MODERN ANALYSIS

(TOPOLOGY OF METRIC SPACES)

Dr. SUDHIR KUMAR PUNDIR
M.Sc., M.Phil., NET, J.R.F., S.R.F. (C.S.I.R.), Ph.D.
HEAD
Department of Mathematics
S.D. (P.G.) College,
Muzaffarnagar (U.P.)

CBS

CBS Publishers & Distributors Pvt Ltd

New Delhi • Bengaluru • Chennai • Kochi • Kolkata • Mumbai
Hyderabad • Jharkhand • Nagpur • Patna • Pune • Uttarakhand

MODERN ANALYSIS
(Topology of Metric Spaces)

ISBN: 978-93-5466-010-8

First Edition: 2021

Published by **Satish Kumar Jain** and produced by **Varun Jain** for

CBS Publishers & Distributors Pvt Ltd
4819/XI Prahlad Street, 24 Ansari Road, Daryaganj, New Delhi 110 002, India.
Ph: 011-23289259, 23266861, 23266867
Fax: 011-23243014

Website: www.cbspd.com
e-mail: delhi@cbspd.com;
cbspubs@airtelmail.in.

Corporate Office: 204 FIE, Industrial Area, Patparganj, Delhi 110 092
Ph: 011-4934 4934   Fax: 011-4934 4935   e-mail: publishing@cbspd.com;
publicity@cbspd.com

Branches

- **Bengaluru:** Seema House 2975, 17th Cross, K.R. Road, Banasankari 2nd Stage, Bengaluru 560 070, Karnataka, India
  Ph: +91-80-26771678/79   Fax: +91-80-26771680   e-mail: bangalore@cbspd.com
- **Chennai:** 7, Subbaraya Street, Shenoy Nagar, Chennai 600 030, Tamil Nadu, India
  Ph: +91-44-26680620, 26681266   Fax: +91-44-42032115   e-mail: chennai@cbspd.com
- **Kochi:** 42/1325, 1326, Power House Road, Opp KSEB, Ernakulum, Kochi 682 018, Kerala, India
  Ph: +91-484-4059061-65,67   Fax: +91-484-4059065   e-mail: kochi@cbspd.com
- **Kolkata:** 6/B, Ground Floor, Rameswar Shaw Road, Kolkata-700014 (West Bengal), India
  Ph: +91-33-2289-1126, 2289-1127, 2289-1128   e-mail: kolkata@cbspd.com
- **Mumbai:** PWD Shed, Gala no 25/26, Ramchandra Bhatt Marg, Next to JJ Hospital Gate no. 2, Opp. Union Bank of India, Noorbaug
  Mumbai-400009, Maharashtra, India
  Ph: +91-22-66661880/89   e-mail: mumbai@cbspd.com

Representatives

- Hyderabad   0-9885175004
- Patna   0-9334159340
- Jharkhand   0-9811541605
- Pune   0-9623451994
- Nagpur   0-9421945513
- Uttarakhand   0-9716462459

Printed at   Glorious Printers, Daryaganj, Delhi, India

Preface

The book entitled **"MODERN ANALYSIS"** meet the needs of UG and PG students of Mathematics and many competitive examinations like NET, GATE, JAM etc. Besides, it will also be very useful for students preparing for various competitive examinations.

The contents of this book are derived from the curricula offered by various universities across the country. This book consists of eleven chapters. In each chapter of the book, an ample amount of theory is given which is supported by solved examples followed by exercises along with their answers. The text is organized around mathematical problems, with each chapter devoted to a single type of problem.

I express my gratitude to the authors and publishers of various books I consulted during the preparation of the book.

I wish to sincerely thank Sh S.K. Jain and Sh Varun Jain, Managing Director, CBS Publishers and Distributors, New Delhi for encouragement and help in bringing out this publication in a present nice form.

My special thanks to Sh. Y.N. Arjuna, Senior director publishing, editorial and publicity and Smt. Ritu Chawla, publishing head, CBS Publishers and Distributors, New Delhi whose encouragement and unstinted support enabled me to complete the book. I also take this opportunity to express my sincere gratitude to Sh. Sunil Dutt, CBS Publishers and Distributors, New Delhi who gave me the inspiration throughout the preparation of the book. Sh. Suresh Sharma, Sh. Ramakant Jha, Sh. Anil Rawat and Sh. Anurag Singh, CBS, New Delhi deserve special mention for their kind support and help in this endeavour and unstinted support enabled me to complete my book. Mr. Jitesh Ahalawat, M/s Tech Yuva Infostat also deserve special mention for nice type setting..

I must also record my appreciation due to my wife Dr. Rimple, daughter Rijuta and son Shrish for their understanding and love during the long period that I have taken to complete this book.

Above all I am thankful to The Almighty God, without whose grace nothing is possible for any one.

Readers are welcomed to point out errors, if any and send their valuable suggestions for improving the quality of the book.

Dr. Sudhir Kumar Pundir
email : skpundir05@yahoo.co.in

Contents

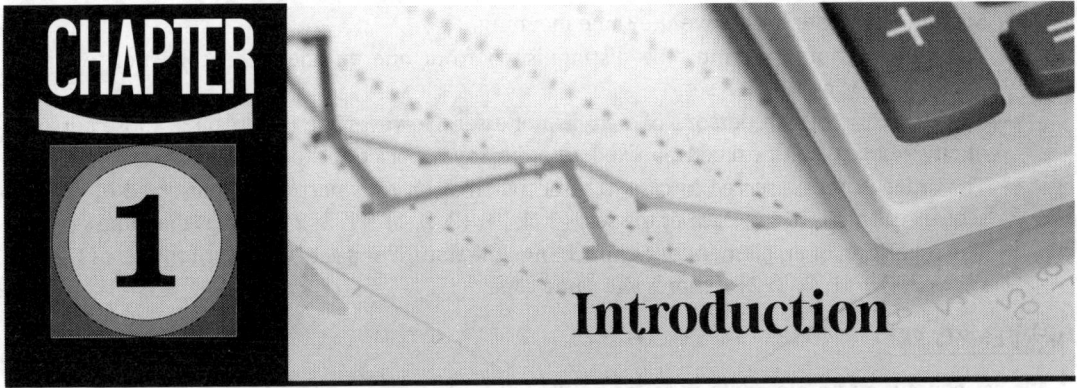

CHAPTER 1

Introduction

1.1 INTRODUCTION

The calculus (differential and integral) briefly called 'the Calculus and mathematical analysis' emerged in 17th century as a powerful tool for bringing about a rapid development of modern mathematical science. The basic notions of calculus are

(i) derivative, and

(ii) integral

These notions can be properly understood without an adequate knowledge of the concept of numbers, limits and functions. We shall therefore deal with the concept of numbers first, and shall thereafter take up the concept of functions, sequence and countability in this chapter.

1.2 CONCEPT OF SETS

The theory of sets is one of the most important tools of pure mathematics. Pure mathematics is the study of sets equipped with assigned structures, knows as mathematical systems. In this section, we shall study some fundamental concepts of set theory.

Definition. *'A set is a well defined collection of objects'.*

The objects of a set are called the elements or members of that set and their membership is defined by certain conditions.

The sets are usually denoted by the capital letters of English alphabets: Say $A, B, C, ..., X, Y, Z$.

For Example

(i) The collection of the letters a, b, c, d, . . .

(ii) The collection of all natural numbers denoted by **N**.

(iii) The students of M.Sc., Mathematics in C.C.S. University, Meerut.

(iv) The collection of vowels in English alphabets. This set containing only five elements, namely a, e, i, o, u.

(v) The collection of all states in Indian union.

If S is a set, an object a in the collection S is called an element of S. This fact is expressed in symbol as $a \in S$ (read as a is in S or a belongs to S). If a is not in S, we write $a \notin S$. For example $4 \in$ **R**, the set of real numbers, but $\sqrt{-2} \notin$ **R** .

Here, Greek letter $\in$ denotes 'belongs to'. It is the abbreviation of the Greek word meaning 'is'.

☛ REMARKS

⟹ By the term 'well defined' we mean that we are given a collection of objects, with certain definite property, so that we are able to determine whether a given objects belongs to our collection or not. Thus, every collection of objects is not a set.

- Set and aggregate both have the same meaning.
- The elements of a set must be distinguished from one another. The collection of sand particles does not form a set.
- The collection of rich persons of a city is not a set. However the collection of those persons of city whose wealth exceeds, a fixed amount, say rupees ten thousands, is a set.
- The order is not preserved in case of a set, whereas order is necessarily preserved in case of sequence. That is to say, each of the sets {1, 2, 3}, {3, 2, 1}, {1, 3, 2} denotes the same set.
- The repetition of an element does not change the nature of a set i.e. each of the sets {1,2,3}, {1, 2, 2, 3}, {1, 3, 3, 2} denotes the same sets.

1.2.1 EMPTY SET

A set containing no elements is called empty set and is denoted by the symbol ϕ.

For Example:

(i) $\phi = \{x : x$ is a negative integer whose square is $-1\}$

(ii) $\phi = \{x : x$ is a natural number lying between 2 and 3 $\}$

(iii) $\phi = \{$the set of such persons, who never die$\}$

(iv) $\phi = \{x : x$ is a real number, $x^2 < 0\}$

(v) $\phi = \{x : x$ is an even prime number greater than five$\}$

(vi) $\phi = \{$the set of real numbers which are solution of equation $x^2 + 1 = 0$ $\}$

(vii) $\phi = \{x : x$ is a straight line passing through three distinct points on a circle$\}$

☛ REMARKS

- The empty set is also known as null set or void set.
- In Roster method, the empty set is denoted by { }.
- To describe the null set, we can use any property, which is not true for any element.
- It is wrong to use the expression 'an empty' or 'a null set' as there is one and only one empty set though, it may have many descriptions. We shall always call 'The empty or the null set'.
- A set consisting of at least one element is called a non-empty or non-void set.
- { ϕ } is not a null set.

1.2.2 SINGLETON SET

Set containing only one element is a Singleton set. The set {a} is a singleton set.

☛ REMARKS

- {0} is not a null set, since it contains 0 as its member. It is a Singleton set.
- A room containing only one man is not same thing as a man. In a similar way, the singleton set {a} is not the same thing as the element a.

FACTS : TO THE POINT

We have the following statements about the real numbers:

- $|x| \geq 0$ and $|x| = 0$ if and only if $x = 0$
- $|x| \leq \varepsilon \Leftrightarrow -\varepsilon \leq x \leq \varepsilon$
- $|x| < \varepsilon \Leftrightarrow -\varepsilon < x < \varepsilon$
- $x = 0$ if and only if either $|x| \leq \varepsilon$ for each $\varepsilon > 0$ or else $|x| < \varepsilon$ for each $\varepsilon > 0$
- $|x + y| \leq |x| + |y|$
- $|x - y| \leq ||x| - |y||$
- $|xy| = |x| \cdot |y|$
- If $x_i \geq y_i$ for each i then
$$\sum_{i=1}^{n} x_i \geq \sum_{i=1}^{n} y_i$$
- If $x_i \geq 0$ for each i and $\sum_{i=1}^{n} x_i = 0$ then $x_1 = x_2 = \ldots = x_n = 0$
- Following are equivalent:
$x \geq y; x - y \geq 0; -y \geq -x; 0 \geq y - x$
- If $x \geq y$ and $z \geq 0$ then $xz \geq yz$
- $x^2 = 0$ if and only if $x = 0$
- If $x > 0$ then $x^{-1} > 0$ and if $x < 0$ then $x^{-1} < 0$
- $x > y > 0 \Leftrightarrow y^{-1} > x^{-1} > 0$
- $\max\{x, y\} \geq 0$ if and only if $x \geq 0$ or $y \geq 0$
- $\max\{x, y\} > 0 \Leftrightarrow x > 0$ or $y > 0$
- $\min\{x, y\} \geq 0 \Leftrightarrow x \geq 0$ or $y \geq 0$
- $\min\{x, y\} > 0 \Leftrightarrow x > 0$ or $y > 0$
- $x > 0 \Leftrightarrow$ there exists a positive integer n such that $n > x^{-1}$
- $x \geq 0 \Leftrightarrow x \geq -\dfrac{1}{n}$ for all positive integer n

1.2.3 FINITE SET

A set is said to be finite if it consists of only finite number of elements.

Here, the process of counting the different elements comes to an end.

For Example:
(i)   Set of natural numbers less than 50.
(ii) Set of all persons in a city.
(iii) Set of English alphabets.
(iv) Set of all persons on the earth.

1.2.4 INFINITE SET

A set which is not finite *i.e.* it contains infinite number of elements. Here, process of counting the different elements never comes to an end.

For Example :
(i)   Set of natural number $N = \{1, 2, 3, ...\}$
(ii) Set of all points of a plane.
(iii) Set of all even integers.
(iv) Set of rational numbers lying between two integers.

1.2.5 EQUAL SETS

Two sets are said to be equal if they contain exactly the same elements. For Example :

$$A = \{x : x \text{ is a letter in the word 'Area'}\} \text{ i.e. } A = \{a, r, e\}$$
and $$B = \{y : y \text{ is a letter in the word 'ear'}\} \quad \text{i.e. } B = \{a, r, e\}$$

Here, A and B are equal sets.

1.2.6 CARDINAL NUMBER OF A SET

The number of distinct elements contained in a finite set A is called cardinal number of A and is denoted by $n(A)$.

1.2.7 EQUIVALENT SETS

Two finite sets are said to be equivalent if they have the same cardinal number.

☛ REMARKS
- Equivalent sets are not always equal but equal sets are always equivalent.
- The number of distinct elements in a finite set is also called the order of the set. If the order of a set is zero, the set is empty.
- If the order of a set is one, the set is singleton.
- The order of an infinite set is never defined.

1.3 SET OF NUMBERS

The number system plays a key role in Mathematics. The real number system R is one of the most important and beautiful mathematical system. There are different ways of introducing the real number system, but the most common way is to start with Peano's Axioms for the natural numbers. The axioms for natural numbers, discovered by the Italian Mathematician Peano are

(i)   1 is a natural number.
(ii) Each natural number n has a successor $(n + 1)$.
(iii) Two natural numbers are equal if their successors are equal.
(iv) Except 1, each natural number is a successor of a natural number.
(v)   Any set of natural numbers which contains 1 and the successor of every natural number k whenever it contain k is the set N of natural number.

☞ REMARK

⟹ Axiom (v) is commonly known as the axiom of induction or principle of finite induction.

The above axioms completely define the set of natural number.

1.3.1 NATURAL NUMBERS

The numbers 1, 2, 3,... are called natural numbers. We represent the set of natural numbers by **N** i.e. **N** = {1, 2, 3, ...}

The Peano's axioms can be used to extend the set **N** of natural numbers to another large system, known as the set of integers.

1.3.2 INTEGERS

The numbers ..., –3, –2, –1, 0, 1, 2, 3, ... are called integers. We represent the set of integers by **Z** i.e. **Z** = {..., –3, –2, –1, 0, 1, 2, 3, ...}

Integers can be used to define the rational numbers.

1.3.3 RATIONAL NUMBERS

Any number of the form p/q, where $p, q \in Z$, $q \neq 0$ and p, q have no common factor (except ± 1) is called a rational number.

The set of rational numbers is denoted by **Q**.

$$\therefore \quad \mathbf{Q} = \{ \frac{p}{q} ; p, q \in \mathbf{Z}, q \neq 0 \}$$

☞ REMKARK

⟹ The set of rational numbers consist of integers and fractions.

Any number which is not rational, is called an Irrational number. For example, $\sqrt{2}$, $\sqrt{3}$, etc. It should be noted that every rational number can be expressed as a terminating or recurring decimal whereas every irrational number can be expressed as a non-terminating infinite decimal.

1.3.4 REAL NUMBERS

A number which is either rational or irrational is called a real number. The set of real numbers is denoted by **R**.

1.4 CARTESIAN PRODUCT OF SETS

Here, we shall study the cartesian product of sets, concept of relation and concept of function.

1.4.1 ORDERED PAIR

An ordered pair is a pair of entries, whose components occur in a specific order. It is written by listing the two components in the specific order, separating them by a comma and enclosing the pair in parenthesis.

Symbolically. If A and B are two non- empty sets, then by ordered pair of elements, we must mean a pair (a,b) such that $a \in A, b \in B$ in the order.

☞ REMARKS

⟹ It may be noted that (a, b) is not the same as $\{a, b\}$. The former denotes an ordered pair where later denotes a set.

⟹ $(a,b) \neq (b,a)$ unless $a = b$.

⟹ Ordered pair may have the same first and second components i.e., two elements of an ordered pair need not be distinct.

1.4.2 CARTESIAN PRODUCT OF TWO SETS

The set of all ordered pair $\{(a,b) : a \in A, b \in B\}$ is called the cartesian product of two sets A and B. It is denoted by $A \times B$.

Symbolically. $A \times B = \{(a,b) : a \in A, b \in B\}$.

For Example. If $A = \{2,3\}$ and $B = \{4,5,6\}$.

Then $A \times B = \{(2,4),(2,5),(2,6),(3,4),(3,5),(3,6)\}$.

☛ REMARKS

⟹ $A \times B = \phi \Leftrightarrow A = \phi$ or $B = \phi$.

⟹ If A and B are finite sets, then $n(A \times B) = n(A) . \, n(B)$.

⟹ If either A or B is an infinite set, then $A \times B$ is an infinite set.

1.5 RELATION

Let us take two sets of natural numbers N_1 and N_2. We define R as a relation between them such that N_2 is a square of N_1. Then

1R1, 2R4, 3R9, 4R16....

In terms of ordered pair, we can write

$R = \{(1,1),(2,4),(3,9),(4,16)...\}$

$= \{(x,y) : x,y \in N$ and $y = x^2\}$.

Then relation from set N_1 to N_2 is a subset of $N_1 \times N_2$ such that $(x,y) \in R$ iff $y = x^2$.

Definition. *Let A and B be two sets. Then a relation R from A to B is a subset of $A \times B$.*

Symbolically. R is a relation from A to $B \Leftrightarrow R \subseteq A \times B$.

☛ REMARKS

⟹ If R is a relation from A to B, then A is called the domain and B the range of R.

⟹ If R is a relation from a non-empty set A to non-empty set B and if $(a,b) \in R$. Then we write $a \, R \, b$ and read as 'a is related to b by relation R'.

⟹ Any subset $A \times A$ defines a relation in A, known as binary relation.

✍ ILLUSTRATIONS

● If $a, b \in N$ and R is defined as "a is a divisor of b", then R is a relation on N.

The subset $N \times N$, which correspond to the relation R is $R = \{(n, nr) : n \in N, r \in N\}$.

● If R is a relation from set $A = \{1,2,3\}$ to the set $B = \{-1,-2\}$ defined by $x + y = 0$ then $R = \{(1,-1),(2,-2)\}$.

Hence, Domain of R is $\{1,2\}$ and range $= \{-1,-2\}$.

● If $A = (a,b,c,d,e\}$ and $B = \{f,g,h,i\}$ and let

$R = \{(a,g),(a,i),(d,h),(e,f)\}$ be a relation from A to B.

Then domain of $R = \{a,d,e\}$

Range of $R = \{g,i,h,f\}$.

● If $a, b \in R$ and S is "$|a-b|$ is a rational number", then S is a relation on R.

1.5.1 TOTAL NUMBER OF RELATIONS

Let A and B be two non-empty finite sets consisting of p and q elements respectively, then $A \times B$ consists of pq ordered pairs. Therefore, total number of subset of $A \times B$ and hence number of relations form A to B is 2^{pq}.

☞ **REMARKS**

⇒ For a non-empty set $A, \phi \subset A \times A$, therefore, it is a relation on A. This relation is called void or empty relation on A.

⇒ The void relation ϕ and the universal relation $A \times B$, are called trivial relation from A to B.

⇒ The void and universal relation on set A are respectively the smallest and the largest relation on A.

1.5.2 IDENTITY RELATION

Let A be a set. The identity relation on A is the relation $I_A = \{(x,x) : x \in A\}$.

For Example. If $A = \{a,b,c\}$, then the relation $I_A = \{(a,a),(b,b),(c,c)\}$ is the identity relation. Here, $R = \{(a,a),(b,b)\}$ is not an identity relation on A as $c \in A$ but as $(c,c) \notin R$.

1.5.3 INVERSE OF A RELATION

Let A, B be two non-empty sets and R be a relation from a set A to B and let (x,y) be element of the subset A of $A \times B$ corresponding to the relation R from A to B.

To the relation R from the set A to the set B, there corresponds a relation from the set B to the set A called the inverse of the relation R, denoted by R^{-1}, such that the subset $B \times A$ corresponding to the relation R^{-1} is equal to $\{(y,x) : (x,y) \in A\}$ i.e., $yR^{-1}x \Leftrightarrow xRy$.

✎ **ILLUSTRATIONS**

● Let $A = \{a, b, c\}$ and $B = \{1, 2, 3\}$ be two sets and let $R = \{(a, 1), (a, 2), (b, 1), (b, 2)\}$ be a relation from A to B, then $R^{-1} = \{(1,a),(2,a),(1,b),(2,b)\}$.

● If $A = \{1,2,3\}, B = \{5,6,7\}$ and let $\{(1,5),(2,5),(2,7)\}$ be a relation from A to B, then $R^{-1} = \{(5,1),(5,2),(7,2)\}$, which is a relation from B to A.

● The inverse of the relation 'is less than' in R is 'is greater than'.

☞ **REMARKS**

⇒ Sometimes, the inverse of a relation coincides with the relation itself. For example, the inverse of the relation 'perpendicular to' in the set of straight lines coincides with itself.

1.5.4 CLASSIFICATION OF RELATIONS

(i) Reflexive relation. Let R be a relation on a set A, then "A relation R is said to be reflexive if $(x,x) \in R$, $\forall x \in A$" i.e., xRx, $\forall x \in A$

For Example :

i. In the set Z of integers, a relation R is defined by xRy iff $x - y$ is divisible by 4. Then, R is a reflexive relation as xRx, $\forall x \in Z$ as $x - x = 0$, which is divisible by 4.

ii. The universal relation on a non-empty set A is reflexive.

iii. The relation 'is less than' (i.e., $<$) in the set of natural number is not reflexive because no number satisfies the relation 'is less than' to itself.

iv. The relation 'is a factor of' in the set of rational number is reflexive, since every rational number is a factor of itself.

v. The relation 'is less than or equal to" (i.e., $\leq$) is in set of natural numbers is reflexive because $n \leq n$ $\forall n \in N$.

(ii) Symmetric relation. A relation R on a set A is said to be symmetric if $(y,x) \in R$ whenever $(x,y) \in R$ $\forall x,y \in R$.

For Example :

(i) Let l_1, l_2 be two lines such that l_1 is perpendicular to l_2 i.e., $l_2 \perp l_1$ then $l_1 \perp l_2 \Rightarrow l_2 \perp l_1$. Therefore, the relation $\perp$ is symmetric.

(ii) The identity and universal relation on a non-empty set are symmetric relation.

(iii) The relation 'less than' for natural number is not symmetric.

(iii) Transitive relation. A relation R on a set A is said to be transitive iff

$$(x,y) \in R \text{ and } (y,z) \in R \Rightarrow (x,z) \in R \forall x,y,z \in A.$$

For Example:

(i) Let a,b,c be three number such that a is a factor of b and b is a factor c, then obviously a is a factor of c. Therefore, 'is a factor of' is a transitive relation.

(ii) If l_1,l_2,l_3 are three lines such that $l_1 \perp l_2$ and $l_2 \perp l_3$ then obviously l_1 is parallel to l_3. Therefore, the relation   is not transitive.

(iv) Anti-symmetric relation. A relation R on a non-empty set A is said to be anti-symmetric iff $(x,y) \in R$ and $(y,x) \in R \Leftrightarrow x = y$ $\forall x, y, \in R$.

☛ REMARKS

⟹ The identity relation R on a set A is an anti-symmetric relation.

⟹ If $(x,y) \in R$ and $(y,x) \in R$, then it may be that $x = y$.

⟹ The universal relation on a set A, containing at least two elements is not anti-symmetric.

(v) Equivalence relation. A relation R on a set A is said to be an equivalence relation if it is

(a) reflexive

(b) symmetric

(c) transitive

For Example: In the set of integers, a relation R is defined by xRy if and only if $x - y$ is divisible by 4, then R is an equivalence relation, since

(a) For xRx, $x - x = 0$ is divisible by 4. Therefore, it is reflexive.

(b) For xRy, let $x - y = 4m$ so $y - x = 4n$, which is also divisible by 4, therefore yRx and hence it is symmetric also.

(c) For xRy let $x - y = 4m$; for yRz, let $y - z = 4n$. By adding these two equation, we get $x - z = 4(m+n)$, which is always divisible by   . Therefore xRz and   hence R is transitive.

1.5.5 COMPOSITION OF RELATION

Let R_1 and R_2 be two relation from sets A to B and B to C respectively. Then we can define a relation $R_1 \, o \, R_2$ from A to C, such that $(x,z) \in R_1 \, o \, R_2$ if and only if there exists $y \in B$ such that

$$(x,y) \in R_1 \text{ and } (y,z) \in R_2.$$

This relation is called composition of R_1 and R_2.

1.6 FUNCTION

Let A and B be two sets, then a rule or correspondence, which associates each element of A to a unique element of B is called a function or mapping from set A to set B.

Symbolically. If f is a function from a set A to set B then we write $f : A \to B$ read as f is a function from A to B or f maps A to B.

1.6.1 RANGE AND DOMAIN OF A FUNCTION

Let an element $y \in B$ be corresponded by an element $x \in A$, then y is called the image of x and is denoted by $f(x)$. Hence, x is defined as the pre-image of y.

'The set A is called the domain and the set B is called the co-domain of the function f'.

The set of all f-image of the element A is called image set or the range of f and is denoted by $f(A)$ or $\{f(x) : x \in A\}$.

Evidently, $f(A) \subseteq B$.

Thus a mapping $f : A \to B$ is the set of ordered pairs $\{(a,b) : a \in A, b \in B\}$.

so that no two ordered pairs have the same first element *i.e.*,

$$f = \{(a,b) : a \in A, b \in B, b = f(x) \quad \forall a \in A\}.$$

For Example. Let $A = \{-2,-1,0,1,2\}$ and B is the set of all integers, then for every $x \in A, f(x) \in B$ where $f(x) = x^2$. Here, A is the domain and B is the co-domain.

Also, $f(a)$ is the value of the function $f(x)$, when x takes the value a. The element of the co-domain which are equal to $f(x)$ for $x \in A$, form the range of f.

When $x = -2$, $f(x) = f(-2) = (-2)^2 = 4$

When $x = -1$, $f(x) = 1$

When $x = 0$, $f(x) = 0$

When $x = 1$, $f(x) = 1$

When $x = 2$, $f(x) = 4$

This can be illustrated in the following figures.

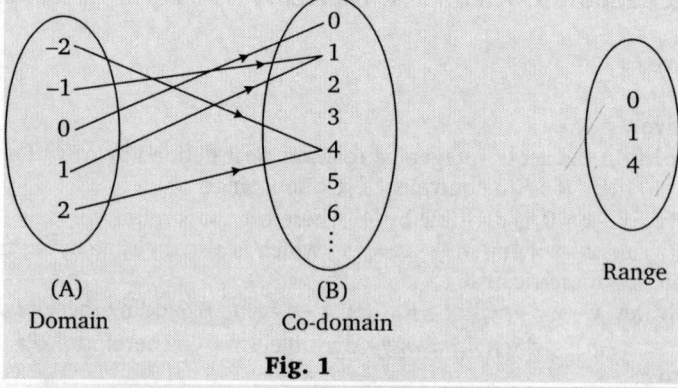

(A)   (B)   Range

Domain   Co-domain

Fig. 1

☛ **REMARKS**

➥ If $f : A \to B$, then a single element in A, can not have more than one image in B. However, two or more element in A may have the same images in B.

➥ Every element in A must have its images in B, but every element in B may not have its pre-image in A.

➥ To each element x in A, there exists a unique element y in B such that $y = f(x)$.

➥ The range of f consists those element in B which appear as the image of at least one element in A. In other words, we can say, range of a function is the image of its domain.

➥ Range is a subset of co-domain.

1.6.2 TYPES OF FUNCTIONS

(a) One-one function. A function f from A to B i.e. $f : A \to B$ is said to be one-one (or injective) if and only if distinct elements of A have distinct images.

Symbolically. f is one-one if for $x_1, x_2 \in A$, we have

$$x_1 \ne x_2 \Rightarrow f(x_1) \ne f(x_2) \quad \forall x_1, x_2 \in A$$

or $f(x_1) = f(x_2) \Rightarrow x_1 = x_2, \quad \forall x_1, x_2 \in A$.

It is also called injective function.

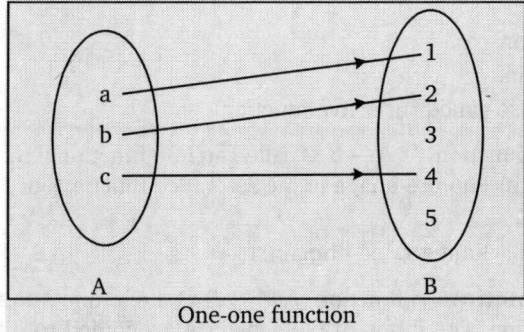

One-one function

Fig. 2

Graphically, a function is one-one if and only if no line parallel to x-axis meets the graph of the function in more than one point.

(b) Many-one function. A function $f : A \rightarrow B$ is called many-one, if at least one element of co-domain B has two or more than two pre-image in domain A.

Symbolically. f is many one, if for some $x_1, x_2 \in A$, we can have $x_1 \neq x_2 \Rightarrow f(x_1) = f(x_2)$. This can be illustrated in the following figure.

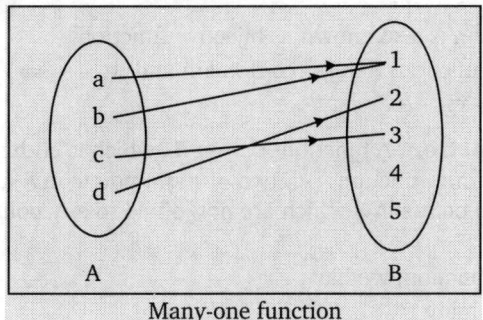

Many-one function

Fig. 3

Graphically, a function is many-one if and only if a line parallel to x-axis meets the graph of the function is more than one point.

☞ **REMARK**

➠ One-many function does not exist.

(c) Onto function. A function $f : A \rightarrow B$ is called an onto function, if there is no element of B, which is not an image of some element of A, *i.e.*, every element of B appears as the image of at least one element of A.

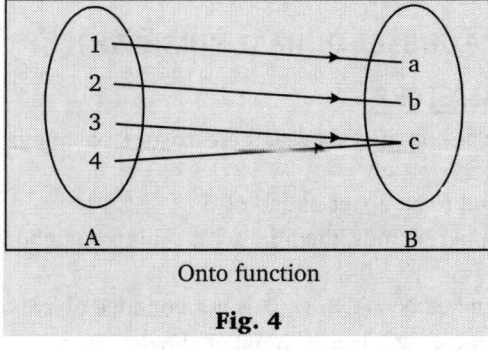

Onto function

Fig. 4

☛ REMARKS

➠ For an onto function
   Range = co-domain.

➠ Onto function is also called surjective function.

(d) Into function. A function $f : A \to B$ is called an into function if there is at least one element of the set B, which has no pre-image in the set A. See functions of Fig. 2 and 3.

☛ REMARK

➠ In an into function : Range $\subset$ co-domain.

(e) One-one into function. A function $f : A \to B$ is a one-one into function if it is both one-one and into function, *i.e.*, different element in A are joined to different element in B and there are some elements in B, which are not joined to any element in A.

Symbolically. One-one into function is defined as
   (i) Range $\subset$ co-domain
   (ii) $f(x_1) \neq f(x_2) \Rightarrow x_1 \neq x_2$

(f) One-one onto function. If a function $f : A \to B$ is both one-one and onto *i.e.*, the different points in A are joined to different points in B and no point in B is left vacant.

☛ REMARKS

➠ One-one onto function is also known as bijective function.

➠ For a one-one onto function, Range=co-domain and $x_1 \neq x_2 \Rightarrow f(x_1) = f(x_2)$.

➠ $f(x_1) = f(x_2) \Rightarrow x_1 = x_2$.

(g) Many-one into function. A function $f : A \to B$, which is both many one and into function is called a many one into function, *i.e.*, two or more points in A are joined to some points in B and there are some points in B which are not joined to any point in A.

☛ REMARK

➠ For many-one into function, we have
   (i) Range $\subset$ Co-domain
   (ii) $x_1 \neq x_2 \Rightarrow f(x_1) = f(x_2)$.

(h) Many-one onto function. If a function $f : A \to B$ is both many one and onto function, then it is many-one and onto function, *i.e.*, in B, one point is joined to at least one point in A and two or more points in A are joined to the same points in B.

☛ REMARK

➠ For many-one onto function, we have
   (i) Range $\subset$ Co-domain
   (ii) $x_1 \neq x_2 \Rightarrow f(x_1) = f(x_2)$.

1.7 BOUNDEDNESS OF A SUBSET OF REAL NUMBERS R

1.7.1 UPPER BOUND OF A SUBSET OF R

A subset S of R is said to be bounded above if there exists a real number u such that $s \leq u$, $\forall s \in S$.

(MEERUT 1990)

The real number u is said to be upper bound of S.

If there exists no such upper bounds, then the set is said to be unbounded above.

For example:

(i) The set of natural number $N = (1, 2, 3, ...)$ is not bounded above or unbounded above.

(ii) The set of positive integers Z^+ is not bounded above.

(iii) The set $S = [1, 2, 3, 4]$ is bounded above by 4.

(iv) The set $\left\{\dfrac{1}{n} : n \in \mathbf{N}\right\}$ is bounded above by 1.

(v) The set of negative integers is bounded above by 0.

1.7.2 LOWER BOUND OF A SUBSET OF R

A subset S of $\mathbf{R}$ is said to be bounded below if there exists a real number l such that $s \leq l, \ \forall s \in S$.

(MEERUT 2000, 2001)

The real number l is said to be the lower bound of S, if there exists no such lower bound, then the set is said to be unbounded below.

For example :

(i) The set of natural number $\mathbf{N}$ is bounded below by 1.

(ii) The set $\left[\dfrac{1}{n} : n \in \mathbf{N}\right]$ is bounded below by 0.

(iii) The set $S = [1, 2, 3, 4]$ is bounded below by 1.

(iv) The set of positive real numbers is bounded below.

1.7.3 BOUNDED SET

A subset S of $\mathbf{R}$ is said to be bounded set if it is bounded below as well as bounded above, i.e., if there exist two real numbers l and u such that $l \leq s \leq u, \ \forall s \in S$.

Equivalently, if there exists an interval $I \,(= [l, u])$ such that $S \subseteq I$.

For example:

(i) Every finite set is bounded.

(ii) The set $\left[\dfrac{1}{n} : n \in \mathbf{N}\right]$ is bounded.

1.7.4 UNBOUNDED SET

A subset S of $\mathbf{R}$ which is not bounded is called an unbounded set.

For example:

(i) The sets $\mathbf{N}, \mathbf{Z}, \mathbf{Q}, \mathbf{R}$ are unbounded sets.

(ii) Set of all prime numbers is an unbounded set.

1.7.5 LEAST UPPER BOUND (OR SUPREMUM)

A real number u is said to be a least upper bound of a set S if

(i) u is an upper bound of S

(ii) if u' is another upper bound of S, then $u \leq u'$, i.e., no real number less than u can be an upper bound of S.

1.7.6 GREATEST LOWER BOUND (OR INFIMUM)

A real number l is called a greatest lower bound of a set S if

(i) l is a lower bound of S

(ii) if l' is another lower bound of S, then $l' \leq l$, i.e., no

FACTS : TO THE POINT

▶ If a set is bounded above, then it has infinitely many upper bounds in as much as every number greater than an upper bound is also an upper bound.

▶ If a set is bounded below, then it has infinitely many lower bounds in as much as every number smaller than a lower bound is also a lower bound.

▶ It is not necessary that lower bounds and upper bounds of a set S are the members of S.

▶ The null set ϕ is bounded but it neither possesses lower bound nor upper bound.

▶ Supremum is defined only for the bounded above sets and infimum for the subset, which are bounded below.

▶ The supremum and infimum of a set may or may not belong to the set. If supremum of a set belongs to the set, then supremum is the largest element of the set.

▶ If infimum of a set belongs to the set, then infimum is the smallest element of the set.

▶ Supremum and infimum of a bounded subset of R, are unique.

real number greater than l can be a lower bound of S.

For example : If $S = \left\{ \dfrac{1}{n} : n \in N \right\}$, then $l.u.b. = 1$ and $g.l.b.$ is 0.

☛ **REMARKS**

⇒ If a real number u is the supremum of a subset S of real numbers then, for every $\varepsilon > 0$, there exists a real number $x \in S$ such that $u - \varepsilon < x < u$.

⇒ If a real number l is the infimum of a subset S of real numbers then, for every $\varepsilon > 0$, there exists a real number $x \in S$ such that $l \le x < l + \varepsilon$.

⇒ In case of singleton set $S = [a], a \in R$, supremum and infimum coincide.

⇒ If u and l are the supremum and infimum of a non-empty subset S of R, then $l \le u$.

THEOREM 1. ***The supremum of a set $S \subset R$, if exists, is unique.*** (MEERUT 1990P, 98)

PROOF. Let S be a non-empty subset of R.

Let if possible, s_1 and s_2 be two supremum of S.

To show $s_1 = s_2$.

Since we assume that s_1 and s_2 are the supremums of S.

⇒ s_1 and s_2 are the upper bounds of S.

Let, first suppose s_1 is a supremum and s_2 is an upper bound of S, then

$$s_1 \le s_2 \qquad \qquad \dots(1)$$

Now, if s_2 is the supremum and s_1 is the upper bound of S, then

$$s_2 \le s_1 \qquad \qquad \dots(2)$$

From (1) and (2)

$$s_1 = s_2.$$

Hence, supremum of a set, if exists is unique.

THEOREM 2. ***The infimum of a set, if exists, is unique.***

PROOF. Proof is similar as theorem 1 and left to the reader.

THEOREM 3. ***If S be a non-empty subset of R, then a real number s is the supremum of S if and only if***

    (i) $x \le s \quad \forall x \in S$

and (ii) for each positive real number ε, there exists a real number $x \in S$ such that $x > s - \varepsilon$.

PROOF. (i) **Necessary Condition** (only if part)

Let us first suppose the given condition is necessary for s to be the supremum of the set S.

Let s be the supremum of $S \Rightarrow s$ is an upper bound of S.

By definition $x \le s \quad \forall x \in S$.

Let $\varepsilon > 0$ be any real number. Then obviously $s - \varepsilon < s$.

⇒ $(s - \varepsilon)$ is not an upper bound of S.   ($\because s$ is l.u.b. of S)

Hence, there must exist some $x \in S$ such that $x > s - \varepsilon$.

(ii) **Sufficient part** (If part)

Let us suppose (i) and (ii) holds.

Then, to show $s = \sup S$.

By condition (i), we have that s is an upper bound of S. To show s is the supremum of S, for this, it is enough to show that no real number less than s can be an upper bound of S.

Let s_1 be any real number less than s

$\Rightarrow \qquad s - s_1 > 0$.

Let us take $\varepsilon = s - s_1 \Rightarrow \varepsilon > 0$.

Then by condition (ii), there exists $x \in S$ such that $x > s - \varepsilon$.

$\Rightarrow \qquad x > s - (s - s_1) \Rightarrow x > s_1, \ x \in S$.

$\Rightarrow \qquad s_1$ is not an upper bound of S.

Hence, we can say that s is an upper bound of S and no real number less than s is an upper bound of S.

$\Rightarrow \qquad s$ is the supremum of S.

THEOREM 4. *Let S be a non-empty subset of R, then a real number t is the infimum of S if and only if*

(i) $x \geq t \quad \forall \ x \in S$

and (ii) for each real number $\varepsilon > 0$, there exists a real number $x \in S$ such that $x > t + \varepsilon$.

PROOF. Proof is similar as theorem 3.

SOLVED EXAMPLES

EXAMPLE 1. *(i) The set R^+ of positive real numbers, is bounded below and unbounded above.*

(ii) The set R^- of negative real numbers, is bounded above and unbounded below.

SOLUTION. (i) Since every member of $R^- \cup [0]$ is a lower bound of R^+, therefore R^+ is bounded below.

To prove R^+ is unbounded above.

Let if possible, suppose u is an upper bound of R^+ we have $u \geq 1$ for $1 \in R^+$. Since $2 \in R^+$, $2 > 0$ and so $u \geq 1, 2 > 0$ gives $u + 2 > 1 + 0$, i.e., $u + 1 > 0$. Thus $(u + 1) \in R^+$ and $(u + 1) > u$ which is a contradiction, that u is an upper bound of R^+.

Hence, R^+ is unbounded above.

(ii) Proof follows in a similar manner.

EXAMPLE 2. *The set R is an unbounded set.*

SOLUTION. From example 1, we conclude that the set R^+ is unbounded above and R^- is always unbounded below.

Also, $R = R^- \cup [0] \cup R^+$

$\Rightarrow \quad R$ is not bounded.

EXAMPLE 3. *The null set ϕ is neither bounded below or above, nor unbounded.*

SOLUTION. Since, there is no member in ϕ, we can not check whether a given real number can be a bound for ϕ or not. Thus, bounds for ϕ do not exist. On the other hand, we can as well say that every real number is a lower or upper bound for there is no member in ϕ which does not satisfy the required property of bounds.

EXAMPLE 4. *Show that every non-empty finite subset of R is bounded.*

SOLUTION. Let S be a non-empty finite subset of R.

$\Rightarrow \quad S$ contains a finite number of elements. Then by the properties of the ordered relation in R, out of these elements one element $a \in S$ shall be the smallest element of S and another element $b \in S$ shall be the greatest element of S.

$$\Rightarrow \qquad a \le x \le b \ \forall \, x \in S$$

Hence, S is always bounded.

EXAMPLE 5.   ***Find the supremum and infimum of the set*** $S = \{x \in Z : x^2 \le 25\}$.   (MEERUT 2001)

SOLUTION.   Since $S = \{x \in \mathbf{Z} : x^2 \le 25\}$

$$= \{-5, -4, -3, -2, -1, 0, 1, 2, 3, 4, 5\}$$

Since S is a finite subset of $\mathbf{R}$, the smallest member of S is –5, which is a lower bound of S, and hence infimum of S is –5. Similarly 5 is the supremum of S.

EXAMPLE 6.   ***Find the supremum and infimum, if they exist, of the following sets***

     (i) $\left\{ \dfrac{1}{n} : n \in N \right\}$ (MEERUT 1994, 95, 97, 2000, 2001)

     (ii) $\left\{ x \in Q : x = \dfrac{n}{n+1}, \ n \in N \right\}$ (MEERUT 1994)

     (iii) $\left\{ 1 + \dfrac{(-1)^n}{n} : n \in N \right\}$ (MEERUT 1990S)

     (iv) $\left\{ \pi + \dfrac{1}{2}, \ \pi + \dfrac{1}{4}, \ \pi + \dfrac{1}{8}, \ ... \right\}$ (MEERUT 1993, 93S)

SOLUTION.   (i)   Here, we have

$$S = \left\{ \frac{1}{n} : n \in \mathbf{N} \right\} = \left\{ 1, \frac{1}{2}, \frac{1}{3}, ... \right\}$$

The set is bounded above by 1, also any member less than 1 is not an upper bound of S, therefore $\sup S = 1$

Also, 0 is a lower bound of S, because $x \ge 0, \forall x \in S$. Let l be any arbitrary positive small number, then there exists $n \in \mathbf{N}$ such that $\dfrac{1}{n} < l$, which shows that l is not an upper bound of S. Thus 0 is a lower bound of S and no other positive real number is a lower bound of S. Therefore, infimum of $S = 0 \notin S$.

(ii)   Let $\ S = \left\{ \dfrac{n}{n+1} : n \in \mathbf{N} \right\} = \left\{ \dfrac{1}{2}, \dfrac{2}{3}, \dfrac{3}{4}, \ ... \right\}$

Then, the set is bounded below by $\frac{1}{2}$ and any number greater than $\frac{1}{2}$ can not be a lower bound of S. Therefore infimum of $S = \dfrac{1}{2}$.

Also, $\left(\dfrac{n}{n+1} \right) < 1, \ \forall n \in \mathbf{N}$, therefore 1 is an upper bound of S, and any number less than 1 not be an upper bound of S. Therefore, supremum of $S = 1$.

(iii)   Let $\ S = \left\{ 1 + \dfrac{(-1)^n}{n} : n \in \mathbf{N} \right\} = \left\{ 0, \dfrac{3}{2}, \dfrac{2}{3}, \dfrac{5}{4}, \dfrac{4}{5}, \dfrac{7}{6}, \dfrac{6}{7}, ... \right\}$

$$= \left\{ \frac{0}{1}, \frac{2}{3}, \frac{4}{5}, \frac{6}{7}, \frac{8}{9}, ..., \frac{2n-2}{2n-1} ... \right\} \cup \left\{ \frac{3}{2}, \frac{5}{4}, \frac{7}{6}, \frac{9}{8}, ..., \frac{2n+1}{2n} ... \right\}.$$

Here, we have that the proper fraction $\dfrac{0}{1}, \dfrac{2}{3}, \dfrac{4}{5}, \dfrac{6}{7}, ...$ are increasing and tending to 1, and the improper fractions begin with $\dfrac{3}{2}$ are decreasing and tending to 1.

Therefore, infimum of $S = 0$ and supremum of $S = \dfrac{3}{2}$.

(iv) Let $S = \left\{ \pi + \dfrac{1}{2}, \ \pi + \dfrac{1}{4}, \pi + \dfrac{1}{8}, ... \right\}$

Here, we have $x \leq \pi + \dfrac{1}{2} \ \forall \ x \in S$

$\Rightarrow \quad \pi + \dfrac{1}{2}$ is an upper bound for S.

Since, $\pi + \dfrac{1}{2} \in S$, therefore no real number less than $\pi + \dfrac{1}{2}$ can be an upper bound for S. Thus, $\pi + \dfrac{1}{2}$ is the least upper bound. Therefore, supremum of

$S = \pi + \dfrac{1}{2}$.

Similarly, we can show that π is the infimum of S.

1.8 SEQUENCES

Let N be the set of natural numbers and S be any set of real numbers. A function, whose domain is the set of natural number s and range is a subset of S, is called a sequence in S.

Symbolically. If we define a function $f : N \rightarrow S$, then f is a sequence. We shall denote a sequence in a number of ways:

(i) Usually, a sequence is denoted by its images. For a sequence f, the image corresponding to $n \in N$ is denoted by f_n or $<f(n)>$ and is called the nth term of the given sequence. For example $< 1, 4, 9, ... >$ is the sequence whose n^{th} term is n^2.

(ii) Using in order, the first few element of a sequence, the rule for writing down different elements becomes clear. For example, $<1, 2, 3, ...>$ is the sequence whose n^{th} term is n.

(iii) Defining a sequence by a recurrence formula $i.e.$, by a rule which express the nth term by $(n-1)^{th}$ term. For example, let $a_1 = 1, a_{n+1} = 2a_n \ \ \forall n \geq 1$.

These above relations define a sequence whose n^{th} term is 2^{n-1}.

☞ **REMARKS**

- A sequence is represented as $< s_n >$ or $\{s_n\}$, when s_n is the nth term of the sequence.
- The set of all distinct terms of a sequence is called the range set of that sequence.
- A sequence, whose range is a subset of R is called a real sequence or a sequence of real numbers.

Constant sequence. A sequence $< s_n >$ defined by $s_n = a, \ \forall n \in N$, is called a constant sequence.

Equality on sequence. Two sequence $< s_n >$ and $< t_n >$ are said to be equal, if $s_n = t_n \ \ \forall n \in N$.

1.8.1 OPERATION ON SEQUENCES

Since, the sequence are real valued functions, therefore, the sum, difference, product etc. of two sequence are defined as follows:

(i) If $< s_n >$ and $< t_n >$ be any two sequence, then the sequence, whose n^{th} terms are $s_n + t_n$, $s_n - t_n$ and $s_n . t_n$ are respectively known as the sum, difference and product of the sequence $< s_n >$ and $< t_n >$ and are denoted by $< s_n + t_n >, < s_n - t_n >$ and $< s_n . t_n >$ respectively.

(ii) If $s_n \neq 0 \ \ \forall n \in N$, then the sequence, whose n^{th} term is $\dfrac{1}{s_n}$ is called reciprocal of the sequence $< s_n >$ and is denoted by $< \dfrac{1}{s_n} >$.

(iii) The sequence , whose n^{th} term $s_n / t_n (t_n \neq 0 \quad \forall n \in \mathbf{N})$ is known as the quotient of the sequence $< s_n >$ by the sequence $< t_n >$ and is denoted by $< \dfrac{1}{s_n} >$.

(iv) The sequence, whose n^{th} term is ks_n, where $k \in \mathbf{R}$ is known as the scalar multiple of the sequence $< s_n >$ by k and is denoted by $< ks_n >$.

1.8.2 BOUNDED SEQUENCE

(i) **Bounded below sequence.** A sequence $< s_n >$ is said to be bounded below is there exists a real number l such that $s_n \geq l \quad \forall n \in \mathbf{N}..$

The number l is known as the lower bound of the sequence $< s_n >$.

(ii) **Bounded above sequence.** A sequence $< s_n >$ is said to be bounded above if there exists a real number u such that $s_n \leq u \quad \forall n \in \mathbf{N}$.

The number u is said to be upper bound of the sequence $< s_n >$.

(iii) **Bounded sequence.** A sequence $< s_n >$ is said to be bounded if it is bounded above as well as bounded below.

(iv) **Unbounded sequence.** A sequence $< s_n >$ is said to be unbounded if it is not bounded.

(v) **Least upper bound.** If a sequence $< s_n >$ is bounded above, then there exists a number u_1 such that

$$s_n \leq u_1 \quad \forall n \in \mathbf{N}. \qquad \ldots(1)$$

This number u_1 is called an upper bound of the sequence $< s_n >$. If $u_1 < u_2$. Then from (1), we find that

$$s_n < u_2 \quad \forall n \in \mathbf{N}$$

which implies, u_2 is also an upper bound of the sequence $< s_n >$. Hence, we can say that any number greater than u_1 is an upper bound of $< s_n >$.

Hence, a sequence has an infinite number of upper bounds, if it is bounded above. Let u be the least of all the upper bound of the sequence $< s_n >$. Then u is defined as the least upper bound (l.u.b) or supremum of the sequence $< s_n >$.

(vi) **Greatest lower bound.** If a sequence $< s_n >$ is bounded below, then there exists a number $l_1 \in \mathbf{R}$ such that

$$l_1 \leq s_n \quad \forall n \in \mathbf{N} \qquad \ldots(2)$$

This number l_1 is known as the lower bound of $< s_n >$. If $l_2 < l_1$, then from (2)

$$l_2 \leq s_n \quad \forall n \in \mathbf{N}$$

which implies, l_2 is also a lower bound of the sequence $< s_n >$. Hence, we can say any number less than l_1 is a lower bound of $< s_n >$.

Hence, a sequence has infinite number of lower bounds, if it is bounded below. Let l be the greatest of

FACTS : TO THE POINT

▶ If $c > 0$ then $<a_n>$ converges to a if and only if for each $\varepsilon > 0$ there exists a positive integer m, depending on ε such that $|a - a_n| \leq c\varepsilon \ \forall \ n \geq m$

▶ If $<a_n>$ is a convergent sequence in R such that $\lim\limits_{n \to \infty} a_n > r$ then $a_n > r$ $\forall$ sufficiently large $n, r \in R$.

▶ Let $r \in R$ and $<a_n>$ be a convergent sequence in R such that $a_n \geq r$ for all sufficiently large n. Then $\lim\limits_{n \to \infty} a_n \geq r$.

▶ If $r > 1$ then $r^n \to \infty$ as $n \to \infty$.

▶ If $a > 1$ then $\log_a n \to \infty$ as $n \to \infty$.

▶ An increasing sequencing of non-negative real numbers diverges to infinitely if and only if it is not bounded above.

▶ The intersection of a nested sequence of closed intervals in R is non-empty.

▶ Let $<s_n>$ be a sequence of real numbers then atleast one of the following hold:

   (i) $<s_n>$ has a convergent subsequence

   (ii) $<s_n>$ has a strictly increasing subsequence

   (iii) $<s_n>$ has a strictly decreasing subsequence

all the lower bounds of the sequence $< s_n >$. Then l is known as greatest lower bound (g.l.b.) or infimum of the sequence $< s_n >$.

✍ ILLUSTRATIONS

* The sequence $< n^2 >$ is bounded below by 1 but not bounded above.

* The sequence $< \dfrac{n}{n+1} >$ is bounded as $\dfrac{1}{2} \leq \dfrac{n}{n+1} < 1 \quad \forall n \in \boldsymbol{N}$.

* The sequence $< \dfrac{1}{n} >$ is bounded since $\left|\dfrac{1}{n}\right| \leq 1 \quad \forall n \in \boldsymbol{N}$.

* The sequence $< 2^n >$ is bounded below and has smallest term as 2. Every member of $]-\infty, 2]$ is a lower bound of the sequence and the sequence is not bounded above.

1.8.3 LIMIT POINT OF THE SEQUENCE

A real number l is called a limit point of a sequence $< s_n >$ if every nbd of l contains infinite number of terms of the sequence.

Thus, $l \in \boldsymbol{R}$ is a limit point of the sequence $< s_n >$ if for given $\varepsilon > 0, s_n \in]l-\varepsilon, l+\varepsilon]$ for infinitely many points.

Here it must be noted that

(i) Limit point of a sequence need not be a member of the sequence.

(ii) A limit point of a sequence may or may not be a limit point of the range of the sequence but the limit point of the range of a sequence is always a limit point of the sequence.

(iii) In case of real number, limit points of a sequence may also be called accumulation, cluster or condensation points.

✍ ILLUSTRATIONS

* The sequence $< \dfrac{1}{n} >$ has one limit point, *i.e.*, 0.

* The sequence $< (-1)^n >$ has two limit points 1 and -1.

* The sequence $< n >$ has no limit point.

* The sequence $< 1 + \dfrac{(-1)^n}{n} >$ has one limit point, *i.e.*, 1.

Sufficient Conditions for number l to be or not be a limit point of the sequence $< s_n >$:

(i) If for every $\varepsilon > 0, \exists m \in \boldsymbol{N}$ such that $s_n \in]l-\varepsilon, l+\varepsilon[\quad \forall n \geq m$ or equivalently $|s_n - l| < \varepsilon \quad \forall n \geq m$, then l is the limit point of the sequence $< s_n >$.

(ii) If for any $\varepsilon = 0, s_n \in]l-\varepsilon, l+\varepsilon[$ for only a finite number of values of n, then l is not a limit point of the sequence $< s_n >$. Such a condition is also necessary for a number l not to be limit point of the sequence $< s_n >$.

1.8.4 BOLZANO-WEIERSTRASS THEOREM FOR SEQUENCE (KANPUR 2000)

STATEMENT. *Every bounded sequence has at least one limit point.*

PROOF. Let $S = \{s_n : n \in \boldsymbol{N}\}$ be the range set of the bounded sequence $< s_n >$.

Then, S is a bounded set. Now, there may be two cases :

(i) Let S be a finite set. Then $s_n = p$ for infinitely many indices n. Here $p \in \boldsymbol{R}$.

Obviously p is a limit point of $< s_n >$.

(ii) Let S be an infinite set. Since, S is bounded, then by Bolzano-Weirstrass theorem for set of real numbers S has a limit point say p. Therefore, every nbd of p contains infinity many distinct points of S, $i.e.$, infinitely many terms of $<s_n>$ and hence p is a limit point of the sequence $<s_n>$.

1.8.5 LIMIT SUPERIOR AND LIMIT INFERIOR

The greatest limit point of a bounded sequence is called the upper limit or limit superior and is denoted by $\overline{\lim}\ s_n$ and the smallest limit point of a bounded sequence is called the lower limit or limit inferior and is denoted by $\underline{\lim}\ s_n$

* By definition, it is obvious that $\underline{\lim}\ s_n \le \overline{\lim}\ s_n$.
* A bounded sequence $<s_n>$ for which the upper limit and lower limit coincide with real number l is said to converge to l.

Limit of a sequence. A sequence $<s_n>$ is said to have a limit l if for a given $\epsilon > 0\ \exists$ a positive integer m such that $|s_n - l| < \epsilon\ \ \forall n \ge m$.

1.8.6 CONVERGENT SEQUENCE

A sequence $<s_n>$ is said to converge to a number l, if for a given $\epsilon > 0$ there exists a positive integer m such that $|s_n - l| < \epsilon\ \ \forall n \ge m$.

☞ REMARK

⇒ A sequence $<s_n>$ is said to be convergent iff it is bounded and has one and only one limit point.

1.8.7 DIVERGENT SEQUENCE

A sequence, which is not convergent, is known as divergent sequence.

1.8.8 OSCILLATORY SEQUENCE

A sequence $<s_n>$ is said to be an oscillatory sequence if it is neither convergent nor divergent.

An oscillatory sequence is said to be oscillate finitely or infinitely according as it is bounded or unbounded.

In other words, we can say

(i) A bounded sequence, which is not convergent is said to be oscillate finitely.

(ii) An unbounded sequence, which does not diverge, is said to be oscillate infinitely.

(iii) A bounded sequence, which does not converge and has at least two limit points is said to be oscillate finitely.

✎ ILLUSTRATIONS

* The sequence $\left\langle 1 + (-1)^n \right\rangle$ oscillate finitely.

* The sequence $\left\langle (-1)^n \right\rangle$ oscillate finitely.

* The sequence $< (-1)^n \left(1 + \dfrac{1}{n} \right) >$ oscillate finitely.

* The sequence $< n(-1)^n >$ oscillate infinitely.

1.9 COUNTABLE SET

A set which can be put in one to one correspondence with the set of all natural numbers or with its subset is said to be a countable set. (MEERUT 2003,04; KANPUR-2000)

For example: If $A = \{2,4,6,8,...\}$, then $\exists$ an one to one mapping $f : A \to \mathbf{N}$ s.t. $f(x) = x/2$. Then A is a countable set.

Further if $B = \{2,4,6,8,10\}$, then this set can be put in one to one correspondence with the set $A = \{1,2,3,4,5\}$, which is a subset of $\mathbf{N}$.

Obviously a countable set can be finite or infinite both.

An infinite countable set is also called as countable infinite or enumerable set or denumerable set.   (MEERUT 2001)

Thus enumerable set always means an infinite countable set.

Further note that if a set $'A'$ is countable then its elements can be put in one to one correspondence with set $\mathbf{N} = \{1,2,3,...\}$. If we denote the elements of A corresponding to the natural numbers $1,2,3,...$, by $a_1, a_2, a_3, ...$ etc., then the set A can be written as $A = \{a_1, a_2, a_3, ...\}$.

Thus a set is countable iff its elements can be written in the form of a sequence.

☛ **REMARK**

⇒ A set which is not countable, is said to be uncountable.

THEOREM 1.   ***Every subset of a countable set is countable.***

(MEERUT 1987,88,94,2003(BP); KANPUR 2000)

PROOF.   Let A be a countable set, then A can be written as a sequence. Let $A = \{a_1, a_2, a_3, ...\}$.

If A is finite then its every subset will also be finite and so will be countable. Now consider that A is enumerable. Let B be any subset of A.

If B is finite or empty then the result is trivial, so let $B \neq \phi$ and let k_1 be the least positive integer s.t. $a_{k_1} \in B$. Again, let k_2 be the least positive integer with $k_2 > k_1$ s.t. $a_{k_2} \in B$. Dealing with the elements of B in this way, we reach to the conclusion that B can be written as $\{a_{k_1}, a_{k_2}, ...\}$ which is a sequence and hence B is a countable set.

THEOREM 2.   ***The union of enumerable collection of enumerable sets is also enumerable.***

(MEERUT 1972,84; GARHWAL 1991,95; KANPUR 2000)

PROOF.   Let $A = \{A_1, A_2, ...,\}$ be an enumerable collection of enumerable sets where $A_i = \{a_{i_1}, a_{i_2}, a_{i_3}, ...\}$, $\forall i \in \mathbf{N}$ denote enumerable sets. For the proof of the theorem, we shall construct a progression in which all the elements of A appear. Such a list is as follows :

$$\bigcup_{i \in N} A_i = \{a_{11} \to a_{12} \to a_{13} \to a_{14} \cdots$$
$$a_{21} \quad a_{22} \quad a_{23} \quad a_{24} \cdots$$
$$a_{31} \quad a_{32} \quad a_{33} \quad a_{34} \cdots$$
$$a_{41} \quad a_{42} \quad a_{43} \quad a_{44} \cdots$$

Choose the path as shown above; then every element of A lies somewhere on the path (*i.e.*, each element occupies some particular position say r^{th} position $r \in \mathbf{N}$) and hence a one-one correspondence between $\cup A_i$ and $\mathbf{N}$ is implied showing thereby that A is enumerable. Thus we can write $\cup A_i = \{a_{11}, a_{12}, a_{21}, a_{31}, a_{22} ...\}$ as a sequence.

$\Rightarrow$ set $\cup A_i$ is enumerable.

Besides the above argument, note that the mapping $f : \bigcup_{i \in N} A_i \to \mathbf{N}$ given by

$$f(a_{pq}) = \frac{(p+q-2)(p+q-1)}{2} + p \quad (= say \ n \in \mathbf{N})$$

shows the enumeration of

$$\cup A_i = \{a_{11}, a_{21}, a_{12}, a_{13}, a_{22}, a_{31}, a_{14}, a_{23}, a_{32}, a_{41}, ...\}.$$

i.e., this mapping assigns a unique place to each element of the set $\cup A_i$ in the above sequence showing that $\cup A_i$ is countable.

COR.I.   *The union of countable collection of countable sets is also countable.*

(MEERUT 1986,87,91,94,2001; AMRAVATI 1997; GARHWAL 2001; KANPUR 2001,02)

COR.2.   *If A_i is countable infinite set, then prove that $\overset{n}{\underset{i=1}{\cup}} A_i$ is also countable infinite.*

(GARHWAL 1995)

THEOREM 3.   *The set $N \times N$ is enumerable.*

(MEERUT 1983,98, GARHWAL 2003)

PROOF.   Here note that $N \times N = \{(u,v) : u,v \in N\}$; then clearly $N \times N$ can be arranged as shown below :

$$N \times N = \{(1,1),(1,2),(1,3),(1,4)...$$
$$(2,1),(2,2),(2,3),(2,4)...$$
$$(3,1),(3,2),(3,3),(3,4)...$$
$$...\quad ...\quad ...\quad ...\} = \cup A_i,$$

where $A_i = \{(i,n) : n \in N\}$

Now obviously each $A_i(i = 1,2,3,...)$ is enumerable which shows that the set $N \times N$, union of enumerable collection of enumerable sets is also enumerable by Theorem 2.

THEOREM 4.   *The set Q of all rational numbers is enumerable.*

(MEERUT 1984,90,95,97,2001(BP), KANPUR 1996)

PROOF.   Recall that a rational number is written as $p/q : p,q \in Z$ and $q \neq 0$.

Let $A_q = \{p/q : p,q \in Z\}$. Then obviously A_q is equivalent to Z and we know that Z is enumerable and so A_q is enumerable. Also the collection $Q = \underset{q \in Z_0}{\cup} A_q$ is enumerable where Z_0 is a subset of Z, because Q is expressed as enumerable union of enumerable sets.

THEOREM 5.   *(a) The set of all real numbers in the open interval $(0,1)$ is not enumerable.*

(MEERUT 1989,96; HIMACHAL 2000; KANPUR 2004)

     (b) Prove that the set of all real numbers in the closed interval $[0,1]$ is not enumerable.

(MEERUT 1998,2004; GARHWAL 1995,97,2003; HIMACHAL 2002)

PROOF.   (a) Suppose contradiction, *i.e.*, let the given interval be enumerable. Then the elements of the interval can be written as a sequence $\{x_1,x_2,x_3,...\}$. Now using the decimal expansion of these x_m's, we can get as follows :

$$x_1 = .a_{11}a_{12}a_{13}...a_{1n}...$$
$$x_2 = .a_{21}a_{22}a_{23}...a_{2n}...$$
$$x_3 = .a_{31}a_{32}a_{33}...a_{3n}...$$
$$...\quad ...\quad ...\quad ...\quad ...$$
$$x_n = .a_{n1}a_{n2}a_{n3}...a_{nm}...$$
$$...\quad ...\quad ...\quad ...\quad ...$$

where a_{ij}'s may be any integer from 0 to 9. Now let

$$x_n = b_1 b_2 b_3 ...$$

where $b_1, b_2,...$ are the digits from 0 to 9 *s.t.* $b_1 \neq a_{11}$, $b_2 \neq a_{22}$, $b_3 \neq a_{33}$ etc. In general $b_m \neq a_{mm}, \forall m \in N$.

Hence, $x \neq x_1, x \neq x_2,..., x \neq x_m,...$.

Obviously, $x \in (0,1)$ and $x \notin \{x_1, x_2, ...\}$; thus we set that besides countable set $\{x_1, x_2, ...\}$ there exists element belonging to interval $(0,1)$, showing that the set of all real numbers lying in the interval $(0,1)$ is not enumerable.

(b) Open interval (0, 1) is a subset of [0, 1]. Since (0, 1) is not denumerable, the interval [0, 1] is also not denumerable.

THEOREM 6. **The set of all real numbers is not enumerable, i.e., R is not enumerable.**

(MEERUT 1991,97; GARHWAL 1996; KANPUR 1996,2000,02)

PROOF. Suppose contradiction, *i.e.*, let **R** be enumerable. Then its every subset must be enumerable (Theorem 1). Consider the set of all real numbers lying in the interval (0, 1) which is not enumerable by preceding theorem 5; which is again a contradiction since it is the subset of **R**. Hence the theorem.

THEOREM 7. **The set of all irrational numbers is not enumerable.**

PROOF. Let the contrary be true. Also by theorem 4, the set of all rational numbers is enumerable. Thus union of rational and irrational numbers would also be enumerable. But we know that **R** is the union of rational and irrational numbers and hence **R** would be enumerable which is a contradiction according to Theorem 6. Hence the set of all irrational numbers is not enumerable.

1.10 ALGEBRAIC NUMBERS

Let

$$P_n(x) = \alpha_n x^n + \alpha_{n-1} x^{n-1} + ... + \alpha_1 x + \alpha_0 (\alpha_n \neq 0)$$

be a polynomial with α_i's as integral numbers, then we define the algebraic number as a root of the polynomial equation $P_n(x) = 0$ of the above form.

THEOREM 1. **The set of all algebraic numbers is enumerable.**

(MEERUT 1996; GARHWAL 1993; KANPUR 1972)

PROOF. Consider the algebraic equation, $\alpha_0 x^n + \alpha_1 x^{n-1} + ... + \alpha_n$ of degree n with $\alpha_0 \neq 0$. Now we define the rank of this equation :

$$|\alpha_0| + |\alpha_1| + |\alpha_2| + ... + |\alpha_n| = m .$$

Clearly, rank is a positive number. Also α_i's are integers; so rank is an integer ≥ 1. Obviously for a given rank the roots of the equation will be finite and therefore will be enumerable.

Again we can put a one-one correspondence in the set of natural numbers with the algebraic equation arranged with respect to rank and hence the set of all algebraic equations is enumerable. Now each algebraic equation has enumerable number of roots and so the set of all algebraic numbers is the enumerable collection of enumerable sets and hence enumerable by Theorem 2.

SOLVED EXAMPLES

EXAMPLE 1. **Show that the set P of all polynomials**

$$P_n(x) = \alpha_n x^n + \alpha_{n-1} x^{n-1} + ... + \alpha_1 x + \alpha_0, (\alpha_n \neq 0)$$

with integral (rational) coefficients is denumerable. (GARHWAL 2001)

SOLUTION. If $|\alpha_n| + |\alpha_{n-1}| + ... + |\alpha_0| = m$, then for each pair of natural numbers (m,n), set P_{mn} of all the polynomials of the form
$$P_n(x) = \alpha_n x^n + \alpha_{n-1} x^{n-1} + ... + \alpha_1 x + \alpha_0$$
is finite and hence countable.

Also the sets $P_{(m,n)} = P_K$, $K = (m,n) \in \mathbf{N} \times \mathbf{N}$ themselves are countable.

Therefore the set $P = \bigcup\limits_{(m,n) \in \mathbf{N} \times \mathbf{N}} P_{m,n}$ is also countable.

1.11 CARDINALLY EQUIVALENT SETS

A set P is said to be cardinally equivalent to a set Q, if there exists at least one one-to-one map from P to Q and is denoted by $P \sim Q$.

THEOREM I. **_Prove that the relation $P \sim Q$ in the family of sets is an equivalence relation._**

PROOF. Recall that a relation R in a set is an equivalence relation in P, iff

  (i) R is reflexive, *i.e.*, aRa, $\forall a \in P$

  (ii) R is symmetric, *i.e.*, $aRb \Rightarrow bRa$, $\forall a, b \in P$,

  (iii) R is transitive, *i.e.*, $aRb, bRc \Rightarrow aRc$ $\forall a, b, c \in P$.

Here we observe that the given relation is

  (i) Reflexive. Since the identity map $I_P : P \to P$ given by $I_P(a) = a, \forall a \in P$ is one-one onto, *i.e.*, $P \sim P$ for every set P.

  (ii) Symmetric. $P \sim Q \Rightarrow \exists$ a one-one map $f : P \xrightarrow{onto} Q$

     $f^{-1} : Q \xrightarrow{onto} P$ is also one-one $\Rightarrow Q \sim P$.

  (iii) Transitive.

     $P \sim Q \Rightarrow \exists$ a one-one map $f : P \xrightarrow{onto} Q$

     $Q \sim R \Rightarrow \exists$ a one-one map $g : Q \xrightarrow{onto} R$

       $\Rightarrow g \circ f : P \xrightarrow{onto} R$ is also one-one

   $\Rightarrow P \sim R$

Thus the given relation is an equivalence relation. It is to be noted that an equivalence relation decomposes the set P into equivalence classes, any two of which are either equal or mutually disjoint. The set of these mutually disjoint equivalence classes is said to be the quotient set of the set for the given equivalence relation. The equivalence relation is also known as equipollent relation.

SOLVED EXAMPLES

EXAMPLE 1. **_If A and B are the sets of all the real numbers of the intervals $[a_1, a_2]$ and $[b_1, b_2]$ respectively, then show that $A \sim B$._**

SOLUTION. Consider a mapping $f : A \to B$ *s.t.*
If $x \in A$, then

$$f(x) = \frac{b_2 - b_1}{a_2 - a_1} x - a_1 \frac{b_2 - b_1}{a_2 - a_1} + b_1$$

$$= \left(\frac{b_2 - b_1}{a_2 - a_1} \right)(x - a_1) + b_1 \, .$$

Mapping f is one-one. For if $f(x) = f(y); x, y \in A$, then
$$\frac{(b_2 - b_1)}{(a_2 - a_1)}(x - a_1) + b_1 = \frac{(b_2 - b_1)}{(a_2 - a_1)}(y - a_1) + b_1$$

$$\Rightarrow \qquad \left(\frac{b_2-b_1}{a_2-a_1}\right)(x-a_1)=\left(\frac{b_2-b_1}{a_2-a_1}\right)(y-a_1)\Rightarrow x=y$$

$\Rightarrow \quad f$ is one-one.

Mapping f in onto. Let $z\in B\Rightarrow b_1\le z\le b_2$.

If $z=\left(\dfrac{b_2-b_1}{a_2-a_1}\right)(x-a_1)+b_1$, then $x=\dfrac{(z-b_1)(a_2-a_1)}{(b_2-b_1)}+a_1$.

Obviously $a_1\le x\le a_2$ and hence $x\in A$.

Thus for any $z\in B$, $\exists x\in A$ s.t. $f(x)=z$ and hence f is onto.

$\Rightarrow \quad f$ is one-one onto and hence $A\sim B$.

☛ **REMARK**

➡ If $A=(a_1,a_2)$, $B=(b_1,b_2)$ then $A\sim B$.

1.12 CARDINAL NUMBER OF A SET

As we have already seen that the relation of equivalence decomposes any collection of sets into equivalence classes containing the equivalent sets. Each equivalent class has a cardinal number which we shall use to represent the property the equivalent sets have in common. It will be in some sense a measure of the points in sets. It is denoted by the card X, i.e., cardinal number of X. The basic property of cardinal number is that card $X=$ card Y iff $X\sim Y$.

Definition. *Any set which is equivalent to the set {1, 2, 3, ..., n} is said to have the cardinal number n.*

In this way we have defined the cardinal number of the finite sets. The null set is said to have the cardinal number zero. Obviously for finite sets, cardinal number is just the number of element in the sets.

(MEERUT 2003,04; GARHWAL 2002)

The cardinal number of N (set of natural numbers) is denoted by a and hence all enumerable sets will have the cardinal number a.

The cardinal number of the set of real numbers is denoted by c and all the sets equivalent to the set R are said to have the cardinal number c. The set of all real numbers in the interval [0, 1] also has the cardinal number c. Since we have proved that all intervals (open or closed) are equivalent to [0,1], hence every interval also has cardinal number c. The cardinal number of the set of all real-valued functions defined in the interval [0, 1] is denoted by f.

The cardinal number of an infinite set is called a transfinite cardinal number a is considered to be the first (smallest) transfinite cardinal.

Since every finite set has an enumerable subset which is equivalent to be the set of natural number N and so every infinite set has a subset with cardinal number a.

Note that the set of cardinal number $\{0,1,2,3,...,a,c,f\}$ is a superset of set of all natural numbers.

1.12.1 SUM OF CARDINAL NUMBERS

Let A and B be any two disjoint sets; then

$$\text{card } A+\text{card } B=\text{card } A\cup B.$$

In general $\displaystyle\sum_{\alpha\in\Delta} card\Lambda_\alpha = card\left(\bigcup_{\alpha\in\Delta} A_\alpha\right)$

where $A_\alpha\cap A_\beta=\phi$, $\forall\alpha,\beta,\in\Delta$ (index set) such that $\alpha\ne\beta$.

1.12.2 PRODUCT OF CARDINAL NUMBERS

We define the product of two cardinal numbers as card $P\times$ card $Q=card(P\times Q)$ where P,Q are any sets.

In general, card $P_1\times$ card $P_2... =$ card [Cartesian product of sets $P_i(i=1,2,3,...)$]

1.12.3 COMPARISON OF CARDINAL NUMBERS

Let P and Q be any two sets, then

(i) card $P <$ card Q, if $\exists$ a set $R \subset Q$ s.t. $P \sim R$ and if $P \sim Q$ then card $P =$ card Q.

(ii) card $P \leq$ and Q, if $\exists$ a set $R \subseteq Q$ s.t. $P \sim R$, *i.e.*, there exists a one-one map

$$f : P \xrightarrow{\text{onto}} R \subseteq Q.$$

(iii) Evidently, card $(P \cup Q) \geq$ card P or card Q.

(iv) $\boldsymbol{a}$ is the smallest infinite cardinal number.

THEOREM 1. **Prove that card $(P \times Q) = \boldsymbol{card\, Q} + \boldsymbol{card\, Q} + ...\boldsymbol{card\, P}$ terms.**

PROOF. We have $P \times Q = \{(x,y) : x \in P, y \in Q\} = \{ \bigcup_{x \in P} (x,y) : y \in Q\}$.

$$\therefore \quad card(P \times Q) = card\left(\{ \bigcup_{x \in P} (x,y) : y \in Q\}\right). \qquad \text{...(1)}$$

Let $x \in P$ be arbitrary but fixed, Consider the map

$$f : Q \to \{(x,y) : y \in Q\}$$

defined by $f(y) = (x,y) \quad \forall y \in Q$.

This show that f is one-one. Therefore

$$card\, Q = card((x,y) : y \in Q))$$

Observing (1) we get the required result.

EXAMPLE 1. **Show that $\alpha \leq \boldsymbol{card\, P} \leq \alpha \quad \Rightarrow \quad \boldsymbol{card\, P} = \alpha$**

SOLUTION. $\alpha \leq card\, P \quad \Rightarrow \quad \exists$ a set Q such that card $Q \leq$ card P where card $Q = \alpha$

$\Rightarrow \quad Q$ is equivalent to a subset of P $\qquad \text{...(1)}$

Again $\qquad card\, P \leq \alpha \quad \Rightarrow \quad card\, P \leq Q$

$\qquad\qquad\qquad\qquad \Rightarrow \quad P$ is equivalent to a subset of Q $\qquad \text{...(2)}$

Combining (1) and (2), we get

$$card\, P = card\, Q \text{ or } card\, P = \alpha \qquad\qquad (\because card\, Q = \alpha)$$

THEOREM 2. **Let P and Q be any two sets; then show that**

(i) card $P +$ card Q is unique,

(ii) card P.card Q is unique.

PROOF. Let $P \sim P_1, Q \sim Q_1, P_1 \cap Q_1 = \phi$. Now

(i) $P \sim P_1 \Rightarrow \exists$ a one-one map $f : P \xrightarrow{\text{onto}} P_1$

$Q \sim Q_1 \Rightarrow \exists$ a one-one $g : Q \xrightarrow{\text{onto}} Q_1$.

Define a function $\psi : P \cup Q \to P_1 \cup Q_1$ s.t.

$$\psi(x) = \begin{cases} f(x) & , \quad \forall x \in P, \\ g(x) & , \quad \forall x \in Q. \end{cases}$$

Evidently ψ is one-one onto since f and g are one-one onto. Hence $P \cup Q \sim P_1 \cup Q_1 \Rightarrow card\ (P \cup Q) = card\ (P_1 \cup Q_1)$

This shows that $card\, P + card\, Q$ is unique.

(ii) $x \in P, y \in Q \quad \Rightarrow \quad (x,y) \in P \times Q$

$$\Rightarrow (f(x), g(y)) \in P_1 \times Q_1.$$

Define a function $\psi : P \times Q \to P_1 \times Q_1$ s.t.,

$$\psi(x.y) = (f(x), g(y)) \quad \forall (x,y) \in P \times Q.$$

Again f and g one-one onto, therefore ψ is one-one onto.

$\Rightarrow \quad P \times Q \sim P_1 \times Q_1 \quad \Rightarrow \quad card(P \times Q) = card(P_1 \times Q_1)$.

This shows that card P. card Q is unique.

EXAMPLE 2.   *Show that* $(P \times Q) \sim (Q \times P)$ *i.e.,* $(P \times Q)$ *is cardinally equivalent to* $(Q \times P)$.

SOLUTION.   We have $(P \times Q) = \{p,q\} : p \in P, q \in Q\}$

Define a map $f : (P \times Q) \to (Q \times P)$ by the formula $f(p,q) = (q,p)$.

f is one-one. Let $f((p_1,q_1)) = f((p_2,q_2))$

$\Rightarrow \quad (q_1,p_1) = (q_2,p_2) \Rightarrow q_1 = q_2$ and $p_1 = p_2$

$\Rightarrow \quad (p_1,q_1) = (p_2,q_2)$.

Obviously f is onto.

Hence, $(P \times Q) \sim (Q \times P)$.

THEOREM 3.   *Let* card $P = p$, card $Q = q$, card $R = r$, *then show that :*

(i) $p + q = q + p$,

i.e., addition of cardinal numbers is commutative.

(ii) $p \cdot q = q \cdot p$ (GARHWAL 2002)

i.e., multiplication of cardinal numbers is commutative.

(iii) $p \cdot (q + r) = p \cdot q + p \cdot r$,

i.e., multiplication is distributive over addition.

(iv) $p \cdot (q \cdot r) = (p \cdot q) \cdot r$

i.e., associative law for multiplication holds.

(v) $p + (q + r) = (p + q) + r$

i.e., associative law for addition holds.

PROOF.   (i) Let $P \cap Q = \phi$. The elements of an arbitrary set may be in any order, therefore

$P \cup Q = Q \cup P \Rightarrow card(P \cup Q) = card(Q \cup P)$

$\Rightarrow \quad card\, P + card\, Q = card\, Q + card\, P$.

(ii) We know that $P \times Q \sim Q \times P$ (See Ex. 2 above)

$\Rightarrow \quad card\,(P \times Q) = card\,(Q \times P)$

$\Rightarrow \quad p \cdot q = q \cdot p$.

(iii) Let Q, R be disjoint sets. Then

$p.(q + r) = card\,[P \times (Q \cup R)] = card\,[(P \times Q) \cup (P \times R)]$

$= card\,(P \times Q) + card\,(P \times R) = p \cdot q + p \cdot r$.

(iv) $\because P \times (Q \times R) \sim (P \times Q) \times R$ under the map $f(x,(y,z)) = ((x,y),z)$,

Hence $card\,[P \times (Q \times R)] = card\,[(P \times Q) \times R]$

or $p \cdot (q \cdot r) = (p \cdot q) \cdot r$.

(v) Suppose that P, Q, R are pairwise disjoint sets. Also we know that

$(P \cup Q) \cup R = P \cup (Q \cup R)$

$\Rightarrow \quad card\,[(P \cup Q) \cup R] = card\,[P \cup (Q \cup R)]$

$\Rightarrow \quad (p \mid q) + r = p + (q + r)$

THEOREM 4.   (i) *If* A_i *is enumerable set for* $i = 1,2,3,...,n$ *then* $\bigcup\limits_{i=1}^{n} A_i$ *is enumerable*

and hence deduce that $n.a = a$.   (GURUKUL KANGRI 2001; GARHWAL 1995)

(ii) *If* A_i *is enumerable set for* $i = 1,2,3,...,n$, *then* $\bigcup\limits_{i=1}^{n} A_i$ *is enumerable*

and hence deduce that $a + a + a + ...$ *to a terms* $= a$

PROOF.   (i) Let $A = \bigcup\limits_{i=1}^{n} A_i$. We know that countable union of enumerable set is enumerable, hence A is enumerable.

DEDUCTION.   Let $A_i \cap A_j = \phi$ for $i \neq j$; then by the definition of sum of cardinal numbers,

$$card\ A_1 + card\ A_2 + ... + card\ A_n = card\ A \quad \left(since\ \bigcup\limits_{i=1}^{n} A_i = A\right) \qquad ...(1)$$

We know that cardinal number of an enumerable set is $\boldsymbol{a}$ and since each $A_i(i = 1, 2, 3, ..., n)$ is enumerable, so we have

$$card\ A_1 = card\ A_2 = ...card\ A_n = \boldsymbol{a}$$

Also A is enumerable so card $A = \boldsymbol{a}$. So from (1), $n \cdot \boldsymbol{a} = \boldsymbol{a}$

       (ii) Let $A = \bigcup\limits_{i=1}^{\infty} A_i$; then A is enumerable being enumerable union of enumerable sets and hence card $A = \boldsymbol{a}$

DEDUCTION.   Let $A_i \cap A_j = \phi$ for $i \neq j$; then card $A = \boldsymbol{a}$ gives

$$\sum\limits_{i=1}^{\infty} card\ A_i = \boldsymbol{a}$$

$$\Rightarrow \quad \boldsymbol{a} + \boldsymbol{a} + \boldsymbol{a} + \boldsymbol{a} + ... \text{ to } \boldsymbol{a} \text{ terms} = \boldsymbol{a}.$$

THEOREM 5.   ***(i) If A_i is non-enumerable set for $1 \leq i \leq n$, then $\bigcup\limits_{i=1}^{n} A_i$ is non-enumerable and hence deduce that***

$$\boldsymbol{c} + \boldsymbol{c} + \boldsymbol{c} + ... \text{to } n \text{ terms} = \boldsymbol{c}.$$

(ii) If A_i is non-enumerable, $\forall i \in N$, then $\bigcup\limits_{i=1}^{\infty} A_i$ is non-enumerable and hence deduce that

$$\boldsymbol{c} + \boldsymbol{c} + \boldsymbol{c} + ... \text{ to } \boldsymbol{a} \text{ terms} = \boldsymbol{c}. \qquad \text{(GARHWAL 1991)}$$

PROOF.   (i) Each A_i is non-enumerable for $1 \leq i \leq n$, therefore card $A_i = \boldsymbol{c}$ for $1 \leq i \leq n$.

This imply $A_i \sim [a_i, a_{i+1})$ where $a_i, a_{i+1} \in \boldsymbol{R}$ for $i = 1, 2, 3, ..., n$.

Let $a_i < a_{i+1}$ for $i = 1, 2, 3, ..., n$. Thus, we have

$$A_1 \sim \{a_1, a_2\},$$
$$A_2 \sim \{a_2, a_3\},$$
$$... \quad ... \quad ...$$
$$... \quad ... \quad ...$$
$$A_n \sim [a_n, a_{n+1}).$$

At first suppose that $A_i \cap A_j = \phi$ for $i \neq j$; then $\bigcup\limits_{i=1}^{n} A_i$ is equivalent to some subset of $[a_1, a_{n+1})$.

$$\therefore \quad card \left(\bigcup\limits_{i=1}^{n} A_i \right) \leq \boldsymbol{c}. \qquad ...(1)$$

Now evidently

$$A_i \subseteq \bigcup\limits_{i=1}^{n} A_i \quad \Rightarrow \quad card\ A_i \leq card \left(\bigcup\limits_{i=1}^{n} A_i \right)$$

$$\Rightarrow \quad c \le card \left(\bigcup_{i=1}^{n} A_i \right).$$ ...(2)

Combining (1) and (2), we have

$$c \le card \left(\bigcup_{i=1}^{n} A_i \right) \le c, \text{ i.e., } card \left(\bigcup_{i=1}^{n} A_i \right) = c.$$

$$\therefore \quad \bigcup_{i=1}^{n} A_i \text{ is non-enumerable.}$$

Now suppose that $A_i \cap A_j = \phi$ for $i \ne j$, then

$$\bigcup_{i=1}^{n} A_i \sim [a_1, a_{n+1}) \Rightarrow card \left(\bigcup_{i=1}^{n} A_i \right) = card[a_1, a_{n+1})$$

$$= card \left(\bigcup_{i=1}^{n} A_i \right) = c.$$

DEDUCTION. If we assume that $A_i \cap A_j = \phi$ for $i \ne j$, then as proved above, we have

$$\sum_{i=1}^{n} card (A_i) = c \quad \left(\because card \bigcup_{i=1}^{n} A_i = \sum_{i=1}^{n} card \, A_i \text{ when } A_i \cap A_j = \phi \text{ for } i \ne j \right)$$

i.e., $c + c + c + \dots$ to n terms $= c$.

(ii) Let $A = \bigcup_{i=1}^{\infty} A_i$, where card $A_i = c$, $\forall i \in N$. Now

card $A_i = c \Rightarrow A_i \sim \left[1 - \dfrac{1}{2^{i-1}}, 1 - \dfrac{1}{2^i} \right)$.

Thus, we have

$$A_1 \sim \left[0, \frac{1}{2}\right), A_2 \sim \left[\frac{1}{2}, \frac{3}{4}\right), A_3 \sim \left[\frac{3}{4}, \frac{7}{8}\right), \dots \quad \dots A_i \sim \left[1 - \frac{1}{2^{i-1}}, 1 - \frac{1}{2^i}\right) \dots.$$

Now assume that $A_i \cap A_j = \phi$ for $i \ne j$.

Then obviously $\bigcup_{i=1}^{\infty} A_i \sim [0, 1)$.

$$\therefore \quad card \left(\bigcup_{i=1}^{\infty} A_i \right) = card \, [0, 1) \text{ or card } A = c.$$

Next suppose that $A_i \cap A_j \ne \phi$ for $i = j$, then $\bigcup_{i=1}^{\infty} A_i$ is cardinally equivalent to some subset of $[0, 1)$.

So, $card \left(\bigcup_{i=1}^{\infty} A_i \right) \le c.$ ...(3)

Again $A_i \subset \bigcup_{i=1}^{\infty} A_i \Rightarrow card \, A_i \le card \left(\bigcup_{i=1}^{n} A_i \right)$

$$\Rightarrow c \le card \left(\bigcup_{i=1}^{\infty} A_i \right)$$ ...(4)

Combining (3) and (4) we get card $\left(\bigcup_{i=1}^{\infty} A_i \right) = c$.

Hence $\bigcup_{i=1}^{\infty} A_i$ is non-enumerable.

DEDUCTION. Suppose that $A_i \cap A_j = \phi$ for $i \neq j$, we have

$$card\left(\bigcup_{i=1}^{\infty} A_i\right) = c \implies \sum_{i=1}^{\infty} card\, A_i = c$$

or $\quad c + c + c + \dots$ to $\boldsymbol{a}$ terms $= \boldsymbol{c}$.

THEOREM 6. **Prove $\boldsymbol{a} + \alpha = \alpha, \alpha$ being any transfinite cardinal number.**

(GURUKUL KANGRI 2001)

PROOF. If an enumerable set is added to an infinite set. Since A is an infinite set and therefore $\exists$ a subset B of A s.t. B is enumerable.

$\therefore \quad$ card $B = \boldsymbol{a}$.

Now we can write $A = (A - B) \cup B$.

$\therefore \quad A \cup N = (A - B) \cup B \cup N = (A - B) \cup (B \cup N)$.

B and N are enumerable sets $\implies B \cup N$ is enumerable

$\implies \quad B \cup N \sim N$.

Now $B \cup N \sim N, N \sim B \implies B \cup N \sim B$.

$\therefore \quad (A - B) \cup (B \cup N) \sim (A - B) \cup B$ i.e., $A \cup N \sim A$

or $\quad card(A \cup N) = card\, A$ i.e., $\quad \alpha + \boldsymbol{a} = \alpha$.

THEOREM 7. **Show that $\boldsymbol{c} \cdot \boldsymbol{c} = \boldsymbol{c}$.** (BANARAS 1977)

PROOF. Let $A = \{x : 0 \leq x \leq 1\}$. Then card $A = \boldsymbol{c}$.

Now, $\quad$ let $B = \{(0, x) : x \in A\}$

Then obviously $B \subset A \times A \qquad \therefore card\, B \leq card(A \times A)$

Again $A \sim B$ under the map f s.t.

$$f(x) = (0, x), \quad \forall x \in A, \quad \therefore card\, A \leq card\, B \ .$$

$\therefore \qquad card\, A \leq card\,(A \times A)$. $\qquad \qquad \dots(1)$

Let x, y be any two real numbers in the closed interval $[0, 1]$. Then x and y can uniquely be expanded in the form of infinite decimals which contain non-zero digits. Now define a map $g : (A \times A) \to A$ by writing

$$g(x, y) = 0. \quad x_1 y_1 x_2 y_2 x_3 y_3 \dots \ .$$

Obviously g is one-one. So by definition

$\qquad card\,(A \times A) \leq card\, A$. $\qquad \qquad \dots(2)$

From (1) and (2), we get

$\qquad card\, A \leq card\,(A \times A) \leq card\, A \ ; \ card\,(A \times A) = card\, A$

or $\qquad \boldsymbol{c} \cdot \boldsymbol{c} = \boldsymbol{c}$

1.13 TRANSCENDENTAL NUMBER

Definition. *A real number which is not an algebraic number is called Transcendental number. Thus the numbers e and π which are real but not algebraic numbers, are transcendental numbers.*

(MEERUT 2003)

All rational numbers are algebraic number hence every rational number is not transcendental, implying that every transcendently number must be irrational, for

$\qquad \boldsymbol{R} = $ (rational numbers) $\cup$ (irrational numbers).

It must be noted by the readers that there are so many irrational numbers which are algebraic e.g., $(n)^{1/r}$. Therefore every irrational number is not transcendental number.

THEOREM 1. **Prove that every monotonic function in a closed interval is discontinuous at a countable number of points of that interval.**

PROOF.   Let $f(x)$ be a monotonic function in the closed interval $[a, b]$. Also let it be a monotonically increasing function and be discontinuous at an arbitrary point x. Then

$$\delta(x) = f(x+0) - f(x-0) > 0 \qquad\qquad \text{...(1)}$$

where $\qquad f(a) = f(a-0), f(b) = f(b+0)$.

Let $\xi_1, \xi_2, ..., \xi_{m-1}$ be numbers in the intervals $x_k < \xi_k < x_{k+1}$ where $a < x_1 < x_2 < ... x_m < b$, where $\xi_0 = a$ and $\xi_m = b$.

$$\therefore \quad f(\xi_k) - f(\xi_{k-1}) \geq f(x_k + 0) - f(x_k - 0) = \delta(x_k) \qquad \text{[by (1)]} \qquad\qquad \text{...(2)}$$

Therefore $\quad f(b) - f(a) = \sum_{k=1}^{m} [f(\xi_k) - f(\xi_{k-1})] \geq \sum_{k=1}^{m} \delta(x_k)$.

Let $\quad \delta(x_k) > \dfrac{1}{n}, \ \forall k$.

Then by last inequality, we have

$$f(b) - f(a) > \frac{m}{n} \quad \text{or} \quad [f(b) - f(a)]n > m.$$

This shows that m which is the number of points of discontinuity x with $\delta(x) > \dfrac{1}{n}$ is bounded above, *i.e.*, the number of points of discontinuity x with $\delta(x) > \dfrac{1}{n}$ are finite in the closed interval $[a, b]$. Since $n \in \mathbf{N}$, we see that the number of points of discontinuity x with $\delta(x) > \dfrac{1}{n}$ are finite in the closed interval $[a, b]$. Since every finite set is countable and for every $x \ \exists \ n \in \mathbf{N}$ therefore the number of points of discontinuity in the closed interval $[a, b]$ will be an enumerable union of countable sets and hence countable. Hence the theorem.

THEOREM 2.   **(Cantor's Theorem).** *Prove that card* $A <$ *card* $P(A)$, $P(A)$ *being power set of the set A.*
(MEERUT 1983,91; GURUKUL KANGRI 2001)

Or

Show that for every cardinal number $n, 2^n > n$.

PROOF.   Let $\quad B^* = \{\{x\} : x \in A\};$ then obviously $B^* \subset P(A)$. Now define a map $f : A \to B^* \ s.t. f(x) = \{x\}$. Obviously $A \sim B^*$. Hence

$$card \ A \leq card \ P(A).$$

Thus we are only to show that $card \ A \neq card \ P(A)$,

i.e., $\qquad\qquad A \propto P(A) \qquad\qquad \text{...(1)}$

Suppose contradiction, *i.e.,* $A \sim P(A)$. So $\exists$ a one-one map $f : A \xrightarrow{\text{onto}} P(A)$.

Let $\qquad\qquad B = \{x \in A : x \notin f(x)\}$.

Clearly, $\qquad\qquad B \subset A \ \Rightarrow \ B \in P(A)$.

Since the mapping f is onto, there must exist $x \in A \ s.t. \ f(x) = B$. Now if $x \in B$ then by definition of $B, x \in f(x)$ which is not possible.

Consider the second possibility that $x \notin B$, then $x \in f(x) = B$ which is again impossible.

It means that our assumption is wrong. Hence (1) is true.

DEDUCTION.   (i) From the last theorem, it following that $n < 2^n$ where card $A = n$ and

$$cardP(A) = 2^{cardA} = 2^n.$$
(MEERUT 1988)

(ii) Also $2^a > a$.

THEOREM 3. **(Equivalence Theorem).** *If* $A_1 \subset B \subset A$ *and* $A \sim A_1$, *then* $A \sim B$.

Or

If $A_1 \subset B \subset A$ *and* card $A =$ card A_1, *then* card $A =$ card B.

PROOF. $A \sim A_1 \Rightarrow \exists$ a one-one map $f : A \xrightarrow{\ onto\ } A_1$.

As $B \subset A$ so f_B is a one-one onto where f_B is the restriction of f to B. This means that $B \sim B_1 \subset A_1$. Similarly, $A_1 \sim A_2 \subseteq B_1$.

Continuing in this way we get equivalent sets

$$A, A_1, A_2, \ldots \text{ and } B, B_1, B_2, \ldots$$

s.t. $\qquad A \supset B \supset A_1 \supset B_1 \supset A_2 \supset B_2 \supset A_3 \supset B_3 \supset \ldots$

Let $\qquad S = A \cap B \cap A_1 \cap B_1 \cap A_2 \cap B_2 \cap \ldots$

Then we can write

$$A = (A - B) \cup (B - A_1) \cup (A_1 - B_1) \cup \ldots \cup S,$$
$$B = (B - A) \cup (A_1 - B_1) \cup (B_1 - A_2) \cup \ldots \cup S$$

Define a map $\psi : A \to B$ s.t.

$$\psi(A - B) = A_1 - B_1,$$
$$\psi(A_1 - B_1) = A_2 - B_2,$$
$$\psi(A_2 - B_2) = A_3 - B_3,$$
$$\ldots \qquad \ldots \qquad \ldots \qquad \ldots$$
$$\ldots \qquad \ldots \qquad \ldots \qquad \ldots$$
$$\psi(B - A_1) = B - A_1,$$
$$\psi(B_1 - A_2) = B_1 - A_2,$$
$$\ldots \qquad \ldots \qquad \ldots \qquad \ldots$$
$$\ldots \qquad \ldots \qquad \ldots \qquad \ldots$$
$$\ldots \qquad \ldots \qquad \ldots \qquad \ldots$$
$$\psi(S) = S.$$

Above definition of ψ makes the mapping ψ one-one and showing that $A \sim B$.

THEOREM 4. **(Schorder-Bernstein Theorem).** *If* card $A \le$ card B *and* card $B \le$ card A, *then* card $A =$ card B.

(MEERUT 1989)

Or

If each of the sets A *and* B *is equivalent to a subset of other, then* $A \sim B$.

PROOF. Let f and g be one-to-one mapping from A onto B and from B into A respectively. Let $f(A) = B_i \subset B$ and $g(B) = A_2$ and $g(B_1) = A_3$, then we have $A \supset A_2 \supset A_3$. Further $g(B_1) = A_3, f(A) = B_1$ implies $g(f(A)) = A_3$, giving gof is a one-to-one mapping from A to $A_3 \Rightarrow$ card $A =$ card A_3. Hence by the above theorem, card $A =$ card A_2. Also existence of g s.t. $g(B) = A_2$ shows that card $B =$ card A_2. Hence card $B =$ card A.

THEOREM 5. **Show that** $2^a = c$.

PROOF. We know that card $[0,1] = c$. On the other hand each $x \in [0,1]$ can be written in the form of binary expansion as $x \equiv 0.x_1 x_2 x_3 \ldots$, where each $x_i = 0$ or 1.

But selecting each x_i in two ways (either 0 or 1) we can form at most 2^a numbers. So the card $[0,1] = 2^a$ implying that $2^a = c$.

THEOREM 6. *Every superset of an uncountable set is uncountable.*

PROOF. Suppose contradiction, *i.e.*, if B is the superset of an uncountable set A, then B is countable. But we know that every subset of a countable set is countable and so A must be countable which is a contradiction and so B is uncountable. That is to say that every superset of an uncountable set is uncountable.

THEOREM 7. *Union of two enumerable sets is also enumerable.*

PROOF. Let A and B be the two enumerable sets.

CASE I. When $A \cap B = \phi$.

Let $A = \{a_1, a_2, ...\}, B = \{b_1, b_2, ...\}$. Now establish correspondence
$$f : A \cup B \to N$$
s.t. $\qquad f(a_n) \to 2n - 1 \qquad \qquad$ (odd positive integer),
$$f(b_n) \to 2n \qquad \qquad \text{(even positive integer)}.$$
Evidently this mapping is one-one from $A \cup B$ onto N.

CASE II. When $A \cap B \neq \phi$, then we can write $A \cup B = A \cup (B - A)$. Taking $B_1 = B - A$ we have $A \cap B_1 = \phi$. As already proved $A \cup B_1$ is countable where B_1 is countable where B_1 is countable being the subset of countable set B and hence $A \cup B = A \cup B_1$ is countable when B_1 is countable infinite but if B_1 is finite say
$$B_1 = \{e_1, e_2, ..., e_m\} .$$
Then $\qquad A \cup B = A \cup B_1 = \{e_1, e_2, e_3, ..., e_m, a_1, a_2, a_3, ...\}$.

Now set a correspondence
$$f : N \to A \cup B_1$$
s.t. $\qquad f(i) = e_i, \qquad 1 \leq i \leq m ,$
$$f(m + i) = a_i, \qquad \forall i$$
$\Rightarrow \qquad A \cup B$ is enumerable.

We can generalize the result that union of two countable sets is countable (whether each is countably infinite or finite).

THEOREM 8. *Every infinite set is equivalent to its proper subset.*

PROOF. **CASE I.** When A is countably infinite, then A can be written as a sequence. Let $A = \{a_1, a_2, a_3 ...\}$. Then the function $f(a_n) = a_{n+1}$ establishes a one to one correspondence between the set A and $A - \{a_1\}$ which is a proper subset of A.

CASE II. When A is uncountably infinite, then it has an enumerable subset say B where $B = \{a_1, a_2, a_3 ...\}$. We shall show that $A \sim (A - \{a_i\})$.

Write $C = A - B$; then $A = B \cup C$ and $B \cap C = \phi$. Also $A - \{a_i\} = B - \{a_i\}) \cup C$. Let $e(x)$ be the identity mapping which associates each $x \in A$ onto itself. Let f be the function: $f(a_n) = a_{n+1}$. Now define a function h:
$$h(x) = \begin{cases} e(x) : x \in C, \\ f(x) : x \in B. \end{cases}$$

Then the range of h is $(B - \{a_i\}) \cup C$ which is a proper subset of $B \cup C = A$. Thus the result follows.

THEOREM 9. *If* α *and* β *are cardinal numbers such that* $\alpha \leq \beta$ *and* $\beta \leq \alpha$, *then* $\alpha = \beta$.

PROOF. Let card $A = \alpha$, card $B = \beta$
$$\alpha \leq \beta \Rightarrow \text{card } A \leq \text{card } B$$
$\Rightarrow \qquad A \sim B$ or $A \sim$ to a subset of B $\qquad \qquad \qquad \qquad$ ...(1)
$\qquad \beta \leq \alpha \Rightarrow B \sim A$ or $B \sim$ to a subset of A . $\qquad \qquad \qquad$ ...(2)

(1) and (2) give the required result.

THEOREM 10. *If an enumerable set is subtracted from an enumerable set, the remaining set will be enumerable.*

PROOF. Suppose contradiction, *i.e.,* $A - B$ is non-enumerable where A and B are enumerable sets. We can write $A = (A - B) \cup B$. Now $A - B$ is non-enumerable gives A to be non-enumerable which is a contradiction. Hence the result follows.

THEOREM 11. *If we subtract an enumerable set from a non-enumerable set, then the remaining set is non-enumerable.*

PROOF. Suppose contradiction, *i.e.,* $A - B$ is enumerable where A is non-enumerable and B is an enumerable set. We can write $A = (A - B) \cup B$. Since both $A - B$ and B are enumerable sets, it implies A is enumerable which a contradiction as A is non-enumerable. Hence the result.

SOLVED EXAMPLES

EXAMPLE 1. *Prove that the set Z of all integers is countable.*

(MEERUT 1993,98,2004; KANPUR 2001)

SOLUTION. Write

$$\mathbf{Z}^+ = \{1,2,3,...\},$$

$$\mathbf{Z}^- = \{-1,-2,-3,...\}.$$

Then, we have $\mathbf{Z}^+ \sim \mathbf{N}$ under the mapping $n \to n$.

Now $\mathbf{Z}^+ \sim \mathbf{N}$ under the mapping $n \to n$,

$$\mathbf{Z}^- \sim \mathbf{N} \text{ under the mapping } (-n) \to n.$$

Also singleton set {0} is finite, so countable. Thus **Z** is the countable union of countable sets and hence countable.

Alternatively, define the mapping $f : \mathbf{N} \to \mathbf{Z}$

$$s.t. f(x) = (-1)^x \left[\frac{x}{2}\right],$$

where $\left[\frac{x}{2}\right]$ represents the integral value of $\frac{x}{2}$, *i.e.,* represents the largest integer less than or equal to $\frac{x}{2}$.

Establish that this mapping gives a one-to one correspondence between **N** and **Z** implying that **Z** is countable.

Alternatively the $f : \mathbf{N} \to \mathbf{Z}$ s.t.

$$f(n) = \frac{n-1}{2} \text{ if } n \text{ is odd} \in \mathbf{N}$$

$$= -\frac{n}{2} \text{ if } n \text{ is even} \in \mathbf{N}.$$

This is also a one-to one correspondence $\Rightarrow \mathbf{Z} \sim \mathbf{N}$.

☞ **REMARK**

It shows that an infinite set can be equivalent to its proper subset, *e.g.,* $\mathbf{Z} \sim \mathbf{N}$.

EXAMPLE 2. *Find the power of an aggregate of numbers given by $\dfrac{M}{2^m}$, M and m being positive and integral.* (GARHWAL 1994,98)

(Power of a set means the cardinal number of that set).

SOLUTION. Let us suppose that

$$B = \left\{ \frac{M}{2^m} : M, \; m \in \boldsymbol{N} \right\}.$$

Write $\quad B_M = \left\{ \dfrac{M}{2^m} : m \in \boldsymbol{N} \right\}.$

Then we have,

$$B_1 = \left\{ \frac{1}{2^1}, \frac{1}{2^2}, ..., \frac{1}{2^n}, ... \right\}$$

$$B_2 = \left\{ \frac{2}{2^1}, \frac{2}{2^2}, ..., \frac{2}{2^n}, ... \right\}$$

$$B_3 = \left\{ \frac{3}{2^1}, \frac{3}{2^2}, ..., \frac{3}{2^n}, ... \right\}$$

$$\cdots \qquad \cdots \qquad \cdots \qquad \cdots$$

$$B_n = \left\{ \frac{n}{2^1}, \frac{n}{2^2}, ..., \frac{n}{2^n}, ... \right\}$$

$$\cdots \qquad \cdots \qquad \cdots \qquad \cdots$$

Evidently,

(i) B_i is enumerable $\forall i \in \boldsymbol{N}$ under the mapping $\dfrac{i}{2^n} \to n$,

(ii) B_i's are pairwise disjoint.

(iii) $B = \overset{\infty}{\underset{i=1}{\bigcup}} B_i$.

Thus B is enumerable being the enumerable union of enumerable sets. Hence card $B = a$, *i.e.*, the power of the given set is a.

EXAMPLE 3. *Prove that $a < c$.*

SOLUTION. Since $\boldsymbol{N} \subset \boldsymbol{R} \Rightarrow$ card N < card R

$$\Rightarrow \quad \boldsymbol{a} < \boldsymbol{c}.$$

EXAMPLE 4. *Prove that $\operatorname{card} P(A) = 2^{\operatorname{card} A}$ for any finite set A.* (MEERUT 1988)

SOLUTION. Let $\operatorname{card} A = \alpha$.

Then $\operatorname{card}(P(A)) = 1 + {}^{\alpha}c_1 + {}^{\alpha}c_2 + ... + {}^{\alpha}c_{\alpha} = (1+1)^{\alpha} = 2^{\alpha} = 2^{\alpha} = 2^{\operatorname{card} A}$.

EXAMPLE 5. *Prove that $\alpha \le \alpha$ for any cardinal number α.*

SOLUTION. Let card $P = \alpha$.

Define an identity map $f : P \to P$ written by

$$f(x) = x, \forall x \in P.$$

Obviously, f is one-one. Hence by definition,

$$\operatorname{card} P \le \operatorname{card} P \quad \textit{i.e.,} \quad \alpha \le \alpha.$$

EXAMPLE 6. *If each $a_i(i = 1, 2, ..., n)$ is a rational number, then the point $a = (a_1, a_2, ..., a_n) \in R^n$ is called a rational point. Show that the set of all rational points in R^n is denumerable.*

SOLUTION. We know that set of rational numbers is countable. So varying $a_1, a_2, ..., a_n$, we can form a^n rational points.

But we know that $a . a = a$ and hence $a^n = a$.

$\Rightarrow$ set of all rational points has the same cardinal number as that of **N**.

$\Rightarrow$ set of rational points is countable.

EXAMPLE 7. *Prove that $c^c = 2^c$.*

SOLUTION. We have $c^c = (2^a)^c = 2^{ac} = 2^c$.

EXAMPLE 8. *Let X be any non-empty set and let C be the family of functions $f : x \rightarrow \{0, 1\}$. Then show that the family of subset of X i.e., the power set of X is equivalent to C.*

SOLUTION. Let $A \in P(X)$ where $P(X)$ denotes the power set of X. Also let ϕ_A denote the characteristic function of A relative to X.

Now define a map $f : P(X) \rightarrow C$ by the formula $f(A) = \phi_A$.

Obviously f as defined above is one-one onto. Hence, $P(X) \sim C$.

EXAMPLE 9. *Prove that $]0, 1] \sim]0, 1[$.*

SOLUTION. Denote the points of $]0, 1]$ by x and of $]0, 1[$ by y. Now define a correspondence

$$y = \frac{3}{2} - x \text{ for } \frac{1}{2} < x \leq 1 \text{; then } \frac{1}{2} \leq y < 1,$$

$$y = \frac{3}{4} - x \text{ for } \frac{1}{4} < x \leq \frac{1}{2} \text{; then } \frac{1}{4} \leq y < \frac{1}{2},$$

$$y = \frac{3}{8} - x \text{ for } \frac{1}{8} < x \leq \frac{1}{4} \text{; then } \frac{1}{8} \leq y < \frac{1}{4},$$

and so on.

From the above correspondence, we see that for every $x \in]0, 1]$, there corresponds one and only one y of $]0, 1[$. Hence by definition $]0, 1] \sim]0, 1[$.

EXAMPLE 10. *Show that for every real number x, the real number in the semi-open interval [x, x + 1) form an uncountable set.*

SOLUTION. Let x be any real number. Define a function

$$f : [x, x + 1) \rightarrow [0, 1) \text{ given by } f(y) = y - x.$$

Then f is well defined, for obviously,

$$f(x) = x - x = 0, f(x + 1) = x + 1 - x = 1.$$

Again $f(y_1) = f(y_2) \Rightarrow y_1 - x = y_2 - x \Rightarrow y_1 = y_2 \Rightarrow f$ is one-one.

Also f is a continuous map which implies that f is an onto map.

Hence $[x, x + 1) \sim [0, 1)$

$\therefore$ $card [x, x + 1) = card [0, 1)$.

Now we know that the set of all real numbers in the set semi-open interval [0, 1) is uncountable and hence the set of all real numbers in [x, x + 1) which is cardinally equivalent to [0, 1) is uncountable.

EXAMPLE II. *If α is any transfinite cardinal number, then $a \leq \alpha$.*

SOLUTION. Let A be an infinite arbitrary set *s.t.* card $A = \alpha$.

Now, A is infinite set $\Rightarrow \exists$ an enumerable subset B of $A \Rightarrow card\ B = \boldsymbol{a}$

$$B \subseteq A \quad \Rightarrow \quad card\ B \leq card\ A \quad \Rightarrow \quad \boldsymbol{a} \leq \alpha.$$

EXAMPLE 12. *Show that the set of all transcendental numbers in any interval is non-enumerable.*   (MEERUT 1995, 2003; PUNJAB UNIV. 2002)

SOLUTION. We know that the set of all algebraic numbers and transcendental numbers is the set of all real numbers which is known to be uncountable. Also we know that the set of algebraic numbers in an interval is enumerable.

But we have already proved that if an enumerable set is removed from a non-enumerable set, the remaining set is non-enumerable. Therefore the complement of the set of all algebraic numbers in any interval relative to the set of all real numbers in that interval is uncountable. But this is the set of all transcendental numbers. Hence the result.

EXAMPLE 13. *Show that the interval (0, 1) is equivalent to the set R of all real numbers and hence show that $card\ (0,1) = card\ R$.*

SOLUTION. Define a function $f : (0,1) \to \boldsymbol{R}$ *s.t.*

$$f(x) = \begin{cases} \dfrac{2x-1}{x}, x \in \left(0, \dfrac{1}{2}\right) \\ \dfrac{2x-1}{1-x}, x \in \left[\dfrac{1}{2}, 1\right) \end{cases}.$$

which show that this function is one-one and onto implying that $(0,1) \sim \boldsymbol{R}$ and hence $card\ (0,1) = card\ R = \boldsymbol{c}$.

Also since $(0,1)$ is uncountable, the set $\boldsymbol{R}$ is also uncountable.

Above property supports our idea of defining the same cardinal numbers $\boldsymbol{c}$ of the two sets $(0,1)$ and $\boldsymbol{R}$. Then $\boldsymbol{c}$ is called the Cardinal number of continuum.

EXAMPLE 14. *Prove that $a < c < f$ where a, c and f denote the cardinal numbers of set of all natural numbers, and real numbers and set of all real valued functions defined over [0,1] respectively.*   (GARHWAL 1997; MEERUT 1983,89)

SOLUTION. We have already proved that $\boldsymbol{a} < \boldsymbol{c}$. Now it remains to prove that $c < f$. Let F be set of all real valued functions defined over [0, 1].

Now consider the mapping $f_k : [0,1] \to \boldsymbol{R}$ defined as $f_k(x) = k, \quad \forall x \in [0,1]$ and k being a real number in [0, 1].

All these functions are real valued and so that set $F^* = \{f_k : 0 \leq k_1\}$ is a proper subset of the set F.

We can set up a one-to-one correspondence between [0,1] and F^* (*s.t.* $F^* \subset F$). Hence $card\ [0,1] = card\ F^* < card\ F$ or $\boldsymbol{c} < \boldsymbol{f}$.

EXAMPLE 15. *Show that a countable set is a Borel set.*

SOLUTION. Let $A = \{a_1, a_2, a_3, \ldots\}$ be a countable set. Now note that

$$\{x : x = a_r\} = \bigcap_{n=1}^{\infty} \left\{x : a_r \leq a_r + \frac{1}{n}\right\}$$

and

$$A = \bigcup_{r \in N} \{a_r\}$$

$\Rightarrow$ A is obtained by the formation of countable union and intersection of closed and open sets and hence A is a Borel set.

1.14 CONTINUUM HYPOTHESIS

It is assumed that there is no cardinal number between **a** and **c**. Thus **c** is assumed to be the second transfinite cardinal number. However, there are other cardinal number greater than **c**. For instance *card P(R)* > **c** .

REVIEW QUESTIONS AND ARCHIVE

1. (a) Define an enumerable set. Show that the set of real numbers can not be enumerable, although the set of rationales is enumerable.

   (b) If $f : A \to B$ and the range of f is uncountable, prove that domain of f is also uncountable.

2. Define cardinal number of a set. Show that $n < 2^n$ for any cardinal number n.

3. Prove that the set of all real numbers in the closed interval [0, 1] is uncountable.

4. Prove that if A and B are enumerable then $A \times B$ is also enumerable.

5. Prove that $\alpha + \alpha = \alpha$ for any infinite cardinal number α.

6. Find the cardinal number of the set $\{x\}$ of those numbers in the interval [0,1] whose ternary expansion does not have the digit 1.

7. Prove the following:

   (i) $[0,1] \sim]0,1[$,

   (ii) $[0,1] \sim [2,5]$,

   (iii) $[0,1[\sim]0,1[$.

8. If $\{E_n\}$ be a sequence of countable sets and $S = \bigcup_{n=1}^{\infty} E_n$, then prove that S is countable.
   (KANPUR 2003)

9. Let α and β be any two cardinal numbers such that $\alpha \leq \beta$ and $\beta \leq \alpha$, then prove that $\alpha = \beta$.

10. (i) Prove that the set of all numbers in any interval can not be enumerable.

    (ii) Show that the set of all characteristic functions on R is uncountable.

11. Set of real numbers is ....
    (KANPUR 2003)

12. Every isolated set of point is ....
    (KANPUR 2002)

13. By an example show that cancellation law does not hold in case of cardinal numbers?

14. Set of integers $\{0,1,2,\ldots\}$ is uncountable.
    (KANPUR 2003)

15. Show that the family of all finite subsets of the natural numbers is countable infinite.

16. Prove that

    (i) $\boldsymbol{a} + \boldsymbol{c} = \boldsymbol{c}$ (ii) $\boldsymbol{a} + \boldsymbol{a} = \boldsymbol{a}$

    (iii) $\boldsymbol{c} + \boldsymbol{c} = \boldsymbol{c}$ (iv) $\boldsymbol{a} \cdot \boldsymbol{c} = \boldsymbol{c}$

    (v) $\boldsymbol{a} \cdot \boldsymbol{a} = \boldsymbol{a}$

17. State and prove Schroder-Bernsteing theorem.

18. Prove that the set of complex numbers is uncountable.

19. Exhibit a $1-1$ correspondence between the points of the closed interval $[0,1]$ of R and the points of the half closed interval $(0,1]$ of R.

20. Show that the set of all polynomial functions with integer (rational) coefficients is countable (or say has the cardinal a)

21. Prove that card $P(A) = 2^{card\ A}$, where A is any finite set. (GARHWAL 1992)

22. Using the mapping $f : N \times N \to N$ given by $f(x,y) = 2^x(x^y + 1) - 1$, show that set $N \times N$ is countable equivalent.

23. Show that the set of points in the closed interval $[2,4]$ and in the open interval $(1,2)$ are cardinally equivalent. (MEERUT 2004)

24. If a finite set of elements in added to an enumerable set, the executing set is also enumerable.

25. If B is a countable subset of an uncountable set A. Then $A - B$ is .... (MEERUT 2001)

26. Show that the set of all sequence whose elements are the digit 0 and 1 is uncountable. (KANPUR 2001; PUNJAB 2003)

27. If $\{E_n\}$ be a sequence of countable sets and $S = \overset{\infty}{\underset{n=1}{\cup}} E_n$ then prove that S is countable.

CHAPTER SUMMARY

This chapter introduced the fundamental notions and basic concepts. The important points discussed in this chapter are as follows :

➭ The numbers 1, 2, 3,... are called natural numbers. We represent the set of natural numbers by N, i.e., $N = \{1, 2, 3, ...\}$

➭ The numbers ... ..., –3, –2, –1, 0, 1, 2, 3, ... are called integers. We represent the set of integers by Z, i.e., $Z = \{..., -3, -2, -1, 0, 1, 2, 3, ...\}$

➭ Any number of the form p/q, where $p, q \in Z$, $q \neq 0$ and p, q have no common factor (except ± 1) is called a rational number.

➭ A number which is either rational or irrational is called a real number. The set of real numbers is denoted by R.

➭ A set is a well defined collection of objects

➭ A set containing no elements is called empty set and is denoted by the symbol ϕ.

➭ Set containing only one element is a Singleton set. The set {a} is a singleton set.

➭ The supremum of a set $S \subset R$, if exists, is unique.

➭ The infimum of a set, if exists, is unique.

➭ A sequence is represented as $< s_n >$ or $\{s_n\}$, when s_n is the nth term of the sequence.

➭ The set of all distinct terms of a sequence is called the range set of that sequence.

➭ A sequence, whose range is a subset of R is called a real sequence or a sequence of real numbers.

\FOR ADVANCED LEARNERS/

➡ If $x > 0$, $y > 0$ then $xy > 0$; if $x > 0$, $y > 0$ then $0 > xy$. If $0 > x$, $0 > y$ then $xy > 0$ and these results hold with $>$ replaced everywhere by $\geq$.

➡ If S is a non-empty set of integers then $m = \sup S$ is an integer.

➡ There exists $n \in \mathbf{Z}$ such that $n - 1 \leq x < n$.

➡ If $x > 0$, $y \geq 0$ then there exists $n \in \mathbf{N}$ such that $nx > y$.

➡ If $q \in \mathbf{Q}$ such that $x < q < y$. If we choose integers $n > \dfrac{1}{y-x}$ and $k \geq ny$ and let m be the least integer such that $y \leq \dfrac{m}{n}$. Then

$$x < \frac{(m-1)}{n} < y$$

➡ If S and T are non-empty set of positive numbers then

$$\sup\{st : x \in S, t \in T\} = \sup S \times \sup T$$

➡ Following conditions are equivalent on non-empty subset X and Y of $\mathbf{R}$

(i) $x \leq y \ \forall \ x \in X, y \in Y$

(ii) $\exists \ r \in \mathbf{R}$ such that $x \leq r < y \ \forall \ x \in X$ and $y \in Y$

➡ If $0 < a \neq 1$ and $a^x = 1$ then $x = 0$

➡ If $a > 0$ and $x > y$. If $a > 1$ then $a^x > a^y$ and that if $a < 1$ then $a^x < a^y$.

➡ If $a > 0$ then for each $x > 0$ there exists a unique $y \in \mathbf{R}$ such that $a^y = x$.

➡ If $a > 1$ then the function $\log_a x$ is strictly increasing and that if $0 < a < 1$ then $\log_a x$ is strictly decreasing.

➡ If in the system of axioms for $\mathbf{R}$, the least upper bound principle is replaced by the axioms of Acrhimedes, then the nested interval principle is equivalent to the completeness of $\mathbf{R}$.

❋❋❋❋

CHAPTER 2

Open and Closed sets in Real Numbers

2.1 INTRODUCTION

The real number system has many important properties. One of these properties says that there is a one-to-one correspondence between real numbers and the set of points on a line and there is a one-to-one correspondence between the real numbers and the points on a directed line. For this reason, the directed line is called the real line or real axis and a real number is called a point of the real line. In this chapter, we shall discuss the concept of neighbourhood, open set, closed set and their properties in real numbers.

2.2 NEIGHBOURHOOD OF A POINT

Here, we are going to introduce the concept of neighbourhood of a point.

Definition 1. *A subset N of $\mathbf{R}$ is said to be a neighbourhood of a point $p \in \mathbf{R}$, if there exists a real, however small number $\varepsilon > 0$ such that*

$$p \in]p-\varepsilon, \; p+\varepsilon[\; \subset \mathbf{N}$$

Definition 2. *A subset N of $\mathbf{R}$ is called a neighbourhood of a point $p \in \mathbf{R}$ if $\exists$ an open interval I such that*

$$p \in I \subset \mathbf{N}$$

Definition 3. *A set T is called a neighbourhood of a set S if T is a neighbourhood of each point of S.*

2.2.1 GEOMETRICAL INTERPRETATION OF ε-NBD

Any ε-nbd viz., $]p-\varepsilon, \; p+\varepsilon[$ (for $\varepsilon > 0$) of p is the set of all those points which are within an ε distance from p on either side of p.

☛ REMARKS

⟹ For brevity in writing, now onwards, the word 'neighbourhood' is shortened to 'nbd'.

⟹ The open interval $]p-\varepsilon, p+\varepsilon[$, $\varepsilon > 0$ is sometimes referred to as ε-nbd of p or simply as nbd of p. An open interval containing a point p is often denoted by I_p. This is also used for a nbd of p.

2.2.2 DELETED NEIGHBOURHOOD OF p

In case the point p is specially excluded from a nbd N_p of p, then such a nbd is called a deleted nbd of p.

Thus, if N is a nbd of p, then the set $N - \{p\}$ is a deleted nbd of p.

☛ REMARK

⇨ A symmetric deleted nbd of p is the union of two open intervals
$$]p-\varepsilon,\ p[\quad \text{and}\quad]p, p+\varepsilon[$$

ILLUSTRATIONS

* **An open interval is a nbd of each of its points.**

  Let $]a, b[$ be an open interval and let p be any arbitrary point of interval $]a, b[$. To show $]a, b[$ is a nbd of p.

  If we take ε as the minimum of two positive numbers $p-a$ and $b-p$ then obviously $\varepsilon > 0$, and $p \in]p-\varepsilon, p+\varepsilon[\ \subset\ ]a, b[\ \Rightarrow\ ]a, b[$ is a nbd of p.

  Since, p is arbitrary point of $]a, b[$, the interval $]a, b[$ is a nbd of each of its points.

* **The set Q of rational numbers is not the nbd of any of its point.**

  Let $p \in \mathbf{Q}$. Therefore any arbitrary positive real number ε, $p-\varepsilon$ and $p+\varepsilon$ are two distinct real numbers contains infinite irrational numbers between them, which are not the members of $\mathbf{Q}$. Therefore
  $$]p-\varepsilon,\ p+\varepsilon] \not\subset \mathbf{Q}\ \ \forall\, \varepsilon > 0$$
  $\Rightarrow\ Q$ is not a nbd of p.

* **A non-empty finite set cannot be a nbd of any of its points.**

  Since, a nbd of a point p contains an open interval containing p. Also, an interval contains infinite points. Therefore, for a nbd of a point, the set must contain infinite points.

  $\Rightarrow$ A non-empty finite set can not be a nbd of any of its points.

* **A closed interval [a, b] is a nbd of each of its points except the ends points.**

  Since, we know that "an open interval is a nbd of each of its points, therefore $x \in]a, b[\ \Rightarrow\ ]a, b[$ is a nbd of x.

  Also $]a, b[\ \subset\ [a, b] \Rightarrow [a, b]$ is a nbd of each point of $]a, b[$.

  Now, it is remains to show $[a, b]$ is not a nbd of a and b. Take a positive real number ε. Then
  $$a \in]a-\varepsilon,\ a+\varepsilon[\not\subset [a, b]$$
  $\Rightarrow$ There is no $\varepsilon > 0$ such that $]a-\varepsilon,\ a+\varepsilon[\ \subset [a, b]$

  $\Rightarrow\ [a, b]$ is not a nbd of a.

  Similarly we can show that $[a, b]$ is not a nbd of b.

* **The null set ϕ is trivally a nbd of each of its points**

  The null set ϕ is a nbd of each of its points, because there is no point in ϕ, therefore we can say there is no point in ϕ which is not a nbd.

* **The set Z^+ of all positive integers or a set of natural number is not a nbd of any of its points.** (MEERUT 1996, 2001)

  Use the argument of (2).

* A non-empty subset N of $\mathbf{R}$ is a nbd of $p \in \mathbf{R}$ iff there exist a positive integer n such that
  $$\left]p-\frac{1}{n},\ p+\frac{1}{n}\right[\ \subset N$$

  Let us first suppose a non-empty subset N of $\mathbf{R}$ be a nbd of point $p \in \mathbf{R}$, then there exist a positive number $\varepsilon > 0$ such that
  $$p \in]p-\varepsilon,\ p+\varepsilon[\ \subset N.$$

  Now, for a given positive real number ε, we can choose a positive integer n such that $\dfrac{1}{n} < \varepsilon$
  $$\frac{1}{n} < \varepsilon \Rightarrow p+\frac{1}{n} < p+\varepsilon.$$

Also $\qquad \dfrac{1}{n} < \varepsilon \;\Rightarrow\; -\dfrac{1}{n} > -\varepsilon$

$\Rightarrow \qquad p - \dfrac{1}{n} > p - \varepsilon.$

Therefore, $\left] p - \dfrac{1}{n},\ p + \dfrac{1}{n} \right[\subset\ ]p - \varepsilon,\ p + \varepsilon[$

$\Rightarrow$ If N is a nbd of p, then there exist a positive integer n such that

$$\left] p - \dfrac{1}{n},\ p + \dfrac{1}{n} \right[\subset N$$

Conversely, let there exist a positive integer n such that

$$\left] p - \dfrac{1}{n},\ p + \dfrac{1}{n} \right[\subset N$$

Then, $\left] p - \dfrac{1}{n},\ p + \dfrac{1}{n} \right[$ is an open interval containing p and contained in **N**.

$\Rightarrow\ N$ is a nbd of p.

* **The intersection of two nbds of a point is also a nbd of that point.**

(ALLAHABAD 2008, 2010; KANPUR 2007, 10; IAS 2003, 07)

Let N_1 and N_2 be two nbds of a point p. Then, by definition, there exist $\varepsilon_1 > 0$ and $\varepsilon_2 > 0$ such that

$\qquad\qquad]p - \varepsilon_1, p + \varepsilon_1[\subset N_1$ ...(1)

and $\qquad\qquad]p - \varepsilon_2,\ p + \varepsilon_2[\subset N_2$ ...(2)

Take $\varepsilon = \min\{\varepsilon_1, \varepsilon_2\}$, then from (1) and (2), we have

$\qquad\qquad]p - \varepsilon, p + \varepsilon[\subset N_1$

and $\qquad\qquad]p - \varepsilon, p + \varepsilon[\subset N_2$

$\Rightarrow\ ]p - \varepsilon, p + \varepsilon[\subset N_1 \cap N_2$

$\Rightarrow\ N_1 \cap N_2$ is a nbd of p.

* **Any superset of a nbd of a point is also nbd of that point.**

(GARHWAL 2007, 09; PURVANCHAL 2005, 08)

Let us suppose N be a nbd of a point $p \in R$ and M is the super set of N, i.e., $N \subset M$.

To show M is also a nbd of p. Since N is a nbd of p, by definition $\exists \varepsilon > 0$ such that $p \in]p - \varepsilon, p + \varepsilon[\subset N$.

Also $N \subset M$

$\Rightarrow\ p \in]p - \varepsilon,\ p + \varepsilon[\subset N \subset M$

$\Rightarrow\ p \in]p - \varepsilon, p + \varepsilon[\subset M$

$\Rightarrow\ M$ is a nbd of p.

* **On the real line R, for each point $p \in R$, there exist at least one nbd of p.**

If $p \in R$, then for each $\varepsilon > $, we have

$\qquad\qquad p \in]p - \varepsilon,\ p + \varepsilon[\subset R.$

$\Rightarrow\ R$ is always a nbd of p.

* On the real line **R**, for each point $p \in R$ and each nbd N of p there exist a nbd M of p such that $M \subset N$ and M is a nbd of each of its points.

Since, we have that N is a nbd of p, therefore $\exists \varepsilon > 0$ such that

$$p \in]p-\varepsilon, \; p+\varepsilon[\; \subset N$$

Take $\qquad M = \;]p-\varepsilon, \; p+\varepsilon[\; .$

As $a \cup]p-\varepsilon, \; p+\varepsilon[$ is an open interval containing p, so it is a nbd of p and also a nbd of each of its points.

$\Rightarrow \exists$ a nbd M of p such that $M \subset N$ and M is a nbd of each of its points.

* The set $R - N$ is a nbd of all its points.
* The set $R - Z$ is a nbd of all its points.
* The set $R - Q$ is not a nbd of any of its points.
* The set $R - I$ is not a nbd of all of its points.
* Empty set ϕ is a nbd of all its points as there is no point in ϕ of which ϕ is not a neighbourhood.

SOLVED EXAMPLES

EXAMPLE 1.   ***Show that [1, 3] is a nbd of 2 but not of 1.***

SOLUTION. $2 \in]1,3[$ and $]1,3[\subset [1,3] \quad \Rightarrow \quad [1,3]$ is a nbd of 2.

Since $[1,3]$ does not contain any open interval containing 1

$\Rightarrow [1,3]$ is not a nbd of 1.

EXAMPLE 2.   ***Which of the following subsets of R are nbd of 3?***

       (i)]2, 4[(ii)]2, 4]   (iii) [2, 4[(iv) [2, 4]

       (v)]3, 8[(vi)]3, 8]   (vii) [3, 8[(viii) [3, 8]

       (ix) [2, 5] $-[4\frac{1}{2}]$.

SOLUTION.   (i)   since $]2,4[$ is an open interval and $3 \in]2,4[$

             $\Rightarrow \qquad]2,4[$ is a nbd of 3.

      (ii)   Since, there exist an open interval $]2,4[$ such that

$$3 \in]2,4[\; \subset]2,4]$$

             $\Rightarrow \qquad]2,4]$ is a nbd of 3.

      (iii)   Since, there exist an open interval $]2,4[$ such that

$$3 \in]2,4[\; \subset [2,4[$$

             $\Rightarrow \qquad [2,4[$ is a nbd of 3.

      (iv)   Since, there exist an open interval $]2,4[$ such that

$$3 \in]2,4[\; \subset [2,4]$$

             $\Rightarrow \qquad [2,4]$ is a nbd of 3.

      (v) $]3,8[$ is not a nbd of 3 as $3 \notin]3,8[$.

      (vi) $]3,8]$ is not a nbd of 3 as $3 \notin]3,8]$.

      (vii) $[3,8[$ is not a nbd since $[3,8[$ does not contain any open interval containing 3.

      (viii) $[3,8]$ is not a nbd of 3 since $[3,8]$ does not contain any open interval containing 3.

      (ix) $[2,5] -[4\frac{1}{2}]$ is a nbd of 3 since there exists an open interval $]2,4\frac{1}{2}[$ such that

$$3 \in]2,4\tfrac{1}{2}[\; \subset [2,5]-[4\tfrac{1}{2}].$$

EXAMPLE 3. *Give an example of a set which is a nbd of:*
  (i) each of its points.
  (ii) not any of its points.
  (iii) each of its point, except the end point.
  (iv) each of its point, except one point.

SOLUTION.
  (i) The open interval $]a, b[$ is a nbd of each of its points.
  (ii) A finite set is not a nbd of any of its points.
  (iii) The closed interval $[a, b]$ is a nbd of each of its points, except the end points.
  (iv) The right half closed interval $]a, b]$ is a nbd of each of its points except b.

EXAMPLE 4. *Let $I_n = \left]-\dfrac{1}{n}, 1+\dfrac{1}{n}\right[$ be an open interval for each $n \in N$. Prove that*

$\overset{\infty}{\underset{n=1}{\cap}} I_n$ is not a nbd of each of its points.

SOLUTION. Since $n \in N$

$$\Rightarrow \quad \frac{1}{n} \to 0 \quad \text{as} \quad n \to \infty$$

$$\Rightarrow \quad -\frac{1}{n} \to 0 \quad \text{as} \quad n \to \infty \quad \text{and} \quad 1+\frac{1}{n} \to 1 \quad \text{as} \quad n \to \infty .$$

Therefore, we can find that

$$0 \in \left]-\frac{1}{n}, 1+\frac{1}{n}\right[\quad \forall n \in N$$

$$1 \in \left]-\frac{1}{n}, 1+\frac{1}{n}\right[\quad \forall n \in N .$$

Also, each point lying between 0 and 1 is an element of the open interval $\left]-\frac{1}{n}, 1+\frac{1}{n}\right[, \forall n \in N$.

But $-\left(\frac{1}{n}\right)$, whatever be the value of n, is not an element of the open interval.

$\left]-\frac{1}{n}, 1+\frac{1}{n}\right[, n \in N \Rightarrow$ All numbers less than 0 are not in $\overset{\infty}{\underset{n=1}{\cap}} I_n$.

Similarly, We can show that all numbers greater than 1 are not in $\overset{\infty}{\underset{n=1}{\cap}} I_n$, $n \in N$.

Now, $[0, 1] = \overset{\infty}{\underset{n=1}{\cap}} \left\{\left]-\frac{1}{n}, 1+\frac{1}{n}\right[\right\} = \overset{\infty}{\underset{n=1}{\cap}} I_n$

and $[0, 1]$ being a closed interval is a nbd of each of the points of the interval $[0, 1]$, except the end point 0 and 1.

Hence, $\overset{\infty}{\underset{n=1}{\cap}} I_n$ is not a nbd of each of its points.

2.3 INTERIOR OF A SET

In this section, we shall discuss the concept of interior points and interior of a set.

2.3.1 INTERIOR POINT OF A SET

A point p is called an interior point of a set S if S is a nbd of p, i.e., p is called an interior point of S if there exist an open interval I such that $p \in I \subset S$.

2.3.2 INTERIOR OF A SET

The set of all interior point of a set S, denoted by S^o or S^i is called interior of a set.

2.3.3 OPEN SETS

A set $S \subset R$ is said to be open if it is a neighbourhood of each of its points.

(or)

A set $S \subset R$ is said to be open if for each $p \in S$, there exist $\varepsilon > 0$ such that $]p-\varepsilon,\ p+\varepsilon[\subset S$.

✎ ILLUSTRATIONS

* Every open interval $]a,b[$ is an open set.
* The interval $[a,b[$ is not an open set, because, it is not a neighbourhood of a.
* The interval $]a,b]$ is not an open set, because it is not a neighbourhood of b.
* The closed interval $[a,b]$ is not an open set, because it is not the nbd of a and b.
* R is an open set, because, if p be any point of R, then the open interval $]p-1,p+1[\subset R$ and consequently R is a neighbourhood of p. Since p is arbitrary, therefore R is a neighbourhood of each of its points.
* The empty set ϕ is an open set, because there is no point at all in ϕ, and consequently, there is no point in ϕ, of which it is not a neighbourhood.
* The open rays $]a,\infty[$ and $]-\infty,a[$ are open sets.
* The closed rays $[a,\infty[$ and $]-\infty,a]$ are not open sets.
* Every point of an open set is an interior point. Therefore for an open set S, $S^o = S$. Thus S is open iff $S^o = S$.

FACTS : TO THE POINT

▶ The open interval is a nbd of all of its points.
▶ The closed interval is a nbd of all of its points except the end points.
▶ The finite set is not a nbd of any real number.
▶ The set N of natural number is not a nbd of any real numbers ($N^o = \phi$)
▶ The set Z of integers is not a nbd of any real numbers ($Z^o = \phi$)
▶ The set Q of rationals is not a nbd of any real numbers ($Q^o = \phi$)
▶ The set I of irrationals is not a nbd of any real numbers ($I^o = \phi$)
▶ The set R of real numbers is a nbd of all real numbers ($R^o = R$)
▶ The set $R - N$ is a nbd of all its points $((R-N)^o = R-N)$
▶ The set $R - Z$ is a nbd of all its points $((R-Z)^o = R-Z)$
▶ The set $R - Q$ is not a nbd of any of its points $((R-Q)^o = \phi)$
▶ The set $R - I$ is not a nbd of any real number $((R-I)^o = \phi)$
▶ The empty set ϕ is a nbd of all its points $(\phi^o = \phi)$
▶ The sets N, Z, Q, I are not open but R is open
▶ The sets $R-N, R-Z$ are open but $R-Q$ and $R-I$ are not open.

THEOREM 1. *Every open interval is an open set.*

(PATNA-2003, 07)

PROOF. Let p be any point of the given interval $]a,b[$. Therefore, $a < p < b$.

Consider two numbers c and d such that
$$a < c < p < d < b$$
$\Rightarrow \qquad p \in]c,d[\subset]a,b[$
$\Rightarrow \qquad]a,b[$ is a nbd of p.

Since, p is arbitrary, therefore $]a,b[$ is a nbd of each of its points.
$\Rightarrow \qquad]a,b[$ is open.

THEOREM 2. *The union of an arbitrary family of open sets is open.*

(AGRA-2008; MEERUT-2011)

PROOF. Let G be the union of an arbitrary family $[G_\lambda : \lambda \in \Lambda]$ of open sets in R, i.e., $G = \cup \{G_\lambda : \lambda \in \Lambda\}$. To show that G is open.

Let $p \in G$, since G is the union of members of $\{G_\lambda : \lambda \in \Lambda\}$, therefore, there must exist an open set $H \in G_\lambda$ such that $p \in H \subset G$. Since, H is an open set and $p \in H$, therefore there must exist $\varepsilon > 0$ such that $]p-\varepsilon,\ p+\varepsilon[\subset H \subset G$. Again, since $]p-\varepsilon, p+\varepsilon[$ is contained in G, therefore, G is a nbd of p. Now, since p is arbitrary, therefore G is a nbd of each of its points.

$\Rightarrow$ G is an open set.

$\Rightarrow$ Arbitrary union of open sets is open.

THEOREM 3. ***The intersection of two open sets is open.*** (AGRA-2006)

PROOF. Let $G_1 \subset R$ and $G_2 \subset R$ be two open sets. To show $G_1 \cap G_2$ is open.

CASE I. If $G_1 \cap G_2 = \phi$, then it is open. ($\because \phi$ is an open set)

CASE II. If $G_1 \cap G_2 \neq \phi$,

Let $p \in G_1 \cap G_2$

$\Rightarrow$ $p \in G_1$ and $p \in G_2$

$\Rightarrow$ G_1 and G_2 are nbds of p. ($\because G_1$ and G_2 both are open)

$\Rightarrow$ $G_1 \cap G_2$ is a nbd of p.

Now, since p is arbitrary, therefore, it follows that $G_1 \cap G_2$ is a nbd of each of its points

$\Rightarrow$ $G_1 \cap G_2$ is open.

✍ ILLUSTRATIONS

* The intersection of an arbitrary family of open sets is not necessarily open. For example: For each $n \in N$, let

$$G_n = \left]a - \frac{1}{n},\ a + \frac{1}{n}\right[$$

Then, each G_n is an open set, but $\cap[G_n : n \in N] = [a]$ is not open.

* Every open set is a union of open intervals. This follows from the fact that, if G is open, then to each $x \in G$ $\exists$ an open interval G_x such that $x \in G_x \subset G$ and so

$$G = \bigcup_{x \in G} \{x\} \subset \bigcup_{x \in G} G_x \subset G$$

$$\Rightarrow \qquad G = \bigcup_{x \in G} G_x$$

* Every non-empty open set consists necessarily infinitely many points.
* Interior of any set S is an open set.
* For any set S, $(S^o)^o = S^o$.
* For two sets S and T, $S \subset T \Rightarrow S^o \subset T^o$.
* Interior of a set S is the largest open set contained in S.

2.4 LIMIT POINT OF A SET

2.4.1 ADHERENT POINT

A point $p \in R$ is said to be an adherent point of a set $A \subset R$ if every nbd of p contains a point of A.

The set of all adherent point of A is called the adherence of A, denoted by Adh.(A).

🏆 FACTS : TO THE POINT

▶ The limit point is also known as accumulation point, limiting point, cluster point or a point of condensation.

▶ The limiting point of the set A may or may not belong to the set A.

▶ A finite set can not have a limit point.

▶ A set may have no limit point, or a finite number of limit points or infinite number of limit points.

▶ If A be a non-empty set bounded above and if supremum, say, x does not belong to A, then l is a limit point of A if,

for $\varepsilon > 0\ \exists\ y \in A : l - \varepsilon < y < l$

$\Rightarrow$ nbd $]l - \varepsilon,\ l + \varepsilon[$ of l contains a point y of A, other than l, as $l \notin A$.

If A be a non-empty set bounded below and if its infimum does not belong to A, then it is a limit point of A.

▶ A point p is not a limit point of a set A, iff there is a nbd N of p such that $N \cap A$ is finite.

▶ A point p is not a limit point of a set A, then it is not a limit point of any of its subsets.

▶ If a point p is a limit point of a set A, then it is a limit point of every superset of A.

☞ REMARK
⟶ An adherent point is also called closure point.

2.4.2 LIMIT POINT

Definition 1. *A point* $p \in R$ *is said to be a limit point of a set* $A \subset R$ *if every nbd of p contains at least one point of A other than p.*

In symbols, the above definition means that a point $p \in R$ is a limit point of $A \subset R$ iff for each nbd N of p

$$(N \cap A) \sim [p] \neq \phi.$$

Definition 2. *A point* $p \in R$ *is said to be a limit point of a set* $A \subset R$ *if for each* $\varepsilon > 0$*, the open interval* $]p - \varepsilon, \ p + \varepsilon[$ *contains a point of A, other than p.*

☞ REMARK
⟶ In order to show that a point p is not a limit point of a set A, it is enough to find a nbd N of p, such that either $N \cap A = [p]$ or $N \cap A = \phi$.

✍ ILLUSTRATIONS

* Adh. $A = A \cup D(A)$.
* Since $]a, b[\cap \phi = \phi \ \forall \ ]a, b[$, the empty set has no limit point.
* The only limit points of $]a, b[,]a, b], [a, b[$ or $[a, b]$ are points of $[a, b]$ and every point of R is a limit point of R.
* Every limit point of A is also an adherent point of A, but converse is not always true. For example, 1 is an adherent point of the set $\left\{ \dfrac{1}{n} : n \in N \right\}$ but it is not the limit point of the set.

2.4.3 DERIVED SET

The set of all limit points of a set $A \subset R$ is said to be the derived set and it is denoted by $D(A)$ or A'.

$$A' = D(A) = \{p : p \text{ is a limit point of } A\}$$

The derived set of A', *i.e.*, $(A')'$, we shall denote by A'' and so on.

2.4.4 ISOLATED POINT

A point $p \in R$ is said to be an isolated point of A, if it is not the limit point of A, *i.e.*, if there exist a nbd of p which contains no point of A, other than p.

2.4.5 DISCRETE SET

A set A is called discrete if all its points are isolated points.

☞ REMARK
⟶ Each point of a set A is either an isolated point of A or a limit point of A.

2.4.6 DENSE-IN-ITSELF SET

A subset A of R is said to be dense-in-itself if it possesses no isolated points, *i.e.*, every point of A is the limit point of A.

2.4.7 CLOSED SET

Definition 1. *A set A is said to be closed if it contains all its limit points.*　　　　(MEERUT 2003)

Definition 2. *A set A is said to be closed if its compliment is open.*

✍ ILLUSTRATIONS

* A finite set is always closed.
* The null set ϕ is closed.
* The set N, Z and R are closed sets.
* The set Q is not a closed set.
* The singleton is always a closed set.

2.4.8 PERFECT SET

A set A is said to be perfect if it is dense-in-itself and contains all its limit points.

(or)

A set A is said to be perfect if it is dense-in-itself and closed.

☛ REMARK

⟹ A set A is said to be perfect if it is identical with its derived set.

2.4.9 CLOSURE OF A SET

The smallest closed set containing A, is called closure of A and denoted by $\bar{A}$. (KANPUR 2005, 2009)

2.4.10 DENSE SET

(i) *A set A is said to be dense or everywhere dense in R if $\bar{A} = R$.*

(ii) *A set A is said to be no where dense in R if interior of the closure of A is empty, i.e., $(\bar{A})^\circ = \phi$.*

2.4.11 FIRST AND SECOND SPECIES

A set A is said to be first species if it has only a finite number of derived sets. It is said to be of second species if the number of its derived sets is infinite.

☛ REMARKS

⟹ A set A is of first species, then its last derived set must be empty.

⟹ A set whose n^{th} derived set is a finite set (so that its $(n + 1)^{\text{th}}$ derived set is empty) is called a set of nth order.

⟹ If every sub-interval of an interval I has a part which does not contain any point of A, then A is called no where dense in I.

⟹ If no sub-interval of I is free from points of A then A is called everywhere dense in I.

✍ ILLUSTRATIONS

* For sets ϕ, N, Q and R it evidently follows that

  (i) $\phi' = \phi$ (ii) $N' = \phi$

  (iii) $Q' = R$ (iv) $R' = R$

  (i) $\bar{\phi} = \phi$ (ii) $\bar{N} = N$

  (iii) $\bar{Q} = R$ (iv) $\bar{R} = R$

  (i) $(\bar{\phi})^\circ = \phi$ (ii) $(\bar{N})^\circ = \phi$

  (iii) $(\bar{Q})^\circ = R$ (iv) $(\bar{R})^\circ = R$

THEOREM 1.   *Interior of a set is an open set.*

PROOF.   Let A be any given set and A° is the interior of A.

     CASE I. If $A^\circ = \phi$, then A° is open.

     CASE II. If $A^\circ \neq \phi$. Let $p \in A^\circ$.

     $\Rightarrow$ p is an interior point of A.

FACTS : TO THE POINT

▶ Some examples of neither open nor closed sets

   (i) The set Q of rationals is neither open nor closed

   (ii) The set I of irrationals is neither open nor closed

   (iii) The set $R - Q$ of rationals is neither open nor closed

   (iv) The set $R - I$ of rationals is neither open nor closed

   (v) Semi closed intervals are neither open nor closed

▶ The sets $R - N$ and $R - Z$ are not closed.

▶ ϕ and whole set R are two sets which are open as well as closed.

$\Rightarrow \;\; \exists$ an open interval I_p such that $p \in I_p \subset A$.

Since I_p is a nbd of each of its point and $I_p \subset A$.

$\Rightarrow$ Each point of I_p is an interior point of A, *i.e.*, $p \in I_p \subset A^\circ$.

Hence, A° is an open set.

THEOREM 2. *A point $p \in R$ is a limit point of a set $A \subset R$ if and only if every neighbourhood of p contains infinitely many points of A.*

PROOF. Let us first suppose that every nbd of p contains infinitely many points of A.

$\Rightarrow$ Every nbd of p contains a point of A, which is different from p.

$\Rightarrow$ p is a limit point of A.

Conversely, let p is a limit point of A. To show that every nbd of p contains infinitely many points of A.

Let, if possible, there exists a *nbd N* of p which contains finitely many points of A. Then there exists a positive number $\varepsilon > 0$ such that the open interval $]p-\varepsilon, p+\varepsilon[$ must contain only finitely many points of A. Now, if p is the only such point which contained in $]p-\varepsilon, p+\varepsilon[$, then p is not a limit point of A. If $]p-\varepsilon, p+\varepsilon[$ contains points of A, other than p also, then since we assumed them finite in number, let they be $p_1, p_2, ..., p_n$. Out of these n points of A, let p_i be the point which is nearest to p and let $|p_i - p| = \varepsilon_1$

$\Rightarrow \;\; \varepsilon_1 = \min[|p_1 - p|, |p_2 - p|, ..., |p_n - p|]$

Then, the real number ε_1 is such that the open interval $]p-\varepsilon_1, p+\varepsilon_1[$ contains no points of A, other than p.

$\Rightarrow$ there exist a nbd of p which contains no point of A, other than p.

$\Rightarrow$ p is not a limit point of A.

which is a contradiction.

Hence, if p is a limit point of A, then every nbd of p contain infinitely many points of A.

THEOREM 3. *If a non-empty subset A of R which is bounded above and has no maximum member, then its supremum is a limit point of the set A.*

PROOF. Let A be a non-empty subset of $\mathbf{R}$, which is bounded above.

Let $\sup A = u$

Now since A has no maximum member $\Rightarrow u \notin A$.

Let $\varepsilon > 0$. Since $\sup A = u \Rightarrow u - \varepsilon$ cannot be an upper bound of A.

$\Rightarrow \;\; \exists x \in A$ such that $x > u - \varepsilon$.

Also, $u + \varepsilon > u$ and $u = \sup A$.

$\Rightarrow \;\; u + \varepsilon$ is also an upper bound of A and so $x \in A \Rightarrow x < u + \varepsilon$

$\Rightarrow$ For every $\varepsilon > 0 \;\exists x \in A$ such that $u - \varepsilon < x < u + \varepsilon$.

Since, $u \notin A$, therefore $x \neq u$.

$\therefore$ every nbd $]u-\varepsilon, u+\varepsilon[$ of u contains a point x of A which is different from u.

Hence, u is a limit point of A.

THEOREM 4. *If a non-empty subset A of R, which is bounded below has no minimum member, then its infimum is a limit point of the set A.*

PROOF. Proof is similar as Theorem 3.

THEOREM 5. *The finite set has no limit point.*

PROOF. Let A be a finite set. To show that every real number p is not a limit point of the set A.

Since A is finite. Therefore, for $\varepsilon > 0$, the open interval $]p-\varepsilon, p+\varepsilon[$ contains only finitely many points of the set A.

$\Rightarrow$ $]p-\varepsilon,\ p+\varepsilon[$ is a nbd of p, which do not contain infinitely many point of A.

$\Rightarrow$ p is not a limit point of A.

Since p is arbitrary, therefore, we conclude that every real number p is not a limit point of A.

$\Rightarrow$ Finite set has no limit point.

THEOREM 6. **(Bolzano-Weierstrass Theorem)** *Every infinite bounded set of real numbers has a limit point.*

(MEERUT 1990, 90P, 90S, 91, 91P, 92, 92P, 93, 94, 2002, 03, 05, 08, 09, 2010; KANPUR 2001, 06; BHU 2002, 06, 09; PATNA 2006; KOLKATA 2004,08)

PROOF. Let $S \subset R$ be any infinite bounded set so that there exist its infimum and supremum k_1 and k_2 respectively. Define a set T as follows :

" $x \in T$ iff it exceeds at most a finite number of members of S."

Now $T \neq \phi$ $(\because k_1 \in T)$

Also, T is bounded above with k_2 as its upper bound, for no number greater than k_2 belongs to T.

Therefore, T is non-empty bounded subset of R.

$\Rightarrow$ T has supremum in R, say $p = \sup T$ (By ordered completeness property)

Now, we shall show that p is a limit point of S.

$p = \sup T \Rightarrow \exists\, q \in T$ such that $q > p - \varepsilon,\ \varepsilon > 0$.

$q \in T \Rightarrow q$ exceeds atmost a finite number of members of S

$\Rightarrow p - \varepsilon$ exceeds at most a finite number of members of S ...(i)

Also, $p = \sup T \Rightarrow p + \varepsilon \notin T$

$\Rightarrow p + \varepsilon$ exceeds an infinite number of members of S ...(ii)

Now, (i) and (ii) $\Rightarrow \exists$ a nbd if p, *i.e.*, $]p-\varepsilon, p+\varepsilon[$, which contains an infinite number of members of S.

$\Rightarrow$ p is the limit point of S.

☛ REMARK

⟹ This theorem gives a sufficient condition for an infinite set to have a limit point.

2.5 THEOREMS ON CLOSED AND DERIVED SETS

THEOREM I. $D(\phi) = \phi$, *i.e., derived set of an empty set is empty.*

PROOF. We know that, ϕ, the subset of R is a nbd of each of its point.

Therefore $p \in R \Rightarrow R$ is a nbd of p.

Also $R \cap \phi = \phi$

$\Rightarrow$ R contains no point of ϕ $(\because \phi$ has no points)

$\Rightarrow$ p is not a limit point of R.

$\Rightarrow$ no point of R is a limit point of ϕ.

$\Rightarrow$ $D(\phi) = (\phi)$.

THEOREM 2. *If A and B are subsets of R, then $A \subset B \to D(A) \subset D(B)$.*

PROOF. Here, $A \subset B$ is given,

To show $D(A) \subset D(B)$.

Let $x \in D(A) \Rightarrow x$ is the limit point of A.

$\Rightarrow$ every nbd N of x contains at least one point of A, other than x

$\Rightarrow$ every nbd N of x contains at least one point of B, other then x $(\because A \subset B)$

$\Rightarrow$ x is the limit point of B

$\Rightarrow$ $x \in D(B)$

Hence, $A \subset B \Rightarrow D(A) \subset D(B)$

THEOREM 3.   **If A and B are any subsets of R, then $D(A \cap B) \subset D(A) \cap D(B)$.**

PROOF.   From previous theorem, we have

$$A \subset B \Rightarrow D(A) \subset D(B)$$

Now, $\because A \cap B \subset A \Rightarrow D(A \cap B) \subset D(A)$

and $A \cap B \subset B \Rightarrow D(A \cap B) \subset D(B)$.

Hence, $D(A \cap B) \subset D(A) \cap D(B)$

☛ **REMARKS**

⟹ $D(A) \cap D(B) \not\subset D(A \cap B)$.

For example, let $A = \mathbf{Q}$, $B = \mathbf{I}$, the set of irrational numbers then $A \cap B = \phi$ and $D(A \cap B) = \phi$, where as $D(A) = D(B) = \mathbf{R}$ and therefore

$$D(A) \cap D(B) = \mathbf{R}$$

Thus it follows that $D(A) \cap D(B) = \mathbf{R} \not\subset D(A \cap B) = \phi$.

THEOREM 4.   **If A and B are any subsets of R, then**
$$D(A \cup B) = D(A) \cup D(B).$$

PROOF.   From Theorem (2), we have

$$A \subset B \Rightarrow D(A) \subset D(B).$$

Since $A \subset A \cup B$ and $B \subset A \cup B$.

Therefore, $D(A) \subset D(A \cup B)$ and $D(B) \subset D(A \cup B)$

$\Rightarrow$ $D(A) \cup D(B) \subset D(A \cup B)$ ...(1)

Now, to show $D(A \cup B) \subset D(A) \cup D(B)$.

Let x be any point such that

$$x \notin D(A) \cup D(B)$$

Then, we shall show that $x \notin D(A \cup B)$

$x \notin D(A) \cup D(B) \Rightarrow x \notin D(A)$ and $x \notin D(B)$

$\rightarrow$ x is neither a limit point of A, nor a limit point of B.

$\Rightarrow$ $\exists$ a nbd N_1 of x, which contains no point of A, other then x and $\exists$ a nbd N_2 of x, which contains no point of B, other than x.

Since N_1 and N_2 are the nbds of x $\Rightarrow$ $N_1 \cap N_2$ is a nbd of x (being the intersection of two nbds of x)

$\Rightarrow$ $N_1 \cap N_2$ is a nbd of x, which contains no point of A and B, other than x.

$\Rightarrow$ $N_1 \cap N_2$ is a nbd of x, which contains no point of $A \cup B$.

$\Rightarrow$ x is not the limit point of $A \cup B$.

$\Rightarrow$ $x \notin D(A \cup B)$.

Therefore, $x \notin D(A) \cup D(B) \Rightarrow x \notin D(A \cup B)$

$\Rightarrow x \in D(A \cup B) \Rightarrow x \in [D(A) \cup D(B)]$

$\Rightarrow D(A \cup B) \subset (D(A) \cup D(B))$.   ...(2)

Now, from (1) and (2), we get

$$D(A \cup B) = D(A) \cup D(B)$$

THEOREM 5. ***The derived set of a bounded set is again a bounded set.***

PROOF. Let $A \subset R$ be any set.

Let m, M be the bounds (infimum, supremum) of A which exist by the ordered completeness property of R

$$A \subset [m, M].$$

Now, we shall show that $D(A)$ is also bounded, *i.e.*, no limit point of $D(A)$ can be less than m or greater than M.

Let if possible, $p < m$, be a limit point of A and let $\varepsilon = m - p$ since $m = \inf A$.

Therefore, it follows that $\exists$ a nbd $]p - \varepsilon, \ p + \varepsilon[$, containing no point of A, other than p.

$\Rightarrow$ p is not a limit point of A, which is a contradiction.

$\Rightarrow$ no limit point of A can be less than m.

Similarly, we can show that no limit point of A, can be greater than M.

Hence $\quad D(A) \subset [m, M]$

$\Rightarrow$ $D(A)$ is bounded.

THEOREM 6. ***The derived set of any infinite bounded set attains its bounds.***

PROOF. Let A be an infinite bounded set. Since A is bounded, therefore, there exist $h, k \in R$ such that $S \subset [h, k]$

$$S \subset [h, k] \Rightarrow D(S) \subset D([h, k])$$

Also, the derived set of the closed interval $[h, k]$ is $[h, k]$.

i.e., $\qquad D([h, k]) = [h, k]$.

Therefore, $\quad D(A) \subset [h, k]$.

$\Rightarrow$ $D(A)$ is bounded.

Since, A is infinite bounded set, therefore, by Bolzano-Weirstrass theorem, A has at least one limit point and so $D(A) \neq \phi$.

$\Rightarrow$ $D(A)$ is non-empty bounded subset of R.

$\Rightarrow$ $D(A)$ has supremum and infimum in R.

Let inf. $D(A) = m$ and sup. $[D(A)] = M$.

Now, we shall show that both m and M belongs to $D(A)$, *i.e.*, both m and M are the limit point of A.

Let $\varepsilon > 0$ be given.

Now $m = \inf$ of $D(A) \Rightarrow \exists$ some $x \in D(A)$ such that $m \leq x < m + \varepsilon$

$\Rightarrow \quad m - \varepsilon < x < m + \varepsilon$

$\Rightarrow \quad x \in]m - \varepsilon, m + \varepsilon[$

$\Rightarrow \quad]m - \varepsilon, m + \varepsilon[$ is a nbd of some $x \in D(A)$.

$\Rightarrow \quad]m - \varepsilon, m + \varepsilon[$ is a nbd of x, which is a limit point of A.

$\Rightarrow \quad]m - \varepsilon, m + \varepsilon[$ contains infinitely many points of A.

$\Rightarrow \quad$ For every $\varepsilon > 0$, the open interval $]m - \varepsilon, m + \varepsilon[$ contains infinitely many points of A.

$\Rightarrow \quad m$ is the limit point of A.

Similarly, we can show that M is the limit point of A.

$\Rightarrow \quad$ both m and M, *i.e.*, the infimum and supremum of $D(A)$ belongs to $D(A)$

$\Rightarrow \quad m$ is the smallest and M is the greatest member of $D(A)$.

$\Rightarrow \quad$ the set A has smallest and greatest limit points.

THEOREM 7. *Finite union of closed sets is closed.*

PROOF. Let $F_1, F_2, ..., F_n$ be n closed sets. Then, to prove that $\bigcup_{i=1}^{n} F_i$ is closed.

Each F_i (i = 1, 2, ..., n) is closed $\Rightarrow$ Each F_i' (i = 1, 2,..., n) is open.

$\Rightarrow \qquad \bigcap_{i=1}^{n} F_i'$ is open $\qquad\qquad$ ($\because$ finite intersection of closed sets is closed)

$\Rightarrow \qquad \left[\bigcup_{i=1}^{n} F_i \right]' = \bigcap_{i=1}^{n} F_i'$ is open $\qquad\qquad$ (By Demorgan's law)

$\Rightarrow \qquad \bigcup_{i=1}^{n} F_i$ is closed.

☛ **REMARK**

⟹ The union of an arbitrary family of closed sets may fail to be a closed set. For example, let $F_n = \left[\dfrac{1}{n}, 2 \right]$ for each $n \in \mathbf{N}$. Then $\cup F_n = \,]\,0, 2\,[$, which is not a closed set $\forall\, n \in \mathbf{N}$.

THEOREM 8. *The intersection of an arbitrary collection of closed set is closed.*

PROOF. Let $\{F_\lambda : \lambda \in \Lambda\}$ be an arbitrary family of closed sets. To show $\cap [F_\lambda]$ is closed.

Now, since, each F_λ is a closed $\Rightarrow$ each F_λ' is open

$\Rightarrow \qquad \cup \{F_\lambda' : \lambda \in \Lambda\}$ is open.

$\Rightarrow \qquad (\cap\{F_\lambda : \lambda \in \Lambda\})'$ is an open set. $\qquad\qquad$ (By Demorgan's law)

$\Rightarrow \qquad \cap \{F_\lambda : \lambda \in \Lambda\}$ is closed.

THEOREM 9. *Let A be a subset of R. Then A is closed if and only if $D(A) \subset A$, i.e., A is closed iff it contains all its limit points.*

PROOF. Let A be any subset of $\mathbf{R}$, and closed.

$\Rightarrow \quad A'$ is open.

Now if $D(A) = \phi$, then $D(A) \subset A$.

Let $D(A) \neq \phi$, let $p \in D(A)$. To show $p \in A$.

Let if possible, $p \notin A$, then $p \in A'$, since A' is open, there exist $\varepsilon > 0$ such that $]\,p - \varepsilon, p + \varepsilon\,[\, \subset A'$.

$\Rightarrow \qquad]\,p - \varepsilon, p + \varepsilon\,[\, \bigcap A = \phi$

$\Rightarrow \qquad]\,p - \varepsilon, p + \varepsilon\,[$ contain no points of A

which contradicting the fact that p is the limit point of A.

Hence, $p \in A$.

$\Rightarrow \quad$ If A is closed, then $D(A) \subset A$.

Conversely, let us suppose $D(A) \subset A$

A contains all its limit points.

To show A is closed.

Let $p \in A' \Rightarrow p \notin A \Rightarrow p \notin D(A)$ $\qquad\qquad$ ($\because D(A) \subset A$)

$\Rightarrow \quad p$ is not the limit point of A

$\Rightarrow \quad \exists\, \varepsilon > 0$ such that $]\,p - \varepsilon, p + \varepsilon\,[$ contains no point of A, other than p

$\exists \, \varepsilon > 0$ s.t. $]\, p - \varepsilon, p + \varepsilon \,[$ contains no point of A $(\because p \notin A)$

$\Rightarrow$ $\exists \, \varepsilon > 0$ such that $]\, p - \varepsilon, p + \varepsilon \,[\subset A'$

$\Rightarrow$ A' contains a neighbourhood of each of its point

$\Rightarrow$ A' is open $\Rightarrow$ A is closed.

THEOREM 10.   *Let A be any subset of R, then $\bar{A} = A \cup D(A)$, i.e., $\bar{A}$ is the set of all adherent points of A.*

PROOF.   Firstly, we shall show that $A \cup D(A)$ is closed.

Let p be any limit point of $A \cup D(A)$.

$\Rightarrow$ either p is a limit point of A or a limit point of $D(A)$.

If p is the limit point of $A \Rightarrow p \in D(A)$.

If p is the limit point of $D(A) \Rightarrow p \in D(A)$ $(\because D(A)$ is closed$)$

$\Rightarrow$ $p \in A \cup D(A)$.

Since p is arbitrary

$\Rightarrow$ $A \cup D(A)$ contains all its limit points

$\Rightarrow$ $A \cup D(A)$ is closed.

Now, $A \cup D(A)$ is a closed set containing A, and $\bar{A}$ is the smallest closed set containing A, we have

$$\bar{A} \subset A \cup D(A). \qquad \qquad \dots(1)$$

Also, $A \subset \bar{A} \Rightarrow D(A) \subset D(\bar{A})$. $\dots(2)$

Now, since $\bar{A}$ is closed $\Rightarrow$ $D(\bar{A}) \subset \bar{A}$. $\dots(3)$

From (2) and (3), we have

$$D(A) \subset \bar{A}$$

$A \subset \bar{A}$ and $D(A) \subset \bar{A}$ $\Rightarrow$ $A \cup D(A) \subset \bar{A}$. $\dots(4)$

From (1) and (4), we have $\bar{A} = A \cup D(A)$.

SOLVED EXAMPLES

EXAMPLE 1.   *Find the limit points of the interval $]0, 1[$.*   (MEERUT 1991)

SOLUTION.   Let $A = \,]0, 1[$.

Now, firstly we shall show that every point of the closed interval $[0, 1]$ is the limit point of A.

Let $p \in [0, 1]$. Then for $\varepsilon > 0$, the open interval $]\, p - \varepsilon, p + \varepsilon \,[$ must contain infinitely many points of A, therefore, it contains at least one point of A, other than p.

$\Rightarrow$ p is the limit point of $]0, 1[$.

Now, we shall show that no points, other than $[0, 1]$ is the limit point of $]0, 1[$, *i.e.*, $p \notin [0, 1]$, then p is not the limit point of $]0, 1[$.

Let $\varepsilon > 0$ be such that ε is less than the distance of the point p from each of the end points 0 and 1 of the closed interval $[0, 1]$.

$\Rightarrow$ $\varepsilon < |p - 0|$ and $\varepsilon < |p - 1|$.

$\Rightarrow$ Open interval $]\, p - \varepsilon, p + \varepsilon \,[$ does not contain any point of the set A.

$\Rightarrow$ p is not the limit point of A.

Therefore, p is the limit point of $]0, 1[$ if and only if $p \in [0, 1]$.

$\Rightarrow$ $D(\,]0, 1[\,) = [0, 1]$.

EXAMPLE 2.   ***Find the limit points of the closed interval [0, 1].***

SOLUTION.   Let $A = [0, 1]$.

Then, in a similar manner as in Ex. 1.

We have $D([0, 1]) = [0, 1]$.

☞ REMARKS

⟹ In example (1), we observe that the points 0 and 1 do not belong to $]0, 1[$. Also, they are limit points of $]0, 1[$. Therefore, the set $]0, 1[$ is dense-in-itself, which is not prefect.

⟹ In example (2), each point of $[0, 1]$ is the limit point of $[0, 1]$, therefore the set $[0, 1]$ is dense-in-itself. Also $[0, 1]$ contains all its limit point, therefore $[0, 1]$ is a closed set. Since the set $[0, 1]$ is dense-in-itself and closed, therefore it is prefect.

EXAMPLE 3.   ***Find the set of the limit points of the set*** $S = \left[\dfrac{1}{n} : n \in N\right].$ (MEERUT 2001)

SOLUTION.   Here, we have

$$S = \left[\frac{1}{n} : n \in N\right] = \left[1, \frac{1}{2}, \frac{1}{3}, ..., \frac{1}{n}, ...\right].$$

We shall show that the set S has only one limit point, namely 0.

Now, firstly, we shall show that 0 is the limit point of S.

Let $\varepsilon > 0 (\Rightarrow -\varepsilon < 0)$. Then, by Archemedian property of real numbers there exist a positive integer m such that $\dfrac{1}{m} < \varepsilon$ so that $-\varepsilon < 0 < \dfrac{1}{m} < \varepsilon$.

Therefore, for every $\varepsilon > 0$, the open interval $]-\varepsilon, \varepsilon[$ contains a point of S, other than 0, namely $\dfrac{1}{m}$.

$\Rightarrow$ 0 is the limit point of S.

Now, to show that no real number l, other than 0 can be a limit point of S. Now there are following cases.

CASE I.   **If $l < 0$.**

In this case, the open interval $]p-1, 0[$ is a nbd of p which contain no point of S.

$\Rightarrow$ l is not the limit point of S.

CASE II.   **If $l > 1$**

In this case, the open interval $]1, l+1[$ is a nbd of p which contains no point of S.

$\Rightarrow$ l is not the limit point of S.

CASE III.   **If $0 < l < 1$ but $l \notin S$**

In this case, $l < 1 \Rightarrow \dfrac{1}{l} > 1$ and $\dfrac{1}{l}$ is not an integer.

Therefore, there exist a positive integer m such that

$$m < \frac{1}{l} < m+1$$

$\Rightarrow$ $\dfrac{1}{m+1} < l < \dfrac{1}{m}$

$\Rightarrow$ $\left]\dfrac{1}{m+1}, \dfrac{1}{m}\right[$ is a nbd of l which contains no point of S

$\Rightarrow$ l is not the limit point of S.

CASE IV. If $l = 1$.

In this case, the open interval $]\frac{1}{2}, 2[$ is a nbd of 1 which contain no point of S other than 1.

$\Rightarrow$ 1 is not the limit point of S.

CASE V. If $l \neq 1$ and $l \in S$.

If $l = \dfrac{1}{m} : m \in \mathbf{N}$ and $m \neq 1$, then $\left] \dfrac{1}{m+1}, \dfrac{1}{m-1} \right[$ is a nbd of l, which contains no point of S, other than l.

$\Rightarrow$ l is not the limit point of S.

Hence, no real number other than 0 is a limit point of S. Therefore, 0 is the only limit point of S.

☞ **REMARK**

➠ Since 0 does not belongs to S. Hence S is not closed.

EXAMPLE 4. *Find the limit point of the set of natural numbers N.*

SOLUTION. Let p be any element of $\mathbf{R}$. Now, find a positive integer ε, such that the open interval $]p-\varepsilon, p+\varepsilon[$ contains no point of $\mathbf{N}$, other than p.

$\Rightarrow$ p is not the limit point of $\mathbf{N}$

$\Rightarrow$ $D(\mathbf{N}) = \phi$

Hence, the set of natural number has no limit point.

EXAMPLE 5. *Find the limit point of the set of irrational numbers.*

SOLUTION. Since, every open interval containing a real number p also contains irrational numbers distinct from p, so every number $p \in \mathbf{R}$ is a limit point of the set of irrational number.

EXAMPLE 6. *Find the limit points of the set S of rational numbers of the form* $\left\{ \dfrac{n}{n+1} : n \in \mathbf{N} \right\}.$

SOLUTION. Here, we have

$$S = \left[\dfrac{n}{n+1} . n \in \mathbf{N} \right].$$

Also, $\dfrac{n}{n+1} = \dfrac{n+1-1}{n+1} = 1 - \dfrac{1}{n+1}.$

Let $\varepsilon > 0$ be arbitrary small positive number, then the nbd $]1-\varepsilon, 1+\varepsilon[$ of the point 1 contains a point of S, other than 1, because by taking $n > \left[\dfrac{\varepsilon}{1-\varepsilon} \right]$, we have

$$\dfrac{n}{n+1} > \dfrac{\varepsilon/(1-\varepsilon)}{[\varepsilon/(1-\varepsilon)]+1}$$

$\rightarrow$ $\dfrac{n}{n+1} > \varepsilon$

$\Rightarrow$ 1 is a limit point of the given set A.

Now, we check whether there is any other limit point of S other than 1.

Let us suppose $p \in A'$, $p \neq 1$.

Now, there are following cases:

CASE I. If $p > 1$, then choose $\varepsilon < p-1$, then the nbd $]p-\varepsilon, p+\varepsilon[$ of p contains no point of S, other than p.

$\Rightarrow$ p is not the limit point of S.

CASE II. If $p < 1$.

$p \in A'$, then there exist a point of S, which is nearest to p and let p_r be this element of S, which is nearest to p. Choose a positive integer ε such that $\varepsilon < |p_r - p|$, then the nbd $]p - \varepsilon, p + \varepsilon[$ of the point p contains no point of S and so as before, we conclude that p is not the limit point of S.

Suppose that $p \in S$ and let $p = \dfrac{n}{n+1}$.

Then the point just before p is $\dfrac{n-1}{(n-1)+1}$, i.e., $\dfrac{n-1}{n}$

And the point just after p is $\dfrac{n+1}{(n+1)+1}$, i.e., $\dfrac{n+1}{n+2}$

Now, we can find that

$$\frac{n+1}{n+2} - \frac{n}{n+1} = \frac{(n+1)^2 - n(n+2)}{(n+1)(n+2)} = \frac{1}{(n+1)(n+2)}$$

and $\dfrac{n}{n+1} - \dfrac{n-1}{n} = \dfrac{n^2 - (n-1)(n+1)}{n(n+1)} = \dfrac{1}{n(n+1)}$.

Also, $\qquad\qquad n + 2 > n$

$\Rightarrow \qquad\qquad \dfrac{1}{n+2} < \dfrac{1}{n}$

$\Rightarrow \qquad\qquad \dfrac{1}{(n+1)(n+2)} < \dfrac{1}{(n+1)n}$

Hence, we have

$$\frac{n}{n+1} - \frac{n-1}{n} > \frac{n+1}{n+2} - \frac{n}{n+1}.$$

Let us choose a positive number $\varepsilon > 0$ such that

$$\varepsilon < \left(\frac{n+1}{n+2} - \frac{n}{n+1} \right)$$

then the nbd $]p - \varepsilon, p + \varepsilon[$ of p contains no point of S, other than p.

$\Rightarrow \quad p$ is not the limit point of S.

Hence, we find that no real number other than 1 is a limit point of S.

EXAMPLE 7. ***Find the limit points of the set***

$$\left[1, -1, 1\frac{1}{2}, -1\frac{1}{2}, 1\frac{1}{3}, -1\frac{1}{3} \right]$$

(MEERUT 1996)

SOLUTION. Let we have

$$S = \left[1, -1, 1\frac{1}{2}, -1\frac{1}{2}, 1\frac{1}{3}, -1\frac{1}{3}, \ldots \right]$$

Now, let us define the sets A, B and C such that

$$A = \left[1\frac{1}{2}, 1\frac{1}{3}, 1\frac{1}{4}, \ldots \right]$$

$$B = \left[-1\frac{1}{2}, -1\frac{1}{3}, -1\frac{1}{4}, \ldots \right]$$

and $C = [1, -1]$

Then, $S = A \cup B \cup C$

$\Rightarrow D(S) = D(A \cup B \cup C) = D(A) \cup D(B) \cup D(C)$.

Now, since C is the a finite set, it has no limit point

$\Rightarrow D(C) = \phi$

Now, we wish to find $D(A)$ and $D(B)$.

We have, the set A is the sequence $< s_n >$ where

$$s_n = 1 + \frac{1}{n+1} : n \in N$$

$$\Rightarrow \lim_{n \to \infty} s_n = \lim_{n \to \infty} \left(1 + \frac{1}{n+1}\right) = 1$$

Therefore, the sequence $< s_n >$ converges to 1.

Now, we shall show that 1 is the limit point of A.

Let $\varepsilon > 0$, then by Archemedian property of real numbers, there exists a positive integer m such that

$$\frac{1}{m} < \varepsilon \Rightarrow \frac{1}{m+1} < \varepsilon$$

$$\Rightarrow \quad -\varepsilon < 0 < \frac{1}{m+1} < \varepsilon$$

$$\Rightarrow \quad 1 - \varepsilon < 1 + \frac{1}{m+1} < 1 + \varepsilon$$

Therefore, for every $\varepsilon > 0$, the open interval $]1-\varepsilon, 1+\varepsilon[$ contains a point of A, other

than 1, namely $1 + \dfrac{1}{m+1}$.

$\Rightarrow \quad$ 1 is the limit point of A.

Also, we can easily show that 1 is the only limit point of A.

$\Rightarrow \quad D(A) = 1$.

Again, the set B is the sequence $\left[-1 - \dfrac{1}{n+1} : n \in N\right]$.

Then, we have

$$D(B) = -1$$

Hence, $D(S) = D(A) \cup D(B) \cup D(C)$

$$= [1] \cup [-1] \cup \phi = [1, -1]$$

$\Rightarrow$ 1 and -1 are the only limit point of the given set S.

EXAMPLE 8. **Find the limit points of the set $S = \left[\dfrac{3n+2}{2n+1} : n \in N\right]$.**

SOLUTION. Here, we have, the given set S is the sequence $< s_n >$ where

$$s_n = \frac{3n+2}{2n+1} : n \in N$$

$$\Rightarrow \quad \lim_{n \to \infty} s_n = \lim_{n \to \infty} \left(\frac{3+2/n}{2+1/n}\right) = \frac{3}{2}.$$

$\Rightarrow \quad$ The sequence $< s_n >$ converges to $\dfrac{3}{2}$.

Now, we shall show that $\dfrac{3}{2}$ is the limit point of the set S. Let $\varepsilon > 0$. Since the sequence

$< s_n >$ converges to $\dfrac{3}{2}$, therefore for given $\varepsilon > 0$ there exist a positive integer m such that

$$|s_n - \tfrac{3}{2}| < \varepsilon \quad \forall n \geq m$$

i.e., $\quad \dfrac{3}{2} - \varepsilon < s_n < \dfrac{3}{2} + \varepsilon \quad \forall n \geq m.$

For every $\varepsilon > 0$, the open interval $\left] \dfrac{3}{2} - \varepsilon, \dfrac{3}{2} + \varepsilon \right[$ contains infinite terms of the sequence $< s_n >$.

$\Rightarrow \quad \dfrac{3}{2}$ is the limit point of S.

Now, to show $\dfrac{3}{2}$ is the only limit point of the set S.

Let if possible l be any limit point of the set S.

To show $l = \dfrac{3}{2}.$

Let $\varepsilon > 0$. Also $s_n \to \dfrac{3}{2}$, therefore

$$|s_n - \tfrac{3}{2}| < \frac{\varepsilon}{2} \quad \forall n \geq p \qquad \qquad \text{...(1)}$$

Since, l is a limit point of the set S, therefore, the open interval $\left] l - \dfrac{\varepsilon}{2}, \; l + \dfrac{\varepsilon}{2} \right[$ contains infinite distinct points of the set S.

$\Rightarrow$ There **must exist** a positive integer $q > p$ such that

$$l - \frac{\varepsilon}{2} < s_q < l + \frac{\varepsilon}{2}$$

$\Rightarrow \qquad |s_q - l| < \dfrac{\varepsilon}{2}. \qquad \qquad \text{...(2)}$

Now, from (1), we have

$$\left| s_q - \tfrac{3}{2} \right| < \frac{\varepsilon}{2}. \qquad \qquad \text{...(3)}$$

Consider

$$\left| \tfrac{3}{2} - l \right| = \left| (s_q - l) + \left(\tfrac{3}{2} - s_q \right) \right|$$

$$\leq |s_q - l| + \left| \tfrac{3}{2} - s_q \right| = |s_q - l| + \left| s_q - \tfrac{3}{2} \right| \leq \frac{\varepsilon}{2} + \frac{\varepsilon}{2}$$

$\Rightarrow \qquad \left| \tfrac{3}{2} - l \right| < \varepsilon.$

Now, since ε is arbitrary, hence, letting $\varepsilon \to 0$.

We have $\left| \dfrac{3}{2} - l \right| = 0 \Rightarrow \dfrac{3}{2} - l = 0 \Rightarrow l = \dfrac{3}{2}$

$\Rightarrow \quad \dfrac{3}{2}$ is the only limit point of S

$\Rightarrow \quad D(S) = \left[\dfrac{3}{2} \right].$

2.6 INTERIOR, EXTERIOR AND BOUNDARY OF A SET

Let R be the set of real numbers and A be any subset of R. Then we have the following definitions:

Definition 1. *A point p is called an interior point of A if there exists a neighbourhood of p contained in A.*

In other words, a point p is said to be an interior point of A if there exists an $\varepsilon > 0$ such that $]\,p-\varepsilon, p+\varepsilon\,[\subset A$

Definition 2. *The set of all interior points of A is called the interior of A and is denoted by A°.*

Definition 3. *A point p is called an exterior point of A if there exists a neighbourhood of p contained in the compliment A' of A, i.e., an exterior point of a set A is an interior point of A'.*

Definition 4. *The set of all exterior points of A is called exterior of A and is denoted by ext A.*

Definition 5. *A point p is called a boundary point or frontier point of A if it is neither an interior nor an exterior point of A.*

Definition 6. *The set of all boundary or frontier points of a set A is called boundary or frontier of the set.*

It is denoted by $b(A)$ or $\mathrm{Fr}(A)$.

THEOREM I. **Let A be a subset of real numbers R, then**

  (i) A° is an open set.

  (ii) A° is the largest open set contained in A.

  (iii) A is open if and only if $A^\circ = A$.

> ## FACTS : TO THE POINT
> ▶ Interior A° of a set A is the set of all those points of A which are not the limit point of A'.
> ▶ A point x is an exterior point of a set A if and only if x is not an adherent point of A.
> ▶ A point x is a boundary point of a set A if and only if every nbd of x intersect both A and A'.

PROOF.

(i) Let $x \in A^\circ$

$\Rightarrow$ x is an interior point of A.

$\Rightarrow$ A is a nbd of x.

$\Rightarrow$ $\exists$ an open set G such that $x \in G \subset A$.

Since G is open, so G is a nbd of each of its points. Also, since $G \subset A$, so A is also a nbd of each point of G.

$\Rightarrow$ Every point of G is an interior point of A.

$\Rightarrow$ $G \subset A^\circ$

$\Rightarrow$ To each $x \in A^\circ$, $\exists$ an open set G such that $x \in G \subset A^\circ$.

$\Rightarrow$ A° is the nbd of each of its points.

$\Rightarrow$ A° is open.

(ii) Let G be an open subset of A and let $x \in G$.

Then $x \in G \subset A$

Since G is open $\Rightarrow$ A is a nbd of x.

$\Rightarrow$ x is an interior point of A.

$\Rightarrow$ $x \in A^\circ$.

Thus, $x \in G \Rightarrow x \in A^\circ$ $(\because G \subset A^\circ)$

$\Rightarrow$ A° contains every open subset of A and it is therefore the largest open subset of A.

(iii) Let us first suppose $A = A^\circ$.

Since A° is open $\Rightarrow$ A is open.

Conversely, let A is open. We have to show that $A = A^o$.

Then A is surely identical with A^o.

Also, A^o is the largest open subset of A. Hence, $A = A^o$.

THEOREM 2.   *Let A, B be any two subsets of R, then*

     (i) $A^o \subset A$ *(ii)* $A \subset B \Rightarrow A^o \subset B^o$

   (iii) $(A \cap B)^o = A^o \cap B^o$ *(iv)* $A^o \cup B^o \subset (A \cup B)^o$

     (v) $A^{oo} = A^o$

PROOF.   (i) Let $x \in A^o$

          $\Rightarrow$ x is an interior point of A

          $\Rightarrow$ A is a nbd of x.

          $\Rightarrow$ $x \in A$.

     Since x is arbitrary, therefore $A^o \subset A$.

     (ii) Let $A \subset B$ and $x \in A^o$

          $\Rightarrow$ x is an interior point of A.

          $\Rightarrow$ x is an interior point of B. $(\because A \subset B)$

          $\Rightarrow$ $x \in B^o$.

     Since x is arbitrary, therefore $A^o \subset B^o$.

    (iii) We know that

$$A \cap B \subset A \Rightarrow (A \cap B)^o \subset A^o \qquad \text{(using (ii))}$$

     and $(A \cap B) \subset B \Rightarrow (A \cap B)^o \subset B^o$

          $\Rightarrow$ $(A \cap B)^o \subset A^o \cap B^o$ ...(1)

     Further, since

$$A^o \subset A \quad \text{and} \quad B^o \subset B$$

          $\Rightarrow$ $A^o \cap B^o \subset A \cap B$

     Also, $A^o \cap B^o$ is open.

          $\Rightarrow$ $(A^o \cap B^o)$ is the largest open set contained in $A \cap B$.

     Hence, $A^o \cap B^o \subset (A \cap B)^o$ ...(2)

     From (1) and (2), we conclude that

$$(A \cap B)^o = A^o \cap B^o$$

    (iv) By set theory, we can write

$$A \subset A \cup B \Rightarrow A^o \subset (A \cup B)^o$$

$$B \subset A \cup B \Rightarrow B^o \subset (A \cup B)^o$$

          $\Rightarrow$ $A^o \cup B^o \subset (A \cup B)^o$

     (v) Since A^o is always open and we know that A is open if and only if $A^o = A$. Using this argument for A^o, we get $(A^o)^o = A^o$

✎ ILLUSTRATIONS

● Let $A =]0, 1[$. Clearly A is open, therefore using $A^o = A$, for A is open.

$\Rightarrow A^o = A =]0, 1[$

Further $A' =]-\infty, 0[\cup [1, \infty]$

and $\text{ext}(A) = (A')^o =]-\infty, 0[\cup]1, \infty[$ ($\because$ 0 and 1 are not the interior point of A')

Since boundary of A consists of those points of A which are neither interior nor exterior points of A.

Therefore, $b(A) = [0, 1]$.

● Since no point of **Q**, the set of rational numbers can have an nbd contained in **Q**.

Now, $\text{ext}(\mathbf{Q}) = (\mathbf{Q}')^o = \phi$

Since Q' consists of all irrational points and no nbd of an irrational point can be contained in Q'. Hence $b(\mathbf{Q}) = \mathbf{R}$

2.7 DENSE AND PERFECT SETS

Let A, B be any two subsets of **R**, then we have the following definitions :

1. A is said to be dense in B if $B \subset \bar{A}$.

2. A is said to be dense in **R** or everywhere dense if $\bar{A} = \mathbf{R}$.

3. A is said to be non-dense or nowhere dense if the interior of the closure is empty, *i.e.,* $(\bar{A})^o = \phi$

4. A is said to be dense in itself if $A \subset D(A)$, *i.e.,* if every point of A is limit point of A.

5. A is said to be perfect if A is dense in itself and closed.

☞ REMARK

➡ Set A is not dense if A contains no open interval. Thus, we conclude that a closed set is non-dense if it contains no open interval.

2.7.1 ISOLATED POINT

Let A be any subset of **R** and p be any arbitrary point of A, then p is said to be isolated point of A if there exists a nbd of p which contain no point of A other than p itself.

In other words, we can say that p is said to be isolated point of A if p is not the limit point of A.

☞ REMARK

➡ Set of all isolated points is called **Discrete** set.

2.7.2 PERFECT SET

A set $A \subset \mathbf{R}$ is said to be perfect if it is closed and has no isolated points.

THEOREM I.   ***A set is perfect if and only if $A - D(A)$***

PROOF.   Let A be any subset of **R**. Then

   A is perfect

   $\Leftrightarrow A$ is dense in itself and closed.

   $\Leftrightarrow$ Every point of A is a limit point of A and each limit point of A belongs to A.

   $\Leftrightarrow A \subset D(A)$ and $D(A) \subset A$

   $\Leftrightarrow A = D(A)$

2.8 SETS OF FIRST AND SECOND CATEGORY

Definition 1. *A subset A of $\mathbf{R}$ is said to be of first category if A can be written as a countable collection of no where dense sets.*

Definition 2. *A subset A of $\mathbf{R}$ is said to be of second category if A is not of first category.*

📖 ILLUSTRATIONS

* The set $\mathbf{Q}$ of all rational points is everywhere dense. Also, since $\mathbf{Q} \subset D(\mathbf{Q}) = \mathbf{R}$, therefore it is dense in itself also. Now, since $\mathbf{Q}$ is not closed, so it is not perfect. Hence, $\mathbf{Q}$ is of first category.

* For any closed interval $A = [0, 1]$, since $A = D(A)$, therefore A is perfect.

* Consider the set $A = \left[\dfrac{1}{n} : n \in \mathbf{N}\right]$. Then every point of A is an isolated point since it is not a limit point of A. It is also non-dense, since $\bar{A} = A \cup \{0\}$ and no open interval can be a subset of A.

THEOREM 1. *Every countable set is of first category.*

PROOF. Let A be a countable subset of $\mathbf{R}$.

Then by definition of countable set, we can write

$$A = \{x_1, x_2, x_3, ..., x_n, ...\}$$

$$\Rightarrow \qquad A = \bigcup_{i=1}^{\infty} \{x_i\}$$

Since every one point set $\{x_n\}$ is nowhere dense.

Hence, A is of first category.

THEOREM 2. *Union of two sets of first category is of first category.*

PROOF. Let A and B be two sets of first category.

Then by definition, we can write

$$A = \bigcup_{n=1}^{\infty} A_n \quad \text{and} \quad B = \bigcup_{n=1}^{\infty} B_n$$

such that each A_n and B_n is nowhere dense.

Clearly, $A \cup B = \left[\left(\bigcup_{n=1}^{\infty} A_n\right) \cup \left(\bigcup_{n=1}^{\infty} B_n\right)\right]$

Since all the sets A_n and B_n form a countable collection.

$\Rightarrow$ $A \cup B$ is a countable collection of no-where dense subsets of $\mathbf{R}$.

Hence, $(A \cup B)$ is of first category.

THEOREM 3. **(Baire Category Theorem)** *The real line R is of second category.*

PROOF. Let if possible $\mathbf{R}$ is of first category.

Then by definition, we can write

$$\mathbf{R} = \bigcup_{n=1}^{\infty} A_n$$

where each A_n is nowhere dense (By def. of first category)

$$\Rightarrow \qquad (\bar{A}_n)^{\circ} = \phi \qquad \qquad \text{(By definition of no where dense subset)}$$

Let us assume that each A_n is closed.

Choose any x_1 which is not in A_1.

Since A_1 is closed.

$\Rightarrow$ x_1 is not the limit point of A_1.

$\Rightarrow$ $\exists$ an open interval I_1 with centre x_1, which contains no point of A_1.

Let B_1 be a closed interval with $0 < |B_1| < 1$ such that $B_1 \subset A_1$ such that $B_1 \cap A_1 = \phi$.

Let B_1' be the open interval with the same end points as those of B_1. Now A_2 is nowhere dense so it can not contain the open interval B_1'. Then choose any x_2 in B_1' such that $x_2 \notin A_2$. Then again, there is an open interval I_2 with centre x_2 which contain no point of A_2. Since B_1' is open, we can choose I_2 such that $I_2 \subset B_1' \subset B_1$

Further, let B_2 be a closed interval with $0 < |B_2| < \dfrac{1}{2}$ such that $B_2 \subset B_1$.

Then, $B_2 \cap A_2 = \phi$.

Continuing the process, we can construct a sequence of non-empty closed intervals

$B_1 \supset B_2 \supset B_3 \dots$ such that $0 < |B_n| < \dfrac{1}{n}$

and $B_n \cap A_n = \phi$

Then, by Cantor's intersection theorem, there exists a point $x_0 \in \mathbf{R}$ such that

$\displaystyle\bigcap_{n=1}^{\infty} B_n = \{x_0\}$

$\Rightarrow$ For each n, x_0 is in B_n.

$\Rightarrow$ $x_0 \not\subset A_n$ for any n $(\because B_n \cap A_n = \phi)$

$\Rightarrow$ $x_0 \notin \displaystyle\bigcup_{n=1}^{\infty} A_n$

which is a contradiction because $\displaystyle\bigcup_{n=1}^{\infty} A_n = \mathbf{R}$.

Hence, $\mathbf{R}$ is of second category.

THEOREM 4.   ***The set of all irrational numbers is of second category.***

PROOF.   Let I be the set of irrational numbers.

Let if possible I is of first category. Using illustration 1, we conclude that set of rational numbers Q is of first category.

Clearly, $I \cup Q = \mathbf{R}$ ...(1)

In previous theorem, we have proved that $\mathbf{R}$ is of second category. But from (1), $\mathbf{R}$ is of first category, being the union of two sets of first category.

Thus, we get a contradiction.

Hence, the set I of irrational numbers is of second category.

THEOREM 5.   ***The derived set of a dense in itself set is perfect.***

PROOF.   Let A be a set which is dense in-itself.

We have to show that $D(A)$ is perfect.

Since A is dense in itself

$\Rightarrow$ $A \subset D(A)$

$\Rightarrow$ $A \cup D(A) = D(A)$

$\therefore$ $D(A \cup D(A)) = D(D(A))$

$\Rightarrow$ $D(A) \cup D(D(A)) = D(D(A))$

$\Rightarrow$ $D(A) \cup D^2(A) = D^2(A)$ ...(1)

Now, since $D(A)$ is closed.

$\Rightarrow$ $D^2(A) \subset D(A)$

> **FACTS : TO THE POINT**
> - A countable set is never perfect and perfect set is never countable.
> - Every isolated set of points is countable.
> - If we are given on a straight line a set of non-overlapping intervals, then set is said to be countable.

$$\Rightarrow \quad D(A) \cup D^2(A) = D(A)$$

Then, from (1)

$$D(A) = D^2(A)$$
$$= D(D(A))$$

$\Rightarrow \quad D(A)$ is perfect. $\qquad\qquad (\because A \text{ is perfect if } A = D(A))$

2.9 COMPACTNESS OF A SET

2.9.1 OPEN COVERING OF A SET

Let $C = \{A_\lambda : \lambda \in \Lambda\}$ be a collection of sets of real numbers. Then C is said to be cover of a set A of real numbers. If each point of A belongs to A_λ for some $\lambda \in \Lambda$.

i.e., $\qquad\qquad$ if $A \subset \cup \{A_\lambda : \lambda \in \Lambda\}$

If each A_λ is open set then C is said to be an open cover of A.

2.9.2 SUBCOVER

Let C be the open cover of A. If C' is a cover of A such that $C' \subset C$, then C' is said to be subcover of A.

2.9.3 COMPACT SET

A set A is said to be compact if each open cover of A has a finite subcover.

THEOREM 1. **(Lindelof Theorem).** *If C is the set of collection of open sets of real numbers then there exists a countable subcollection $\{G_i\}$ of C such that*

$$\cup\{G : G \in C\} = \bigcup_{i=1}^{\infty} G_i.$$

PROOF. Let us write

$$P = \cup\{G : G \in C\}$$

and $x \in P$.

$\Rightarrow$ There exist at least one $G \in C$ such that $x \in G$ and since G is open

$\Rightarrow$ There exist an open interval I_1 such that $I_1 \subset G$

Further, since every interval I_1 contains infinitely many rational points, so, we can find an open interval J_1 with rational end points such that $x \in J_1 \subset I_1$.

Also, the collection of all open intervals with rational end points is countable.

$\Rightarrow$ The collection $\{J_i\}$, $i \in P$ is countable.

and $P = \cup\{J_i : i \in P\}$.

Now, select a set G in C (for each interval $\{J_i\}$) which contains it.

Thus we get a countable subcollection $\{G_i\}_{i=1}$ of C

and $P = \bigcup_{i=1}^{\infty} G_i$

$\Rightarrow \quad \cup\{G : G \in C\} = \bigcup_{i=1}^{\infty} G_i.$

THEOREM 2. **(Lindelof Covering Theorem).** *If A is a set of real numbers and C be an open cover of A, then there exists a countable subcollection of C which also covers A.*

PROOF. By definition of an open cover, we can write

$$A \subset \cup\{G : G \in C\}$$

Using previous theorem, we can write that there exists a countable subcollection $\{G_i\}_{i=1}^{\infty}$ of C such that

$$\bigcup[\,G : G \in C = \bigcup G_i\,]$$

which implies $A \subseteq \overset{\infty}{\underset{i=1}{\bigcup}} G_i$.

☛ **REMARK**

⟹ Above theorem states that from any open covering of an arbitrary set A of real numbers, we can pull out a countable covering.

2.10 HEINE-BOREL THEOREM

This theorem tells us that if we know that A is closed and bounded, we can reduce the covering to a finite covering.

THEOREM I.   **(Heine-Borel Theorem). *Let F be a closed and bounded set of real numbers. Then each open covering of F has a finite subcovering.***

or

Every closed and bounded set of real numbers is compact.

PROOF.   Let C be a collection of open sets such that

$$F \subset \cup\{G : G \in C\}$$

Then, we have to show that there exists a finite subcollection $[G_1, G_2, ..., G_n]$ of C such that $F \subset \overset{n}{\underset{i=1}{\bigcup}} G_i$.

Let us define C^* such that

$$C^* = C \cup \{F'\}$$

Since F is closed $\Rightarrow F'$ is open.

$$\Rightarrow C^* \text{ is a collection of open sets.}$$

Further $F \subset \cup[G : G \in C]$

Therefore $R = F \cup F' \subset [\cup\{G : G \in C\}] \cup F = \cup[G \cup F : G \in \ \ ]$

$$= \cup[G : G \in C^*]$$

$\Rightarrow$ C^* is open covering of R.

$\Rightarrow$ C^* is open covering of closed and bounded interval $[a, b]$ of R. Now we have to show that $[a, b]$ is covered by a finite sub-collection of C.

Let if possible, it is not true.

Then, if $I_0 = [a, b]$ can not be covered by a finite subcollection of C^*, then one of the intervals $\left[a, \frac{1}{2}(a+b)\right], \left[\frac{1}{2}(a+b), b\right]$ can not be so covered.

Denote this interval by $[a_1, b_1]$.

In a similar argument, one of the interval $\left]a_1, \frac{1}{2}(a_1+b_1)\right[, \ \left]\frac{1}{2}(a_1+b_1), b_1\right[$ can not be covered by a finite subcollection of C^*. We denote this interval by $[a_2, b_2]$.

Continuing in the same way, we get a sequence of closed intervals

$$\{I_n\} = \{[a_n, b_n]\}$$

such that

$$I_n \supset I_{n+1} \supset I_{n+2} \supset \cdots$$

and $\qquad |I_n| = b_n - a_n$

$$= \frac{b-a}{2^n}$$

$$\to 0 \text{ as } n \to \infty$$

Hence, by Cantor's intersection theorem, we can say that $\overset{\infty}{\underset{n=1}{\cap}} I_n$ consists of a single point, say p, which must belong to one of the members say G_0 of $\mathbf{C}$.

Now, since G_0 is an open set and $p \in G_0$, there exists $\varepsilon > 0$ such that $]p-\varepsilon, p+\varepsilon[\subset G$. Choose k so large such that

$$\frac{b-a}{2^k} < \varepsilon$$

Then $I_k \subset [p-\varepsilon, p+\varepsilon[\subset G$.

$\Rightarrow I_k$ is covered by a single member of $\mathbf{C}$

which is a contradiction.

Thus $[a, b]$ must be covered by a finite subcollection of C^* and thus F must also be covered by a finite subcollection of C^*. Further, if this finite subcollection does not contain F, it is a subcollection of $\mathbf{C}$ and our theorem is proved. If this subcollection contain F denote it by $\{G_1, G_2, ..., G_n, F'\}$

$\Rightarrow \qquad F \subset F' \cup G_1 \cup G_2 \cup ... \cup G_n$

Further, since no point of F is contained in F', we have

$$F \subset G_1 \cup G_2 \cup ... \cup G_n$$

and the collection $\{G_1, G_2, ..., G_n\}$ is a finite subcollection of $\mathbf{C}$ which covers F.

THEOREM 2. ***Compact subsets of real numbers R are closed and bounded.***

PROOF. Consider a compact subset A of real numbers $\mathbf{R}$.

Let us define $A_n =]-n \; n[$

Clearly a collection $\mathbf{C} = \{A_n : n \in \mathbf{N}\}$ is an open cover of $\mathbf{R}$, so $\mathbf{C}$ is an open cover of A also because $A \subset \mathbf{R}$. Since A is given to be compact, so there exists finitely many positive integers $n_1, n_2, ..., n_k$ such that the subcollection $[A_{n_1}, A_{n_2}, ..., A_{n_k}]$ of $\mathbf{C}$ covers A.

Define $\qquad n_0 = \max \{n_1, n_2, ..., n_k\}$

$\Rightarrow \qquad A \subset A =]-n_0, n_0[$

$\Rightarrow \qquad A$ is bounded.

Further, we have to show that A is closed. For this, it is sufficient to prove that no point of $\mathbf{R} - A$ can be limit point of A.

Let $a \in \mathbf{R} - A$ be arbitrary.

$\Rightarrow \qquad a \notin A$.

Consider the family of closed sets

$$F_n = \left[a - \frac{1}{n}, a + \frac{1}{n}\right] \quad \forall n \in \mathbf{N}$$

Then, $\qquad \mathbf{C} = [\mathbf{R} - F_n : n \in \mathbf{N}]$ is a collection of open sets.

Also, $\overset{\infty}{\underset{n=1}{\cap}} F_n = \{a\}$

Now, since $a \notin A$, we have

$A \subset [\mathbf{R} - \cap\{F_n : n \in \mathbf{N}\}]$

$= \cup \{ \boldsymbol{R} - F_n : n \in \boldsymbol{N} \}$, which work as a swubcollection which covered A.

Then by definition of compactness of A, there exists finitely many positive integers $m_1, m_2, ..., m_k$ such that every point of A is contained in one of the open sets $\boldsymbol{R} - F_{m_1}, \boldsymbol{R} - F_{m_2}, ..., \boldsymbol{R} - F_{m_k}$.

Therefore if $x \in A$, then for some $i \in [1, 2, 3, ..., k]$ and $x \in \boldsymbol{R} - F_m$.

$\Rightarrow$ No point of A is contained in $F_{m_i} = \left[a - \dfrac{1}{m_i}, a + \dfrac{1}{m_i} \right]$

$\Rightarrow$ a is not the limit point of A.

$\Rightarrow$ A contains all its limit points.

Hence, A is closed.

☛ **REMARK**

⇒ Using above two theorems, we conclude the following result : "A subset of real numbers is compact if and only if it is closed and bounded."

THEOREM 3. ***The set of real numbers R is not compact.***

PROOF. Let us define $A_n =]-n, n[$.

Also, let $\boldsymbol{C} = \{ A_n : n \in \boldsymbol{N} \}$ is an open cover of $\boldsymbol{R}$.

Now, if $\{ A_{n_1}, A_{n_2}, ..., A_{n_k} \}$ be any finite subfamily of $\boldsymbol{C}$.

Define $n_0 = \max \{ n_1, n_2, ..., n_k \}$.

Then, $n_0 \notin A_{n_i}$ for any $i = 1, 2, ..., k$.

$\Rightarrow$ No finite subfamily of $\boldsymbol{C}$ can cover $\boldsymbol{R}$. Hence, $\boldsymbol{R}$ is not compact.

THEOREM 4. ***Open intervals on R are not compact.***

PROOF. Let us define an open interval $I =]a, b[$ on $\boldsymbol{R}$.

If $A_n = \left] a + \dfrac{1}{n}, b \right[$

Then, $\boldsymbol{C} = \{ A_n : n \in \boldsymbol{N} \}$ be an open cover of $]a, b[$. $\qquad \left(\because \overset{\infty}{\underset{n=1}{\cup}} A_n =]a \ b[\right)$

Then it is not possible to find a finite subcollection of $\boldsymbol{C}$ which covers A because if

$\boldsymbol{C'} = \{ A_{n_1}, A_{n_2}, ..., A_{n_k} \}$ be any finite subcollection of $\boldsymbol{C}$. If $n_0 = \max \{ n_1, n_2, ..., n_k \}$.

Then it is clear that the subset $\left] a, a + \dfrac{1}{n_0} \right[$ of A is not covered by $\boldsymbol{C'}$.

$\Rightarrow$ $\exists$ an open cover of A which does not have a finite subcover.

$\Rightarrow$ $]a, b[$ is not compact.

Hence, open intervals on real numbers is not compact.

2.11 CANTOR TERNARY SET

Consider the closed interval $[0, 1]$. Divide this closed interval into three equal parts and remove the middle one, *i.e.*, remove the open interval $\left(\dfrac{1}{3}, \dfrac{2}{3} \right)$. This is our first step of construction of Cantor ternary set.

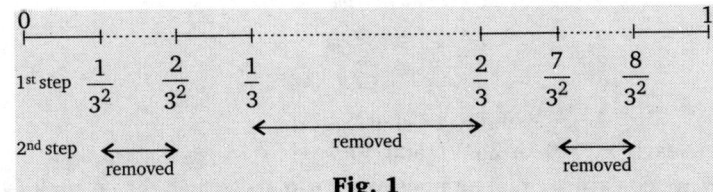

Fig. 1

Again divide each of the remaining two intervals into three equal parts and remove the middle one, i.e., remove the open intervals $\left(\dfrac{1}{3^2}, \dfrac{2}{3^2}\right)$ and $\left(\dfrac{7}{9}, \dfrac{8}{9}\right)$. This is our second step of construction.

Continuing in this way infinitely many times, we have that at p^{th} step, 2^{p-1} open intervals are removed each of length $\dfrac{1}{3^p}$. The remaining set constitutes Cantor ternary set. Let us denote this set by F. Then clearly F^c is open being enumerable union of mutually disjoint open intervals $\Rightarrow$ Being the complement of an open set, set F is closed.

Here, $F = [0, 1] - F^c$, where $F^c = \left(\dfrac{1}{3}, \dfrac{2}{3}\right) \cup \left(\dfrac{1}{9}, \dfrac{2}{9}\right) \cup \left(\dfrac{7}{9}, \dfrac{8}{9}\right) \cup \dots$

Thus Cantor ternary set surely contains the points 0, 1, $\dfrac{1}{3}$, $\dfrac{2}{3}$, $\dfrac{1}{9}$, $\dfrac{2}{9}$, $\dfrac{7}{9}$, $\dfrac{8}{9}$, etc.

Definition. *The Cantor set F is the set of all numbers in the interval $[0, 1]$ which have a ternary expansion without the digit 1, i.e., ternary expansion involves only two digits 0 or 2.*

Obviously Cantor set contains no open interval.

2.11.1 PROPERTIES OF CANTOR SET

PROPERTY 1.   ***Cantor set is obviously bounded as $F \subset [0, 1]$***

PROPERTY 2.   ***By Heine-Borel theorem, being closed and bounded set F is compact also.***

PROPERTY 3.   ***Cantor ternary set is perfect.***   (GARHWAL, 1997)

PROOF.   Set F is closed. Therefore every limit point of F belongs to F. Further, we shall show that every number of F is a limit point of numbers having terminating ternary expansions.

Let x_0 be any element of F. We can write the ternary expression of x_0 as $x_0 = .a_1\, a_2\, a_3 \dots a_n \dots$ where each $a_n = 0$ or 2.

Now construct a sequence $<x_n>$ of elements of F given as follows :

$$x_1 = .a_1'\, a_2\, a_3 \dots a_n \dots,$$
$$x_2 = .a_1\, a_2'\, a_3 \dots a_n \dots$$
$$\dots \qquad \dots \qquad \dots \qquad \dots$$
$$\dots \qquad \dots \qquad \dots \qquad \dots$$
$$x_n = .a_1\, a_2\, a_3 \dots a_n' \dots,$$

where, $\forall\, n \in \mathbb{N}$, $a_n' = 0$, if $a_n = 2$ and $a_n' - 2$, if $a_n = 0$,

$\Rightarrow$ $x_n \neq x_0$, $\forall\, n$.

Also, $\lim\limits_{n \to \infty} x_n = x_0 \Rightarrow x_0$ is a limit point of F.

Therefore set F is dense-in-itself and hence it is a perfect set.

PROPERTY 4.   ***Cantor set is uncountable.***

PROOF.   Suppose contradiction, *i.e.*, F is countable. Then the elements of F can be arranged to form a sequence. Let $F = \{x_n : n \in \mathbf{N}\}$.

Expressing each x_i's in ternary expression, let

$$x_1 = 0.x_{11}\, x_{12}\, x_{13} \dots x_{1m} \dots,$$
$$x_2 = 0.x_{21}\, x_{22}\, x_{23} \dots x_{2m} \dots,$$
$$\dots \qquad \dots \qquad \dots \qquad \dots$$
$$\dots \qquad \dots \qquad \dots \qquad \dots$$
$$x_n = 0.x_{n1}\, x_{n2}\, x_{n3} \dots x_{nm} \dots,$$

where each $x_{nr} = 0$ or 2, $\forall\, r$ and $\forall\, n$.

Now, we construct a real number $x = .\alpha_1 \alpha_2 \alpha_3 \dots$ such that $\alpha_m = 2$, if $x_{mm} = 0$, and $\alpha_m = 0$, if $x_{mm} = 2$.

In either case, $x_{mm} \neq \alpha_m \; \forall m$ giving $x_m \neq x$, $\forall m \in N$. Since $\alpha_m = 0$ or 2, $\forall m$ it follows that $x \in F$.

Thus F contains other points also besides x_m and hence F is uncountable. It proves the required result.

PROPERTY 5. *The cardinal number of the Cantor set is the cardinal number* **c** *of the linear continuum.*   (MEERUT 1990)

PROOF. Let us consider a mapping $f : F \to [0, 1]$. If $x \in F$ and its ternary expansion is $x = 0 . x_1 \, x_2 \, x_3 \ldots$, where each $x_i = 0$ or 2, then we define $f(x)$ as the number whose binary (base 2) expansion is $0 . k_1 \, k_2 \, k_3, \ldots$, where $k_i = \frac{1}{2} x_i$, $\forall i$.

Thus $f(0 . x_1 \, x_2 \, x_3 \ldots) = 0 . k_1 \, k_2 \, k_3 \ldots$, where $k_i = \frac{1}{2} x_i$, $\forall i$

This function f is one-one.

For, let $x = 0 . x_1 \, x_2 \, x_3 \ldots$, $y = 0 . y_1 \, y_2 \, y_3 \ldots$ be any two elements of F s.t. each x_i and $y_i = 0$ or 2. Then

$$f(x) = f(y) \Rightarrow f(0 . x_1 \, x_2 \, x_3 \ldots) = f(0 . y_1 \, y_2 \, y_3 \ldots)$$
$$\Rightarrow 0 . k_1 \, k_2 \, k_3 \ldots = 0 . p_1 \, p_2 \, p_3 \ldots \text{ where } k_i = \frac{1}{2} x_i \text{ and } p_i = \frac{1}{2} y_i$$
$$\Rightarrow \text{ each } k_i = p_i \text{ and hence each } x_i = y_i.$$
$$\Rightarrow x = y \Rightarrow f \text{ is one-one.}$$

Function f is onto also.

Let $z \in [0, 1]$. Let its binary expansion be $z = 0 . l_1 \, l_2 \, l_3 \ldots$ where each $l_i = 0$ or 1. If $2l_i = x_i$, then each $x_i = 0$ or 2 and hence the number $x = 0 . x_1 \, x_2 \, x_3 \ldots$ where

$x_i = 2l_i$, belongs to the set F s.t. $f(x) = f(0 . x_1 x_2 x_3 \ldots) = 0 . l_1 l_2 l_3 \ldots = z$, where $l_i = \dfrac{1}{2} x_i$.

Thus, for each $z \in [0, 1], \exists$ a preimage x in F under the mapping f. Hence f is an onto mapping.

Thus, $f \sim [0, 1]$, showing that card $F =$ card $[0, 1] =$ **c**.

Alternatively, note that F contains numbers whose ternary expansions contain only two digits 0 or 2. Thus F contains as many elements as the number of ways of forming the numbers of the form $0 . x_1 \, x_2 \, x_3 \, x_4 \ldots$.

For each place, x_i can be chosen in two ways (either $x_i = 0$ or $x_i = 2$). So we can form 2^n such numbers by the digits 0 and 2.

Hence, card $F = 2^n$ which is equal to **c** implying that card $F =$ **c**.

☛ **REMARK**

⇒ Above function proves that the Cantor set can be put into a one-to-one correspondence with the interval $[0, 1]$.

REVIEW QUESTIONS AND ARCHIVE

1. Define the following :
   (i) Neighbourhood
      (MEERUT–1990, 92, 94, 2000, 01, 11, 13, 15, 16, 18,
      BHOPAL–2005, 09, HIMACHAL–2003, 07, 09)
   (ii) Open and closed sets
   (iii) Adherent and limit points
   (iv) Interior and closure of a set
   (v) Dense and non-dense set
   (vi) Discrete and derived sets

   (vii) Limit point   (MEERUT–2001)
2. Define Cantor Ternary Set.(MEERUT-2009, 10, 12, 15,
   GARHWAL–2003, 16, GURUKUL KANGRI–2001,
   ROHTAK–2000, 03, HIMACHAL–2002)
3. Define compact subsets of **R**.
4. Give an example of each of the following :
   (i) An open set which is not an interval
   (ii) An interval which is not open
   (iii) A closed set which is not an interval

(iv) An interval which is not closed

(v) A set which is neither open interval nor an open set

(vi) A set which is neither closed interval nor closed set.

5. Write 'T' for True and 'F' for False statement:

(i) A finite set can be open **(T/F)**

(ii) A finite non-empty set can be open. **(T/F)**

(iii) Every infinite set is open. **(T/F)**

(iv) Union of arbitrary family of closed sets is closed. **(T/F)**

(v) Set of irrational numbers neither open nor closed. **(T/F)**

6. Show that an open interval is a nbd of each of its points.

7. Show that the set **Q** of rational numbers is not a nbd of any of its points.

8. Show that a non-empty finite set can not be a nbd of any of its points.

9. Show that the set of integers is not a nbd of any of its points.

10. Show that every open interval is an open set.

11. Show that interior of a set is open.

12. Show that if $A \subset B$, then $D(A) \subset D(B)$.

13. Show that the set of rational numbers is not an open set.

14. Show that $D(A \cup B) = D(A) \cup D(B)$

15. Show that $D(A \cap B) \subset D(A) \cap D(B)$.

16. Show that the union of any collection of open sets is open.

17. Show that the intersection of a finite collection of open sets is open.

18. State and prove Bolzano-Weirstrass theorem for real numbers.

19. State and prove Baire's category theorem for real numbers.

20. Show that the set of irrational numbers is of second category.

21. Show that compact subset of real numbers are closed and bounded.

22. Show that open intervals on real numbers are not compact.

23. Show that every non-empty set contains both rational and irrational points.

24. Show that every closed set in **R** is the intersection of a countable collection of open sets.

25. Show that p be an adherent point of a set A if and only if there is a sequence $< s_n >$ with $s_n \in A$ and $p = \lim s_n$.

OBJECTIVE EVALUATION

▶ FILL IN THE BLANKS

1. Every _____ set is a union of open intervals.

2. The interior of a set A is the largest open set contained in _____.

3. Finite intersection of open set is _____.

4. An adherent point is also called _____ point.

5. A set is said to be _____ if it is dense in itself and closed.

6. A set A is said to be everywhere dense if $\bar{A} =$ _____.

7. A finite set has _____ limit points.

8. Every infinite bounded set of real numbers has a _____.

9. The set of natural numbers has _____ limit point.

10. $D[\,]a, b[\,] =$ _____.

▶ TRUE OR FALSE

1. Every open interval is a nbd of each of its points. **(T/F)**

2. Every closed interval is a nbd of each of its points. **(T/F)**

3. The closed interval is a nbd of each of its points except the end points. **(T/F)**

4. Every finite set is closed. **(T/F)**

5. Every infinite bounded set has at least one limit point. **(T/F)**

6. The empty set ϕ is open as well as closed. **(T/F)**

7. The only limit points of $]a, b[$ is a and b. **(T/F)**

8. A finite set $S = [1, 2, 3, 4]$ is an open set. **(T/F)**

9. The set of rational numbers is a nbd of each of its points. **(T/F)**

10. The derived set of a finite set is always finite. **(T/F)**

▶ MULTIPLE CHOICE QUESTIONS (CHOOSE THE MOST APPROPRIATE ONE)

1. The closed interval $[a, b]$ is a nbd of:
- (a) each of its points
- (b) each of its end points
- (c) each of its points except the end points
- (d) none of these

2. The set of rational numbers is a nbd of:
- (a) each of its points
- (b) each $n \in N$
- (c) not a nbd of any of its points
- (d) none of these

3. The number of limit points of a set of natural numbers N is:
- (a) 0
- (b) 1
- (c) 2
- (d) 4

4. If a and b are any two distinct real numbers then there exists nbds of a and b which are disjoint. This property is called:
- (a) Housedroff property
- (b) Denseness property
- (c) Completeness
- (d) None of these

5. If a point $p \in R$ is limit point of $A \subset R$, then which of the following is not true:
- (a) $(N \cap A) - \{p\} \neq \phi$ for every nbd N of p
- (b) $N \cap [A - \{p\}] = \phi$
- (c) $N - \{p\} \cap A \neq \phi$ for every nbd N of p
- (d) None of these

6. A point $p \in A \subset R$ is said to be an isolated point of A if:
- (a) p is an adherent point of A
- (b) p is not an adherent point of A
- (c) p is not the limit point of A
- (d) p is the limit point of A

7. The limit point of the set $S = \left\{ \dfrac{3n+2}{2n+1} : n \in N \right\}$ is:
- (a) $\dfrac{5}{2}$
- (b) $\dfrac{3}{2}$
- (c) $\left[\dfrac{3}{2}, \dfrac{5}{2} \right]$
- (d) None of these

8. The only limit point of the set $S = \left\{ \dfrac{1}{n} : n \in N \right\}$ is:
- (a) 0
- (b) 1
- (c) 2
- (d) 3

ANSWERS

▶ FILL IN THE BLANKS

1. open	**2.** A	**3.** open	**4.** closure	**5.** perfect	**6.** R	**7.** no
8. limit point	**9.** no	**10.** $[a, b]$				

▶ TRUE OR FALSE

1. T	**2.** F	**3.** T	**4.** T	**5.** T	**6.** T	**7.** F	**8.** F	**9.** F
10. T								

▶ MULTIPLE CHOICE QUESTIONS

1. (c)	**2.** (c)	**3.** (a)	**4.** (a)	**5.** (b)	**6.** (c)	**7.** (b)	**8.** (a)

＼ CHAPTER SUMMARY ／

This chapter introduced the fundamental notions and concepts of open and closed sets of real numbers and showed that these are the main ideas behind the metric space. Typical examples of limit point, adherent points, etc. were presented as well. The important points discussed in this chapter are as follows :

➲ A subset N of R is said to be a nbd of a point $p \in R$ if for given $\varepsilon > 0$ ∃ an open interval $]p - \varepsilon, p + \varepsilon[\subset N$.

➲ An open interval is a nbd of each of its points.

➲ A closed interval is a nbd of each of its points except the end points.

➲ A set is said to be open if it is a nbd of each of its points.

➲ The empty set ϕ and whole set R are open.

➲ Arbitrary union of open sets is open.

➲ Finite intersection of open sets is open.

⊃ Singleton set is always closed.

⊃ Finite union of closed sets is closed.

⊃ Arbitrary intersections of closed sets is closed.

⊃ A point $p \in \mathbf{R}$ is called a limit point of a set $A \subset \mathbf{R}$ if every nbd N of p contains a point other than p.

⊃ Every point of the closed interval $[a, b]$ is a limit point of the set of points in open interval $]a, b[$.

⊃ Every real number is an accumulation point of the set of rational numbers.

⊃ Every limit point of A is an adherent point of A but not conversely.

⊃ An adherent point of A may or may not belong to A.

⊃ A bounded infinite set of real numbers has at least one limit point.

⊃ A set is said to be closed if it contains all its limit points.

⊃ If N is a nbd of p then p is said to be an interior point of N.

⊃ The interior of a set $A \subset \mathbf{R}$ is the largest open set contained in A.

⊃ If $A, B \subseteq \mathbf{R}$, then

  (i) A is said to be dense in B if $B \subseteq \bar{A}$.

  (ii) A is said to be everywhere dense if $\bar{A} = \mathbf{R}$.

  (iii) A is said to be nowhere dense if $(\bar{A})^\circ = \phi$.

  (iv) A is said to be dense-in-itself if $A \subset D(A)$

⊃ A set $\subset$ is said to be perfect if A is dense-in-itself and closed.

⊃ A point p is said to be isolated point of $A \subseteq \mathbf{R}$ if p is not the limit point of A.

⊃ The set of real numbers is of second category.

⊃ A countable set is never perfect and conversely.

⊃ The set of isolated points is countable.

⊃ Compact subset of real numbers are closed and bounds.

⊃ The set $\mathbf{R}$ of real numbers is not compact.

⊃ The Cantor ternary set is always closed, bounded and compact.

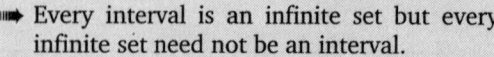

FOR ADVANCED LEARNERS

➡ Every interval is an infinite set but every infinite set need not be an interval.

➡ The set $\mathbf{N}, \mathbf{Z}, \mathbf{Q}$ and $\mathbf{R}$ all are bounded above sets.

➡ The set $\mathbf{Z}, \mathbf{Q}$ and $\mathbf{R}$ are not bounded below.

➡ The set $\mathbf{Q}$ of rationals is not ordered complete.

➡ Every open interval is an open set but converse is not necessarily true.

➡ Arbitrary intersection of nbds of point p is not a nbd of p.

➡ Limit point of a set may or may not belong to the set.

➡ A set may or may not have a limit point.

➡ Every limit point of a set is also the adherent point of that set.

➡ An adherent point may or may not be a limit point of the set.

➡ The supremum of a non-empty bounded set is neither the greatest member of the set or is a limit point of the set.

➡ A non-empty bounded and closed set contains its supremum and infimum.

➡ The supremum and infimum of a set A are also the supremum and infimum of $\bar{A}$ and are contained in $\bar{A}$ according as A is bounded above or below.

➡ Every infinite bounded set of real numbers has a limit point. This result does not hold for the system of rational numbers.

➡ If a set $A \subseteq \mathbf{R}$ satisfies the Heine-Borel property then any closed subset of A also satisfies the Heine-Borel property.

➡ Every set satisfying the Heine-Borel property is a compact set.

➡ The union of finite family of compact sets is compact.

➡ The intersection of any family of compact sets is compact.

➡ The sets $\mathbf{N}, \mathbf{Z}, \mathbf{Q}$ and $\mathbf{R}$ are not compact.

●●●●

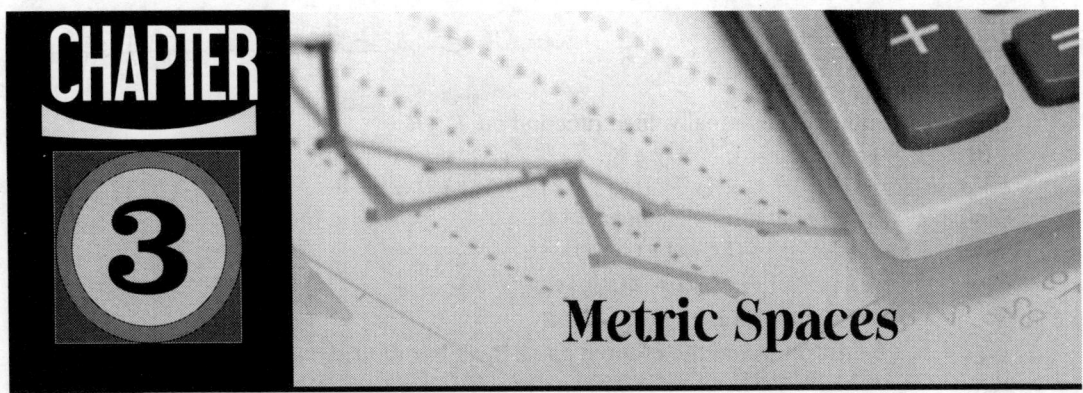

CHAPTER 3

Metric Spaces

3.1 INTRODUCTION

In 1905, the French mathematician M. Frechet introduced the concept of a metric on a set. The subject of real numbers has two type of properties. The first type consists of the algebraic which deals with addition, multiplication, etc. The second type consists of properties having to do with the notion of distance between two numbers and with the concept of limit. The second type of properties are called topological or metric. In this chapter, we shall study these properties in a general space in which the notion of distance is defined. ∞∞

3.2 METRIC SPACE

Let X be a non-empty set. Then a mapping $d : X \times X \to R$ is said to be metric, if it satisfying the following conditions :

M(i) $d(x, y) \geq 0 \quad \forall x, y \in X$

M(ii) $d(x, y) = 0$ iff $x = y \quad \forall x, y \in X$

M(iii) $d(x, y) = d(y, x) \quad \forall x, y \in X$ (Symmetric property)

M(iv) $d(x, y) \leq d(x, z) + d(z, y) \quad \forall x, y, z \in X$ (Triangle inequality)

If d is a metric on X, then ordered pair (X, d) is said to be **metric space**.

☞ REMARKS

⟹ The function d is also known as distance function, i.e., d is a real valued function denotes the distance between x and y.

⟹ In 1905, the French mathematician Maurice Frechet thought of generalizing the notion of distance and extending it to arbitrary set, which seems the beginning of metric space.

3.2.1 DESCRIPTIONS OF THE PROPERTIES OF METRIC SPACES

* $d(x, y) \geq 0 \Rightarrow$ The distance between any two points of X is non-negative real number.
* $d(x, y) = 0$ **iff** $x = y \Rightarrow$ If the two points coincide then the distance is zero and if the distance is zero, then two points are same.
* $d(x, y) = d(y, x) \Rightarrow$ The distance does not depend on the order of the points x and y.
* $d(x, y) \leq d(x, z) + d(z, y) \Rightarrow$ The sum of the length of two sides of a triangle is greater than or equal to the length of the third side. The sign of equality holds when three points lie on the straight line.

3.2.2 EXAMPLES ON METRIC SPACES

(a) The real line R. Let R be the set of real numbers and let $d : R \times R \to R$ be the function

defined by
$$d(x, y) = |x - y| \; \forall \; x, y \in \boldsymbol{R}$$
(MEERUT 2004,07; RAJ. 2004,06,08)

Then clearly, we have

(i) d is a non-negative real valued function on $\boldsymbol{R} \times \boldsymbol{R}$

(ii) $d(x, y) = 0 \Leftrightarrow x = y \; \forall \; x, y \in \boldsymbol{R}$

(iii) $d(x, y) = |x - y| = |-(y - x)| = |y - x| = d(y, x) \; \forall x, y \in \boldsymbol{R}$

(iv) Let x, y, z be any three elements of $\boldsymbol{R}$.

Then, $d(x, y) = |x - y| = |(x - z) + (z - y)|$
$$\leq |x - z| + |z - y| = d(x, z) + d(z, y)$$

$\Rightarrow \quad d(x, y) \leq d(x, z) + d(z, y) \; \forall \; x, y, z \in \boldsymbol{R}$

$\Rightarrow \quad d$ is a metric on $\boldsymbol{R}$ and the ordered pair $(\boldsymbol{R}, d)$ is a metric space.

☛ **REMARK**

➡ The metric defined above is called usual metric on $\boldsymbol{R}$.

(b) The Euclidean plane $\boldsymbol{R}^2$. Let $\boldsymbol{R}^2$ be the set of all ordered pairs of real numbers. Define a mapping $d : \boldsymbol{R}^2 \times \boldsymbol{R}^2 \to \boldsymbol{R}$ such that
$$d(\boldsymbol{x}, \boldsymbol{y}) = \sqrt{(x_1 - y_1)^2 + (x_2 - y_2)^2} \;\; \forall x, y \in \boldsymbol{R}^2$$
where $\boldsymbol{x} = (x_1, x_2)$, $\boldsymbol{y} = (y_1, y_2)$

Then clearly we have

(i) d is a non-negative real value function on $\boldsymbol{R}^2 \times \boldsymbol{R}^2$.

(ii) $d(\boldsymbol{x}, \boldsymbol{y}) = 0 \Leftrightarrow \sqrt{\{(x_1 - y_1)^2 + (x_2 - y_2)^2\}} = 0$

$\qquad\qquad\qquad \Leftrightarrow x_1 - y_1 = 0$ and $x_2 - y_2 = 0$

$\qquad\qquad\qquad \Leftrightarrow x_1 = y_1$ and $x_2 = y_2 \Leftrightarrow \boldsymbol{x} = \boldsymbol{y}$.

(iii) For all $\boldsymbol{x}, \boldsymbol{y} \in \boldsymbol{R}^2$

$$d(\boldsymbol{x}, \boldsymbol{y}) = \sqrt{\{(x_1 - y_1)^2 + (x_2 - y_2)^2\}}$$
$$= \sqrt{\{(y_1 - x_1)^2 + (y_2 - x_2)^2\}} = d(\boldsymbol{y}, \boldsymbol{x}).$$

(iv) Let $\boldsymbol{x} = (x_1, x_2)$, $\boldsymbol{y} = (y_1, y_2)$, $\boldsymbol{z} = (z_1, z_2)$ be any three elements of $\boldsymbol{R}^2$. Now, by triangle inequality for real numbers, we have

$$\sqrt{\{(a_1 + b_1)^2 + (a_2 + b_2)^2\}} \leq \sqrt{(a_1^2 + a_2^2)} + \sqrt{(b_1^2 + b_2^2)}, \text{ for } a_1, b_1, a_2, b_2 \in \boldsymbol{R}$$

Put $a_1 = x_1 - z_1$, $a_2 = x_2 - z_2$, $b_1 = z_1 - y_1$, $b_2 = z_2 - y_2$ in above inequality, we have

$$\sqrt{\{(x_1 - y_1)^2 + (x_2 - y_2)^2\}} \leq \sqrt{\{(x_1 - z_1)^2 + (x_2 - z_2)^2\}} + \sqrt{\{(z_1 - y_1)^2 + (z_2 - y_2)^2\}}$$

$\Rightarrow \quad d(\boldsymbol{x}, \boldsymbol{y}) \leq d(\boldsymbol{x}, \boldsymbol{z}) + d(\boldsymbol{y}, \boldsymbol{z})$

Hence, d is a metric on $\boldsymbol{R}^2$.

☛ **REMARK**

➡ The metric space $(\boldsymbol{R}^2, d)$ defined above is called Euclidean metric space.

(c) Discrete metric space. Let X be any non-empty set. The mapping $d : X \times X \to \mathrm{R}$ defined by
$$d(x, y) = \begin{cases} 0 & \text{if } x = y \\ 1 & \text{if } x \neq y \end{cases}$$

PROOF. $d(x, y)$ is a metric because it satisfies all the following properties such that

(i) $d(x, y) = 0$ or $d(x, y) = 1$

$\Rightarrow d(x, y) \geq 0 \;\; \forall \, x, y \in X$

(ii) If $x = y \Rightarrow d(x, y) = 0$.

Conversely, if $d(x, y) = 0$, then $x = y$.

If it is not possible, *i.e.*, $x \neq y$. Then, $d(x, y) = 1$

It contradicts the fact that $d(x, y) = 0$

$\Rightarrow d(x, y) = 0$ iff $x = y$.

(iii) If $x = y$ $\Rightarrow$ $d(x, y) = 0 = d(y, x)$

If $x \neq y$ $\Rightarrow$ $d(x, y) = 1 = d(y, x) \;\; \forall \, x, y \in X$

Hence, $d(x, y) = d(y, x)$.

(iv) Let $x, y, z \in X$

CASE I. $x = y$

(i) $x = z \Rightarrow z = y$

then, $d(x, y) = 0, \; d(x, z) = 0, \; d(y, z) = 0$

$\Rightarrow d(x, y) \leq d(x, z) + d(z, y)$

(ii) $x \neq z \Rightarrow z \neq y$

$\Rightarrow d(x, y) = 0, \; d(x, z) = 1 = d(y, z)$

then, $d(x, z) + d(z, y) = 1 + 1 = 2 \geq 0$

$\Rightarrow d(x, y) \leq d(x, z) + d(z, y)$

CASE II. $x \neq y$

(i) $x = z \Rightarrow z \neq y$

$d(x, y) = 1, \; d(x, z) = 0, \; d(y, z) = 1$

$\Rightarrow d(x, z) + d(z, y) = 1 \geq d(x, y)$

$\Rightarrow$ $d(x, y) \leq d(x, z) + d(z, y)$

(ii) $x \neq z$, then either $z = y$ or $z \neq y$

(a) $x \neq z, \; z = y$

then, $d(x, y) = 1, \; d(x, z) = 1, \; d(z, y) = 0$

$\Rightarrow d(x, z) + d(z, y) = 1 + 0 = 1 \geq d(x, y)$

$\Rightarrow d(x, z) + d(z, y) \geq d(x, y)$

(b) $x \neq y \neq z$

$d(x, y) = 1, \; d(x, z) = 1, \; d(z, y) = 1$

$1 \leq 1 + 1$

$\Rightarrow d(x, y) \leq d(x, z) + d(z, y)$

For any combination of x, y, z, triangular inequality is satisfied.

Hence, d is metric and (X, d) is a metric space.

The metric space defined above is called **Discrete metric space**.

(d) Let (X_1, d_1) and (X_2, d_2) be two metric spaces.

Define $X = X_1 \times X_2$

and $d(x, y) = d_1(x_1, y_1) + d_2(x_2, y_2) \;\; \forall \, \boldsymbol{x}, \boldsymbol{y} \in X$,

when $\boldsymbol{x} = (x_1, y_1)$ and $\boldsymbol{y} = (x_2, y_2)$

Then (X, d) is a metric space. Hence, Cartesian product of two metric spaces is a metric space.

☞ REMARK

⟶ The space (X, d) defined above is called Product metric space.

(e) The Postman metric for R^2. Consider a well-planned city, in which the roads are either parallel or perpendicular to each other and there are rectangular blocks of housing complexes. Suppose someone wants to go from point A to point B. How do we find the minimum distance that he has to travel.

Since he can not go as a cross flies. Therefore, the Euclidean metric is useless.

The product metric is also useless. Then what? He has to go along one road till he reaches a road on which B is situated and then move along the perpendicular road till he reaches B.

If the coordinates of A and B with reference to a pair of rectangular axes, one of which is parallel to one set of roads, and the other is perpendicular to it, be (x_1, x_2) and (y_1, y_2) respectively, then he will have to move a distance $|x_1 - y_1| + |x_2 - y_2|$

Therefore, we can define a metric as follows :

Let R^2 be the set of all ordered pairs of real numbers and let

$$d : R^2 \times R^2 \to R$$

defined by $d(x, y) = |x_1 - y_1| + |x_2 - y_2|$

where $x = (x_1, x_2)$ and $y = (y_1, y_2)$.

Then, clearly we have

(i) d is a non-negative real-valued function on $R^2 \times R^2$

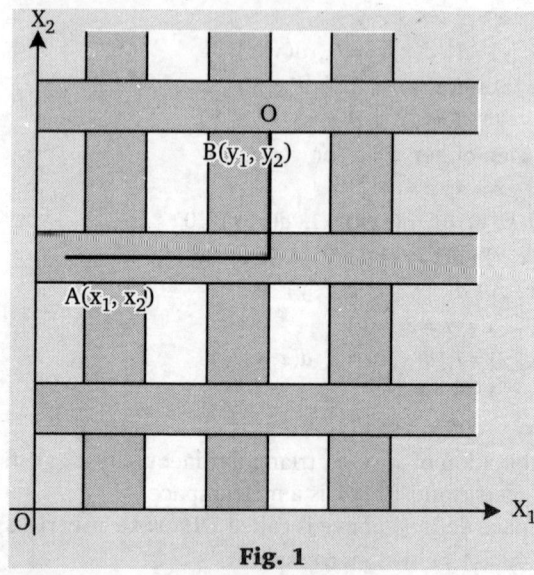

Fig. 1

(ii) $d(x, y) = 0 \Leftrightarrow |x_1 - y_1| + |x_2 - y_2| = 0$
$\Leftrightarrow |x_1 - y_1| = 0$ and $|x_2 - y_2| = 0$
$\Leftrightarrow x_1 = y_1$ and $x_2 = y_2 \Leftrightarrow x = y$.

(iii) For all $x, y \in R^2$
$d(x, y) = |x_1 - y_1| + |x_2 - y_2|$
$= |y_1 - x_1| + |y_2 - x_2| = d(y, x)$.

(iv) Let $\mathbf{x} = (x_1, x_2)$, $\mathbf{y} = (y_1, y_2)$, $\mathbf{z} = (z_1, z_2)$ be any three points of $\mathbf{R}^2$.

Consider $d(x, y) = |x_1 - y_1| + |x_2 - y_2|$

$$= |(x_1 - z_1) + (z_1 - y_1)| + |(x_2 - z_2) + (z_2 - y_2)|$$
$$\leq (|x_1 - z_1| + |z_1 - y_1|) + (|x_2 - z_2| + |y_2 - z_2|)$$
$$= (|x_1 - z_1| + |x_2 - z_2|) + (|z_1 - y_1| + |z_2 - y_2|)$$
$$= d(x, z) + d(z, y)$$
$$\Rightarrow \qquad d(x, y) \leq d(x, z) + d(z, y).$$

Hence, d is a metric on $\mathbf{R}^2$.

☞ **REMARK**

⟹ The metric defined above is called the postman metric because it measures the distances the way a postman would cover when he will be distributing mail.

3.2.3 SUBSPACE

Let (X, d) be a metric space and Y be a proper subset of X.

Let d_1 be the restriction of d on Y, *i.e.*

$$d_1(x, y) = d(x, y) \ \forall x, y \in X \times Y.$$

Then (Y, d_1) is called subspace of (X, d).

3.2.4 PSEUDO-METRIC

A mapping $d : X \times X \to \mathbf{R}$ is called a pseudo-metric or semi-metric for X if and only if

(i) $d(x, y) \geq 0; \ \forall x, y \in X$

(ii) $d(x, x) = 0; \ \forall x \in X$

(iii) $d(x, y) = d(y, x); \ \forall x, y \in X$

(iv) $d(x, y) \leq d(x, z) + d(z, y); \ \forall x, y, z \in X.$

☞ **REMARKS**

⟹ For a pseudo metric $x = y$ implies $d(x, y) = 0$ but converse is not true. Therefore, every metric is a pseudo metric but a pseudo metric is not necessarily a metric.

⟹ The pseudo metric is said to be finite if
$$d(x, y) < \infty; \ \forall x, y \in X.$$

⟹ The pseudo metric d differ from metric in the sense that :
(a) $d(x, y)$ may be equal to zero even if $x \neq y$, *i.e.*, distance between a pair of distinct points may be zero.
(b) $d(x, y) = \infty$, for some $x, y \in X$, *i.e.*, ∞ is defined as a measure of a distance between a pair of points.

✎ ILLUSTRATIONS

• Consider a set X of real valued functions defined over the closed interval $[-1, 1]$. Let $f(x)$ and $g(x)$ be two arbitrary real-valued functions defined over $[-1, 1]$. Let us define

$$d(f, g) = \int_{-1}^{1} \{[f(x) - g(x)]^2 dx\}^{1/2}$$

It is easy to verify that d is a pseudo metric on X.   (MEERUT 1996)

• Let (X, d) be a pseudo metric and let '~' be a relation in X defined by setting

$x \sim y$ iff $d(x, y) = 0$.

The relation $\sim$ is an equivalence relation in X. In fact

(i) Since $d(x, x) = 0, \ \forall x \in X$

$$\Rightarrow x \sim x, \forall x \in X.$$

Therefore, the relation '$\sim$' is reflexive.

(ii) $x \sim y \Leftrightarrow d(x, y) = 0 \Leftrightarrow d(y, x) = 0 \Leftrightarrow y \sim x$

$\Rightarrow$ '$\sim$' is symmetric.

(iii) Since $x \sim y$ and $y \sim z \Rightarrow d(x, y) = 0$ and $d(y, z) = 0$

$$\Rightarrow d(x, y) + d(y, z) = 0$$

$$\Rightarrow d(x, z) = 0$$

$$\Rightarrow x \sim z$$

$\Rightarrow$ '$\sim$' is transitive.

3.2.5 NORM

The size of an element x is a real number denoted by $||x||$ and is called norm. Norm is a generalization of the real valued functions, which satisfying the following conditions

(i) $||x|| \geq 0$

(ii) $||x|| = 0 \Leftrightarrow x = 0$

(iii) $||cx|| = |c| \, ||x||, c \in R$

(iv) $||x + y|| \leq ||x|| + ||y||.$

☞ **REMARK**

⟾ The metric defined with the help of norm as follows :

$$d(x, y) = ||x - y||$$

This metric is known as metric induced by norm.

For example:

Consider the set of all bounded real valued functions defined on $[0, 1]$.

Define norm of a function as follows :

$$||f|| = \sup \{|f(x)| : x \in [0, 1]\}.$$

Here, it can be easily verified that

$$d(f, g) = ||f - g|| = \sup |f(x) - g(x)|$$

is a metric on $[0, 1]$. We denote this space by $\mathbf{C}[0, 1]$.

3.3 DISTANCE BETWEEN TWO SETS AND DIAMETER OF A SET

Definition 1. *Let (X, d) be a metric space and let A be a non-empty subset of X. Then diameter of the set A, denoted by $\delta(A)$, is defined by $\delta(A) = \sup\{d(x, y) : x, y \in A\}$.*

Definition 2. *The distance between a point $x \in X$ and a set A is denoted by $d(x, A)$ and is defined by $d(x, A) = \inf \{d(x, y) : y \in A\}$.*

Definition 3. *The distance between two non-empty subsets A and B of a metric space X is denoted and defined by $d(A, B) = \inf \{d(x, y) : x \in A, y \in B\}$.*

☞ **REMARKS**

⟾ The diameter of a set is always non-negative.

⟾ The diameter of the empty set is $-\infty$.

⟾ A set A is said to be bounded if $\delta(A) < \infty$.

⟾ If A is a closed sphere of radius r, then $d(A) = 2r$.

⟾ The diameter of a finite set is finite and diameter for an infinite set is infinite.

⟾ The diameter of the set A is the supremum of the set of all distances between points of A.

⟾ For the empty set ϕ, the distance of ϕ from point x is denoted by $d(x, \phi) = \infty$.

⟾ $d(x, A) = 0$ if $x \in A$.

⟾ The distance of set A and empty set is ∞, *i.e.,* $d(A, \phi) = \infty$.

⟾ $d(A, B) = 0$ iff $A \cap B \neq \phi$.

THEOREM 1. *Let (X, d) be a metric space and let x, y, z be any three points of X, then*
$$d(x, y) \geq |d(x, z) - d(z, y)|.$$
(MEERUT 2000,03)

PROOF. By definition of metric space, we have
$$d(x, z) \leq d(x, y) + d(y, z)$$
$$= d(x, y) + d(z, y) \qquad \text{(By symmetric property)}$$
$$\Rightarrow \quad d(x, z) - d(z, y) \leq d(x, y) \qquad \qquad \dots(1)$$
By M (iv), we have
$$d(z, y) \leq d(z, x) + d(x, y)$$
$$= d(x, z) + d(x, y) \qquad (\because d(x, z) = d(z, x))$$
$$\Rightarrow \quad d(z, y) - d(x, z) \leq d(x, y) \qquad \qquad \dots(2)$$
From (1) and (2), we conclude that
$$d(x, y) \geq |d(x, z) - d(z, y)|.$$

☞ **REMARK**

➡ The above inequality states that the difference of the lengths of any two sides of a triangle is less than or equal to the third side. Also, the sign of equality occurs, when three points lie on the straight line.

THEOREM 2. *Let (X, d) be a metric space, then*
$$|d(x, y) - d(x', y')| \leq d(x, x') + d(y, y'), \quad \forall x, x', y, y' \in X \qquad \text{(MEERUT 1994, 2016)}$$

PROOF. By M (iv), we have
$$d(x, y) \leq d(x, x') + d(x', y)$$
$$\leq d(x, x') + d(x', y') + d(y', y) \qquad \text{(By } M \text{ (iv))}$$
$$= d(x, x') + d(x', y') + d(y, y') \qquad [\because d(y, y') = d(y', y)]$$
$$\Rightarrow \quad d(x, y) - d(x', y') \leq d(x, x') + d(y, y'). \qquad \dots(1)$$
Also, we have
$$d(x', y') \leq d(x', x) + d(x, y')$$
$$\leq d(x', x) + d(x, y) + d(y, y') \qquad [\because d(x, y') \leq d(x, y) + d(y, y')]$$
$$= d(x, x') + d(x, y) + d(y, y') \qquad [\because d(x, x') = d(x', x)]$$
$$\Rightarrow \quad d(x', y') - d(x, y) \leq d(x, x') + d(y, y'). \qquad \dots(2)$$
From (1) and (2), we conclude that
$$|d(x, y) - d(x', y')| \leq d(x, x') + d(y, y').$$

THEOREM 3. *Let $X \neq \phi$. Then a mapping $d : X \times X \to R_*$ is a metric if and only if the following conditions hold :*

(i) $d(x, y) = 0 \iff x = y; \quad \forall x, y \in X$

(ii) $d(x, y) \leq d(x, z) + d(y, z); \quad \forall x, y, z \in X$.

(MEERUT 1992, 2001,02; AVADH-2001,05)

PROOF. Let us first suppose d is a metric on R.
Then by M (iv), we have
$$d(x, y) \leq d(x, z) + d(z, y) \qquad \dots(1)$$
Also, $\qquad d(z, y) = d(y, z) \qquad \dots(2)$
From (1) and (2), we conclude that
$$d(x, y) \leq d(x, z) + d(y, z); \quad \forall x, y, z \in X.$$
Now, suppose condition (i) and (ii) holds : To show d is a metric.

Let x, y be any two elements of X, then by (ii), we have

$$d(x, x) \leq d(x, y) + d(x, y) \quad \text{[Replacing } x, x, y \text{ for } x, y, z \text{ respectively]}$$

i.e., $\qquad 2d(x, y) \geq d(x, x) \qquad \qquad \qquad \qquad \qquad \qquad \qquad$ ...(3)

But from (i) $d(x, x) = 0$, therefore from (3), we have

$$d(x, y) \geq 0 \quad \forall \ x, y \in X \implies M \text{ (i) is satisfied.}$$

Now, apply condition (ii) for the points x, y, x, we have

$$d(x, y) \leq d(x, x) + d(y, x) = 0 + d(y, x) \qquad \qquad (\because d(x, y) = 0)$$

$\implies \qquad d(x, y) \leq d(y, x). \qquad \qquad \qquad \qquad \qquad \qquad \qquad$ ...(4)

Again applying condition (ii) for the points y, x, y, we get

$$d(y, x) \leq d(y, y) + d(x, y) = 0 + d(x, y) \qquad \qquad (\because d(y, y) = 0)$$

$\implies \qquad d(y, x) \leq d(x, y). \qquad \qquad \qquad \qquad \qquad \qquad \qquad$ ...(5)

From (4) and (5), we conclude that

$$d(x, y) = d(y, x) \implies M \text{ (iii) is satisfied.}$$

Finally, for any x, y, z, we have

$$d(x, y) \leq d(x, z) + d(y, z) = d(x, z) + d(z, y)$$

$\implies \qquad d(x, y) \leq d(x, z) + d(z, y)$

$\implies \qquad M$ (iv) is satisfied.

Hence, d is a metric on X.

THEOREM 4. **Let (X_1, d_1) and (X_2, d_2) be any two metric spaces $\ d$ is defined as**

$$d[(x_1, x_2), (y_1, y_2)] = \sqrt{d_1^2(x_1, y_1) + d_2^2(x_2, y_2)}$$

where $x_1, y_1 \in X_1$, $x_2, y_2 \in X_2$. Show that d is a metric for $X_1 \times X_2$.

(MEERUT 1996, 2018)

PROOF. Since (X_1, d_1) and (X_2, d_2) are two metric spaces. Therefore, by definitions, we have

(i) $d_1(x_1, y_1) \geq 0$, $d_2(x_2, y_2) \geq 0$

(ii) $d_1(x_1, y_1) = 0 \Leftrightarrow x_1 = y_1$ and $d_2(x_2, y_2) = 0 \Leftrightarrow x_2 = y_2$.

(iii) $d_1(x_1, y_1) = d_1(y_1, x_1)$ and $d_2(x_2, y_2) = d_2(y_2, x_2)$.

(iv) $d_1(x_1, y_1) \leq d_1(x_1, z_1) + d_1(z_1, y_1)$

$\qquad d_2(x_2, y_2) \leq d_2(x_2, z_2) + d_2(z_2, y_2)$.

$\qquad \forall \ x_1, y_1, z_1 \in X_1, \ x_2, y_2, z_2 \in X_2$

To show $d[(x_1, x_2), (y_1, y_2)] = \sqrt{d_1^2(x_1, y_1) + d_2^2(x_2, y_2)}$ is a metric.

(a) Since $\quad d_1(x_1, y_1) \geq 0 \quad$ and $\quad d_2(x_2, y_2) \geq 0$

Therefore $d_1^2(x_1, y_1) \geq 0$ and $d_2^2(x_2, y_2) \geq 0$

$\implies \quad d_1^2(x_1, y_1) + d_2^2(x_2, y_2) \geq 0$

$\implies \quad \sqrt{d_1^2(x_1, y_1) + d_2^2(x_2, y_2)} \geq 0$

$\implies \quad d[(x_1, x_2), (y_1, y_2)] \geq 0$.

(b) Here, we have

$$d_1(x_1, y_1) = 0 \Leftrightarrow x_1 = y_1$$

and $d_2(x_2, y_2) = 0 \Leftrightarrow x_2 = y_2$

Now, $\quad d[(x_1, x_2), (y_1, y_2)] = 0$

$$\Leftrightarrow \sqrt{[d_1^2(x_1, y_1) + d_2^2(x_2, y_2)]} = 0$$

$$\Leftrightarrow d_1^2(x_1, y_1) = 0, \ d_2^2(x_2, y_2) = 0$$

$\Leftrightarrow d_1(x_1, y_1) = 0 \Leftrightarrow x_1 = y_1$ and $d_2(x_2, y_2) = 0 \Leftrightarrow x_2 = y_2$

$\Rightarrow (x_1, x_2) = (y_1, y_2)$.

(c) Here, we have

$$d_1(x_1, y_1) = d_1(y_1, x_1)$$

and $d_2(x_2, y_2) = d_2(y_2, x_2)$.

$$\Rightarrow d_1^2(x_1, y_1) = d_1^2(y_1, x_1)$$

and $d_2^2(x_2, y_2) = d_2^2(y_2, x_2)$

$$\Rightarrow d_1^2(x_1, y_1) + d_2^2(x_2, y_2) = d_1^2(y_1, x_1) + d_2^2(y_2, x_2)$$

$$\Rightarrow \sqrt{d_1^2(x_1, y_1) + d_2^2(x_2, y_2)} = \sqrt{d_1^2(y_1, x_1) + d_2^2(y_2, x_2)}$$

$$\Rightarrow d[(x_1, x_2), (y_1, y_2)] = d[(y_1, y_2), (x_1, x_2)]$$

(d) Let $(x_1, x_2), (y_1, y_2), (z_1, z_2) \in X_1 \times X_2$.

Then, we have

$$d(x, y) = \sqrt{|d_1^2(x_1, y_1) + d_2^2(x_2, y_2)|}$$

$$\leq \sqrt{[d_1(x_1, z_1) + d_1(z_1, y_1)]^2 + [d_2(x_2, z_2) + d_2(z_2, y_2)]^2}$$

(By triangle inequality)

$$\leq [d_1^2(x_1, z_1) + d_2^2(x_2, z_2)]^{1/2} + [d_1^2(z_1, y_1) + d_2^2(z_2, y_2)]^{1/2}$$

$$\leq d(x, z) + d(z, y).$$

Hence, from (a), (b), (c) and (d), we conclude that

$$d[(x_1, x_2), (y_1, y_2)] = \sqrt{[d_1^2(x_1, y_1) + d_2^2(x_2, y_2)]} \text{ is a metric on } X_1 \times X_2.$$

THEOREM 5. ***Let d be a metric on X then $\sqrt{d}$ is a metric on X.***

PROOF. For $x, y \in X$, clearly, we have

$$\sqrt{d(x, y)} \geq 0 \qquad\qquad (\because d(x, y) \geq 0 \ \forall \ x, y \in X)$$

and $\quad \sqrt{d(x, y)} = 0 \ \Leftrightarrow \ x = y \qquad\qquad (\because d(x, y) = 0 \Leftrightarrow x = y)$

Also, $\quad \sqrt{d(x, y)} = \sqrt{d(y, x)} \qquad\qquad (\because d(x, y) = d(y, x))$

Finally, since

$$d(x, y) \leq d(x, z) + d(z, y) \quad \forall x, y, z \in X$$

Therefore,

$$\sqrt{d(x, y)} \leq \sqrt{d(x, z) + d(z, y)}$$

$$\leq \sqrt{d(x, z)} + \sqrt{d(y, z)} \qquad\qquad (\because \sqrt{a+b} \leq \sqrt{a} + \sqrt{b})$$

Hence, $\sqrt{d}$ is a metric on X.

☞ **REMARK**

⟹ In a similar manner we may prove the following result

"If d is a metric on X then d^2 is a metric on X."

3.4 SOME INEQUALITIES

3.4.1 THE TRIANGLE INEQUALITY : $|z + w| \leq |z| + |w|, (z, w \in C)$ (KANPUR 2008; LUCKNOW 2007)

PROOF. Consider

$$|z + w|^2 = (z + w)(\bar{z} + \bar{w}) = z\bar{z} + z\bar{w} + \bar{z}w + w\bar{w}$$

$$= |z|^2 + 2\operatorname{Re}(z\bar{w}) + |w|^2 \qquad\qquad [\because z\bar{w} \text{ and } \bar{z}w \text{ are conjugates}]$$

$$\leq |z|^2 + 2|z\bar{w}| + |w|^2$$

$$= |z|^2 + 2|z||w| + |w|^2 \qquad\qquad [\because |\overline{w}| = |w|, |z\overline{w}| = |z| \cdot |\overline{w}|]$$

$$= (|z| + |w|)^2$$

Hence, $|z + w| \leq |z| + |w|$

☞ REMARK

⟹ The generalization of this inequality gives

$$|z_1 + z_2 + \dots + z_n| \leq |z_1| + |z_2| + \dots + |z_n|; \quad (z_i \in \mathbf{C}, \; i = 1, 2, \dots, n)$$

3.4.2 $\dfrac{|z + w|}{1 + |z + w|} \leq \dfrac{|z|}{1 + |z|} + \dfrac{|w|}{1 + |w|} (z, w \in C)$ (HIMACHAL 2007, 10)

PROOF.   Let $f(t) = t(1 + t)^{-1}$, for $t > -1$. Then $f'(t) = (1 + t)^{-2}$.

Obviously $f(t)$ is increasing function. Hence

$$|z + w| \leq |z| + |w|$$

$$\Rightarrow \qquad f(|z + w|) \leq f(|z| + |w|)$$

$$\frac{|z + w|}{1 + |z + w|} \leq \frac{|z| + |w|}{1 + |z| + |w|}$$

$$= \frac{|z|}{1 + |z| + |w|} + \frac{|w|}{1 + |z| + |w|}$$

$$\leq \frac{|z|}{1 + |z|} + \frac{|w|}{1 + |w|}.$$

3.4.3   IF $p > 1, \dfrac{1}{p} + \dfrac{1}{q} = 1, a \geq 0, b \geq 0,$. THEN, $ab \leq \dfrac{a^p}{p} + \dfrac{b^q}{q}$, WITH EQUALITY IF AND IF ONLY $a^p = b^q$

(ROHTAK 2004,06; ROHILKHAND 2007)

PROOF.   Let $f(t) = 1 - \lambda + \lambda t - t^{\lambda}$, where $\lambda = 1/p$ and $t \geq 0$. Then using principle of maxima and minima, it is easy to verify that $f'(t) < 0$ for $0 < t < 1$ and $f'(t) > 0$ for $t > 1$. Hence $f(t) \geq f(1) = 0$ with equality if and only if $t = 1$.

Thus we have $1 - \lambda + \lambda t - t^{\lambda} \geq 0$, that is $t^{\lambda} \leq (1 - \lambda) + \lambda t$.   ...(1)

If $b = 0$, then $ab = 0 \leq a^p / p$. If $b > 0$, we put $t = a^p \, b^{-q}$ in (1) and we get

$$(a^p \, b^{-q})^{1/p} \leq (1 - 1/p) + (1/p) \, a^p \, b^{-q} \qquad\qquad [\because \lambda = 1/p]$$

Using the relation $\dfrac{1}{p} + \dfrac{1}{q} = 1$ and simplifying, we get

$$ab \leq \frac{a^p}{p} + \frac{b^q}{q} \qquad\qquad\qquad ...(2)$$

3.4.4   HOLDER'S INEQUALITY

If $a_i, b_i \; (i = 1, 2, \dots, n)$ are negative real numbers, then

$$\sum_{i=1}^{n} a_i \, b_i \leq \left(\sum_{i=1}^{n} a_i^p \right)^{1/p} \left(\sum_{i=1}^{n} b_i^q \right)^{1/q} \qquad\qquad ...(3)$$

where $p > 1$ and $\dfrac{1}{p} + \dfrac{1}{q} = 1$

(MEERUT 2001; KURUKSHETRA 2006; PATNA 2004,05, 07,2010; ALLAHABAD 2007; BHU-2009)

PROOF.   Let $A = \left(\Sigma \, a_i^p \right)^{1/p}$, $B = \left(\Sigma b_i^q \right)^{1/q}$.

If $AB = 0$, then either $A = 0$ or $B = 0$ and so in this case, we see that both sides of (3) are equal to zero. If $AB > 0$, then putting $a = \dfrac{a_i}{A}$, $b = \dfrac{b_i}{B}$ in (2), we get

$$\frac{a_i}{A} \cdot \frac{b_i}{B} \le \frac{1}{p} \left(\frac{a_i}{A} \right)^p + \frac{1}{q} \left(\frac{b_i}{B} \right)^q$$

or $\Sigma a_i b_i \le AB \left[\dfrac{1}{pA^p} \Sigma a_i^p + \dfrac{1}{qB^q} \Sigma b_i^q \right]$

$$= AB \left[\frac{1}{pA^p} . A^p + \frac{1}{qB^q} B^q \right] = AB \left[\frac{1}{p} + \frac{1}{q} \right] = AB \qquad \left[\because \frac{1}{p} + \frac{1}{q} = 1 \right]$$

Thus, $\displaystyle\sum_{i=1}^{n} a_i b_i \le \left(\sum_{i=1}^{n} a_i^p \right)^{1/p} \left(\sum_{i=1}^{n} b_i^q \right)^{1/q}$ ...(4)

☛ **REMARK**

➠ If $p = 2$, then $q = 2$ and (4) becomes

$$\sum_{i=1}^{n} a_i b_i \le \left(\sum_{i=1}^{n} a_i^2 \right)^{1/2} \left(\sum_{i=1}^{n} b_i^2 \right)^{1/2} \qquad \text{...(5)}$$

If f, g are non-negative real valued integrable functions defined on $[a, b]$, then

$$\int_a^b fg dx < \left\{ \int_a^b (f(x))^p \, dx \right\}^{1/p} \left\{ \int_a^b (g(x))^q \, dx \right\}^{1/q} \qquad \text{...(6)}$$

It is called Holder's inequality for integrals.

3.4.5 CAUCHY-SCHWARZ INEQUALITY   (KANPUR 2001; OSMANIA 2004,06; RAJASTHAN 2005,09)

Let $z = (z_1, z_2, ..., z_n)$ and $w = (w_1, w_2, ..., w_n)$ be two n-tuples of real or complex

numbers. Then $\displaystyle\sum_{i=1}^{n} |z_i w_i| \le \left(\sum_{i=1}^{n} |z_i|^2 \right)^{1/2} \left(\sum_{i=1}^{n} |w_i|^2 \right)^{1/2}$ ...(7)

or in terms of norm,

$$\sum_{i=1}^{n} |z_i w_i| \le ||z|| \, ||w|| \qquad \text{...(8)}$$

PROOF.   If $z = 0$ or $w = 0$, the inequality reduces to equality. We therefore assume that $z \ne 0$ and $w \ne 0$. We know that the geometric mean of two non-negative quantities a, b is less than or equal to their arithmetic mean, that is

$$a^{1/2} b^{1/2} \le \frac{a+b}{2}$$

Setting $a = (|z_i| / ||z||)^2$ and $b = (|w_i| / ||w||)^2$

$$\frac{|z_i| |w_i|}{||z|| \, ||w||} \le \frac{|z_i|^2 / ||z||^2 + |w_i|^2 / ||w||^2}{2}$$

Summing these inequalities as i varies from 1 to n , we obtain

$$\left(\sum_{i=1}^{n} |z_i w_i| \right) / ||z|| \, ||w|| \le \frac{1+1}{2} = 1 \qquad \left[\because \sum_{i=1}^{n} |z_i|^2 = ||z||^2 \text{ and } \sum_{i=1}^{n} |w_i|^2 = ||w||^2 \right]$$

Thus, $\displaystyle\sum_{i=1}^{n} |z_i w_i| \le ||z|| \, ||w||$.

3.4.6 MINKOWSKI'S INEQUALITY

If $p \geq 1$ and a_i, b_i $(i = 1, 2, ...)$ are non-negative real numbers, then

$$\left(\sum_{i=1}^{n} (a_i + b_i)^p \right)^{1/p} \leq \left(\sum_{i=1}^{n} a_i^p \right)^{1/p} + \left(\sum_{i=1}^{n} b_i^p \right)^{1/p}$$

(KANPUR 2002,05; LUCKNOW 2007; ALLAHABAD 2008; PATNA 2004,06,09)

PROOF. If $p = 1$, the case is trivial. Now, suppose $p > 1$. Let $\dfrac{1}{q} = 1 - \dfrac{1}{p}$ so that $q > 1$.

Then, $\sum_{i=1}^{n} (a_i + b_i)^p = \sum_{i=1}^{n} a_i (a_i + b_i)^{p-1} + \sum_{i=1}^{n} b_i (a_i + b_i)^{p-1}$.

$$\leq \left(\sum_{i=1}^{n} a_i^p \right)^{1/p} \left(\sum_{i=1}^{n} (a_i + b_i)^{q(p-1)} \right)^{1/q} + \left(\sum_{i=1}^{n} b_i^p \right)^{1/p} \left(\sum_{i=1}^{n} (a_i + b_i)^{q(p-1)} \right)^{1/q}$$

[Using Holder's inequality]

$$= \left(\sum_{i=1}^{n} a_i^p \right)^{1/p} \left(\sum_{i=1}^{n} (a_i + b_i)^p \right)^{1/q} + \left(\sum_{i=1}^{n} b_i^p \right)^{1/p} \left(\sum_{i=1}^{n} (a_i + b_i)^p \right)^{1/q}$$

$[\because q(p-1) = p]$

or $\left(\sum_{i=1}^{n} (a_i + b_i)^p \right)^{1-1/q} \leq \left(\sum_{i=1}^{n} d_i^p \right)^{1/p} + \left(\sum_{i=1}^{n} b_i^p \right)^{1/p}$

or $\left(\sum_{i=1}^{n} (a_i + b_i)^p \right)^{1/p} \leq \left(\sum_{i=1}^{n} a_i^p \right)^{1/p} + \left(\sum_{i=1}^{n} b_i^p \right)^{1/p}$ ...(9)

If $a_i = \lambda b_i$, the inequality (9) reduces to equality provided λ is independent of i.
Putting $p = 2$ in (9), we get

$$\left(\sum_{i=1}^{n} (a_i + b_i)^2 \right)^{1/2} \leq \left(\sum_{i=1}^{n} a_i^2 \right)^{1/2} + \left(\sum_{i=1}^{n} b_i^2 \right)^{1/2}$$

☞ **REMARK**

⇒ If f, g are non-negative real valued functions defined on $[a, b]$, then

$$\left\{ \int_a^b (f+g)^p \, dx \right\}^{1/p} \leq \left\{ \int_a^b (f(x))^p \, dx \right\}^{1/p} + \left\{ \int_a^b (g(x))^p \, dx \right\}^{1/p}, \ (p \geq 1) \qquad ...(10)$$

which is called Minkowski's inequality for integrals.

3.4.7 MINKOWSKI'S INEQUALITY IN TERMS OF NORM

Let $z = (z_1, z_2, ..., z_n)$ and $w = (w_1, w_2, ..., w_n)$ be two n-tuples of real or complex numbers. Then

$$\left(\sum_{i=1}^{n} |z_i + w_i|^2 \right)^{1/2} \leq \left(\sum_{i=1}^{n} |z_i|^2 \right)^{1/2} + \left(\sum_{i=1}^{n} |w_i|^2 \right)^{1/2} \qquad ...(11)$$

or in terms of norm,

$$||z + w|| \leq ||z|| + ||w|| \qquad ...(12)$$

PROOF. We have

$$||z + w||^2 = \sum_{i=1}^{n} |z_i + w_i|^2 = \sum_{i=1}^{n} |z_i + w_i| \, |z_i + w_i|$$

$$\leq \sum_{i=1}^{n} |z_i + w_i| (|z_i| + |w_i|)$$

$$= \sum_{i=1}^{n} |z_i + w_i| \, |z_i| + \sum_{i=1}^{n} |z_i + w_i| \, |w_i|$$

$$\leq ||z+w|| \, ||z|| + ||z+w|| \, ||w|| \qquad\qquad\text{by (8)}$$

Thus, $||z+w||^2 \leq ||z+w||(||z||+||w||)$ ...(13)

If $||z+w||=0$, the inequality (12) holds trivially; otherwise it follows from (13) on dividing it by $||z+w||=0$ that

$$||z+w|| \leq ||z|| + ||w||$$

☞ **REMARK**

⟹ If a_i, b_i $(i=1, 2, ..., n)$ are non-negative real number and $0 < p \leq 1$, then

$$\sum_{i=1}^{n}(a_i+b_i)^p \leq \sum_{i=1}^{n}a_i^p + \sum_{i=1}^{n}b_i^p \qquad\qquad\text{...(14)}$$

3.4.8 **MINKOWSKI'S INEQUALITY FOR COMPLEX NUMBERS**

If z_i, w_i $(i = 1, 2, ..., n)$ are complex numbers, then

$$\left(\sum_{i=1}^{n}|z_i + w_i|^p\right)^{1/p} \leq \left(\sum_{i=1}^{n}|z_i|^p\right)^{1/p} + \left(\sum_{i=1}^{n}|w_i|^p\right)^{1/p}, \; (p \geq 1)$$

and $\displaystyle\sum_{i=1}^{n}|z_i + w_i|^p \leq \sum_{i=1}^{n}|z_i|^p + \sum_{i=1}^{n}|w_i|^p, \; (0 < p \leq 1)$

PROOF.   We have

$$\left(\sum_{i=1}^{n}|z_i + w_i|^p\right)^{1/p} \leq \left(\sum_{i=1}^{n}|z_i|^p + |w_i|^p\right)^{1/p}$$

$$\leq \left(\sum_{i=1}^{n}|z_i|^p\right)^{1/p} + \left(\sum_{i=1}^{n}|w_i|^p\right)^{1/p}$$

and $\displaystyle\sum_{i=1}^{n}|z_i + w_i|^p \leq \sum_{i=1}^{n}(|z_i|+|w_i|)^p \leq \sum_{i=1}^{n}|z_i|^p + \sum_{i=1}^{n}|w_i|^p$

SOLVED EXAMPLES

EXAMPLE I.   **Show   that   the   mapping $d : R^2 \times R^2 \to R$ defined   by**
$d(x, y) = \max(|x_1 - y_1|, |x_2 - y_2|)$, where $x = (x_1, x_2)$, $y = (y_1, y_2) \in R$
is metric on R^2. (ROHILKHAND 1994; GARHWAL 1998)

SOLUTION.   (i) Since $|x_1 - y_1| \geq 0$ and $|x_2 - y_2| \geq 0$

which implies $\max\{(x_1 - y_1), (x_2 - y_2)\} \geq 0 \; \Rightarrow \; d(x, y) \geq 0$

(ii) Let $d(x, y) = 0 \Rightarrow \max(|x_1 - y_1|, |x_2 - y_2|) = 0$

Maximum of two positive numbers will be zero only when both the numbers are zero

$\Rightarrow \quad |x_1 - y_1| = 0$ and $|x_2 - y_2| = 0 \Rightarrow x_1 = y_1$ and $x_2 = y_2$

$\Rightarrow \quad (x_1, x_2) = (y_1, y_2) \Rightarrow x = y$

Conversely, $x = y \Rightarrow x_1 = y_1$

and $\qquad x_2 = y_2 \Rightarrow |x_1 - y_1| = 0 = |x_2 - y_2|$

$\Rightarrow \qquad \max(|x_1 - y_1|, |x_2 - y_2|) = 0 \Rightarrow d(x, y) = 0$

i.e., $\quad d(x, y) = 0$ iff $x = y$.

(iii) We know that

$$|x_1 - y_1| = |y_1 - x_1| \text{ and } |x_2 - y_2| = |y_2 - x_2|$$

$$\Rightarrow \quad \max(|x_1 - y_1|, |x_2 - y_2|) = \max(|y_1 - x_1|, |y_2 - x_2|)$$

$$\Rightarrow \quad d(x, y) = d(y, x).$$

(iv) Consider

$$d(x, y) = \max\{|x_1 - y_1|, |x_2 - y_2|\}$$

$$= \max\{|x_1 - z_1 + z_1 - y_1|, |x_2 - z_2 + z_2 - y_2|\}$$

$$\leq \max\{|x_1 - z_1| + |z_1 - y_1|, |x_2 - z_2| + |z_2 - y_2|\}$$

$$\leq \max\{|x_1 - z_1|, |x_2 - z_2|\} + \max\{|z_1 - y_1|, |z_2 - y_2|\}.$$

$$\Rightarrow \quad d(x, y) \leq d(x, z) + d(z, y).$$

From (i), (ii), (iii) and (iv), we conclude that d is a metric on R^2.

EXAMPLE 2. *Let X be the set of real valued bounded continuous function defined on the closed interval $[0, 1]$. We define the norm of a function $f \in X$ by*
$$||f|| = \int_0^1 |f(x)|\, dx.$$

Define a mapping $d : X \times X \to R$ by

$$d(f, g) = ||f - g|| = \int_0^1 |f(x) - g(x)|\, dx \; ; \quad \forall f, g \in X.$$

Show that d is a metric on X.

(MEERUT 1994,95,96,2005; PURVANCHAL 2008; GARHWAL 2004,07,08)

SOLUTION. (i) Since, we have
$$|f(x) - g(x)| \geq 0.$$

Therefore, $\int_0^1 |f(x) - g(x)|\, dx \geq 0$

$$\Rightarrow \qquad\qquad ||f - g|| \geq 0$$

$$\Rightarrow \qquad\qquad d(f, g) \geq 0.$$

(ii) Since $|f(x) - g(x)| = 0$ iff $f(x) = g(x)$

Therefore, $\int_0^1 |f(x) - g(x)|\, dx = 0$ iff $f(x) = g(x)$

$$\Rightarrow \quad ||f - g|| = 0 \text{ iff } f(x) = g(x)$$

$$\Rightarrow \quad d(f, g) = 0 \text{ iff } f(x) = g(x).$$

(iii) Since $|f(x) - g(x)| = |g(x) - f(x)|.$

Therefore, $\int_0^1 |f(x) - g(x)|\, dx = \int_0^1 |g(x) - f(x)|\, dx$

$$\Rightarrow \quad ||f - g|| = ||g - f||$$

$$\Rightarrow \quad d(f, g) = d(g, f).$$

(iv) Let $f, g, h \in X$. Then, we have

$$||f - g|| = \int_0^1 |f(x) - g(x)|\, dx = \int_0^1 |f(x) - h(x) + h(x) - g(x)|\, dx$$

$$\leq \int_0^1 [|f(x) - h(x)| + |h(x) - g(x)|]\, dx$$

$$= \int_0^1 |f(x) - h(x)|\, dx + \int_0^1 |h(x) - g(x)|\, dx$$

$$= ||f - h|| + ||h - g||$$

$$\Rightarrow \quad d(f, g) \leq d(f, h) + d(h, g).$$

Hence, from (i), (ii), (iii) and (iv), we conclude that d is a metric on R.

EXAMPLE 3. *Let l_∞ denote the set of all bounded sequence. If $x = \langle x_n \rangle$ and $y = \langle y_n \rangle$ are any two points of l_∞ we define*
$$d(x, y) = \sup\{|x_n - y_n| : n \in N\}$$

Show that L_∞ is a metric space under d.

(MEERUT 1992, 93)

SOLUTION. (i) Since $|x_n - y_n| \geq 0$

Therefore, $\sup |x_n - y_n| \geq 0$

$\Rightarrow$ $d(x, y) \geq 0$.

(ii) Let $d(x, y) = 0$

$\Rightarrow$ $\sup \{ |x_n - y_n| : n \in \mathbf{N} \} = 0$

$\Leftrightarrow$ $|x_n - y_n| = 0 \ \forall n \in \mathbf{N}$

$\Leftrightarrow$ $x_n = y_n \ \forall n \in \mathbf{N}$

$\Leftrightarrow$ $<x_n> = <y_n> \ \Leftrightarrow \ x = y$

(iii) Since $|x_n - y_n| = |y_n - x_n|$

Therefore $\sup |x_n - y_n| = \sup |y_n - x_n|$

$\Rightarrow$ $d(x, y) = d(y, x)$.

(iv) Let $z = \langle z_n \rangle$ be an element of L_∞ . Then for any positive integer n, we have

$$|x_n - y_n| = |x_n - z_n + z_n - y_n|$$
$$\leq |x_n - z_n| + |z_n - y_n|$$
$$\leq \sup \{ |x_n - z_n| : n \in \mathbf{N} \} + \sup \{ |z_n - y_n| : n \in \mathbf{N} \}$$
$$= d(x, z) + d(z, y)$$

$\Rightarrow$ $d(x, y) \leq d(x, z) + d(z, y)$.

EXAMPLE 4. *Let (X, d) be a metric space and let M be a positive number, then there exists a metric d_1 on X such that the metric space (X, d_1) is bounded with $\delta(x) \leq M$.* (MEERUT 1996, 97, 2003, 11; AVADH 2001,03,08)

SOLUTION. Define d_1 by

$$d_1(x, y) = \frac{Md(x, y)}{1 + d(x, y)}, \text{ where } x, y \in X.$$

To show d_1 is a metric.

(i) Since $d(x, y) \geq 0$ and $M > 0$.

$\therefore$ $\dfrac{Md(x, y)}{1 + d(x, y)} \geq 0$

$\Rightarrow$ $d_1(x, y) \geq 0; \quad \forall x, y \in X$.

(ii) $d_1(x, y) = 0 \ \Leftrightarrow \ \dfrac{Md(x, y)}{1 + d(x, y)} = 0$

$\Leftrightarrow d(x, y) = 0 \ \Leftrightarrow \ x = y$.

(iii) $d_1(x, y) = \dfrac{Md(x, y)}{1 + d(x, y)} = \dfrac{Md(y, x)}{1 + d(y, x)} = d_1(y, x)$.

(iv) Let $x, y, z \in X$, therefore

$$d_1(x, y) = \frac{Md(x, y)}{1 + d(x, y)} = M - \frac{M}{1 + d(x, y)}$$

$$\leq M - \frac{M}{1 + d(x, z) + d(z, y)} = \frac{M[d(x, z) + d(z, y)]}{1 + d(x, z) + d(z, y)}$$

$$= \frac{Md(x, z)}{1 + d(x, z) + d(z, y)} + \frac{Md(z, y)}{1 + d(x, z) + d(z, y)}$$

$$\leq \frac{Md(x, z)}{1 + d(x, z)} + \frac{Md(z, y)}{1 + d(z, y)}$$

$$\leq d_1(x, z) + d_1(z, y)$$
$$\Rightarrow \qquad d_1(x, y) \leq d_1(x, z) + d_1(z, y)$$

From (i), (ii), (iii) and (iv), we conclude that

$$d_1(x, y) = \frac{Md(x, y)}{1 + d(x, y)}$$

is a metric on X.

Also since $d_1(x, y) = \dfrac{M}{1 + d(x, y)} \leq M$ for every points $x, y \in X$

Therefore, d_1 is a bounded metric for X with $\delta(x) \leq M$.

☛ **REMARKS**

⟹ A metric space (X, d) is said to be bounded if there exist a positive number M such that $d(x, y) \leq M$ for every pair of points x, y of X.

⟹ A metric space which is not bounded is said to be unbounded.

⟹ A metric space X is said to be bounded if its diameter is finite.

EXAMPLE 5. *Let (X, d) be a metric space and let $d_1(x, y) = \min \{1, d(x, y)\}$ Then show that d_1 is a metric on X.*

(MEERUT 1990,91,93,95, 2013; AGRA 2002; AVADH 2004,07; GARHWAL 2001,04,07,09)

SOLUTION. (i) Here, we have

$$d_1(x, y) = 1 \quad \text{or} \quad d_1(x, y) = d(x, y)$$

Clearly, $d(x, y) \geq 0$ [∵ d is a metric]

Therefore, in both the cases $d_1(x, y) \geq 0$.

(i) Since $d(x, y) \geq 0$ and $M > 0$.

$$\therefore \qquad \frac{Md(x, y)}{1 + d(x, y)} \geq 0$$

$$\Rightarrow \qquad d_1(x, y) \geq 0; \quad \forall \, x, y \in X.$$

(ii) If $d_1(x, y) = 0 \Rightarrow \min \{1, d(x, y)\} = 0$

$$\rightarrow \quad d(r, y) = 0 \Rightarrow x = y \qquad\qquad [∵ d \text{ is a metric}]$$

and if $x = y \Rightarrow d(x, y) = 0$

$$\Rightarrow \quad \min \{1, d(x, y)\} = \min \{1, 0\} = 0$$

$$\Rightarrow \qquad d_1(x, y) = 0$$

Hence, $d_1(x, y) = 0 \iff x = y$

(iii) Since either

$$d_1(x, y) = 1 \quad \text{or} \quad d_1(x, y) = d(x, y)$$

Therefore, if $d_1(x, y) = d(x, y)$, then $d(x, y) < 1$

Hence, $d(y, x) = d(x, y) < 1$.

But $d(y, x) < 1$, gives $d_1(y, x) = d(y, x) = d(x, y) = d_1(x, y)$.

If $d_1(x, y) = 1$, then $d(x, y) \geq 1$ and therefore $d(y, x) \geq 1$.

But $d(y, x) \geq 1$ gives $d_1(y, x) = 1$.

Hence, $d_1(x, y) = d_1(y, x)$.

Therefore, in each case $d_1(x, y) = d_1(y, x)$.

(iv) Here, we have to prove that
$$d_1(x, y) \leq d_1(x, z) + d_1(z, y).$$ ...(A)

If $d(x, y) \leq 1$, then if either $d_1(x, z) = 1$ or $d_1(z, y) = 1$.

$\Rightarrow$ Inequality (A) holds good.

If both $d_1(x, z) \neq 1$ and $d_1(z, y) \neq 1$, then $d_1(x, z) = d(x, z)$ and $d_1(z, y) = d(z, y)$

Then we have
$$d(x, y) \leq d(x, z) + d(z, y) = d_1(x, z) + d_1(z, y)$$ ...(B)

But $d_1(x, y) = \min\{1, d(x, y)\} \leq d(x, y)$. ...(C)

From (B) and (C), we have
$$d_1(x, y) \leq d_1(x, z) + d_1(z, y).$$

Hence, from (i), (ii), (iii) and (iv), we conclude that d_1 is a metric on X.

EXAMPLE 6. *Let d be a metric for non-empty set X. Show that d_1 defined by $d_1(x, y) = 2d(x, y)$ is also metric for X. Then show that d_1 is a metric on X.*

(GARHWAL 2009; INDORE 2001,05,07)

SOLUTION. (i) Since $d(x, y)$ is a metric on X, therefore $d(x, y) \geq 0$.

$\Rightarrow \qquad 2d(x, y) \geq 0 \Rightarrow d_1(x, y) = 0$.

(ii) Since $d(x, y)$ is a metric on X, therefore

$\qquad\qquad d(x, y) = 0 \qquad \Leftrightarrow \qquad x = y$

or $\qquad\qquad 2d(x, y) = 0 \qquad \Leftrightarrow \qquad x = y$

or $\qquad\qquad d_1(x, y) = 0 \qquad \Leftrightarrow \qquad x = y$

(iii) Since $d(x, y)$ is a metric on X, therefore

$\qquad\qquad d(x, y) = d(y, x)$

$\Rightarrow \qquad 2d(x, y) = 2d(y, x)$

$\Rightarrow \qquad d_1(x, y) = d_1(y, x)$.

(iv) Since $\qquad d(x, y) \leq d(x, z) + d(z, y)$

$\Rightarrow \qquad 2d(x, y) \leq 2d(x, z) + 2d(z, y)$

$\Rightarrow \qquad d_1(x, y) \leq d_1(x, z) + d_1(z, y)$

Hence, from (i), (ii), (iii) and (iv), we conclude that d_1 is a metric on X.

EXAMPLE 7. *Show that the function d defined by $d(p, q) = ||p - q||$ where p, q are vectors in a normed vector space V is metric on V.*

(GARHWAL 1990; ROHILKHAND 1993)

SOLUTION. Let $p, q, r \in V$ and x is any scalar.

By definition of normed vector space, we have

(a) $||p|| \geq 0$ $\qquad\qquad$ (b) $||p|| = 0 \Leftrightarrow p = 0$

(c) $||xp|| = |x| \, ||p||$ $\qquad$ (d) $||p + q|| \leq ||p|| + ||q||$

To show d is a metric on V.

(i) Since $\qquad d(p, q) = ||p - q||$

Then by (a) $||p - q|| \geq 0$

(ii) Using (b), $\qquad d(p, q) = 0 \qquad \Leftrightarrow \qquad ||p - q|| = 0$

$\qquad\qquad\qquad\qquad\qquad\qquad \Leftrightarrow \qquad p - q = 0 \Leftrightarrow p = q$

(iii) Using (c), we have

$\qquad\qquad d(p, q) = ||p - q||$

$\Rightarrow \qquad d(q, p) = ||q - p|| = ||-(p - q)|| = |-1| \, ||p - q|| = d(p, q)$.

(iv) Consider

$$||p-q|| = ||p-r+r-q|| \le ||p-r|| + ||r-q||$$
$$\Rightarrow \quad d(p,q) \le d(p,r) + d(r,q)$$

Hence, from (i), (ii), (iii) and (iv), we conclude that $d(p,q)$ is a metric on V.

EXAMPLE 8. *Let C be the set of all complex number then mapping $d : C \times C \to R$ is a metric on C if d is defined as*

$$d(z_1, z_2) = |z_1 - z_2|, \; \forall \, z_1, z_2 \in C$$

SOLUTION. Since d is defined as $d(z_1, z_2) = |z_1 - z_2|, \forall z_1, z_2 \in C$.

We have to show that (X, d) is metric space, *i.e.*, d is metric.

(i) Since $|z_1 - z_2| \ge 0, \; \forall z_1, z_2 \in C$

$\Rightarrow \quad d(z_1, z_2) \ge 0, \forall z_1, z_2 \in C$

(ii) Since $|z_1 - z_2| = 0 \Leftrightarrow z_1 - z_2 = 0 \Leftrightarrow z_1 = z_2$

$\Rightarrow \quad d(z_1, z_2) = 0$ iff $z_1 = z_2$

(iii) $|z_1 - z_2| = |z_2 - z_1|, \; \forall z_1, z_2 \in C$

$\Rightarrow \quad d(z_1, z_2) = d(z_2, z_1), \; \forall \, z_1, z_2 \in C$.

(iv) Let $z_1, z_2, z_3 \in C$

$|z_1 - z_2| = |(z_1 - z_3) + (z_3 - z_2)|$

$\le |z_1 - z_3| + |z_3 - z_2|, \; \forall z_1, z_2, z_3 \in C$

$\Rightarrow \quad d(z_1, z_2) \le d(z_1, z_3) + d(z_3, z_2), \; \forall z_1, z_2, z_3 \in C$.

Hence, by above properties, d is metric.

EXAMPLE 9. *Let R be the set of real number the mapping $d : R \times R \to R$ defined as $d(x,y) = |x^2 - y^2|, \; \forall x, y \in R$. Show that it is a pseudo-metric on R which is not a metric on R.* (MEERUT 1996)

SOLUTION. Here 'd' is defined as

$$d(x, y) = |x^2 - y^2|, \; \forall x, y \in R.$$

To show d is a metric.

(i) Since $|x^2 - y^2| \ge 0, \; \forall x, y \in R$

$\Rightarrow \quad d(x, y) \ge 0, \; \forall x, y \in R$

(ii) Since $d(x, x) = |x^2 - x^2| = 0, \; \forall x \in R$

But here $d(x, y) = |x^2 - y^2| = 0$

$\Rightarrow \quad x^2 - y^2 = 0 \Rightarrow x = \pm y$

$\Rightarrow \quad d(x, y) = 0$ if $x = y$ but not conversly.

(iii) Since $d(x, y) = |x^2 - y^2| = |y^2 - x^2|$

$\Rightarrow \quad d(x, y) = d(y, x), \; \forall x, y \in R$.

(iv) Let $x, y, z \in R$

$|x^2 - y^2| = |(x^2 - z^2) + (z^2 - y^2)|$

$\le |(x^2 - z^2)| + |z^2 - y^2|$

$\Rightarrow \quad d(x, y) \le d(x, z) + d(z, y), \; \forall x, y, z \in$

By above properties $d(x, y)$ is pseudo metric space.

EXAMPLE 10. *Let d_1, d_2 be two metric for a non-empty set X. Show that the mapping d defined by $d(x,y) = d_1(x, y) + d_2(x, y), \; \forall x, y \in X$ is also a metric for X.* (GARHWAL 1994)

SOLUTION. Let d_1, d_2 be metrices on a non-empty set X. Then they satisfying the following properties :

(i) $d_1(x, y) \geq 0$, $d_2(x, y) \geq 0$, $\forall x, y \in X$

(ii) $d_1(x, y) = 0$ iff $x = y$ and $d_2(x, y) = 0$ iff $x = y$

(iii) $d_1(x, y) = d_1(y, x)$ and $d_2(x, y) = d_2(y, x)$, $\forall x, y \in X$

(iv) $d_1(x, y) \leq d_1(x, z) + d_1(z, y)$, $\forall x, y \in X$

$d_2(x, y) \leq d_2(x, z) + d_2(z, y)$, $\forall x, y \in X$

Now, we have to show that

$$d(x, y) = d_1(x, y) + d_2(x, y), \ \forall x, y \in X$$

is metric on X.

(i) $\because$ $d_1(x, y) \geq 0$, $d_2(x, y) \geq 0$

$\Rightarrow$ $d_1(x, y) + d_2(x, y) \geq 0$, $\forall x, y \in X$

$\Rightarrow$ $d(x, y) \geq 0$, $\forall x, y \in X$.

(ii) Since $d_1(x, y) = 0$ iff $x = y$

$d_2(x, y) = 0$ iff $x = y$

$\Rightarrow$ $d_1(x, y) + d_2(x, y) = 0$ iff $x = y$

$\Rightarrow$ $d(x, y) = 0$ iff $x = y$

(iii) Since $d_1(x, y) = d_1(y, x)$ and $d_2(x, y) = d_2(y, x)$ $\forall x, y \in X$

$d(x, y) = d_1(x, y) + d_2(x, y)$

$\Rightarrow$ $d_1(y, x) + d_2(y, x) \Rightarrow d(y, x)$ $\forall x, y \in X$.

(iv) Let x, y, z be elements of X

$d_1(x, y) \leq d_1(x, z) + d_1(z, y)$

and $d_2(x, y) \leq d_2(x, z) + d_2(z, y)$

$\Rightarrow$ $d_1(x, y) + d_2(x, y) \leq d_1(x, z) + d_1(z, y) + d_2(x, z) + d_2(z, y)$

$\Rightarrow$ $d(x, y) \leq [d_1(x, z) + d_2(x, z)] + [d_1(z, y) + d_2(z, y)]$

$\Rightarrow$ $d(x, y) \leq d(x, z) + d(z, y)$ $\forall x, y, z \in X$

So, $d(x, y) \leq d(x, z) + d(z, y)$ $\forall x, y, z \in X$

So, $d(x, y)$ is a metric on X.

EXAMPLE II.   ***Let (X_1, d_1) and (X_2, d_2) be two metric spaces. For any pair of points $x = (x_1, x_2)$, $y = (y_1, y_2)$ in $X = X_1 \times X_2$ and d is defined as***

$$d(x, y) = d_1(x_1, y_1) + d_2(x_2, y_2)$$

Then prove that d is metric for $X = X_1 \times X_2$.   (MEERUT 1993)

SOLUTION. Since d_1 and d_2 are two metric spaces then it satisfying all properties of metric space. Proceeding same as above example, we have to show that d is metric for $X = X_1 \times X_2$ defined as $d(x, y) = d_1(x_1, y_1) + d_2(x_2, y_2)$.

(i) Since $d_1(x_1, y_1) \geq 0$, $\forall x_1, y_1 \in X_1$

$d_2(x_2, y_2) \geq 0$, $\forall x_2, y_2 \in X_2$

$\Rightarrow$ $d_1(x_1, y_1) + d_2(x_2, y_2) \geq 0$, $\forall x_1, y_1 \in X_1$, $x_2, y_2 \in X_2$

$\Rightarrow$ $d(x, y) \geq 0$, $\forall x, y \in X$, where $x = (x_1 x_2)$, $y = (y_1, y_2)$.

(ii) $d_1(x_1, y_1) = 0$ iff $x_1 = y_1$

$d_2(x_2, y_2) = 0$ iff $x_2 = y_2$

$\Rightarrow$ $d_1(x_1, y_1) + d_2(x_2, y_2) = 0$ iff $x_1 = y_1$ and $x_2 = y_2$

$\Rightarrow d(x, y) = 0$ iff $(x_1, x_2) = (y_1, y_2)$
$\Rightarrow d(x, y) = 0$ iff $x = y$.

(iii) Here, we have $d_1(x_1, y_1) = d_1(y_1, x_1), \ \forall x_1, y_1 \in X_1$

$$d_2(x_2, y_2) = d_2(y_2, x_2), \ \forall \ x_2, y_2 \in X_2$$

$$d(x, y) = d_1(x_1, y_1) + d_2(x_2, y_2)$$

$$= d_1(y_1, x_1) + d_2(y_2, x_2) = d(y, x), \ \forall \ x, y \in X$$

(iv) Let $\quad x = (x_1, x_2), \quad y = (y_1, y_2), \quad z = (z_1, z_2)$

$$X = X_1 \times X_2, \ x_1, y_1, z_1 \in X_1, \ x_2, y_2, z_2 \in X_2$$

$d_1(x_1, y_1) \leq d_1(x_1, z_1) + d_1(z_1, y_1)$

$d_2(x_2, y_2) \leq d_2(x_2, z_2) + d_2(z_2, y_2)$

$d_1(x_1, y_1) + d_2(x_2, y_2) \leq d_1(x_1, z_1) + d_1(z_1, y_1) + d_2(x_2, z_2) + d_2(z_2, y_2)$

$d_1(x_1, y_1) + d_2(x_2, y_2) \leq d_1(x_1, z_1) + d_2(x_2, z_2) + d_1(z_1, y_1) + d_2(z_2, y_2)$

$d(x, y) \leq d(x, z) + d(z, y), \ \forall x, y, z \in X$

Hence, d is metric space.

EXAMPLE 12. ***Let R be the set of all real number and let R^2 denote the set of all ordered pairs of real number, then the function $d : R^2 \times R^2 \rightarrow R$ where d is defined as $d(x, y) = [(x_1 - y_1)^2 + (x_2 - y_2)^2]^{1/2}$ is a metric on R^2.***

(ROHILKHAND–1994)

SOLUTION. Here we shall show that d is a metric space defined as $d : R^2 \times R^2 \rightarrow \mathbf{R}$

such that $\quad d(x, y) = [(x_1 - y_1)^2 + (x_2 - y_2)^2]^{1/2}$

(i) Since $\quad (x_1 - y_1)^2 \geq 0, \ (x_2 - y_2)^2 \geq 0$

$$\Rightarrow \qquad (x_1 - y_1)^2 + (x_2 - y_2)^2 \geq 0$$

$$\Rightarrow \quad [(x_1 - y_1)^2 + (x_2 - y_2)^2]^{1/2} \geq 0$$

$$\Rightarrow \qquad\qquad\qquad d(x, y) \geq 0, \ \forall x, y \in \mathbf{R}^2.$$

(ii) Let $d(x, y) = 0 \Leftrightarrow [(x_1 - y_1)^2 - (x_2 - y_2)^2]^{1/2} = 0$

$$\Leftrightarrow (x_1 - y_1)^2 + (x_2 - y_2)^2 = 0$$

$$\Leftrightarrow (x_1 - y_1)^2 = 0 \ \text{ and } \ (x_2 - y_2)^2 = 0$$

$$\Leftrightarrow x_1 - y_1 = 0 \ \text{ and } \ x_2 - y_2 = 0$$

$$\Leftrightarrow x_1 = y_1 \ \text{ and } \ x_2 = y_2 \ \Leftrightarrow \ x = y$$

Hence, $d(x, y) = 0 \Leftrightarrow x = y$

(iii) $\quad d(x, y) = [(x_1 - y_1)^2 + (x_2 - y_2)^2]^{1/2}$

$$= [(y_1 - x_1)^2 + (y_2 - x_2)^2]^{1/2} = d(y, x).$$

(iv) $\quad d(x, y) = [(x_1 - y_1)^2 + (x_2 - y_2)^2]^{1/2}$

$$= [\{(x_1 - z_1) + (z_1 - y_1)\}^2 + \{(x_2 - z_2) + (z_2 - y_2)\}]^{1/2}$$

$$\leq [(x_1 - z_1)^2 + (x_2 - z_2)^2]^{1/2} + [(z_1 - y_1)^2 + (z_2 - y_2)^2]^{1/2}.$$

On using Minkowski's inequality such that

$$\Rightarrow \qquad\qquad d(x, y) \leq d(x, z) + d(z, y)$$

$$\left[\Sigma(x_i + y_i)^p]^{1/p}\right] \le \left(\sum_{i=1}^{n} x_i^p\right)^{1/p} + \left[\sum_{i=1}^{n} y_i^p\right]^{1/p}$$

∴ $d(x, y) \le d(x, z) + d(z, y), \; \forall x, y, z \in \mathbf{R}^2$.

Hence, we conclude that d is a metric on $\mathbf{R}^2$.

EXAMPLE 13.   *In R^n let us define $d(x,y) = \left[\sum_{i=1}^{n}(x_i - y_i)\right]^{1/2}$ where $x = (x_1, x_2, ..., x_n)$ and*

$y = (y_1, y_2, ..., y_n)$.
Show that d is a metric on R^n.   (MEERUT–2011, KANPUR–2003, 11)

SOLUTION.   (i)   Clearly we have $d(x,y) = \left[\sum_{i=1}^{n}(x_i - y_i)\right]^{1/2} \ge 0$

(ii)   We have $d(x,y) = 0 \quad \Leftrightarrow \quad \left[\sum_{i=1}^{n}(x_i - y_i)\right]^{1/2} = 0$

$$\Leftrightarrow \quad (x_i - y_i)^2 = 0 \;\; \forall i = 1,2,...,n$$

$$\Leftrightarrow \quad x_i = y_i \;\; \forall i = 1,2,...,n$$

$$\Leftrightarrow \quad (x_1, x_2 ..., x_n) = (y_1, y_2 ..., y_n)$$

$$\Leftrightarrow \quad \mathbf{x = y}$$

(iii)   We have $d(x,y) = \left[\sum_{i=1}^{n}(x_i - y_i)^2\right]^{1/2} = \left[\sum_{i=1}^{n}(y_i - x_i)^2\right]^{1/2} = d(y,x)$

(iv)   To prove that the triangle inequality, let us take $a_i = x_i - y_i$ and $b_i = y_i - z_i$ and $p = 2$ in Minkowski's inequality, we get

$$\left[\sum_{i=1}^{n}(x_i - z_i)^2\right]^{1/2} \le \left[\sum_{i=1}^{n}(x_i - y_i)\right]^{1/2} + \left[\sum_{i=1}^{n}(y_i - z_i)\right]^{1/2}$$

$$\Rightarrow \quad d(x,z) \le d(x,y) + d(y,z)$$

Hence, we conclude that d is a metric on $\mathbf{R}^n$.

☛ **REMARK**
⇒ The metric discussed above is known as **usual metric on R^n**.

EXAMPLE 14.   *Let $p \ge 1$ and l_p denote the set of all sequences $<x_n>$ such that $\sum_{n=1}^{\infty} |x_n|^p$ is*

convergent. Define

$d(x,y) = \left[\sum_{n=1}^{\infty} |x_n - y_n|^p\right]^{1/p}$ where $x = (x_1, x_2, ..., x_n)$ and $y = (y_1, y_2, ..., y_n)$.

Show that d is a metric on l_p.

SOLUTION.   Let $a, b \in l_p$. Firstly, we shall prove that $d(a, b)$ is a real number. By Minkowski's inequality, we have

$$\left[\sum_{i=1}^{n}|a_i + b_i|^p\right]^{1/p} \le \left[\sum_{i=1}^{n}|a_i|^p\right]^{1/p} + \left[\sum_{i=1}^{n}|b_i|^p\right]^{1/p} \qquad ...(1)$$

Since $a, b \in l_p$ then RHS of (1) has a finite limit as $n \to \infty$. Therefore $\left(\sum_{i=1}^{n}|a_i + b_i|^p\right)^{1/p}$

is a convergent series.

Similarly, we can prove that $\left(\sum\limits_{i=1}^{n} |a_i - b_i|^p\right)^{1/p}$ is also a convergent series. Thus, $d(a, b)$ is a real number.

Taking limit as $n \to \infty$ in both the sides of (1), we get

$$\left[\sum_{i=1}^{\infty} |a_i + b_i|^p\right]^{1/p} \leq \left[\sum_{i=1}^{\infty} |a_i|^p\right]^{1/p} + \left[\sum_{i=1}^{\infty} |b_i|^p\right]^{1/p} \qquad ...(2)$$

Clearly, $d(x, y) \geq 0$ and $d(x, y) = 0$ if and only if $x = y$. Also, $d(x, y) = d(y, x)$.

Now, it remains to prove that it satisfies triangle inequality. Let $x, y, z \in l_p$.

Let us take $a_i = x_i - y_i$ and $b_i = y_i - z_i$ in (2), we get

$$\left(\sum_{i=1}^{\infty} |x_i - z_i|^p\right)^{1/p} \leq \left(\sum_{i=1}^{\infty} |x_i - y_i|^p\right)^{1/p} + \left(\sum_{i=1}^{\infty} |y_i - z_i|^p\right)^{1/p}$$

$$\Rightarrow \qquad d(x, z) \leq d(x, y) + d(y, z) \quad \forall \, x, y, z \in l_p$$

Hence, from above we conclude that d is a metric on l_p.

EXAMPLE 15. *Let X be the set of sequences in R. Let $x, y \in X$ and $x = (x_n)$, $y = (y_n)$. Define*

$$d(x, y) = \sum_{n=1}^{\infty} \frac{|x_n - y_n|}{2^n (1 + |x_n - y_n|)}$$

Show that d is a metric on X.

SOLUTION. (i) Let $x, y \in X$.

Since, we have $\dfrac{|x_n - y_n|}{2^n (1 + |x_n - y_n|)} \leq \dfrac{1}{2^n} \quad \forall n$

Also, $\sum\limits_{n=1}^{\infty} \dfrac{1}{2^n}$ is a convergent series.

Therefore, $\sum\limits_{n=1}^{\infty} \dfrac{|x_n - y_n|}{2^n (1 + |x_n - y_n|)}$ is a convergent series. (By comparison test)

$\Rightarrow \; d(x, y)$ is a real number and $d(x, y) \geq 0$.

(ii) Now $d(x, y) = 0 \quad \Leftrightarrow \quad \sum\limits_{n=1}^{\infty} \dfrac{|x_n - y_n|}{2^n (1 + |x_n - y_n|)} = 0$

$\Leftrightarrow \; |x_n - y_n| = 0 \quad \forall n$

$\Leftrightarrow \; x_n = y_n \quad \forall n \; i.e. \; x = y$

(iii) Consider $d(x, y) = \sum\limits_{n=1}^{\infty} \dfrac{|x_n - y_n|}{2^n (1 + |x_n - y_n|)}$

$= \sum\limits_{n=1}^{\infty} \dfrac{|y_n - x_n|}{2^n (1 + |y_n - x_n|)} = d(y, x) \qquad$ (By definition of modulus)

(iv) Let $x, y, z \in X$. Then consider

$$\frac{|x_n - y_n|}{1 + |x_n - y_n|} = 1 - \frac{1}{1 + |x_n - y_n|} \leq 1 - \frac{1}{(1 + |x_n - z_n| + |z_n - y_n|)}$$

$$= \frac{|x_n - z_n| + |z_n - y_n|}{1 + |x_n - z_n| + |z_n - y_n|}$$

$$= \frac{|x_n - z_n|}{1 + |x_n - z_n| + |z_n - y_n|} + \frac{|z_n - y_n|}{1 + |x_n - z_n| + |z_n - y_n|}$$

$$\leq \frac{|x_n - z_n|}{1 + |x_n - z_n|} + \frac{|z_n - y_n|}{1 + |z_n - y_n|}$$

Multiplying both sides of the above inequality by $\dfrac{1}{2^n}$ and taking the sum from $n = 1$ to ∞, we get

$$d(x,y) \le d(x,z) + d(z,y) \quad \forall x,y,z \in X$$

Hence, d is a metric on X.

EXAMPLE 16. *Let $(X_1, d_1), (X_2, d_2), \ldots, (X_n, d_n)$ are metric spaces then show that $X_1 \times X_2 \times \ldots \times X_n$ is a metric space with d defined by*

$$d(x,y) = \sum_{i=1}^{n} d_i(x_i, y_i) \text{ where } x = <x_n>, \ y = <y_n>.$$

SOLUTION. (i) Clearly, $d(x,y) = \sum\limits_{i=1}^{n} d_i(x_i, y_i) \ge 0$ $\hspace{1.5cm}$ ($\because$ Each d_i is a metric)

(ii) We have $d(x,y) = 0$ $\hspace{1cm}$ $\Leftrightarrow \sum\limits_{i=1}^{n} d_i(x_i, y_i) = 0$

$\hspace{4.5cm}$ $\Leftrightarrow d_i(x_i, y_i) = 0 \ \ \forall i = 1, 2, \ldots, n$

$\hspace{4.5cm}$ $\Leftrightarrow x_i = y_i \ \ \forall i = 1, 2, \ldots, n$

$\hspace{4.5cm}$ $\Leftrightarrow (x_1, x_2, \ldots, x_n) = (y_1, y_2, \ldots, y_n)$

$\hspace{4.5cm}$ $\Leftrightarrow x = y$

(iii) We have $\quad d(x,y) = \sum\limits_{i=1}^{n} d_i(x_i, y_i) = \sum\limits_{i=1}^{n} d_i(y_i, x_i)$ $\hspace{0.8cm}$ ($\because$ Each d_i is a metric on X)

$\hspace{4cm}$ $= d(y, x) \ \ \forall x, y \in X$

(iv) Consider $\quad d(x,y) = \sum\limits_{i=1}^{n} d_i(x_i, y_i)$

$\hspace{4cm}$ $\le \sum\limits_{i=1}^{n} [d_i(x_i, z_i) + d_i(z_i, y_i)]$ $\hspace{0.8cm}$ ($\because$ Each d_i is a metric on X)

$\hspace{4cm}$ $= \sum\limits_{i=1}^{n} d_i(x_i, z_i) + \sum\limits_{i=1}^{n} d_i(z_i, y_i) = d(x,z) + d(z,y)$

$\Rightarrow \hspace{2cm}$ $d(x,y) \le d(x,z) + d(z,y) \ \ \forall x, y, z \in X$

Hence, we conclude that d is a metric on X.

EXAMPLE 17. *Let $p = <p_n>$ be a bounded sequence of strictly positive numbers so that $0 < p_n < \sup. \ p_n = H < \infty$. Further, let $l(p)$ be the set of all sequence $x = <x_n>$ such that*

$$\sum_{n=1}^{\infty} |x_n|^p < \infty \text{ i.e. let } l(p) = \left\{ x = <x_n> : \sum_{n=1}^{\infty} |x_n|^{p_n} < \infty \right\}$$

Define $d(x,y) = \left(\sum\limits_{n=1}^{\infty} |x_n - y_n|^{p_n} \right)^{1/m}$ where $m = max(1, H)$

Show, that d is a metric on $l(p)$.

SOLUTION. (i) Since $m = \max\{1, H\}$ then

$$\sum_{n=1}^{\infty} (|x_n - y_n|^{p_n})^{1/m} \ge 0 \ \Rightarrow \ d(x,y) \ge 0$$

(ii) Since, $|x_n - y_n| = 0$ iff $x_n = y_n$

$\Rightarrow (|x_n - y_n|^{p_n})^{1/m} = 0$ iff $x_n = y_n$

$$\Rightarrow (\sum |x_n - y_n|^{p_n})^{1/m} = 0 \text{ iff } x_n = y_n$$

$$\Rightarrow d(x,y) = 0 \text{ iff } x_n = y_n \text{ i.e. iff } x = y$$

(iii) Since $|x_n - y_n| = |y_n - x_n|$

$$\Rightarrow |x_n - y_n|^{p_n} = |y_n - x_n|^{p_n}$$

$$\Rightarrow (|x_n - y_n|^{p_n})^{1/m} = (|y_n - x_n|^{p_n})^{1/m}$$

$$\Rightarrow \sum(|x_n - y_n|^{p_n})^{1/m} = \sum(|y_n - x_n|^{p_n})^{1/m}$$

$$\Rightarrow d(x,y) = d(y,x) \quad \forall x \in <x_n>, y \in <y_n>$$

(iv) Let $x = <x_n>$, $y = <y_n>$ and $z = <z_n>$ be any three elements of $l(p)$. Put $a_n = x_n - z_n$, $b_n = z_n - y_n$. Further, since $H = \sup p_n$ and $m = \max\{1, H\}$, we have $0 < \dfrac{\quad}{\quad} \le 1$. Therefore,

$$|a_n + b_n|^{t_n} \le |a_n|^{t_n} + |b_n|^{t_n} \qquad\qquad ...(1)$$

where $t_n = \dfrac{p_n}{m}$

Now,
$$\begin{aligned}
d(x,y) &= (\sum |x_n - y_n|^{p_n})^{1/m}\\
&= (\sum |x_n - z_n + z_n - y_n|^{p_n})^{1/m}\\
&= (\sum |a_n + b_n|^{p_n})^{1/m} = (\sum |x_n - z_n + z_n - y_n|^{p_n})^{1/m}\\
&= (\sum |a_n + b_n|^{p_n})^{1/m} = (\sum (|a_n + b_n|^{t_n})^m)^{1/m}\\
&\le (\sum (|a_n|^{t_n} + |b_n|^{t_n})^m)^{1/m}\\
&\le (\sum |a_n|^{mt_n})^{1/m} + (\sum |b_n|^{mt_n})^{1/m} \quad \text{(By Minkowski's inequality)}\\
&\le (\sum |a_n|^{p_n})^{1/m} + (\sum |b_n|^{p_n})^{1/m} \quad (\because t_n = p_n/m)\\
&= (\sum |x_n - z_n|^{p_n})^{1/m} + (\sum |z_n - y_n|^{p_n})^{1/m}\\
&= d(x,z) + d(z,y)
\end{aligned}$$

Hence, from above, we conclude that d is a metric on $l(p)$.

EXAMPLE 18. ***Let l_p be the set of the all sequences, $x = <x_n>$ such that $\sum\limits_{n=1}^{\infty} |x_n|^p < \infty$ where p is a positive constant. Define***

(i) $d(x,y) = (\sum |x_n - y_n|^p)^{1/p}, p \ge 1$

(ii) $d^*(x,y) = (\sum |x_n - y_n|^p), 0 < p < 1$

Show that d and $d*$ are both metrices on l_p.

SOLUTION. Clearly, this space is a particular case of the space $l(p)$. Considering in the previous example. If $p \ge 1$ then

$$m = \max\{1, p\} = p$$

and when $0 < p < 1$, we have $m = 1$

However, due to the importance, we give independent proof here by checking only 4th properties of metric spaces in each case.

(i) Let $x = <x_n>, y = <y_n>, z = <z_n>$

Put $a_n = x_n - z_n, b_n = z_n - y_n$

Then $(\sum |x_n - y_n|^p)^{1/p} = (\sum |x_n - z_n + z_n - y_n|^p)^{1/p}$

$$\leq (\Sum |x_n - z_n|^p)^{1/p} + (\Sum |z_n - y_n|^p)^{1/p}$$

(By Minkowski's inequality)

$$= d(x,z) + d(z,y)$$

In order that d is well defined, we must show that $d(x,y) < \infty$. We have

$$d(x,y) = (\Sum |x_n - y_n|^p)^{1/p}$$
$$\leq (\Sum |x_n|^p)^{1/p} + (\Sum |-y_n|^p)^{1/p}$$

(By Minkowski's inequality)

$$< \infty \qquad (\because \Sum |x_n|^p < \infty, \Sum |y_n|^p < \infty)$$

(ii) We have $d^*(x,y) = \Sum |x_n - y_n|^p = \Sum |x_n - z_n + z_n - y_n|^p$

$$\leq \Sum |x_n - z_n|^p + \Sum |z_n - y_n|^p$$
$$= d^*(x,z) + d^*(z,y)$$

Further, $d^*(x,y) = \Sum |x_n - y_n|^p \leq \Sum |x_n|^p + \Sum |-y_n|^p < \infty$

$\Rightarrow d^*$ is well defined.

EXERCISE 3.1

1. Give an example of a pseudo-metric which is not metric. Is every metric a pseudo metric? (MEERUT 1996; ROHTAK 2006,08; RAJASTHAN 2002,05; GARHWAL 2007)

2. Let (X, d) be a metric space and let x, y, z be any three points of X, then show that $d(x,y) \geq |d(x,z) - d(z,y)|$.

3. Let (X, d) be a metric space and let $x, x', y, y' \in X$. Then show that $|d(x,y) - d(x',y')| \leq d(x,x') + d(y,y')$. (MEERUT 1994)

4. Let A, B be subsets of a metric space (X,d), then show that $\delta(A \cup B) \leq \delta(A) + \delta(B) + d(A,B)$.

5. Define the diameter of subset A of the metric space X. What is the diameter of empty set? What do you mean by the distance between two non-empty subsets A and B of metric space X. If x is a point of X and A is a subset of X, write the distance of x from A. (MEERUT 1992,95)

6. Let $R[0,1]$ denote the classes of all Reimann integrable function f from $[0,1]$ into R. Consider the mapping $d: R[0,1] \times R[0,1] \to R$ defined by
$$d(f,g) = \int_0^1 |f - g|(x)dx = \int_0^1 |f(x) - g(x)|dx$$
Show that d is a pseudo-metric but not metric on R. (MEERUT 1996)

7. Show that $d: R^2 \times R^2 \to R$ defined by
$$d(x,y) = |x_1 - y_1| + |x_2 - y_2|,$$
$$x = (x_1, x_2), y = (y_1, y_2),$$
where $x \in R^2$ is a metric on .

8. Let $X = R^n$ denote the set of all ordered n-tuples of real numbers for a fixed $n \in N$. Let $\mathbf{x} = (x_1, x_2, ..., x_n), \mathbf{y} = (y_1, y_2, ..., y_n)$ Define the mapping d_1, d_2 and d_3 of $R^n \times R^n$ into R by

(i) $d_1(x,y) = \left[\sum_{i=1}^{n} (x_i - y_i)^2 \right]^{1/2}$ (KANPUR 2003; MEERUT 1997, 2011)

(ii) $d_2(x,y) = \sum_{i=1}^{n} |x_i - y_i|$. (KANPUR 2004; MEERUT 1993)

(iii) $d_3(x,y) = \max\{|x_1 - y_1|, |x_2 - y_2|, ..., |x_n - y_n|\}$.

Show that d_1, d_2, d_3 are metrices on R^n.

9. Show that the set C of all complex numbers is a metric space under
$$d(z_1, z_2) = \frac{|z_1 - z_2|}{[(1 + |z_1|^2)(1 + |z_2|^2)]^{1/2}}$$ (ROHILKHAND 1983)

10. If (X, d) be a metric space, then show that d_1 defined by $d_1(x,y) = \frac{d(x,y)}{1 + d(x,y)}$ is also a metric on X. (MEERUT 1992,94,95,2000,03, 10; KANPUR 1995)

11. Show that the sum of two metric spaces is again a metric space.

12. If $d(x, y) = 2|x - y|$. Then show that d is a metric on R.

13. Let (X, d) be a metric space and let A, B be subsets of X, then show that $A \subset B \Rightarrow \delta(A) \leq \delta(B)$.

14 Let $d(x, y) = \min\{2, |x - y|\}$. Show that d is a metric for $\boldsymbol{R}$. (MEERUT 2001)

15. Let $C[0, 1]$ denote the collection of all real valued bounded continuous functions defined on the closed interval $[0, 1]$, we define the norm of $f \in C[0, 1]$ by

$$||f|| = \sup\{|f(x) : x \in [0, 1]\}$$

where d is defined as

$$d(f, g) = ||f - g|| = \sup\{|f(x) - g(x)| : x \in [0, 1]\}$$

then show that d is a metric for $C[0, 1]$. (KANPUR 1992, 2004,06,08)

16. Let $\boldsymbol{R}$ be the set of all real number and let

$$d(x, y) = \frac{|x - y|}{1 + |x - y|}, \quad \forall x, y \in \boldsymbol{R}$$

Show that d is a metric for $\boldsymbol{R}$. (MEERUT 1991; ROHILKHAND 2005)

17. Let (X, d) be any metric space then show that function d^* defined as

$$d^*(x, y) = \min\{2, d(x, y)\}, \quad \forall x, y \in X$$

is a metric on X.

HINTS TO SELECTED PROBLEMS

1. $d(x, y) = |x^2 - y^2| \quad \forall x, y \in \boldsymbol{R}$.

12. Since we know that $d(x, y) = |x - y|$ is a metric, use this result to prove $d(x, y) = 2|x - y|$ is a metric.

13. By definition of the diameter of a set, we have $\delta(A) = \sup\{d(x, y) : x, y \in A\}$

Since $A \subset B$

$$x, y \in A \Rightarrow x, y \in B$$

$\therefore \{d(x, y) : x, y \in A\}$

$\Rightarrow \{d(x, y) : x, y \in B\}$

$\Rightarrow \sup\{d(x, y) : x, y \in A\}$

$\qquad \leq \sup\{d(x, y) : x, y \in B\} \qquad (\because A \subset B)$

$\therefore \delta(A) \leq \delta(B)$.

16. Do same as example 4.

17. Do same as example 5.

ANSWERS

1. $d(x, y) = |x^2 - y^2|$, Yes.

REVIEW QUESTIONS AND ARCHIVE

1. Define the following:

(i) Metric Spaces (MEERUT–2002, 03, 04, 10, 11, 18, GARHWAL–2000, 05, 06, 07, 08, 10, 11, 15, 17, AGRA–1994, 2003, 05, 11, 12, 17, KANPUR–1995, 2002, 03, 05, 11, 14, 15, AVADH–2005)

(ii) Discrete metric spaces (MEERUT–2005, 18, GARHWAL–2001, 02, 05, 08, PURVANCHAL–2004, 06, 09, 11, 12)

(iii) Usual metric spaces (MEERUT–20004, 07, 11, 12, 14, RAJASTHAN–2004, 06, 08, 11)

(iv) Pseudo metric spaces (MEERUT–1996, 2000, 03, 11, 12, 17, KANPUR–2008, 12, GARHWAL–2002, 05, 07, 09, 12, ROHILKHAND–2004, 07, 11, BHOPAL–2004, 06, 09)

(v) Diameter of a set (MEERUT–2002, 05,)

(vi) Triangle inequality (KANPUR–2008, LUCKNOW–2007, 12)

(vii) Holder's inequality (MEERUT–2001, 11, 12, ALLAHABAD–2007, BHU–2009, KURUKSHETRA–2006, PATNA–2004, 05, 07, 10, 12)

(viii) Cauchy-Schwarz inequality (KANPUR–2001, RAJASTHAN–2005, 09, OSMANIA–2004, 06, 12)

(ix) Minkowski's inequality (KANPUR–2002, 05, LUCKNOW–2007, ALLAHABAD–2008, PATNA–2004, 06, 09, 11)

2. Let $(\boldsymbol{R}^2, d)$ be the Euclidean plane. Define d_1 in terms of d by

$$d_1(x, y) = \begin{cases} d(x, y); & \text{if } x, y \text{ and origin } 0 \\ & \text{are collinear} \\ d(x, 0) + d(0, y), & \text{otherwise} \end{cases}$$

Show that d_1 is a metric on $\boldsymbol{R}^2$.

3. Let (X, d) be a metric space and c be any fixed positive real number. Show that d^* defined by

$$d^*(x, y) = kd(x, y) \quad \forall x, y \in X$$

is a metric on X.

4. If $x_1, x_2, ..., x_n$ are any n points in a metric space (X, d) show that

$d(x_1, x_n) \le d(x_1, x_2) + d(x_2, x_3)$
$$+ ... + d(x_{n-1}, x_n)$$

5. Let (X, d) be a metric space and f be any real valued function on X. Show that d^*

defined by setting.
$d^*(x, y) = d(x, y) + |f(x) - f(y)| \; \forall \, x, y \in X$
is also a metric on X.

6. If $d(x, y) = \min\{2, |x - y|\}$
Show that d is a metric on $\mathbf{R}$.

OBJECTIVE EVALUATION

▶ FILL IN THE BLANKS

1. In a metric space (X, d), $d(x, y) = 0$ iff _____.

2. In a metric space (X, d), $x, y \in X$, then $d(x, y) \le d(x, z) +$ _____.

3. Let (X, d) be a metric space and $A \ne \phi$, $A \subseteq X$, then diameter of A is defined by $d(A) =$ _____.

4. Let $X = \mathbf{R}^n$, the set of all ordered n-tuples of real numbers and let $x = (x_1, x_2, ..., x_n)$,

$y = (y_1, y_2, ..., y_n)$ then $d(x, y) =$ _____.

5. If S be any subset of X, then diameter $\delta(S) = 0$ iff S contains at most _____ point.

6. If S, T be any two subsets of X, then $S \subseteq T \Rightarrow \delta(S)$ _____ $\delta(T)$.

7. If $S \cap T \ne \phi$, then $\delta(S \cap T)$ _____ $\delta(S) + \delta(T)$.

▶ TRUE OR FALSE

1. Every pseudo-metric is a metric. **(T/F)**

2. Every metric is a pseudo metric. **(T/F)**

3. $\delta(A) =$ supremum $\{d(x, y) : x, y \in A\}$. **(T/F)**

4. The distance $d(A, B)$ between two non-

empty sets A and B is given by
$d(A, B) =$ infimum $\{d(x, y) : x \in A, y \in B\}$ **(T/F)**

5. $d(A, B) = 0 \Leftrightarrow A \cap B = \phi$. **(T/F)**

6. Distance between a set A and empty set is ∞. **(T/F)**

▶ MULTIPLE CHOICE QUESTIONS (CHOOSE THE MOST APPROPRIATE ONE)

1. The diameter of a X in a discrete metric space (X, d) is:
  (a) 1   (b) 0
  (c) ∞ (d) $-\infty$

2. The diameter of an empty set ϕ is:
  (a) 0   (b) ∞
  (c) $-\infty$ (d) 1

3. The metric space defined by
$$d(x, y) = \begin{cases} 0 & \text{if } x = y \\ 1 & \text{if } x \ne y \end{cases} \text{ is called:}$$

  (a) Discrete metric space
  (b) Indiscrete metric space
  (c) Usual metric space
  (d) None of these

4. The metric space defined on real numbers is called:
  (a) Discrete metric space
  (b) Indiscrete metric space
  (c) Usual metric space
  (d) None of these

ANSWERS

▶ FILL IN THE BLANKS

1. $x = y$ **2.** $d(z, y)$ **3.** sup $\{d(x, y) : x, y \in A\}$ **4.** $[\Sigma(x_i - y_i)^2]^{1/2}$ **5.** one   **6.** $\le$ **7.** $\le$

▶ TRUE OR FALSE

1. T   **2.** F   **3.** T   **4.** T   **5.** F   **6.** T

▶ **MULTIPLE CHOICE QUESTIONS**

1. (a)   **2.** (b)   **3.** (a)   **4.** (c)

■■□□ \ **CHAPTER SUMMARY** / ■■□□

In this chapter, we have introduced the fundamental notions and concept of metric space. In 1905, the French mathematician Maurice Frechet thought of generalizing the notion of distance and extending it to arbitrary set S. In his doctoral dissertation 'Les Espaces Abstrail', he introduced the concept of metric on a set which is going to be the theme of the chapter. Important points discussed in this chapter are as follows :

⮑ A non-negative real valued function $d : X \times X \to R^*$ is said to be metric on X if following four conditions are satisfied :

  (i) $d(x, y) \geq 0 \ \forall \, x, y \in X$

  (ii) $d(x, y) = 0 \Leftrightarrow x = y \ \forall x, y \in X$

  (iii) $d(x, y) = d(y, x) \ \forall x, y \in X$

  (iv) $d(x, y) \leq d(x, z) + d(z, y) \ \forall x, y, z \in X$

⮑ If $d(x, y) = \begin{cases} 0 & if \ x = y \\ 1 & if \ x \neq y \end{cases}$

Then d is a metric on X, called discrete metric space.

⮑ The metric space defined on real numbers is called usual metric space.

⮑ Every metric space is a pseudo metric but converse is not necessarily true.

⮑ The distance of a point from a set may be zero even if point belongs to the set.

⮑ $d(S, T) = $ infimum $\{ d(x, y) : x \in S, y \in T \}$.

We construct the metric space from pseudo metric space.

⮑ Discrete metric spaces, in which X is uncountable, are rich sources of examples.

⮑ $d(A, B) = d(B, A)$

⮑ The concept of norm for Euclidean space can be extended to abstract metric space. Thus the norm of an element x of a norm in a way gives the size of the element k.

⮑ $|z + w| \leq |z| + |w|$ **(Triangle inequality)**

⮑ $\dfrac{|z + w|}{1 + |z + w|} \leq \dfrac{|z|}{1 + |z|} + \dfrac{|w|}{1 + |w|}$, $z, w \in C$.

⮑ $\sum\limits_{i=1}^{n} a_i b_i \leq \left(\sum\limits_{i=1}^{n} a_i^p \right)^{1/p} \left(\sum\limits_{i=1}^{n} b_i^q \right)^{1/q}$

(Holder's inequality)

⮑ $\sum\limits_{i=1}^{n} |z_i \, w_i| \leq ||z|| . || w ||$

(Minkowski's inequality)

■■□□ \ **FOR ADVANCED LEARNERS** / ■■□□

➡ Let $x_1, x_2, ..., x_n$ be elements of metric space (X, d) then generalised triangle inequality is given by

$d(x_1, x_n) \leq d(x_1, x_2) + d(x_2, x_3) + d(x_{n-1}, x_n)$

➡ The mapping $d : R^n \times R^n \to R$ is a metric on R^n defined by

$d(x, y) = \sum\limits_{i=1}^{n} |x_i - y_i|$

This metric is called **Taxicab metric**.

➡ **(Metric on Riemann integral).** The metric $d(f, g) = R\int_b^a |f - g|$ defined on the set of continuous real valued mappings on the compact interval $[a, b]$ is called metric on Riemann integral.

➡ The mapping $d(f, g) = \int ||f - g|$ defines a pseudometric on the set of Lebesgue integrable function on R.

➡ If S is any non-empty subset of X and x, y are two points of X then

$|d(x, S) - d(y, S)| \leq d(x, y)$

➡ For non-empty subsets A, B of X, $d(A, B) = 0$ and $d(A, B) = d(B, A)$.

➡ The standard metric on R and the metric induced on R as a subset of the extended real line R give rise to the same topology on R.

➡ The union of finitely many bounded subsets of a metric space is bounded.

●●●●

CHAPTER 4

Open and Closed sets in Metric Space

4.1 INTRODUCTION

In this chapter, we shall discuss some basic types of set, viz., open set, closed set with their fundamental properties. The closure, interior, exterior and boundary points for a set will be discussed in details also.

4.2 OPEN AND CLOSED SPHERES IN A METRIC SPACE

Let (X, d) be a metric space. Let $p \in X$ and $r > 0$ be given. Then, the set of all points $x \in X$ such that $d(p, x) < r$ is called the open ball or open sphere of radius r and centre p in X and is denoted by $S(p, r)$ or $B(p, r)$.

i.e., $\qquad S(p, r) = \{x \in X : d(x, p) < r\}$.

The set of all points $x \in X$ such that $d(p, x) \le r$ is called closed ball of radius r and centre p and is denoted by $S^*(p, r)$

i.e., $\qquad S^*(p, r) = \{x \in X : d(x, p) \le r\}$ $\hfill$ (JABALPUR 2002,03)

ILLUSTRATIONS

In the case of real line:

* $S(p, r)$ is the open interval $]p - r, p + r[$.

* The open interval $]a, b[$ is the open ball with centre at the points $p = \frac{1}{2}(a + b)$ and radius $r = \frac{1}{2}(b - a)$.

* $S^*(p, r)$ is the closed interval $[p - r, p + r]$.

* The closed interval $[a, b]$ is the closed ball $S^*(p, r)$ with centre $p = \frac{1}{2}(a + b)$ and radius $r = \frac{1}{2}(b - a)$.

4.2.1 OPEN SET

Definition. *Let (X, d) be a metric space. A subset A of X is said to be open iff to each $x \in A$, there exists $r > 0$ such that*

$\qquad\qquad S(x, r) \subseteq A$ $\hfill$ (MEERUT 1993,94; GARHWAL 2004,06,08,10)

For example:

(i) The subset $]0, 3[$ is open in $X = [0, 3]$ under the metric d given by

$\qquad\qquad d(x, y) = |x - y|$.

Since for every $x \in]0, 3[$, we can find $\varepsilon_x > 0$ s.t. $S(x, \varepsilon_r) \subseteq]0, 3[$.

4.2.2 INTERIOR POINT

Let $< X, d >$ be a metric space.

and $A \subseteq X$, $x \in A$ is said to be an interior point of A iff $\exists\, r > 0$ s.t.

$$S(x, r) \subseteq A$$

For example: In $< R, d >$ usual metric space we have

(i) $A = (0, 1)\,[S(1, r) - \{1\}] \cap (0, 1) \neq \phi \ \ \forall\, r \in R^{+}$

(ii) $A = (0, 1) \cup \{3\}$

$[S(3, r) - \{3\}] \cap A = \phi$ for $r = 1/2$

Also, $S(3, r) \cap A \neq \phi \ \forall\, r > 0$.

4.2.3 LIMIT POINT

Let (X, d) be a metric space and $A \subset X$. A point $x \in X$ is called a limit point or limiting point or accumulation point or cluster point if every open sphere centered at x contains a point of A, other than x i.e., $x \in X$ is called limit point of A if $[S(x, r) - \{x\}] \cap A \neq \phi$, $r \in R^{+}$.

4.2.4 ADHERENT POINT

Let (X, d) be a metric space and A be any subset of X. A point $x \in X$ is said to be adherent point of A if every open sphere centered at x contains at least one point of A not necessarily different from x.

There are two types of adherent points :

(a) Limit point. A point $x \in X$ is called limiting point of A if

$$[S(x, r) - \{x\}] \cap A \neq \phi \ \ \forall\, r \in R^{+},$$

i.e., every open sphere contains a point of A, other than x.

(b) Isolated point. A point $x \in X$ is called isolated point of A if

$$[S(x, r) - \{x\}] \cap A = \phi$$

for some $r \in R^{+}$, *i.e.*, some open sphere contains only x from A.

i.e., if x is not the limit point of A.

4.2.5 DERIVED SET

Set of all limit points of a set A is called derived set of A and is denoted by $D(A)$.   (MEERUT 1994)

4.2.6 CLOSED SET

Let (X, d) be a metric space and $A \subset X$. Then A is called closed set if the derived set of A contains in A, i.e., $D(A) \subset A$
(MEERUT 2006; RAJ. 2005,07,08,10; GARHWAL 2009)

(or) if every limit point of A belongs to the set itself.

4.2.7 PERFECT SET

A closed set which has no isolated point is called perfect set.

4.2.8 DIFFERENCE BETWEEN LIMIT AND LIMITING POINT

The limit and limit point both are different terms. For example, consider the sequence $\langle 1, 1, 1, ... \rangle$ in which every element is identically equal to 1. The limit of the sequence is

---

FACTS : TO THE POINT

▶ Open sphere is defined as spherical neighbourhood of a point p.

▶ An open (or closed) sphere is always non-empty, since it contains its centre at least.

▶ Let X be a non-empty set and let d be the discrete metric space on X. The open sphere centered at a point $p \in X$ and radius r is $\{p\}$ if $r \leq 1$ and X if $r > 1$, The closed sphere centered at p, and of radius $r < 1$ is $\{p\}$. Also, the only closed ball of radius $r \geq 1$ is X.

▶ $S(x, r)$ is always a bounded set.

▶ For usual metric on R if $a \in R$ then $S(a, r) =]a - r, a + r[$.

▶ If we consider C, with usual metric and $a \in C$ then

$S(a, r) = \{z \in C : d(a, z) < r \text{ i.e. } |z - a| < r\}$

This is the interior of the circle with centre a and radius r.

▶ In R^{2} with usual metric, $S(x, r)$ is the interior of the circle with centre x and radius r.

1, but the set of all points of this sequence is the singleton set [1] and hence its limit point does not exist, because finite set has no limit point.

Here, it is also possible that the set of points of a sequence may have a limit point, but cannot have a limit.

✍ ILLUSTRATIONS

● Consider R with usual metric, let $a \in R$, then

$$S(x,r) = \{x \in R : d(a,x) < r\} = \{x \in R : |a - x| < r\}$$
$$= \{x \in R : a - r < x < a + r\}$$
$$=]a - r, a + r[$$

● Consider C with usual metric. Let $a \in C$ then

$$S(x,r) = \{z \in C : d(a,z) < r\}$$
$$= \{z \in C : |z - a| < r\}$$

which is the interior of the circle with centre a and radius r.

● The d be the discrete metric on X then

$$S(a,r) = \begin{cases} X; & \text{if } r > 1 \\ (a); & \text{if } r \le 1 \end{cases}$$

● Consider $X = \{0, 1\}$ with usual metric $d(x, y) = |x - y|$.

Here, $$S\left(0, \frac{1}{2}\right) = \left\{x \in (0,1) : d(0,x) < \frac{1}{2}\right\}$$

$$= \left\{x \in (0,1) : |x| < \frac{1}{2}\right\} = \left[0, \frac{1}{2}\right[$$

● Consider R^2 with the metric d given by

$d((x_1, y_1), (x_2, y_2)) = |x_1 - x_2| + |y_1 - y_2|$
Then $S((0, 0), 1) = \{(x, y) \in R^2 : |x - 0| + |y - 0| < 1\}$
$$= \{(x, y) \in R^2 : |x| + |y| < 1\}$$

which is the interior of the square bounded by the four lines $x + y = 1$, $-x + y = 1$, $x + y = -1$ and $x - y = 1$.

● Consider R^2 with the metric $d((x_1,y_1),(x_2,y_2)) = \max\{|x_1 - x_2|, |y_1 - y_2|\}$

Then $S((0,0),1) = \{(x,y) \in R^2 : \max\{|x - 0|, |y - 0|\} < 1\}$

$$= \{(x,y) \in R^2 : \max\{|x|, |y|\} < 1\}$$

$$= \{(x,y) \in R^2 : |x| < 1 \text{ and } |y| < 1\}$$

which is the interior of the square with vertices $(1, 1)$, $(-1, 1)$, $(-1, -1)$ and $(1, -1)$.

● In R, with usual metric, the set $\{0\}$ is not an open set since any open sphere with centre 0 is not contained in $\{0\}$.

● In R, with usual metric, any finite non-empty subset A of R is not an open set.

THEOREM 1. *In a metric space (X, d), the empty set ϕ and the whole space X, are open sets.*

PROOF. Let (X, d) be a metric space. To show that ϕ and X are open sets.

To prove that ϕ is open in X, it suffices to show that

$$x \in \phi \implies \exists \varepsilon > 0 : S(x, \varepsilon) \subset \phi.$$

Since ϕ does not contain any element and hence this condition is automatically fulfilled.

Now to show X is open.

Since, corresponding to every point $x \in X, \exists$ an open sphere with its centre at x which is contained in X. Hence, X is an open set.

THEOREM 2. *In a metric space (X, d), ϕ and X are closed sets.*

PROOF. Let (X, d) be a metric space.

To show ϕ and X are closed sets.

We know that

$$D(\phi) = \phi \subset \phi$$

Therefore $D(\phi) \subset \phi$.

$\Rightarrow$ ϕ contains all its limit points.

$\Rightarrow$ ϕ is closed.

Now to show X is closed.

Since all the limit points of X belongs to X.

i.e., $x \in D(X) \Rightarrow x \in X$

$\therefore$ $D(X) \subset X$

$\Rightarrow$ X contains all its limit points.

$\Rightarrow$ X is a closed set.

THEOREM 3. *In a metric space, every open sphere is an open set.*

(MEERUT 1991,93, 94;2010, 11(SEM-II); KANPUR 2000,06,07,09; GARHWAL 2001,04,05,07,10)

PROOF. Let (X, d) be a metric space and let $S(p, r)$ be any open sphere.

We wish to show that to each point $x_0 \in S(p, r)$ there exists an open sphere centred at x_0 and contained in $S(p, r)$.

Let $x_0 \in S(p, r)$ be arbitrary.

Now, $x_0 \in S(p, r) \Rightarrow d(x_0, p) < r \Rightarrow r - d(x_0, p) > 0$.

Let us define $\rho = r - d(x_0, p)$.

Then clearly, $\rho > 0$.

Define an open sphere $S(x_0, \rho)$.

We claim that $S(x_0, \rho) \subset S(p, r)$.

Let $x \in S(x_0, \rho)$. Then by definition of open sphere, we have

$$d(x, x_0) < \rho = r - d(x_0, p)$$

Now, $d(x, p) < d(x, x_0) + d(x_0, p)$ (By triangle inequality)

$$< r - d(x_0, p) + d(x_0, p) = r$$

$\Rightarrow$ $d(x, p) < r \Rightarrow x \in S(p, r)$

Since x is arbitrary.

Hence, $S(x_0, \rho) \subset S(p, r)$

$\Rightarrow$ $S(p, r)$ is an open set.

THEOREM 4. *In a metric space (X, d), arbitrary union of open sets is open.*

(MEERUT 1990,92,98,2001,02,11; ROHILKHAND 2002,04,09; KANPUR 2003,07,08)

PROOF. Let (X, d) be a metric space and let $\{G_\lambda : \lambda \in \Lambda\}$ be an arbitrary collection of open subset of X.

Define $G = \cup \{G_\lambda : \lambda \in \Lambda\}$.

To show G is open.

Let $x \in G$

$\Rightarrow$ $x \in G_\lambda$ for some λ.

Since G_λ is open, therefore there exist $r > 0$ such that
$$S(x, r) \subset G_\lambda. \qquad \qquad \ldots(1)$$
By definition of G, we have
$$G_\lambda \subset G. \qquad \qquad \ldots(2)$$
From (1) and (2), we have
$$S(x, r) \subset G_\lambda \subset G$$
$$\Rightarrow \quad S(x, r) \subset G.$$
Since x is arbitrary, therefore, we have shown that, to each $x \in G$, there exists a number $r > 0$ such that $S(x, r) \subset G$
$$\Rightarrow \quad G \text{ is open.}$$
Hence, the arbitrary union of open sets in a metric space, is open.

THEOREM 5. *In a metric space (X, d), the intersection of a finite number of open sets is open.*
(MEERUT 1990,95,96,2000,04,05,08,10,18; KANPUR-1992,94; GARHWAL 2006,08; BHOPAL 2001,03,07; LUCKNOW 2002,04,07)

PROOF. Let (X, d) be a metric space and G_i $(i = 1, 2, ..., n)$ be a finite collection of open subsets of X.

Define $\quad H = \cap [G_i : i = 1, 2, ..., n]$.

To show H is open.

Let $x \in H$. Then by definition of H
$$x \in G_i \text{ for each } i = 1, 2, ..., n$$
Also each G_i is open, therefore there exists $r_i > 0$ such that
$$S(x, r_i) \subset G_i, \text{ for each } i$$
Let $\quad r = \min \{r_1, r_2, ..., r_n\}$
Then, $S(x, r) \subset S(x, r_i) \subset G_i, \qquad \forall i = 1, 2, ..., n$
$$\Rightarrow \quad S(x, r) \subset G_i, \qquad \forall i = 1, 2, ..., n$$
$$\Rightarrow \quad S(x, r) \subset \cap \{G_i : i = 1, 2, ..., n\}$$
$$\Rightarrow \quad S(x, r) \subset H.$$
Since x is arbitrary, therefore it is shown that to each $x \in H$, there exists $r > 0$, such that $S(x, r) \subset H$. Hence H is open.

THEOREM 6. *In a metric space (X, d), for every pair of distinct points $x, y \in X$, there exist disjoint open sets U and V such that $x \in U$ and $y \in V$.*
(MEERUT 1993; KANPUR 1996)

PROOF. Let (X, d) be a metric space and x, y be two distinct points of X such that $x \neq y$
Then clearly $\quad d(x, y) > 0$.

Define $\quad \varepsilon = \dfrac{1}{3} d(x, y)$, then clearly $\varepsilon > 0$.

Also, define $U = S(x, \varepsilon)$ and $V = S(y, \varepsilon)$, then $x \in U$ and $y \in V$.

Since every open sphere is open set, therefore U and V both are open sets.

Now to show U and V are distinct, *i.e.*, $U \cap V = \phi$.

Let if possible $U \cap V \neq \phi$ and $p \in U \cap V$.

Therefore, $\quad p \in U \cap V \Rightarrow p \in U$ and $p \in V$.

Now $\qquad p \in U \Rightarrow d(p, x) < \varepsilon \qquad \qquad (\because U \text{ is an open sphere})$
and $\qquad p \in V \Rightarrow d(p, y) < \varepsilon \qquad \qquad (\because V \text{ is an open sphere})$
By triangle inequality, we have
$$d(x, y) \leq d(x, p) + d(p, y)$$
$$= d(p, x) + d(p, y) \qquad \qquad [\because d(p, x) = d(x, p)]$$

$$< \varepsilon + \varepsilon = 2\varepsilon$$
$$\Rightarrow \quad d(x, y) < 2\varepsilon$$
$$\Rightarrow \quad d(x, y) < 2 \cdot \frac{1}{3} d(x, y)$$
$$\Rightarrow \quad d(x, y) < \frac{2}{3} d(x, y)$$

which is absurd.

Therefore, we have a contradiction

$$\therefore \quad U \cap V = \phi$$

Hence, we can find two disjoint open sets U and V such that $x \in U$ and $y \in V$.

THEOREM 7. *A subset of a metric space is open if and only if it is the union of a family of open sphere.*

(MEERUT 1998; KANPUR 1997, 2000,05; GARHWAL 2000,08)

PROOF. Let (X, d) be a metric space and A be any subset of X.

Let us first suppose A is open. To show A can be written as the union of family of open spheres.

If $A = \phi$, then it can be written as the union of empty family of open sphere.

If $A \neq \phi$, let $x \in A$, $\exists r_x > 0$ s.t. $S(x, r_x) \subset A$

Define $\bigcup_{x \in A} S(x, r_x)$, a family of open spheres. Now we claim that

$$A = \bigcup_{x \in A} S(x, r_x),$$

Let
$$\alpha \in A \Rightarrow \alpha \in S(\alpha, r_\alpha) \Rightarrow \alpha \in \bigcup_{x \in A} S(x, r_x)$$

Also, if
$$\beta \in \bigcup_{x \in A} S(x, r_x) \Rightarrow \beta \in S(x_0, r_{x_0}) \Rightarrow \beta \in S(x_0, r_{x_0}) \subset A$$

$$\Rightarrow \quad \beta \in A \Rightarrow A = \bigcup_{x \in A} S(x, r_x)$$

$$\Rightarrow \quad A = \bigcup \{S(x_i, r) : x_i \in A\}$$

Hence A can be written as the union of a family of open spheres.

Conversely, let A can be written as the union of open spheres, *i.e.*, let $A = \bigcup_{\lambda \in \Lambda} S_\lambda$,

Λ is index set.

To show A is open set.

Since, we know that every open sphere in a metric space is an open set. Therefore each S_λ is an open set.

Also, arbitrary union of open sets is again open.

$\Rightarrow \quad A$ is open.

Since $A = \bigcup \{S(x_i, r) : x_i \in A\}$.

Also the right hand side is the union of open sets.

Therefore, A is open.

THEOREM 8. *Every non-empty open set on the real line is the union of a countable collection of pairwise disjoint open intervals.*

(MEERUT 1991,95; KANPUR 1993,2005,08; GARHWAL 2002,05,08;BANGALURU 2002,07)

PROOF. Let G be an open subset of $\boldsymbol{R}$. Let $x \in G$.

Since G is open, there exists an open interval $S(x, r)$, centered at x such that $S(x, r) \subset G$.

Define I_x to be the union of all open intervals, which contain x and are contained in G. Then we have

(a) I_x is open interval containing x and contained in G.

(b) I_x contains each open interval which contains x and is contained in G.

(c) If y is any other point in I_x, then $I_x = I_y$.

If x and y are two distinct points of G, then either $I_x = I_y$ or $I_x \cap I_y = \phi$.

If $z \in I_x \cap I_y$, then we have $z \in I_x$ and $z \in I_y$

$\Rightarrow \quad I_x = I_z$ and $I_y = I_z \Rightarrow I_x = I_y$.

Now, let I be the collection of all distinct sets of the form I_x for points x belonging to G.

Then clearly, we have that I is the collection of open intervals and G is the union of this collection.

Now, to show I is countable.

Let G_i denotes the set of all rational point in G.

Then, obviously $G_i \neq \phi$.

Now, define a map f of G_i onto I such that for each $i \in G_i$, let $f(i)$ be the unique interval in I to which i belongs.

$\Rightarrow \quad G_i$ is countable.   ($\because G_i$ is a non-empty subset of the countable set

                                              $\mathbf{Q}$ of all rational numbers)

$\Rightarrow \quad I$ is countable.

Hence, we can say that every non-empty open set on the real line is the union of a countable collection of pairwise disjoint open intervals.

THEOREM 9.   *Let (X, d) be a metric space and A is any subset of X. Then A is closed if and only if its complement (i.e., $X - A$) is open.*

                          (MEERUT 1993,95,99; KANPUR 1994,2004; ALLAHABAD 2008; ROHILKHAND 2010)

PROOF.   Let (X, d) be a metric space. Let us first suppose A is closed.

To show its complement $(X - A)$ is open.

Let $x \in X - A$, then $x \notin A$.

Since A is closed and $x \notin A$.

$\Rightarrow \quad x$ is not the limit point of A.

Then by definition of limit point $\exists$ an open sphere $S(x, r)$ such that

$$S(x, r) \cap A = \phi$$

$\Rightarrow \quad S(x, r) \subset X - A$ for some $r > 0$.

Since $x \in X - A$, is arbitrary, therefore each point of $X - A$ is the centre of some open sphere which is contained in $X - A$.

$\Rightarrow \quad X - A$ is open.

Conversely, let $X - A$ is open. $x \in X$

To prove that A is closed, we have to show that A contains all its limit points.

For this it is sufficient to show that if $x \notin A \Rightarrow x$ is not a limit point of A.

So let $x \notin A$, then $x \in X - A$, also since $X - A$ is open, therefore, there exists an open sphere $S(x, r)$ which contained in $X - A$, i.e.,

$$S(x, r) \subset X - A$$

and $S(x, r) \cap A = \phi$ for some $r > 0$.

$\Rightarrow \quad x$ is not the limit point.

Hence, A is closed.

THEOREM 10. *In a metric space, every closed sphere is a closed set.*

(MEERUT 1990,92,2008; KANPUR 2002,09; GARHWAL 2000,02; ROHTAK 2006)

PROOF. Let (X, d) be a metric space. Consider a closed sphere $S(x_0, r)$ in X. To show $S(x_0, r)$ is a closed set.

For this, we shall show that the complement $S'(x_0, r)$ of $S(x_0, r)$ is open.

Let $x \in S'(x_0, r)$. Then $x \notin S(x_0, r)$.

Then by definition of open sphere $d(x, x_0) > r$.

Define $\qquad \rho = d(x, x_0) - r$. $\qquad\qquad\qquad\qquad$ ...(1)

Obviously $\quad \rho > 0$.

Now, we take an open sphere $S(x, \rho)$ of radius ρ centered at x.

We claim that $S(x, \rho) \subset S'(x_0, r)$

Let $\qquad\qquad y \in S(x, \rho) \Rightarrow d(x, y) < \rho$. $\qquad\qquad$ ...(2)

Consider $\quad d(x, x_0) \leq d(x, y) + d(y, x_0)$

or $\qquad\quad d(y, x_0) \geq d(x, x_0) - d(x, y)$

$\qquad\qquad\qquad\qquad > d(x, x_0) - \rho \qquad\qquad$ [using (2)]

$\qquad\qquad\qquad\qquad = d(x, x_0) - [d(x, x_0) - r] \qquad$ [using (1)]

$\qquad\qquad\qquad\qquad = r$

$\Rightarrow \qquad d(y, x_0) > r$

$\Rightarrow \qquad\qquad y \in S'(x_0, r)$

$\Rightarrow \qquad\qquad S(x, \rho) \subset S'(x_0, r) \qquad\qquad$ ($\because y$ is arbitrary)

$\Rightarrow \quad x$ is an interior point of $S'(x_0, r)$.

$\Rightarrow \quad S'(x_0, r)$ is open.

Hence, $S(x_0, r)$ is closed.

☞ REMARK

⟹ Let (X, d) be a metric space and S be any subset of X defined by

$\qquad S = \{x \in X : d(x, x_0) = r\}$.

where $r > 0$ and $x_0 \in X$. Then S is a closed set. $\qquad\qquad$ (KANPUR 2005)

THEOREM 11. *In a metric space (X, d), the intersection of an arbitrary family of closed sets is closed.* (MEERUT 1993,2003; RAVISHANKAR-2005,07; KANPUR 2004,07,09; GARHWAL 2010)

PROOF. Let (X, d) be a metric space and let $(H_\lambda : \lambda \in \Lambda)$ be an arbitrary collection of closed subset of X. Then to show $\bigcap [H_\lambda : \lambda \in \Lambda]$ is also a closed set.

Since H_λ is closed for each $\lambda \in \Lambda$.

$\Rightarrow \quad X - H_\lambda$ is open.

$\Rightarrow \quad \bigcup (X - H_\lambda)$ is open. $\qquad\qquad$ ($\because$ Arbitrary union of open sets is open)

$\Rightarrow \quad X - \bigcap \{H_\lambda : \lambda \in \Lambda\}$ is open. $\qquad\qquad$ (By De-morgan's law)

$\Rightarrow \quad \bigcap \{H_\lambda : \lambda \in \Lambda\}$ is closed. $\qquad$ ($\because$ Complement of an open set is closed)

Hence, the arbitrary intersection of closed sets is closed.

THEOREM 12. *In a metric space (X, d) the finite union of closed sets is closed.* (MEERUT 1991,97,2014,18; KANPUR 2004,06; BHOPAL 2006; GARHWAL 2010)

PROOF. Let (X, d) be a metric space and $H_i : i = 1, 2, ..., n$ be closed subsets of X. To show $\bigcup_{i=1}^{n} H_i$ is closed.

Since each H_i is closed.

$\Rightarrow \quad H_i'$ is open. $\qquad\qquad$ ($\because$ Complement of a closed set is open)

$\Rightarrow \quad \overset{n}{\underset{i=1}{\cap}} H_i'$ is open. $\qquad\qquad$ ($\because$ Finite intersection of open sets is open)

$\Rightarrow \quad \left[X - \overset{n}{\underset{i=1}{\cup}} H_i \right]$ is open. $\qquad\qquad$ (By De-Morgan's law)

$\Rightarrow \quad \overset{n}{\underset{i=1}{\cup}} H_i$ is closed.

THEOREM 13. *Show that in a discrete metric space, every set is open as well as closed.*

PROOF. Let A be any non-empty subset of a discrete metric space (X, d).

(i) If $A = \phi \Rightarrow A$ is open.

If $A \neq \phi$ and $x \in A$, where x is any arbitrary point of A.

and $S\left(x, \dfrac{1}{2}\right) = \{x\} \subseteq A \Rightarrow x$ is an interior point of A.

$\Rightarrow A$ is open.

Hence in a discrete metric space every set is open.

(ii) Let $F \subseteq X$.

Since in discrete metric space every set is open

$\Rightarrow (X - F)$ is open.

$\Rightarrow F$ is closed. $\qquad\qquad$ (Arbitrary subset of X)

$\Rightarrow$ Every set is closed in discrete metric space. Hence, the result.

☛ **REMARKS**

➠ In a metric space, we see that openness and closedness of a set is not like a 'door', *i.e.*, if t is open it doesn't mean it is not closed, or if a set is closed it doesn't mean it is not open. We have so many examples where sets are neither open nor closed and open as well as closed.

➠ The union of infinite number of closed sets may or may not be closed.

For example : Let $F_x = \left[0, \dfrac{n}{n+1} \right]$. Then, $\overset{\infty}{\underset{n=1}{\cup}} F_n = [0, 1[=$ semi open, which is not closed

as '1' is a limit point of $[0, 1[$ and $1 \notin [0, 1[$. $\qquad\qquad$ (MEERUT 2011 (SEM-I))

SOLVED EXAMPLES

EXAMPLE. *Show that the intersection of an infinite number of open sets is not necessarily open. by giving a suitable example.* (MEERUT 2011 (SEM-I))

SOLUTION. Let us consider an open interval collection $\left\{ \left] -\dfrac{1}{n}, \dfrac{1}{n} \right[: n \in \boldsymbol{N} \right\}$ in $\boldsymbol{R}$ with usual metric

$d(x, y) - |x - y|$.

So the intersection of this type collection will be

$$\cap \left\{ \left] -\dfrac{1}{n}, \dfrac{1}{n} \right[: n \in \boldsymbol{N} \right\} = \{0\}$$

which is not open as $0 \in \{0\}$ is not an interior point of $\{0\}$.

Hence in a metric space the intersection of an infinite collection of open set is not open.

4.3 NEIGHBOURHOOD

Let (X, d) be a metric space. A set N of X is said to be neighbourhood (nbd) of a point $p \in X$ if there exists an $\varepsilon > 0$ such that $S(x, \varepsilon) \subset N$

Or

Let (X, d) be metric space. A set N of X is said to be nbd of a point $p \in X$ iff $\exists$ an open set G s.t. $p \in G \subset N$

4.3.1 NBD OF SET

Let (X, d) be a metric space. A set N of x is said to be nbd of a set A iff $\exists$ an open set G s.t.
$$A \subset G \subset N$$

✎ ILLUSTRATIONS

* **On the real line :**
  (i) The open interval $]a, b[$ is a nbd of each of its points.
  (ii) R is a nbd of each of its points.
  (iii) The closed interval $[a, b]$ is a nbd of each of its points except the end points.
  (iv) The set of integers Z is not a nbd of any of its points.
  (v) The set of rational numbers Q is not a nbd of any of its points.

* Let (X, d) be a discrete metric space $x \in X$. $\{x\}$ is a nbd of x. Every super set of $\{x\}$ is also a nbd of x.

THEOREM 1. *Let (X, d) be a metric space and A be any subset of X. If N is the neighbourhood of A and $M \supset N$, then M is also a neighbourhood of A.*

In other words "Every super set of a neighbourhood of A is also a neighbourhood of A".

PROOF. Let (X, d) be a metric space and N be a nbd of $A \subset X$.

Since N is a nbd of A, therefore, by definition there exists an open set G such that
$$A \subset G \subset N \qquad ...(1)$$
Given that $\qquad N \subset M \qquad ...(2)$
From (1) and (2), we conclude that
$$A \subset G \subset M$$
$\Rightarrow$ M is a nbd of A.

THEOREM 2. *The intersection of a finite number of neighbourhood of A is also a neighbourhood of A.*

PROOF. Let (X, d) be a metric space and A is any subset of X.

Also let $N_1, N_2, ..., N_k$ are a finite number of neighbourhood of A, then to show that $\bigcap \{N_i : i = 1, 2, ..., k\}$ is also a neighbourhood of A.

🚀 FACTS : To THE POINT

▶ In a discrete metric space (X, d), any subset A of X is closed.

▶ In a metric space, closure of an open sphere $S(x, r)$ need not be equal to the corresponding closed sphere $\overline{S}(x, r)$.

▶ Let (X, d) be a metric space and $A \subseteq X$. Then following are equivalent:
  (i) A is dense in X
  (ii) The only closed set which contains A is X
  (iii) The only open set disjoint from A is ϕ
  (iv) A intersects every non-empty open set
  (v) A intersects every open sphere

▶ If G is an open set and $G \cap A = \phi$. Then $G \cap \overline{A} = \phi$.

▶ If $x \in S$ and $r > 0$. Then $S \cap S(x, r)$ is the open sphere and $S \cap \overline{S}(x, r)$ is the closed sphere with centre x and radius r in the subspace X.

▶ The set X itself, the empty $\phi \subset X$ and open spheres in X are open sets and X, ϕ and closed spheres in X are closed sets.

▶ Following conditions are equivalent:
  (i) Every subset of S that is open in S, is open in X
  (ii) S is open in X

▶ Let $x \in S$ and $U \subset S$. Then U is a nbd of x in S if and only if $U = S \cap N$ for some neighbourhood V of x in X.

Since each $N_i \, (i = 1, 2, ..., k)$ is a nbd of A, then by definition there exist open sets $G_i \, (i = 1, 2, ..., k)$ such that $A \subset G_i \subset N_i \, ; \, i = 1, 2, ..., k$.

Since $A \subset G_i$ for each $i = 1, 2, ..., k$.

Therefore $\quad A \subset \cap \{G_i : i = 1, 2, ..., k\}$

Also, $\qquad G_i \subset N_i \, \forall i = 1, 2, ..., k$

$\Rightarrow \quad \cap \{G_i : i = 1, 2, ..., k\} \subset \cap \{N_i : i = 1, 2, ..., k\}$

$\Rightarrow \qquad A \subset \cap \{G_i : i = 1, 2, ..., k\}$

$\qquad\qquad\qquad \subset \cap \{N_i : i = 1, 2, ..., k\}$ ...(1)

Since each G_i is open, therefore $\cap [G_i : i = 1, 2, ..., k]$ is open.

$[\because$ Finite intersection of open sets is again open$]$

Hence, from (1), we conclude that $\cap \{N_i : i = 1, 2, ..., k\}$ is also a neighbourhood of A.

☞ **REMARK**

⟹ The intersection of an infinite number of neighbourhood of a set A is not necessarily the nbd of A. For example, in the metric space (R, d) where $d(x, y) = |x - y|$, (i.e., usual metric)

$\left] -\dfrac{1}{n}, \dfrac{1}{n} \right[$ is a nbd of $\{0\}$ for each $n \in N$ but

$\cap \left\{ \left] -\dfrac{1}{n}, \dfrac{1}{n} \right[: n \in N \right\} = \{0\}$, which is not a nbd of $\{0\}$.

THEOREM 3. ***In a metric space every open sphere is a neighbourhood of each of points.***

PROOF. Let (X, d) be a metric space and let $S(x_0, r)$ be an open sphere centred at x_0 and of radius r let $y \in S(x_0, r)$.

Now to show that $S(x_0, r)$ is a neighbourhood of y, i.e.,we have to show that there exists an open sphere centred at p, which is contained in $S(x_0, r)$.

Now we have

$$y \in S(x_0, r) \Rightarrow d(y, x_0) < r$$

$\Rightarrow \qquad r - d(y, x_0) > 0$.

Let $\varepsilon = r - d(y, x_0)$, where $\varepsilon > 0$.

Now we have to show that $S(y, \varepsilon)$ contained in $S(x_0, r)$ let $x \in S(y, \varepsilon)$

$$d(x, y) < \varepsilon \Rightarrow d(x, y) < r - d(y, x_0)$$

$$d(x, x_0) \le d(x, y) + d(y, x_0)$$

$$< r - d(y, x_0) + d(y, x_0) < r$$

$\Rightarrow \qquad d(x, x_0) < r \Rightarrow x \in S(x_0, r)$

$\Rightarrow \quad S(y, \varepsilon) \subseteq S(x_0, r)$

$\Rightarrow \quad S(x_0, r)$ is a neigbourhood of y

$\Rightarrow \quad y$ is an arbitrary point of $S(x_0, r)$

$\Rightarrow \quad S(x_0, r)$ is a neighbourhood of each of its points.

Alternative Method: Since every open sphere is open in a metric space, hence

$y \in S(x_0, r) \subset S(x_0, r)$

As is an arbitrary point

$\Rightarrow \quad S(x_0, r)$ is nbd of each of its point.

THEOREM 4. ***A subset in a metric space is open if and only if it is a neighbourhood of each of its points.***

(MEERUT 1992 93, 95, 96, 2011, 18)

PROOF.

Let (X, d) be a metric space and A is any subset of X, i.e., $A \subset X$.

Let us first suppose A is open. To show it is a nbd of each of its points.

Let x be any arbitrary point of A, i.e., $x \in A$.

Since A is open, we can write $x \in A \subset A$.

$\Rightarrow \quad A$ is a nbd of x.

Conversely, let A is a nbd of each of its points. To show A is open.

Let $x \in A$, and A is a nbd of x. Then by definition of nbd, there exists an open set G_x such that $x \in G_x \subset A$.

Let $\qquad G = \bigcup\{G_x : x \in A\}$.

We claim that $G = A$.

If $x \in A$, then by definition $x \in \bigcup\{G_x : x \in A\} = G$

$\Rightarrow \qquad\qquad A \subset G$. $\hspace{4cm}$ ...(1)

Now, if $y \in G$, then $y \in G_x$ for some $x \in A$. But $G_x \subset A$ and hence $y \in A$

$\Rightarrow \qquad\qquad G \subset A$. $\hspace{4cm}$ ...(2)

From (1) and (2) we conclude that $A = G$.

Now since G is open. $\hspace{3cm}$ (Being the union of collection of open sets)

Hence, A is open.

THEOREM 5. *Let (X, d) be a metric space and let $x \in X$. If $\{N_i : i = 1, 2, ..., k\}$ are finite number of neighbourhood of x, then $\bigcap\{N_i : i = 1, 2, ..., k\}$ is also a nbd of X.*

PROOF.

Given that N_i $(i = 1, 2, ..., k)$ is a nbd of x for each i.

Therefore by definition of neighbourhood there exist open set G_i $(i = 1, 2, ..., k)$ such that

$$x \in G_i \subset N_i; \quad i = 1, 2, ..., k$$

$\Rightarrow \qquad x \in \bigcap\{G_i : i = 1, 2, ..., n\} \subset \bigcap\{N_i : i = 1, 2, ..., k\}.$ $\hspace{2cm}$ ...(1)

Since, each G_i is open, therefore $\bigcap\{G_i : i = 1, 2, ..., n\}$ is open.

$\hspace{4cm}$ ($\because$ Finite intersection of open sets is open)

Hence, from (1) we conclude that $\bigcap\{N_i : 1, 2, ..., k\}$ is a nbd of x.

THEOREM 6. *Let (X, d) be a metric space and A be a subset of X. A point $x \in X$ is a limit point of A if and only if every open sphere $S(x, r)$ centered at x contains infinitely many points of A.* (MEERUT-1995,98,2006; KANPUR-2003; GARHWAL 2004,06,07)

PROOF.

Let us first suppose every open sphere $S(x, r)$ centered at x contains infinitely many points of A. Then clearly, we can say that every open sphere $S(x, r)$ centered at x contains at least one point of A, other than x. Therefore, x is the limit point of A.

Conversely, let x is the limit point of A. To show every open sphere $S(x, r)$ centered at x contains infinitely many points of A.

Let if possible there exists a sphere $S(x, r)$ which contains only a finite number of points of A. Let $x_1, x_2, ..., x_n$ be those points of $S(x, r) \cap A$ which are distinct from x.

Define $r_1 = \min\{d(x, x_m) : 1 \leq m \leq n\}$. Then clearly $r_1 > 0$.

Then open sphere $S(x, r_1)$ contains no points of A distinct from x.

$\Rightarrow \quad x$ is not the limit point of A.

which is a contradiction.

Hence, every open sphere centered at x must contain infinitely many points of A.

THEOREM 7. *Let (X, d) be a metric space and $A \subset X$. A point $x \in X$ is an adherent point of A if and only if $d(x, A) = 0$.* (MEERUT 1990,92,94; KANPUR 2000,03,05)

PROOF. Let us first suppose $d(x, A) = 0$.

Then $d(x, A) = \inf \{d(x, y) : y \in A\} = 0$.

Let $S(x, r)$ be any sphere centered at x.

By definition of infimum, there exists a point $y_0 \in A$ such that

$$0 \le d(x, y_0) < r$$

$\Rightarrow \qquad y_0 \in S(x, r)$

$\Rightarrow \quad x$ is an adherent point of A.

Conversely, let us suppose x be an adherent point of A. To show $d(x, A) = 0$.

Since x is an adherent point of A, therefore, every open sphere centered at x must contain a point of A

$\Rightarrow \quad$ To each $r > 0$ there exists a $y \in A$ such that $0 \le d(x, y) < r$.

Since r is arbitrary, let $r \to 0$. (By taking r very small)

$\Rightarrow \qquad \inf \{d(x, y) : y \in A\} = 0$

$\Rightarrow \qquad\qquad d(x, A) = 0$.

THEOREM 8. *A subset A of a metric space X is closed if and only if $D(A) \subset A$, i.e., iff A contains all its limit points.* (MEERUT 2003,05)

PROOF. Let (X, d) be a metric space and $A \subset X$.

Let us first suppose A is closed. To show A contains all its limit point.

Since A is closed $\Rightarrow A'$ is open.

Let $x \in A'$, since A' is open, therefore by definition of open set there exists a nbd N of x such that $N \subset A'$.

Now, since $A \cap A' = \phi$.

$\Rightarrow \quad N$ contains no point of A.

$\Rightarrow \quad x$ is not a limit point of A.

$\Rightarrow \quad$ No point of A' can be a limit point of A.

$\Rightarrow \quad A$ contains all its limit points.

$\Rightarrow \quad D(A) \subset A$.

Conversely, let A contains all its limit point, *i.e.*, $D(A) \subset A$.

To show A is closed.

For this, we shall show that A' is open.

Let $x \in A'$, then $x \notin A$.

Since $D(A) \subset A$ and $x \notin A$, therefore $x \notin D(A)$

$\Rightarrow \quad x$ is not the limit point of A

$\Rightarrow \quad$ there exists a nbd N of x such that $N \cap A = \phi$

$\Rightarrow \quad N \not\subset A$

$\Rightarrow \quad N \subset A'$

$\Rightarrow \quad A'$ contains a neighbourhood of x.

Since x is arbitrary, therefore, we can say A' is a nbd of each of its points

$\Rightarrow \quad A'$ is open.

Hence, A is closed.

THEOREM 9. *Let A be any subset of a metric space (X, d) then derived set $D(A)$ of A is a closed set.*

PROOF. To show $D(A)$ is closed, we show that $D(A)$ contains all its limit point.

Let x be a limit point of $D(A)$ for all $r > 0$, the open sphere $S(x, r)$ contains infinitely many points of $D(A)$.

We know that each point of $D(A)$ is a limit point of A.

$\Rightarrow$ Every open sphere $S(x, r)$ must contain infinitely many points of A

$\Rightarrow$ x is a limit point of A

$\Rightarrow$ $x \in D(A)$

$\Rightarrow$ $D(A)$ contains all its limit points and so $D(A)$ is closed.

THEOREM 10. *Let A and B be subset of a metric space X. Then*

      (i) $A \subset B \Rightarrow D(A) \subset D(B)$

      (ii) $D(A \cap B) \subset D(A) \cap D(B)$ (MEERUT 2004)

      (iii) $D(A \cup B) = D(A) \cup D(B)$.   (MEERUT 2005)

PROOF.   (i) We have

$$A \subset B$$

To show $D(A) \subset D(B)$

Let $x \in D(A)$

$\Rightarrow x$ is the limit point of A

$\Rightarrow$ every nbd of x contains at least one point of A, other than x

$\Rightarrow$ every nbd of x contains at least one point of B, other than x

                                                                        ($\because A \subset B$)

$\Rightarrow x$ is the limit point of B

$\Rightarrow x \in D(B)$.

Since x is arbitrary, therefore $D(A) \subset D(B)$.

   (ii) To show $D(A \cap B) \subset D(A) \cap D(B)$.

Since we know that

$$A \cap B \subset A \Rightarrow D(A \cap B) \subset D(A) \qquad \text{[Using (i)]} \qquad \qquad ...(1)$$

and $A \cap B \subset B \Rightarrow D(A \cap B) \subset D(B)$.   ...(2)

From (1) and (2), we conclude that

$$D(A \cap B) \subset D(A) \cap D(B).$$

   (iii) To show $D(A \cup B) = D(A) \cup D(B)$

Let $x \notin D(A) \cup D(B)$

$\Rightarrow$ $x \notin D(A)$ and $x \notin D(B)$

$\Rightarrow x$ is not the limit point of A and x is not the limit point of B

$\Rightarrow$ $x \notin D(A \cup B)$.

Since x is arbitrary, therefore $D(A \cup B) \subset D(A) \cup D(B)$.   ...(1)

Also we know that

$$A \subset A \cup B \Rightarrow D(A) \subset D(A \cup B) \qquad \qquad \qquad ...(2)$$

and $B \subset A \cup B \Rightarrow D(B) \subset D(A \cup B)$.   ...(3)

Now, (2) and (3) gives

$$D(A) \cup D(B) \subset D(A \cup B) \qquad \qquad \qquad ...(4)$$

From (1) and (4) we conclude that

$$D(A) \cup D(B) = D(A \cup B).$$

4.4 EQUIVALENT METRICS

Two metrics d and d^* on the same set X are said to be equivalent iff every d-open set is d^*-open and every d^*-open set is d-open.

EXAMPLE. *Let (X, d) be a metric space and d^* is a mapping such that*

$d^* : X \times X \to R$ *defined as* $d^*(x, y) = \dfrac{Md(x, y)}{1 + d(x, y)}$, $M > 0$ *is also a metric for*

X. Also show that d and d^* are equivalent. (MEERUT 1992,96,97, 2016)

SOLUTION. We have already shown that $d^*(x, y) = \dfrac{Md(x, y)}{1 + d(x, y)}$ is metric for X as it follows the

four properties of metric.

Now it remains only to show that d and d^* are equivalent.

For this we show that d-open sphere centered at $x \in X$ contains a d^* open sphere centered at x and vice-versa.

Let $S(x, r)$, $r > 0$ be any d open sphere centred at $x \in X$.

Let $S^*(x, \rho)$, $\rho > 0$ be any d^* open sphere centered at $x \in X$ where $\rho = \dfrac{Mr}{1+r}$.

Now we have to show that $S^*(x, \rho) \subseteq S(x, r)$.

Here let $\quad x_1 \in S^*(x, \rho)$

$\Rightarrow \qquad d^*(x, x_1) < \rho$

$\Rightarrow \qquad \dfrac{Md(x, x_1)}{1 + d(x, x_1)} < \dfrac{Mr}{1+r}$

$\Rightarrow \quad d(x, x_1) + rd(x, x_1) < r + rd(x, x_1)$

$\Rightarrow \qquad d(x, x_1) < r \;\Rightarrow\; x_1 \in S(x, r)$

$\Rightarrow \qquad S^*(x, \rho) \subseteq S(x, r)$.

Now, it remains to show that $S(x, r) \subseteq S^*(x, \rho)$.

Here, $\qquad\qquad\qquad r = \dfrac{\rho}{M - \rho}$

Let $\quad x_1 \in S(x, r) \;\Rightarrow\; d(x, x_1) < r$

$\Rightarrow \qquad \dfrac{d^*(x, x_1)}{M - d^*(x, x_1)} < \dfrac{\rho}{M - \rho}$

$\Rightarrow \quad Md^*(x, x_1) - \rho d^*(x, x_1) < \rho M - \rho d^*(x, x_1)$

$\Rightarrow \qquad d^*(x, x_1) < \rho \;\Rightarrow\; x_1 \in S^*(x, \rho)$

$\Rightarrow \qquad S(x, r) \subseteq S^*(x, \rho)$.

Hence d and d^* are equivalent metric.

> **FACTS : TO THE POINT**
>
> ▶ In *R*, with usual metric, any closed interval $[a, b]$ is closed set.
> ▶ In *R*, with usual metric, $[a, b[$ and $]a, b]$ are neither closed nor open.
> ▶ *Z* is closed.
> ▶ *Q* is not closed in *R*.
> ▶ *I* is not closed in *R*.
> ▶ In *R* with usual metric, every singleton is closed.
> ▶ Every subset of a discrete metric space is closed.
> ▶ In *R*, with usual metric, the closed sphere $S(x, r) = [a - r, a + r]$.
> ▶ Any finite subset of a metric space is closed.

SOLVED EXAMPLES

EXAMPLE 1. *Find the closed and open spheres for the usual metric for R.*

SOLUTION. We know that the usual metric for **R** is defined by $d(x, y) = |x - y|$.

Let $x_0 \in \mathbf{R}$. Then the open sphere $S(x_0, r)$, centered at x_0, with radius r is given by

$$S(x_0, r) = \{x \in \mathbf{R} : |x - x_0| < r\}$$
$$= \{x \in \mathbf{R} : x_0 - r < x < x_0 + r\} \;=\;]x_0 - r, \; x_0 + r[$$

Hence, the open spheres on the real line are open intervals.

Similarly, the closed sphere with centre x_0 and radius r is the closed interval $[x_0 - r, \; x_0 + r]$.

EXAMPLE 2. *Consider the usual metric $d(x, y) = |x - y|$ for $[0, 1]$. Describe $S\left(\left]\frac{1}{2}, 1\right[\right)$ and $S\left[\frac{1}{2}, 1\right]$.*

(MEERUT–2016)

SOLUTION.
$$S\left(\left]\tfrac{1}{2}, 1\right[\right) = \left\{ x \in [0,1] : \left|x - \tfrac{1}{2}\right| < 1 \right\}$$
$$= \left\{ x \in [0,1] : \tfrac{1}{2} - 1 < x < \tfrac{1}{2} + 1 \right\}$$
$$= \left[x \in [0,1] : -\tfrac{1}{2} < x < \tfrac{3}{2} \right]$$
$$= [0, 1]. \qquad \text{(Don't go outside the given reason [0, 1])}$$

Similarly, $S\left[\frac{1}{2}, 1\right] = [0, 1]$.

EXAMPLE 3. *Let R be the set of all real number with usual metric $d(x, y) = |x - y|$. Find whether or not the given sets are open such that*

(i) $A = [0, 1[$
(ii) $B =]0, 1[$
(iii) $C =]0, 1]$
(iv) $D = [0, 1]$
(v) $E = \{1\}$
(vi) $F = [1, 2, 3]$.

SOLUTION. To show whether a set is open or not we check that for each point of given set an open interval of type described above exists or not and also contained in given set.

(i) Here $A = [0, 1[$
Let us choose a positive number r.
So the open interval $]0 - r, 0 + r[=]-r, r[\notin A$.
So no open sphere with centre '0' contained in A
$\Rightarrow \quad A$ is not open set.

(ii) Let us take a point x of set $B =]0, 1[$ and let $r = \min\{x - 0, 1 - x\}$ so it is obvious that $]x - r, x + r[\subseteq B$.
$\Rightarrow \quad B$ is an open set.

(iii) Here $C =]0, 1]$. Let us choose a positive number r.
So the open interval $]1 - r, 1 + r[\notin A$.
Thus no open sphere contained in A having radius $(1 + r)$.
$\Rightarrow \quad C$ is not open set.

(lv) $D = [0, 1]$ is not open set because D is an interval $]0 - r, 0 + r[\subset D$

(v) $E = \{1\}$ is not open set because it contains a single point 1 and so it is not possible to find $r > 0$ such that $]1 - r, 1 + r[\subseteq E$.

(vi) F is not open because it consist elements $\{1, 2, 3\}$ but it is not possible to find $r > 0$ such that $]1 - r, 1 + r[\subseteq F$ and other so it is not open.

EXAMPLE 4. *Describe the open spheres of unit radius about (0, 0) for each of the following matrices for R^2.*

(i) $d(z_1, z_2) = \sqrt{(x_1 - x_2)^2 + (y_1 - y_2)^2}$

(ii) $d(z_1, z_2) = \max.[|x_1 - x_2|, |y_1 - y_2|]$

where $z_1 = (x_1, y_1), z_2 = (x_2, y_2)$ are any two points of R^2.

SOLUTION. (i) Let d be the usual metric on R^2 and, here we are given that
$$d(z_1, z_2) = \sqrt{(x_1 - x_2)^2 + (y_1 - y_2)^2}$$
where $z_1 = (x_1, y_1)$ and $z_2 = (x_2, y_2)$ be any two points of R^2.

The open space with centre z_0 and radius r is given by

$$S(z_0, r) = \left\{ \sqrt{(x_1 - x_0)^2 + (y_1 - y_0)^2} \right\} < r.$$

Here, the open spheres $S(z_0, r)$ consists of all points of the cartesian plane which lie within the circle

$$(x - x_0)^2 + (y - y_0)^2 = r^2$$

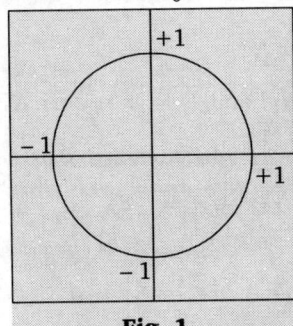

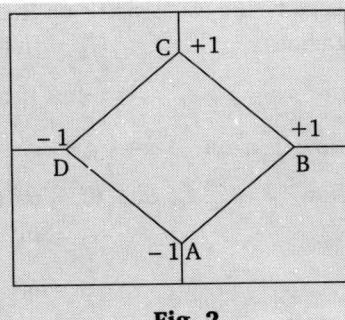

| Fig. 1 | Fig. 2 |

(ii) Here, we have

$$S(z_0, 1) = \{(x, y) \in \mathbf{R}^2 : d(z, z_0) < 1\}$$
$$= \{(x, y) \in \mathbf{R}^2 : \max [\{|x - 0|, |y - 0|\} < 1]\}$$
$$= \{(x, y) \in \mathbf{R}^2 : \max [\{|x|, |y|\} < 1]\}.$$

$\Rightarrow$ The open spheres $S(z_0, 1)$ is the integer space of the square bounded by the lines $x = 1$, $x = -1$, $y = 1$ and $y = -1$.

EXAMPLE 5. *Let (X, d) be any discrete metric space. Describe open sphere for d.*

(MEERUT–2018)

SOLUTION. The discrete metric space is defined by

$$d(x, y) = \begin{cases} 0 & \text{if } x = y \\ 1 & \text{if } x \neq y \end{cases}.$$

Let $x_0 \in X$ and $r > 1$, then we have

$$S(x_0, r) = \{x \in X : d(x, x_0) < r\} = X$$

($\because$ Since $d(x, x_0) = 0$ or 1, each of which is less than r so that $x \in X \Rightarrow x \in S(x_0, r)$)

If $r \leq 1$, then we have

$$S(x_0, r) = \{x \in X : d(x, x_0) < r\} = \{x_0\}$$

($\because d(x, x_0) = 0 < r$ and $d(x, x_0) = 1 \; \forall r$ if $x \neq x_0$)

EXAMPLE 6. *Show that every singleton set in R is closed for the usual metric d for R.*

SOLUTION. Let $a \in \mathbf{R}$. To show $\{a\}$ is closed.

Consider $\mathbf{R} - \{a\} =]-\infty, a[\; \cup \;]a, \infty[$
$$= \text{union of two open sets} = \text{open}$$

$\Rightarrow \quad \mathbf{R} - \{a\}$ is open.

$\Rightarrow \quad \{a\}$ is closed.

EXAMPLE 7. *Show that every closed interval is a closed set for the usual metric for R.*

SOLUTION. Consider a closed interval $[a, b]$ where $a \in \mathbf{R}$, $b \in \mathbf{R}$.

Also, consider $\mathbf{R} - [a, b] = [x \in \mathbf{R} : b < x < a]$
$$= [x \in \mathbf{R} : x < a, \; x > b]$$

$$= [x \in R : x < a] \cup [x \in R : x > b] =]-\infty, a[\cup]b, \infty[$$

(∵ Union of two open sets is again open)

which is open.

Hence, $[a, b]$ is closed.

EXAMPLE 8. *Give an example of two closed subsets A and B of real line R such that*

$$d(A, B) = 0, \quad A \cap B = \phi$$

(MEERUT 1994)

SOLUTION. Let d be the usual metric on R.

Define $$N_1 = [n+1 : n \in N]$$

$$A = [n : n \in N_1] \text{ and } B = \left[n + \frac{1}{n} : n \in N_1 \right]$$

Now if $n \in A \Rightarrow n + \frac{1}{n} \in B \Rightarrow B - A = \frac{1}{n}$, which tends to 0 as $n \to \infty$.

Now $$d(A, B) = \inf \{d(x, y) : x \in A, \ y \in B\}$$

$$= \inf . \{|x - y| : x \in A, \ y \in B\}$$

$$= \inf . \left\{ \frac{1}{n} : n \in A, \ n + \frac{1}{n} \in B \right\} = 0.$$

Hence, $$d(A, B) = 0 \text{ and } A \cap B = \phi.$$

EXAMPLE 9. *Give an example of a set which has*

   (i) no limit point   **(ii) exactly one limit point**

 (iii) exactly two limit points

  (iv) infinite number of limit points

   (v) every point of the set as its limit points.

SOLUTION. (i) The set of rational number Q, has no limit point.

(ii) The set $S = \left[\frac{1}{n} : n \in N \right]$ has exactly one limit point 0.

(iii) The set $S = \left\{ \frac{1}{2} - \frac{1}{2}, \frac{2}{3}, -\frac{2}{3} ... \right\}$ has exactly two limit points 1 and –1.

(iv) The open interval $]1, 2[$ has infinite number of limit points.

(v) The closed interval $[1, 2]$ is a set in which every point is its limit point.

EXAMPLE 10. *Give an example to show that in a metric space the union of an infinite collection of closed set is not necessarily closed.*

SOLUTION. Let us consider an infinite collection $F_n = \left[\frac{1}{n}, 1 \right]$, $n \in N$ of closed intervals for usual metric space (R, d) and we know that in usual metric every closed interval is a closed set.

$$\Rightarrow \quad F_n \text{ is a closed set in } (R, d).$$

Now, $$\cup \{F_n : n \in N\} = \{1\} \cup \left[\frac{1}{2}, 1 \right] \cup \left[\frac{1}{3}, 1 \right] \cup \left[\frac{1}{4}, 1 \right] \cup ... =]0, 1]$$

$\Rightarrow$ which is not closed.

Hence, union of an infinite collection of closed set is not necessarily closed.

EXAMPLE 11. *If A and B are closed sets of a metric space (X, d) then show that*

  (a) $A \cup B$ is closed   **(b) $A \cap B$ is closed.**

SOLUTION. (a) Since A and B are closed set of a metric space X

    $\Rightarrow A' = X - A$ and $B' = X - B$ both are open set of X

    $\Rightarrow A' \cap B'$ is an open set of X

$\Rightarrow$ $(A' \cap B')'$ is closed set of X

$\Rightarrow$ $(A')' \cup (B')'$ is closed set of X (By De-Morgan's law)

$(\because (A')' = A, (B')' = B)$

$\Rightarrow$ $A \cup B$ is closed set of X.

(b) Since A and B are closed set of X.

$\Rightarrow$ A' and B' both are open sets of X.

$\Rightarrow$ $A' \cup B'$ is an open sets of X

$\Rightarrow$ $(A' \cup B')'$ is closed set of X (By De-Morgan's law)

$\Rightarrow$ $(A')' \cap (B')'$ is closed set of X.

$\Rightarrow$ $A \cap B$ is closed set of X.

EXAMPLE 12. *Let (X, d) be a metric space and $S(x_0, r)$ the open sphere with centre x_0 and radius r. Let A be a subset of X which intersects $S(x_0, r)$ and has diameter less than r. Show that $A \subseteq S(2r, x_0)$*   (KANPUR 1996)

SOLUTION. Since $S(r, x_0)$ is open sphere with centre x_0 and radius r and A is any non-empty subset of X which intersects $S(x_0, r)$

Therefore, $A \cap S(x_0, r) \neq \phi$

Let $y \in A \cap S(x_0, r)$

Then $y \in A$ and $y \in S(x_0, r) \Rightarrow d(y, x_0) < r$.

Since $x \in A$ and x is arbitrary point of A and also $y \in A$.

$\Rightarrow$ $d(x, y) < r$ $(\because dia(A) < r)$

$d(x, x_0) \leq d(x, y) + d(y, x_0)$ (By triangular inequality)

$< r + r = 2r$

$\Rightarrow$ $d(x, x_0) < 2r \Rightarrow x \in S(x_0, 2r)$

$\Rightarrow$ $A \subseteq S(x_0, 2r)$.

EXAMPLE 13. *Let (X, d) be a metric space and let $p \notin S(x_0, r)$ where $x_0 \in X$ and $r > 0$ then show that*

$$d(p, S(x_0, r)) \geq d(x_0, p) - r$$

SOLUTION. Let $x \in S(x_0, r)$ where x is any arbitrary point of $S(x_0, r)$, then

$d(x_0, p) \leq d(x_0, x) + d(x, p)$

$d(x, p) \geq d(x_0, p) - d(x_0, x)$ ...(1)

But we consider that $x \in S(x_0, r)$

$\Rightarrow$ $d(x_0, x) < r$.   ...(2)

Using equation (2) in equation (1), we have

$d(x, p) \geq d(x_0, p) - r$

This condition holds for all $x \in S(x_0, r)$

$\Rightarrow$ $d(p, S(x_0, r)) \geq d(x_0, p) - r$.

EXAMPLE 14. *In the usual metric space (R, d) find the derived set of set Q of all rational number.*

SOLUTION. Here we have to show that every real number is limit point of set **Q**.

Let x be any real number and for given $\varepsilon > 0$.

Then $x - \varepsilon$ and $x + \varepsilon$ are two distinct real number and exist infinitely many points between them thus for given $\varepsilon > 0$ open interval $]x - \varepsilon, x + \varepsilon[$ contain at least one point of rational set **Q** other than x

$\Rightarrow$ x is limit point of **Q**

$\Rightarrow$ every real number is a limit point of Q

$\Rightarrow$ the set of the limits of Q is the set of all real number R so $D(Q) = R$.

EXAMPLE 15.   *On the real line R, show that the set of integers, Z has no limit point.*

(MEERUT 1992,93)

SOLUTION.   Let $\qquad\qquad\qquad x \in R$

Consider $\qquad\qquad]x - \varepsilon, x + \varepsilon[$

Then $\qquad\qquad\quad]x - \varepsilon, \ x + \varepsilon[\not\subset Z$

($\because$ By Denseness property of real numbers, "Between any two distinct real numbers there always exist infinitely many rational and irrational numbers")

EXAMPLE 16.   *Let (X, d) be a metric space and x, y be two distinct points of X. Prove that there exists two disjoint open spheres with centre x and y respectively.*

SOLUTION.   We have $x \neq y$ so let $d(x, y) = r > 0$

Consider the open spheres $S\left(x, \dfrac{r}{4}\right)$ and $S\left(y, \dfrac{r}{4}\right)$.

We claim that $S\left(x, \dfrac{r}{4}\right) \cap S\left(y, \dfrac{r}{4}\right) = \phi$.

Let if possible, $S\left(x, \dfrac{r}{4}\right) \cap S\left(y, \dfrac{r}{4}\right) \neq \phi$. Then $z \in S\left(x, \dfrac{r}{4}\right) \cap S\left(y, \dfrac{r}{4}\right)$.

Thus, $z \in S\left(x, \dfrac{r}{4}\right)$ and $z \in S\left(y, \dfrac{r}{4}\right) = \phi$

$\Rightarrow \quad d(x, z) < \dfrac{r}{4}$ and $d(y, z) < \dfrac{r}{4}$ $\qquad\qquad$ (By definition of open spheres)

Now using $d(x, y) \leq d(x, z) + d(z, y)$ we get

$r \leq \dfrac{r}{4} + \dfrac{r}{4} = \dfrac{1}{2}r$, which is a contradiction.

Hence, $S\left(x, \dfrac{r}{4}\right) \cap S\left(y, \dfrac{r}{4}\right) = \phi$.

EXAMPLE 17.   *Let $A = \left\{(a_n) : (a_n) \in l_2 \text{ and } \left[\displaystyle\sum_{n=1}^{\infty} a_n^2\right]^{1/2} < 1\right\}$. Prove that A is an open subset of l_2.*

SOLUTION.   Firstly, we shall prove that $A = S(0, 1)$ where $O = (0, 0, 0, \ldots)$.

Let $x \in A$. Then $\displaystyle\sum_{n=1}^{\infty} x_n^2 < 1$

Therefore, $\qquad d(x, 0) = \left[\displaystyle\sum_{n=1}^{\infty} (x_n - 0)^2\right]^{1/2} = \left[\displaystyle\sum_{n=1}^{\infty} x_n^2\right]^{1/2} < 1$

$\Rightarrow \qquad\qquad d(x, 0) < 1$

$\Rightarrow \qquad\qquad x \in S(0, 1)$

$\Rightarrow \qquad\qquad A \subseteq S(0, 1) \qquad\qquad\qquad\qquad\qquad\qquad\qquad\qquad ...(1)$

Further, let $y \in S(0, 1)$. Then $d(0, y) < 1$.

$\Rightarrow \qquad \left[\displaystyle\sum_{n=1}^{\infty} (y_n - 0)^2\right]^{1/2} < 1$

$\Rightarrow \qquad\qquad \left[\displaystyle\sum_{n=1}^{\infty} y_n^2\right]^{1/2} < 1$

$$\Rightarrow \qquad\qquad y \in A$$

$$\Rightarrow \qquad\qquad S(0,1) \subseteq A \qquad\qquad \ldots(2)$$

From (1) and (2), we conclude that $A = S(0, 1)$

Further, since we know that every open sphere is an open set. Hence, A is an open set.

EXAMPLE 18. *The set of integers Z has no limit point.*

SOLUTION. Let $x \in \mathbf{R}$. If x is an integer then $S = \left(x, \dfrac{1}{2}\right) = \left]x - \dfrac{1}{2}, x + \dfrac{1}{2}\right[$ does not contain any integer

other than x. Hence x is not the limit point of $\mathbf{Z}$. Now let x is not an integer and n be the integer which is closer to x. Choose r such that $0 < r < |x - n|$.

Then, $S(x, r) =]x - r, x + r[$ contains no integer.

Hence, x is not a limit point of $\mathbf{Z}$.

EXERCISE 4.1

1. Consider the usual metric
   $d(x, y) = |x - y|$, for $[0, 1]$

   Describe $S\left(\left[\dfrac{1}{4}, \dfrac{1}{4}\right[\right), S\left(\left[\dfrac{1}{4}, \dfrac{1}{4}\right]\right), \quad S\left(\left]0, \dfrac{1}{8}\right[\right)$

   and $S\left(\left]\dfrac{1}{16}, \dfrac{1}{16}\right[\right)$.

2. Show that the Cantor set c is not open.

3. Show that on the real line, every open interval is an open set.

4. Show that on the real line, every closed interval is a closed set.

5. Show that in a metric space, the intersection of two open spheres need not be an open sphere but it will always contain another open sphere.

   (MEERUT 1993)

6. If A and B are open sets of a metric space X, then show that
   (i) $A \cap B$ is open set.
   (ii) $A \cup B$ is also open set.

7. Let x_1 and x_2 be two distinct elements in the metric space (X, d), show that two disjoint open spheres will exist, which are centred at x_1 and x_2 respectively.

8. If A_1 is open set in metric space (X_1, d_1) and A_2 is open set in metric space (X_2, d_2), then if $X = X_1 \times X_2$, show that $A_1 \times A_2$ is open.

9. Show that the right half open interval $[a, b[$ is neither closed not open (that is say open) for the usual metric on $\mathbf{R}$.

10. In a metric space (X, d), the empty set ϕ and the whole space X are closed as well as open. (MEERUT 1991)

11. Show that every finite subset of $\mathbf{R}$ is closed with respect to the usual metric for $\mathbf{R}$.

12. Show that every subset of X containing x is a neighbourhood of x, where (X, d) is discrete space defined as
    $$d(x, y) = \begin{cases} 0 & \text{if } x = y \\ 1 & \text{if } x \neq y \end{cases}$$

13. Let (X, d) be any metric space and let
    (a) $d^*(x, y) = \dfrac{d(x, y)}{1 + d(x, y)}, \quad \forall x, y \in X$

    Show that d^* is also a metric on X and d and d^* are equivalent.

    (MEERUT 1992,94,95,2011(SEM-I); KANPUR 1995)

    (b) If $d^*(x, y) = \min \{1, d(x, y)\}$, then show that $d^*(x, y)$ is metric on X and also d and d^* are equivbalent.

    (MEERUT 1990,91,93,95)

14. Show that in a discrete metric space, every set is open as well as closed.

15. Show that the closed open interval $[a, b[$ is neither closed nor open for the usual metric on $\mathbf{R}$.

16. Give an example of a set which (i) is both open and closed, (ii) is neither open nor closed.

17. If A' denote the derived set of A, then find a set A such that
    (i) $A \cap A' = \phi$ (ii) $A = A'$
    (iii) $A \subseteq A$ (iv) $A \subseteq A'$

18. On the real line, show that the set of rational numbers $\mathbf{Q}$ has no limit point.

HINTS TO SELECTED PROBLEMS

1. Since we know that the Cantor set is the intersection of closed sets, therefore, being the intersection of closed set, it is again closed.

4. Use the following result
$$R - [a, b] = \{x \in R : x < a\}$$
$$= \{x \in R : x < a\} \cup \{x \in R : x > b\}$$

$$=]\infty, a[\cup]b, \infty[,$$

union of two open sets.

16. (i) ϕ and R

  (ii) Set of rational number.

18. Use the denseness property of real numbers.

ANSWERS

1. $\left]0, \dfrac{1}{2}\right[, \left]0, \dfrac{1}{2}\right[, \left]0, \dfrac{1}{8}\right[, \left]0, \dfrac{1}{8}\right[$ 8. (i) ϕ and R (ii) Q the set of rationals

9. (i) $A = \left[1, \dfrac{1}{2}, \dfrac{1}{3}, ..., \dfrac{1}{n}\right]$ (ii) $A = \{\text{set of closed interval}\}$

4.5 CLOSURE OF A SET

(GARHWAL 2000; KANPUR 2001)

In a metric space (X, d), the set of all adherent points of a set $A \subset X$ is called the closure of A and is denoted by $C(A)$ or $\overline{A}$.

ILLUSTRATIONS

* On the real line, every real number is an adherent point of the set $R \sim Q$ of all irrational numbers. For example, if p be any real number whatever any $\varepsilon > 0$ be given, then $]p - \varepsilon, p + \varepsilon[$ contain infinitely many irrational numbers and hence
$$]p - \varepsilon, p + \varepsilon[\cap (R \sim Q) \neq \phi$$
Hence, p is an adherent point of $R \sim Q$ and $(R \sim Q) = R$.

* On the real line, the closure of each of the sets $]0, 1[,]0, 1]$ and $[0, 1]$ is $[0, 1]$.

* The closure of the set of integers is Z itself.

* Let $A = \left\{\dfrac{1}{n} : n \in Z^+\right\}$. Then closure $A = \overline{A} = \left\{\dfrac{1}{n} : n \in Z^+\right\} \cup \{0\}$.

THEOREM 1. *In a metric space (X, d), the closure of a set $A \subset X$ is a closed superset of A.*

PROOF. Let p be any point of A and N be any nbd of p. Then $p \in N \cap A$, i.e., $N \cap A \neq \phi$. Since every nbd of p intersects A, therefore $p \in \overline{A}$. Since
$$p \in A \Rightarrow p \in \overline{A} \Rightarrow A \subset \overline{A}.$$
Now, $\overline{A}$ is a closed set $\Leftrightarrow X - \overline{A}$ is an open set.
$$\Leftrightarrow X - \overline{A} \text{ is a nbd of each of its points.}$$
Therefore, to show that $\overline{A}$ is a closed set, it is enough to show that $X - \overline{A}$ is a nbd of each of its points.

Consider an arbitrary point $q \in X - \overline{A}$.

Now, $q \in X - \overline{A} \Rightarrow q \notin \overline{A}$
$\Rightarrow$ there exists an $\varepsilon > 0$ such that $S(q, \varepsilon)$ contains no point of A.

It can be easily seen that no point of $S(q, \varepsilon)$ can be in $\overline{A}$. In fact if $r \in S(q, \varepsilon)$, then $S(q, \varepsilon)$ is a nbd of r containing no point of A and therefore $r \notin \overline{A}$. Since r is arbitrary, it follows that no point of $S(q, \varepsilon)$ is in $\overline{A}$. Thus $S(q, \varepsilon) \subset X - \overline{A}$, showing that $X - \overline{A}$ is a nbd of q. Again since q is arbitrary point of $X - \overline{A}$, therefore it follows that $X - \overline{A}$ is a nbd of each of its points.

Hence $\bar{A}$ is closed.

THEOREM 2. *In a metric space* (X, d)*, the closure of a set* $A \subset X$ *is the smallest, closed set containing* A*.*

PROOF. Since we have already shown that the closure of a set A is a closed super set of A, it only remains to show that $\bar{A}$ is the smallest among all closed superset of A, *i.e.*, if B is any closed set containing A, then $\bar{A} \subset B$.

Now, B is closed superset of $A \Rightarrow X - B$ is an open set disjoint from A.

$\Rightarrow$ no point of $X - B$ is an adherent point of A.

$\Rightarrow \quad \bar{A} \subset B$.

☞ **REMARK**

➠ The properties of $\bar{A}$, expressed in the above two theorems is sometimes taken as definition of closure of A. It gives us a description of $\bar{A}$ without using the notion of an adherent point of . Therefore:

(i) The closure of A, *i.e.*, $\bar{A}$ is the smallest closed set containing A.

(ii) The closure of A, *i.e.*, $\bar{A}$ is the intersection of all closed sets containing A, *i.e.*, $\bar{A} = \bigcap [F : F$ is closed and $A \subset F]$.

THEOREM 3. *A subset* A *of a metric space is closed if and only if* $\bar{A} = A$*.*

PROOF. Let (X, d) be a metric space.

Let us first suppose A is closed, then A is the smallest closed set containing $A \Rightarrow \bar{A} = A$.

Conversely if $\bar{A} = A$, then since $\bar{A}$ is closed, therefore A is closed.

THEOREM 4. *Let A be a subset of a metric space, then*

$$\bar{A} = A \cup D(A)$$

(MEERUT 2000,01,05)

PROOF. Let p be the limit point of A. Then p is an adherent point of A also.

i.e., $\qquad\qquad A' \subset \bar{A}$ ...(1)

Now every point of A is an adherent point of A

i.e., $\qquad\qquad A \subset \bar{A}$. ...(2)

From (1) and (2), we conclude that

$$A \cup A' \subset \bar{A}.$$ ...(3)

Now, let p be any point of $\bar{A}$.

If $p \in A$, then clearly $\bar{A} \subset A \cup A'$. Now, let $p \in \bar{A}$, $p \notin A$.

Let N be any nbd of p. Since $p \in \bar{A}$, $N \cap A \neq \phi$. Since $p \notin A$.

It follows that $p \notin N \cap A$ so that $N \cap A$ is a non-empty set, containing a point other than p. Since N is any nbd of p; it follows that $p \in A'$.

Therefore $\qquad\qquad \bar{A} \subset A \cup A'.$ ...(4)

From (3) and (4), we conclude that

$$\bar{A} = A \cup A' \Rightarrow \bar{A} = A \cup D(A)$$

THEOREM 5. *Let* (X, d) *be a metric space and* A, B *be subsets of X. Then*

(i) $\bar{\phi} = \phi$ $\qquad\qquad$ *(ii)* $A \subset \bar{A}$

(iii) $A \subset B \Rightarrow \bar{A} \subset \bar{B}$. (MEERUT 2005)

(iv) $\overline{A \cup B} = \bar{A} \cup \bar{B}$ (MEERUT 2005)

(v) $\overline{(A \cap B)} \subset \bar{A} \cap \bar{B}$ (MEERUT 1992)

(vi) $\bar{\bar{A}} = \bar{A}$ (MEERUT 1992)

PROOF.   (i)   Since ϕ is closed, therefore using theorem (3), we have
$$\bar{\phi} = \phi$$

      (ii)   By definition, $\bar{A}$ is the smallest closed set containing A.

      i.e., $A \subset \bar{A}$

     (iii)   From (ii), we have
$$B \subset \bar{B}.$$

Since $A \subset B$. Therefore $A \subset B \subset \bar{B}$

$\Rightarrow$ $A \subset \bar{B}$.

But $\bar{B}$ is closed.

$\Rightarrow$ $\bar{B}$ is the closed set containing A.

Also, $\bar{A}$ is the smallest closed set containing A.

Therefore, $\bar{A} \subset \bar{B}$.

     (iv)   Since we know that $A \subset A \cup B$ and $B \subset A \cup B$.

Therefore from (iii)
$$\bar{A} \subset \overline{A \cup B} \quad \text{and} \quad \bar{B} \subset \overline{A \cup B}$$

$\Rightarrow$ $\bar{A} \cup \bar{B} \subset \overline{A \cup B}$.   ...(1)

Now, $\bar{A}$ and $\bar{B}$ are closed sets, $\bar{A} \cup \bar{B}$ is also closed (being the union of two closed sets). Also $A \subset \bar{A}$ and $B \subset \bar{B}$.

$\Rightarrow$ $A \cup B \subset \bar{A} \cup \bar{B}$

$\Rightarrow$ $\bar{A} \cup \bar{B}$ is the closed set containing $A \cup B$.

But $\overline{A \cup B}$ is the smallest closed set containing $A \cup B$.

Therefore, $\overline{A \cup B} \subset \bar{A} \cup \bar{B}$ ...(2)

From (1) and (2), we conclude that
$$\overline{A \cup B} = \bar{A} \cup \bar{B}.$$

     (v)   Since we know that
$$A \cap B \subset A \Rightarrow \overline{A \cap B} \subset \bar{A}$$

and $A \cap B \subset B \Rightarrow \overline{A \cap B} \subset \bar{B}$

Hence, $\overline{A \cap B} \subset \bar{A} \cap \bar{B}$.

     (vi)   Since $\bar{A}$ is a closed set, therefore using theorem (3), we have $(\bar{\bar{A}}) = \bar{A}$

☛ **REMARK**

⇒ The inclusion in part (v) of the above theorem can not be replaced by an equality. For example, let

$A = \,]0, 1[\,, \ B = \,]1, 2[\,$. Then, $\bar{A} = [0, 1]$ and $\bar{B} = [1, 2]$

$\Rightarrow$ $\bar{A} \cap \bar{B} = \{1\}$.

Also, $A \cap B = \phi \Rightarrow (\overline{A \cap B}) = \phi$.

Thus we find that in this case $(\overline{A \cap B})$ is a proper subset of $\bar{A} \cap \bar{B}$.

THEOREM 6.   ***A finite set in a metric space has no limit point.***

PROOF.   Let (X, d) be a metric space and A be any finite subset of X.

As $p \in X$ is a limit point of any set B if every open sphere $S(x, r)$ contains an infinite number of points of B, other then p.

But A has finite number of points. Hence, A has no limit point.

THEOREM 7. *Let (X, d) be a metric space and $A \subset X$. Then $\bar{A} = \{x \in X : d(x, A) = 0\}$.*

(MEERUT 1990,92,94)

PROOF.   Let (X, d) be a metric space and A be any subset of X.

To show $\bar{A} = \{x \in X : d(x, A) = 0\}$, where $\bar{A} = A \cup D(A)$.

Let $d(x, A) = \varepsilon > 0$, then $x \notin A$

and $\{S(x, \varepsilon / 3) - [x]\} \cap A = \phi$.

Let $x \notin D(A)$, then

$$d(x, A) = \varepsilon \Rightarrow x \notin A, \; x \notin D(A) \Rightarrow x \notin A \cup D(A)$$
$$\Rightarrow x \notin \bar{A}. \qquad\qquad [\because \bar{A} = A \cup D(A)]$$

Now, $d(x, A) = 0 \Rightarrow x \in A$ and $\{S(x, \varepsilon) - [x]\} \cap A = \phi$
$$\Rightarrow x \in D(A).$$

Finally, $d(x, A) = 0 \Rightarrow x \in A$ or $x \in D(A)$
$$\Rightarrow x \in A \cup D(A).$$
$$\Rightarrow x \in \bar{A} \qquad\qquad [\because \bar{A} = A \cup D(A)]$$

and $d(x, A) = \varepsilon > 0 \Rightarrow x \notin \bar{A}$.

Furthermore, $d(x, A) \geq 0$

and hence the result follows.

THEOREM 8. *In a metric space (X, d), all finite sets are closed.*   (MEERUT 1992)

PROOF.   Since we know that every finite set can be written as the finite union of singletons and each singleton is closed. Also, finite union of closed sets is again closed.

Hence, finite set is always closed.

THEOREM 9. **(Bolzano-Weirstrass Theorem)** *Every bounded sequence in a metric space has a limit point.*

PROOF.   Let $< x_n >$ be a bounded sequence in a metric space (X, d). If there are only a finite number of distinct elements in $\{x_n : n \in N\}$ then at least one of them must occur infinitely often. If this element be x, then x is the limit point of $< x_n >$.

If the sequence $< x_n >$ contains infinitely many distinct element in X, then $S = \{x_n : n \in N\}$ is an infinite bounded set and therefore has a limit point, say y. Since y is the limit point of S, therefore every nbd of y must contains infinitely many points of S. This implies that the sequence $< x_n >$ must be frequently in every nbd of y. Hence, y is a limit point of $< x_n >$.

4.6 INTERIOR POINT AND INTERIOR OF A SET

Definition 1. *Let (X, d) be a metric space and A be any subset of X. A point $p \in A$ is said to be an interior point of A, if A is a nbd of p.*

Definition 2. *Let (X, d) be a metric space and $A \subset X$. A point $p \in A$ is said to be an interior point of A if there exist $\varepsilon > 0$ such that*
$$S(p, \varepsilon) \subset A.$$

Definition 3. *The set of all interior points of a set $A \subset X$ is called the interior of A and is denoted by A°.*

✎ ILLUSTRATIONS

* **On the real line:**
    (i) Every point of $]a, b[$ is an interior point of $]a, b[$ i.e., int $(]a, b[) =]a, b[$.
    (ii) No point of Z is an interior point of Z, i.e., $Z^\circ = \phi$.

(iii) The interior of the closed ray $[0, \infty[$ is the open ray $]0, \infty[$.

(iv) 0 is not an interior point of $[0, 1[$. Every point of $]0, 1[$ is an interior point of $[0, 1[$. The interior of $[0, 1[$ is $]0, 1[$.

(v) The set of rational numbers **Q** has no interior point, *i.e.*, $\mathbf{Q}^\circ = \phi$.

(vi) Let d be the discrete metric on a non-empty set X, and let $p \in X$. Then $\text{int}\{p\}=\{p\}$. In fact, for each set $A \subset X$, $\text{int } A = A$.

THEOREM 1.   **Let (X, d) be a metric space and A be a subset of X then**

     (i) A° is an open set.

     (ii) A° is the largest open set contained in A.

     (iii) A is open if and only if $A^\circ = A$.   (MEERUT 1996,2001)

PROOF.   (i) Let $x \in A^\circ$. Then   is an interior point of A.

     $\Rightarrow$ A is the nbd of x (By definition of interior point)

     $\Rightarrow$ $\exists$ an open set G such that $x \in G \subset A$.   (By definition of neighbourhood)

Since G is open, therefore G is a nbd of each of its points. Also, Since $G \subset A$, therefore A is also a nbd of each point of G

     $\Rightarrow$ Every point of G is an interior point of A.

     $\Rightarrow$ $G \subset A^\circ$

     $\Rightarrow$ to each $x \in A^\circ$, there exists an open set   such that $x \in G \subset A^\circ$

     $\Rightarrow$ A° is the nbd of each of its points.

     $\Rightarrow$ A° is open.

     (ii) Let G be an open subset of A and let $x \in G$. Then $x \in G \subset A$.

Now, since G is open, A is a nbd of x.

     $\Rightarrow$ x is an interior point of $A \Rightarrow x \in A^\circ$.

Therefore, $x \in G \Rightarrow x \in A^\circ$ $(\because G \subset A^\circ)$

     $\Rightarrow$ A° contains every open subset of A and it is, therefore, the largest open subset of A.

     (iii) Let us first suppose $A = A^\circ$.

Since A° is open, therefore A is also open.

Conversely, let A is open. To show $A = A^\circ$.

Then A is surely identical with A°.

Also, A° is the largest open subset of A. Hence, $A = A^\circ$.

THEOREM 2.   **Let X be a metric space and A be a subset of X. Then A° is equal to the set of all those points of A which are not limit point of A'.**

PROOF.   Let x be a point of A which is not a limit point of A'.

Then there exists a nbd N of x which contains no point of A'.

     $\therefore$ $N \subset A$

     $\Rightarrow$ A is also a nbd of x.

     $\Rightarrow$ $x \in A^\circ$.

Conversely, let $x \in A^\circ$.

Since A° is open.   (By definition)

     $\Rightarrow$ A° is a nbd of x.   ($\because$ Every open set is a nbd of each of its points)

Also, A^o contains no point of A'. $(\because A^o \subset A$ and A contains no point of $A')$

$\Rightarrow$ x is not a limit point of A'.

$\Rightarrow$ no point of A^o is a limit point of A'.

Hence, A' consists exactly those points of A which are not limit points of A'.

4.7 EXTERIOR POINTS AND EXTERIOR OF A SET

Definition 1. *Let A be a subset of a metric space* (X, d). *A point* $x \in X$ *is said to be exterior point of A if it is an interior point of the complement of A.*

Definition 2. *The set of all exterior points of A is called the exterior of A and is denoted by* $ext\,(A)$ *or* A^e.

☞ REMARKS

➠ Clearly $ext(A) = (A')^o$ and $ext.(A') = (A'')^o = A^o$.

➠ Also, $A \cap ext(A) = \phi$

➠ Since $ext.(A)$ is the interior of A' therefore $ext.(A)$ is the largest open set containing A'.

4.8 FRONTIER POINTS AND FRONTIER OF A SET

Definition 1. *Let* (X, d) *be a metric space. A point x of a metric space is said to be frontier point of a subset* $A \subset X$ *if and only if it is neither an interior nor exterior point of A.*

Definition 2. *The set of all frontier points of A is called the frontier of A and is denoted by* $F_r(A)$.

4.9 BOUNDARY POINTS AND BOUNDARY OF A SET

Definition 1. *Let* (X, d) *be a metric space. A point* $x \in X$ *is said to be boundary point of a subset* $A \subset X$ *if it is a frontier point of A and belongs to A.*

Definition 2. *The set of all boundary points of A is called the boundary of A and is denoted by* $b(A)$.

☞ REMARKS

➠ $b(A) \subset F_r(A)$

➠ A boundary point of A is essentially a point of A, whereas a frontier point may not belong to A. Thus every boundary point of A is also a frontier point of A but not conversely.

SOME MORE DEFINITIONS

Let (X, d) be a metric space and A, B be subsets of X. Then

1. A is said to be dense in B if $B \subset \overline{A}$

2. A is said to be dense in X or everywhere dense if $\overline{A} = X$.

3. A is said to be nowhere dense in X if $(\overline{A})^o = \phi$, *i.e.,* interior of the closure of A is empty.

4. A is said to be dense in itself if $A \subset D(A)$.

5. A is said to be perfect iff A is dense-in-itself and closed.

☞ REMARKS

➠ A is said to be everywhere dense iff every point of X is an adherent point of A.

➠ A is said to be perfect if $A = D(A)$.

<u>THEOREM 1.</u> **Let (X, d) be a metric space and A be any subset of X. Then**

(i) $A^o = \cup\{G : G$ is open, $G \subset A\}$

(ii) $ext(A) = \cup\{G : G$ is open, $G \subset A'\}$.

PROOF.
(i) Let $x \in A^\circ$, then by definition of interior point, A is a nbd of x.

$\Rightarrow$ there exists an open set G such that $x \in G \subset A$

$\Rightarrow$ $x \in \cup\{G : G$ is open, $G \subset A\}$.

Conversely, let $x \in \cup\{G : G$ is open, $G \subset A\}$. Then $x \in G$ for some open set G such that $G \subset A$.

$\Rightarrow$ A is a neighbourhood of x.

$\Rightarrow$ $x \in A^\circ$.

Therefore, $x \in A^\circ$ if and only if $x \in \cup\{G : G$ is open, $G \subset A\}$.

Since x is arbitrary, therefore

$\Rightarrow$ $A^\circ = \cup\{G : G$ is open, $G \subset A\}$.

(ii) By definition, we have $ext(A) = (A')^\circ$

But by (i), we have $(A')^\circ = \cup\{G : G$ is open, $G \subset A'\}$

which implies $ext(A) = \cup\{G : G$ is open, $G \subset A'\}$

THEOREM 2. *Let A be a subset of a metric space (X, d). Then a point $x \in X$ is an exterior point of A if and only if x is not an adherent point of A i.e. $x \in (\bar{A})'$.*

PROOF.
Let us first suppose x be an exterior point of A.

$\Rightarrow$ x is an interior point of A'.

$\Rightarrow$ A' is a nbd of x containing no point of A.

$\Rightarrow$ x is not an adherent point of A.

$\Rightarrow$ $x \in (\bar{A})'$.

Conversely, let $x \in (\bar{A})'$. To show x is an exterior point of A.

Now, $x \in (\bar{A})' \Rightarrow x$ is not an adherent point of A.

$\Rightarrow$ there exists a nbd N of x which contains no points of A

$\Rightarrow$ $x \in N \subset A'$

$\Rightarrow$ A' is a nbd of x

$\Rightarrow$ x is an interior point of A'

$\Rightarrow$ x is an exterior point of A.

THEOREM 3. *Let (X, d) be a metric space and $A \subset X$. Then, a point $x \in X$ is a frontier point of A if and only if every nbd of x intersects both A and A'.*

PROOF.
Here, we have

$x \in F_r(A) \Leftrightarrow x \notin A^\circ$ and $x \notin ext(A) = (A')^\circ$

$\Leftrightarrow$ neither A nor A' is a nbd of x

$\Leftrightarrow$ no nbd of x can be contained in A or in A'

$\Leftrightarrow$ every nbd of x intersects both A and A'.

☛ **REMARK**

⟼ $F_r(A) = F_r(A')$.

THEOREM 4. *Let A be any subset of a metric space (X, d). Then A°, $ext(A)$ and $F_r(A)$ are disjoint and $X = A^\circ \cup ext(A) \cup F_r(A)$. Also, $F_r(A)$ is a closed set.*

PROOF.
Let (X, d) be a metric space. Since we know that $ext(A) = (A')^\circ$.

Also, we know that $A^\circ \subset A$, $(A')^\circ \subset A'$ and $A \cap A' = \phi$.

$\Rightarrow$ $A^\circ \cap ext(A) = A^\circ \cap (A')^\circ = \phi$.

Now, $x \in F_r(A) \Leftrightarrow x \notin A^o$ and $x \notin ext(A)$

$\Leftrightarrow x \notin A^o \cup ext(A)$

$\Leftrightarrow x \in [A^o \cup ext(A)]'$.

Therefore, $F_r(A) = [A^o \cup ext(A)]'$

$\Rightarrow \quad F(A) \quad A = \phi$ and $F_r(A) \cap ext(A) = \phi$

and $x = A^o \cup ext(A) \cup F_r(A)$.

Finally, since A^o and $ext(A)$ both are open.

Therefore, $(A^o)'$ and $(ext(A))'$ both are closed.

Hence, $F_r(A)$ is closed.

THEOREM 5. **Let (X, d) be a metric space and A, B be any subsets of X. Then**

(i) $X^o = X$, $\phi^o = \phi$ **(ii) $A^o \subset A$**

(iii) $A \subset B \Rightarrow A^o \subset B^o$ **(iv) $(A \cap B)^o = A^o \cap B^o$**

(v) $A^o \cup B^o \subset (A \cup B)^o$ (vi) $A^{oo} = A^o$ (MEERUT 1992; KANPUR 1992,97,98)

PROOF.   (i) Since we know that X and ϕ both are open sets.

Therefore, $X^o = X$ and $\phi^o = \phi$ $(\because A$ is open iff $A^o = A)$

(ii) Let $x \in A^o \Rightarrow x$ is an interior point of A

$\Rightarrow A$ is a nbd of x.

$\Rightarrow \quad x \in A \Rightarrow A^o \subset A$

(iii) Let $A \subset B$ and let $x \in A^o$

$\Rightarrow x$ is an interior point of A

$\Rightarrow x$ is an interior point of B $(A \subset B)$

$\Rightarrow x \in B^o \Rightarrow A^o \subset B^o$.

(iv) Since we know that

$\left. \begin{array}{l} A \cap B \subset A \Rightarrow (A \cap B)^o \subset A^o \\ A \cap B \subset B \Rightarrow (A \cap B)^o \subset B^o \end{array} \right\}$ (using (iii))

This implies

$(A \cap B)^o \subset A^o \cap B^o$.   ...(1)

Now, let $x \in A^o \cap B^o$

$\rightarrow \quad x \subset A^o$ and $x \in B^o$

$\Rightarrow x$ is an interior point of A and x is an interior point of B.

$\Rightarrow x$ is an interior point of $A \cap B$.

$\Rightarrow x \in (A \cap B)^o$.

Therefore $(A^o \cap B^o) \subset (A \cap B)^o$ ...(2)

From (1) and (2), we conclude that

$A^o \cap B^o = (A \cap B)^o$.

(v) We know that

$A \subset A \cup B \Rightarrow A^o \subset (A \cup B)^o$

and $B \subset A \cup B \Rightarrow B^o \subset (A \cup B)^o$.

This implies that $A^{\circ} \cup B^{\circ} \subset (A \cup B)^{\circ}$.

(vi) Since A° is always open and we know that A is open if and only if $A^{\circ} = A$.

Apply the above result for A°, we get

$$(A^{\circ})^{\circ} = A^{\circ}$$

$$\Rightarrow \qquad A^{\circ\circ} = A^{\circ}.$$

☞ **REMARK**

⟹ In result (v), $A^{\circ} \cup B^{\circ} \neq (A \cup B)^{\circ}$.

For example : If $A = [0,1[$ and $B = [1,2[$ $\Rightarrow$ $A^{\circ} =]0,1[$ and $B^{\circ} =]1,2[$

$\Rightarrow$ $A^{\circ} \cup B^{\circ} =]0,1[\; \cup \;]1,2[\; = \;]0,2[-[1]$

also, $A \cup B = [0,2[$ $\Rightarrow$ $(A \cup B)^{\circ} =]0,2[$.

Therefore, $(A \cup B)^{\circ} \neq A^{\circ} \cup B^{\circ}$.

THEOREM 6.   *Let (X, d) be a metric space and let A, B be subsets of X. Then*

      (i) $ext(X) = \phi$, $ext(\phi) = X$ (ii) $ext(A) \subset A'$

   (iii) $ext(A) = ext[(ext(A))']$

   (iv) $A \subset B \Rightarrow ext[B] \subset ext[A]$ (MEERUT 1992)

   (v) $A^{\circ} \subset ext[ext(A)]$ *(vi) $ext(A \cup B) = ext(A) \cap ext(B)$*

PROOF.   (i) $ext(X) = (X')^{\circ} = \phi^{\circ} = \phi$ ($\because X' = \phi$)

                    $ext(\phi) = (\phi')^{\circ} = X^{\circ} = X$ ($\because \phi' = X$)

     (ii) $ext(A) = (A')^{\circ} \subset A'$.

    (iii)   Here, we have

$$ext[(ext(A))'] = ext(A'^{\circ})'] = ext(A'^{\circ\prime})$$

$$= [(A'^{\circ})']^{\circ} = (A'^{\circ\prime\prime})^{\circ} = (A'^{\circ})^{\circ} \qquad (\because A'' = A)$$

$$= A'^{\circ\circ} = A'^{\circ} = (A')^{\circ} = ext(A) \qquad (\because A^{\circ\circ} = A^{\circ})$$

    (iv)   Now $A \subset B \Rightarrow B' \subset A'$

$$\Rightarrow \quad (B')^{\circ} \subset (A')^{\circ}$$

$$\Rightarrow ext(B) \subset ext(A).$$

    (v)   We have $ext(A) \subset A' \Rightarrow ext(A') \subset ext(ext(A))$

$$\Rightarrow A^{\circ} \subset ext(ext(A)) \qquad\qquad (\because A^{\circ} = ext(A'))$$

    (vi)   We have

$$ext(A \cup B) = [(A \cup B)']^{\circ} = [A' \cap B']^{\circ}$$

$$= A'^{\circ} \cap B'^{\circ} = ext(A) \cap ext(B).$$

THEOREM 7.   *Let (X, d) be a metric space and $A \subset X$. Then*

   (i) Closure of the complement of A is the complement of the interior of A.

   (ii) The interior of A is the complement of the closure of the complement of A.

   (iii) The closure of A is the complement of the interior of the complement of A.

PROOF. (i) Since we know that
$$A^\circ = A'^{-\,\urcorner}.$$
Therefore, $(A^\circ)' = (A'^{-\,\prime})' = A'^{-\,\prime\prime} = A^{\circ\prime} = A'^{-}$

(ii) Using(i) $A^{\circ\,\prime} = A'^{-}$

$\Rightarrow$ $A^{\circ\,\prime\prime} = A'^{-\,\urcorner}$ (By taking complement)

$\Rightarrow$ $A^\circ = A'^{-\,\urcorner}$

(iii) Using (i) $A'^{-} = A^{\circ\,\prime}$.

Replacing A by A', we get
$$(A')'^{-} = (A')^{\circ\,\prime}$$
$$A''^{-} = A'^{\circ\,\prime} \Rightarrow \bar{A} = A^{\circ\,\prime}.$$

THEOREM 8. *Let (X, d) be a metric space and $A \subset X$. Then $\bar{A} = A^\circ \cup F_r(A)$.*

PROOF. Since we know that $\bar{A} = \bigcap[F : F$ is closed and $A \subset F]$

$\Rightarrow$ $(\bar{A})' = \bigcup[F' : F'$ is open and $F' \subset A'] = ext\,(A)$

$\Rightarrow$ $(\bar{A})'' = [ext\,(A)]'$ (By taking complement)

$\Rightarrow$ $\bar{A} = A^\circ \cup F_r(A)$.

Hence, we have $\bar{A} = A^\circ \cup F_r(A)$.

THEOREM 9. *Let (X, d) be a metric space and let A, B be subsets of X. Then*

  (i) $F_r(A) = \bar{A} \cap A'^{-} = \bar{A} - A^\circ$ (MEERUT 2004)

  (ii) $A^\circ = A - F_r(A)$ (MEERUT 2005)

  (iii) $[F_r(A)]' = A^\circ \cup A'^\circ$ (MEERUT 2005)

  (iv) $F_r(A^\circ) \subset F_r(A)$

  (v) $F_r(\bar{A}) \subset F_r(A)$

  (vi) $F_r(A \cup B) \subset F_r(A) \cup F_r(B)$

  (vii) $F_r(A \cap B) \subset F_r(A) \cup F_r(B)$

PROOF. Let (X, d) be a metric space and A and B be any two subset of X.

(i) We know that
$$F_r(A) = [A^\circ \cup ext(A)]' = A'^\circ \cap [ext(A)]'$$ (By De-Morgan's law)
$$= A'^{-\,\prime\prime\prime} \cap A^{-\,\prime\prime} = A'^{-} \cap \bar{A}$$ ($\because A'' = A$)

Now, $\bar{A} \cap A'^{-} = \bar{A} - A'^{-\,\urcorner}$ ($\because A \cap B' = A - B$)
$$= \bar{A} - A^\circ$$

which implies
$$F_r(A) = \bar{A} \cap A'^{-} = \bar{A} - A^\circ.$$

(ii) Since $F_r(A) = \bar{A} - A^\circ$ (According to (i))

$\Rightarrow$ $A - F_r(A) = A - (A' - A^\circ) = A^\circ$

(iii) $[F_r(A)]' = (\bar{A} \cap A'^{-})' = A^{i\prime} \cup A'^{-\,\urcorner}$.

Now using $A'^\circ = (A')^\circ = (A')'^{-\,\urcorner} = A''^{-\,\urcorner} = A^{-\,\prime}$

$\Rightarrow$ $F_r(A)' = A'^\circ \cup A^\circ = A^\circ \cup A'^\circ$.

(iv) Now, $F_r(A^o) = A^{o-} \cap A^{o-\prime} = A^{o-} \cap A^{\prime-\prime\prime-}$ $\qquad (\because A^{\prime-\prime} = A^o)$

$\qquad = A^{o-} \cap A^{\prime-} \subset F_r(A)$.

(v) $\qquad F_r(\overline{A}) = \overline{\overline{A}} \cap A^{-\prime-} = \overline{A} \cap A^{-\prime-}$ $\qquad (\because \overline{\overline{A}} = \overline{A})$

Now, $\qquad A \subset \overline{A} \Rightarrow (\overline{A})^\prime \subset A^\prime$

$\qquad\qquad \Rightarrow A^{-\prime-} \subset A^{\prime-}$.

Hence, $F_r(\overline{A}) \subset \overline{A} \cap A^{\prime-} = F_r(A)$

$\qquad\qquad \Rightarrow F_r(\overline{A}) \subset F_r(A)$.

(vi) Consider

$\qquad F(A \cup B) \quad \overline{(A \cup B)} \cap (A \cup B)^\prime$

$\qquad = (\overline{A} \cup \overline{B}) \cap (A^\prime \cap B^\prime)^- \subset (\overline{A} \cup \overline{B}) \cap (A^{\prime-} \cap B^{\prime-})$

$\qquad = [\overline{A} \cap (A^{\prime-} \cap B^\prime)] \cup [\overline{B} \cap (A^{\prime-} \cap B^{\prime-})]$

$\qquad = [(\overline{A} \cup A^{\prime-}) \cap B^\prime] \cup [(\overline{B} \cap B^{\prime-}) \cap A^\prime] \subset F_r(A) \cup F_r(B)$.

(vii) $F_r(\overline{A \cup B}) = \overline{(A \cup B)} \cap (A \cap B)^{\prime-}$

$\qquad \subset (\overline{A} \cap \overline{B}) \cap (A^\prime \cup B^\prime)^- \qquad\qquad (\because \overline{A \cup B} \subset \overline{A} \cap \overline{B})$

$\qquad = (\overline{A} \cap \overline{B}) \cap (A^{\prime-} \cup B^{\prime-})$

$\qquad [(\overline{A} \cap \overline{B}) \cap A^{\prime-}] \cup [(\overline{A} \cup \overline{B}) \cap B^{-\prime}]$

$\qquad = [(\overline{A} \cap A^{\prime-})] \cup [(\overline{A} \cup \overline{B}) \cap B^{-\prime}]$

$\qquad = [(\overline{A} \cap A^{\prime-}) \cap \overline{B}] \cup [\overline{A} \cup (\overline{B} \cup B^{\prime-})]$

$\qquad = [F_r(A) \cap \overline{B}] \cup [\overline{A} \cap F_r(B)] \subset F_r(A) \cup F_r(B)$.

THEOREM 10. *Let (X, d) be a metric space and A be any subset of X. Then*

(i) *If A is open, $F_r(A) = \overline{A} - A^o$*

(ii) *$F_r(A) = \phi$ if and only if A is open as well as closed.*

(iii) *A is open if and only if $A \cap F_r(A) = \phi$.*

(iv) *A is closed if and only if $F_r(A) \subset A$.*

PROOF. (i) We know that $F_r(A) = \overline{A} - A$.

If A is a open then $A = A^o$.

$\therefore \qquad F_r(A) = \overline{A} - A^o$.

(ii) Let us first suppose $F_r(A) = \phi$.

Thus we have

$\qquad F_r(A) = \phi \Rightarrow \overline{A} - A^o = \phi \Rightarrow \overline{A} \subset A^o \Rightarrow \overline{A} \subset A \qquad (\because A^o \subset A)$

$\qquad\qquad \Rightarrow D(A) \subset A \qquad\qquad\qquad\qquad (\because \overline{A} = A \cup D(A))$

$\qquad\qquad \Rightarrow A$ is closed.

Also, $\quad F_r(A) = \phi \Rightarrow \overline{A} - A^o = \phi \Rightarrow \overline{A} \subset A^o$

$\qquad\qquad\qquad \Rightarrow A \cup D(A) \subset A^o \Rightarrow A \subset A^o$.

But $A^o \subset A$. Therefore $A^o = A$.

$\qquad\qquad\qquad \Rightarrow A$ is open. $\qquad (\because A$ is open if and only if $A^o = A)$

Hence, $F_r(A) = \phi$ then A is closed as well as open.

Conversely, let A is open as well as closed. To show

$$F_r(A) = \phi$$

Since we know that

$$F_r(A) = \bar{A} - A^\circ \qquad \qquad ...(1)$$

Also, if A is closed, then

$$\bar{A} = A$$

and if A is open, then

$$A^\circ = A.$$

Put the above values in (1), we get $F_r(A) = \phi$.

(iii) We know that

$$F_r(A) = \bar{A} \cap A'^-.$$

If A is open, then A' is closed.

$\Rightarrow \qquad A'^- = A'.$

Now, $A \cap F_r(A) = A \cap [\bar{A} \cap A'^-] = A \cap [\bar{A} \cap A']$
$$= [A \cap \bar{A}] \cap A' = A \cap A' = \phi$$

Conversely, let $A \cap F_r(A) = \phi$.

Then, $A \cap F_r(A) = \phi \Rightarrow A \cap (\bar{A} \cap A'^-) = \phi$

$\Rightarrow (A \cap A') \cap A'^- = \phi$

$\Rightarrow A \cap A'^- = \phi \Rightarrow A \subset A'^{-\prime} \Rightarrow A \subset A^\circ$

Now, $A^\circ \subset A$ gives $A^\circ = A$.

$\Rightarrow \qquad A$ is open.

(iv) Let A is closed. Then $\bar{A} = A$. $\qquad \qquad (\because A$ is closed iff $\bar{A} = A)$

$\Rightarrow \qquad F_r(A) = \bar{A} \cap A'^- = A \cap A'^- \subset A.$

Conversely, let $F_r(A) \subset A$. Then $A \cup F_r(A) = A$.

But $A \cup F_r(A) = \bar{A} \Rightarrow A = \bar{A} \Rightarrow A$ is closed. $\qquad (\because A$ is closed iff $\bar{A} = A)$

SOLVED EXAMPLES

EXAMPLE 1. *Consider the usual metric space (R, d). Find the closure of the following sets*

(i) $A = \left[\dfrac{1}{n} : n \in N\right]$ (ii) *Z, the set of integers*

(iii) *Q, the set of rationals*   (iv) $]0, 1[$

SOLUTION. (i) Since 0 is the only limit point of A.

$\therefore \qquad D(A) = [0]$

and so, $\qquad \bar{A} = \left\{\dfrac{1}{n} : n \in N\right\} \cup [0].$

(ii) Since $D(\mathbf{Z}) = \phi$, therefore $\bar{\mathbf{Z}} = \mathbf{Z} \cup \phi = \mathbf{Z}$

(iii) $\qquad \qquad D(\mathbf{Q}) = \mathbf{R}$

$\therefore \qquad \bar{\mathbf{Q}} = \mathbf{Q} \cup D(\mathbf{Q}) = \mathbf{Q} \cup \mathbf{R} = \mathbf{R}.$

(iv) $\qquad \qquad \overline{]0, 1[} = [0, 1].$

EXAMPLE 2. *Give an example to show that in a metric space, it is not necessary that $\overline{(A \cap B)} = \overline{A} \cap \overline{B}$.*

SOLUTION. Let us consider the usual metric space (R, d)

$$A = [0, 1[, \ B =]1, 2]$$
$$A \cap B = \phi$$
$$\overline{(A \cap B)} = \overline{\phi} = \phi$$
$$\overline{A} = [0, 1] \text{ and } \overline{B} = [1, 2]$$
$$\overline{A} \cap \overline{B} = \{1\}.$$

Hence, $\overline{(A \cap B)} \neq \overline{A} \cap \overline{B}$.

EXAMPLE 3. *Define a dense set and give an example.*

SOLUTION. By definition of dense set we have a subset A of metric space (X, d) is said to be dense in X iff $\overline{A} = X$.

Example of dense set. Let (R, d) be a usual metric space on R and Q be set of rational number then $Q \subseteq R$

$$\overline{Q} = Q \cup D(Q) = Q \cup R = R$$
$$\Rightarrow \qquad \overline{Q} = R$$

$\Rightarrow \quad Q$ is dense in R.

EXAMPLE 4. *Define a nowhere dense set by giving a suitable example.*

SOLUTION. Let A be non-empty subset A of X, then A is said to be nowhere dense in X iff $(\overline{A})^o = \phi$.

Let (R, d) be usual metric on R and Z is the set of integers $Z \subseteq R$.

$$\overline{Z} = Z \cup D(Z) = Z \cup \phi = Z$$

$\Rightarrow \quad (\overline{Z})^o = Z^o = \phi \Rightarrow Z$ is nowhere dense in R.

EXAMPLE 5. *Show that Cantor's set E is a non-dense set.* (MEERUT 1995)

SOLUTION. Since we know that Cantor set is always a closed set.

$$\therefore \qquad \overline{E} = E \qquad\qquad (\because A \text{ is closed iff } \overline{A} = A)$$
$$\Rightarrow \qquad (\overline{E})^o = (E)^o = \bigcup [G \subset [0, 1] : G \subset E] = \phi$$
$$\Rightarrow \qquad (\overline{E})^o = \phi$$

$\Rightarrow \quad E$ is non-dense.

EXAMPLE 6. *In any metric space, show that $(\overline{A})' = (A')^o$.* (MEERUT 1990)

SOLUTION. Let (X, d) be a metric space and A be any subset of X.

Now, $(\overline{A})' = X - \overline{A} = X -$ intersection of all closed set F_i

$$= X - \cap F_i, \text{ where } F_i \text{ is closed and } A \subset F_i$$
$$= \cup [X - F_i], \text{ where } X - F_i \text{ is open.}$$

Since $X - F_i \subset X - A$.

Therefore, $(\overline{A})' =$ Union of open subsets of $X - A = A' = (A')^o$.

EXAMPLE 7. *Consider the usual metric $d(x, y) = |x - y|$ on R and find (i) interior, (ii) exterior, (iii) frontier, (iv) and boundary of each of the following subsets of R*

(a) $A =]0, 1[$ *(b)* $B = [0, 1[$

(c) $C = \left\{ \dfrac{1}{n} : n \in N \right\}$ *(d)* N

SOLUTION. (a) (i) Since A is an open set.

$\Rightarrow A$ is a nbd of each of its points.

$\Rightarrow$ every point of A is an interior point of A.

$\Rightarrow$ $A^o = A =]0, 1[$.

(ii) $A' =]-\infty, 0[\cup [1, \infty[$.

Here A' is a nbd of each of its points except 0 and 1.

$\Rightarrow$ $ext(A) = (A')^o =]-\infty, 0[\cup]1, \infty[$.

(iii) $F_r(A) = [A^o \cup ext(A)]' = [0, 1]$.

$\Rightarrow F_r(A)$ contains two points 0 and 1.

(iv) $b(A) = \phi$ ($\because$ no frontier point is a point of A)

(b) (i) $B^o =]0, 1[$ (same as in (a))

  (ii) $B' =]-\infty, 0[\cup [1, \infty]$

$\Rightarrow$ $ext(B) = (b')^o =]-\infty, 0[\cup]1, \infty[$

(iii) $F_r(B) = [0, 1]$

(iv) $b(B) = \{0\}$ ($\because$ 0 is the only frontier point which belong to B)

(c) (i) Since C can not be a nbd of any of its point $\dfrac{1}{n}$, $n = 1, 2, \ldots$, therefore $\exists \, \varepsilon > 0$ such that

$$\left]\dfrac{1}{n} - \varepsilon, \; \dfrac{1}{n} + \varepsilon\right[\subset D$$

$\Rightarrow$ no point of D can be its interior point so that $D^o = \phi$.

(ii) Clearly $D' = R - D$ is a nbd of each of its point except 0. Hence $ext(D) = (D')^o = R - \{D \cup (0)\}$.

(iii) $F_r(D) = [D^o \cup ext(D)]' = D \cup \{0\}$

(iv) $b(D) = D$. Since all the points of D are frontier point of D.

(d) (i) Since N is not a nbd of any of its point $\Rightarrow N^o = \phi$.

(ii) Since N has no limit points, therefore we have $\bar{N} = N$

      $\Rightarrow N$ is a closed set.

      $\Rightarrow N'$ is open set.

      $\Rightarrow ext(N) = (N')^o = N' = R - N$

(iii) $F_r(N) = [N^o \cup ext(N)]' = [\phi \cup ext(N)]' = [ext(N)]' = (N')' = N$.

(iv) $b(N) = N$ (because every point of N is a frontier point of N).

EXAMPLE 8.   *Show that a subset F of a metric space X is closed iff $\{x \in X : d(x, F) = 0\} \subseteq F$.*

SOLUTION.   Let (X, d) be a metric space and $F \subseteq X$.

We know that $\bar{F} = \{x \in X_1 : d(x, F) = 0\}$.

Let F be closed. Now we have to show that $\{x \in X : d(x, F) = 0\} \subseteq F$.

Since we have F is closed.

$\Rightarrow$ $\bar{F} = F \Rightarrow \bar{F} \subseteq F$ (by the definition of closure)

$\Rightarrow$ $\{x \in X : d(x, F) = 0\} x \subseteq F$.

Conversely, let $\{x \in X : d(x, F)\} = 0 \subseteq F$, then we have to prove that F is closed.

Since we have $\{x \in X : d(x, F) = 0\} \subseteq F \Rightarrow \bar{F} \subseteq F$.

But $F \subseteq \bar{F} \Rightarrow F = \bar{F}$.

4.10 SEPARABLE SPACES

4.10.1 SEPARABLE SPACES

A metric space (X, d) is said to be separable if X contains a countable dense subset, *i.e.*, there exists a countable subset A of X such that $\bar{A} = X$. (MEERUT 1995)

For Example: The usual metric space (R, d) is separable, because the set Q of rational numbers is a countable dense subset of R.

4.10.2 BASES FOR THE NEIGHBOURHOOD SYSTEM OF A POINT

Let (X, d) be a metric space and $N(x)$ denotes the family of all neighbourhoods of a point $x \in X$. A sub family $\mathfrak{B}(x)$ of $N(x)$ is said to be base for $N(x)$ if to each $N \in N(x)$ there exist $B \in \mathfrak{B}(x)$ such that $B \subset N$.

☛ REMARKS

➠ Here, $\mathfrak{B}(x)$ is said to be local base at x or a fundamental system of neighbourhoods of x.

➠ The set of all open interval in R form a base for the family of open subsets of R.

4.10.3 BASES FOR THE OPEN SETS OF A METRIC SPACE   (MEERUT 2002,03)

Let (X, d) be a metric space and let G be the family of all open subsets of a metric space (X, d). Then a subfamily $\mathfrak{B}$ of G is said to be base if for each point $x \in X$ and each nbd N of x $\exists B \in \mathfrak{B}$ such that $x \in B \subset N$.

4.10.4 FIRST AND SECOND COUNTABLE SPACES   (MEERUT 2002,03)

(a) A metric space (X, d) is said to satisfy the first axiom of countability if each point $x \in X$ possesses a countable local base. Such a space is said to be first countable.

(b) A metric space (X, d) is said to satisfy the second axiom of countability if there exist a countable local base for G, where G denotes the family of all open subsets of X. Such a space is said to be second countable space.

☛ REMARKS

➠ A second countable space is also said to be completely separable space.

➠ The usual metric (R, d) is a first as well as a second countable spaces.

➠ Every metric space (X, d) is first countable.   (MEERUT 2002,03)

4.10.5 INDUCED METRIC

Let (X, d) be a metric space and Y be a proper subset of X. Let d^* denote the restriction of d on $Y \times Y$, *i.e.*, $d^*(x, y) = d(x, y)$.

Whenever x and y are points of Y. Then d^* is a metric for Y called the induced metric. Then (Y, d^*) is said to be subspace of (X, d).

4.10.6 HEREDITARY PROPERTY

A property of a metric space is said to be hereditary if and only if every sub space of that space has that property.

THEOREM 1.   *Every metric space is first countable.*

PROOF.   Let (X, d) be a metric space and $x \in X$ be arbitrary. We have to show that X is first countable. For this, we will have to show that there exists a countable local base for $x \in X$.

Let us consider the collection of open spheres $\mathcal{B}(x) = \left\{ S\left(x, \frac{1}{n}\right) : n \in \mathbf{N} \right\}$

We claim that $\mathcal{B}(x)$ forms a countable local base if there exists an open set G such that $\exists$ a nbd $\mathbf{N}$ of x.

$\Rightarrow \qquad x \in G \subseteq \mathbf{N}$

Since G is open, so by definition, we have for $\varepsilon > 0$

$$S(x, \varepsilon) \subset G \subset N$$

Let us choose n so large such that $\frac{1}{n} < \varepsilon$

Then, clearly

$$S\left(x, \frac{1}{n}\right) \subset S(x, \varepsilon) \subset G \subset N$$

$\Rightarrow$ Every nbd of x contains a number of G.

$\Rightarrow \quad \mathcal{B}(x)$ forms a countable local base at x.

Hence, (X, d) is first countable.

THEOREM 2. **A metric space is separable if and only if it is second countable.**

(KANPUR 1991,94; AGRA 1993)

PROOF. Let us first suppose (X, d) be a separable space.

Define $\qquad A = \{x_n : n \in \mathbf{N}\}$

Let us suppose A be a countable dense subset of X.

$\Rightarrow \qquad\qquad \bar{A} = X$.

We have to show that the collection

$$\mathcal{B} = \left[S\left(x_n, \frac{1}{m}\right) : n \in \mathbf{N}, \ m \in \mathbf{N} \right]$$

is a countable dense for the family G if all open subsets of X.

Now, since $\mathcal{B}$ is a countable family of open countable subsets.

$\Rightarrow \quad \mathcal{B}$ is a countable set.

Further, since $S\left(x_n, \frac{1}{m}\right)$ is an open set. $\qquad$ ($\because$ Every open sphere is an open set)

Thus, it remains to prove that $\mathcal{B}$ is a base for G.

Let $x \in G$ be an open set, then for given $\varepsilon > 0$, we can write

$$S(x, \varepsilon) \subset G$$

Choose a positive integer $m_0 > \frac{2}{\varepsilon}$.

Now, $\bar{A} = X \Rightarrow$ Every point of X is an adherent point of A

$\qquad$ ($\because$ By definition the closure of a set is the set of all adherent points of this set)

By definition of adherent point $\exists$ an open sphere centred at x must contain a point of A.

Accordingly $\exists$ an open sphere $S\left(x, \frac{1}{m_0}\right)$ must contain a point of A, say x_i

But $\qquad\qquad x_i \in S(x, \frac{1}{m_0})$

$\Rightarrow \qquad\qquad d(x, x_i) < \frac{1}{m_0}$

$\Rightarrow$ $\qquad\qquad x \in S(x_i, \frac{1}{m_0})$

Now, we want to show that

$$S(x_i, \frac{1}{m_0}) \subset S(x, \varepsilon)$$

Let $y \in S(x_i, \frac{1}{m_0}) \Rightarrow d(y, x_i) < \frac{1}{m_0}$ $\qquad$ (By definition of open sphere)

Now, $\qquad d(x, y) \le d(x, x_i) + d(x_i, y)$

$$< \frac{1}{m_0} + \frac{1}{m_0} = \frac{2}{m_0} < \varepsilon$$

$\Rightarrow \qquad\qquad\qquad y \in S(x, \varepsilon)$

Since y is arbitrary, therefore

$$S(x_i, \frac{1}{m_0}) \subset S(x, \varepsilon)$$

$\Rightarrow \qquad\qquad x \subset S(x_i, \frac{1}{m_0}) \subset S(x, \varepsilon) \subset G$

$\Rightarrow$ For every point x of G, there exists a member $S(x_i, \frac{1}{m_0})$ of $\mathfrak{B}$ containing x and contained in G.

$\Rightarrow$ $\mathfrak{B}$ is a base for G.

Hence, (X, d) is second countable.

Conversely, let us suppose that (X, d) be a second countable metric space.

We have to show that (X, d) is separable.

By definition of second countable space $\exists$ a countable base β such that $\mathfrak{B} = [B_n : n \in \mathbf{N}]$

For each $n \in \mathbf{N}$, select a point $b_n \in B_n$. Therefore, we get a set

$$B = [b_n : n \in \mathbf{N}]$$

Clearly B is countable.

Now, we wish to show that B is dense in X.

Let $x \in X$ be arbitrary and G be any open nbd of x. Now, since β is a base, so there exists at least $B_{n_0} \in \mathfrak{B}$ such that $x \in B_{n_0} \subset G$

Now, by definition of B, $b_{n_0} \in B$ such that $b_{n_0} \in B_{n_0} \subset G$

$\Rightarrow \qquad G$ contains a point of B.

Since G is an open nbd of x.

$\Rightarrow \qquad$ Every nbd of x contains a point of B.

$\Rightarrow \qquad x$ is an adherent point of B so that $x \in \bar{B}$.

Thus, we show that $x \in X \Rightarrow x \in \bar{B}$

$\Rightarrow \qquad\qquad\qquad X \subset \bar{B}$

But $\qquad\qquad\qquad \bar{B} \subset X$

$\therefore \qquad\qquad\qquad \bar{B} = X$

$\Rightarrow \qquad B$ is a countable dense subset of X.

Hence, (X, d) is separable.

THEOREM 3. **Let (X, d_1) be a metric space and (Y, d_2) be a subspace of (X, d_1). A subset A of Y is d_2-open if and only if there exists a d_1-open subset G of X such that $A = G \cap Y$.**

PROOF. Let $S^*(y_0, r)$ be the d_2-open sphere in Y with centre y_0 and radius r such that

$$S^*(y_0, r) = \{y \in Y : d_2(y, y_0) < r\}$$

Clearly, $\qquad S^*(y_0, r) = Y \cap S(y_0, r)$

where, $S(y_0, r)$ is d_1-open sphere in X.

Let B be any d_2-open subset of Y, then to each $y \in B$, there exists a d_2-open sphere $S^*(y, r(y))$ with centre y and radius $r(y) > 0$ such that

$$S^*(y, r(y)) \subset B \qquad\qquad \text{(radius } r(y) \text{ depends on } y\text{)}$$

Then,
$$\begin{aligned} B &= \cup [S^*(y, r(y)) : y \in B] \\ &= \cup [Y \cap S(y, r(y)) : y \in B] \\ &= Y \cap [\cup \{S(y, r(y))\} : y \in B] \\ &= Y \cap G, \text{ where } G = \cup [S(y, r(y)) : y \in B] \end{aligned}$$

Since G is the union of open sphere $\Rightarrow$ G is open.

Conversely, let us write $B = G \cap Y$

where G is a d_1-open subset of X. Let $y \in B$ be arbitrary, then y is also a point of G since $B = G \cap Y$.

Since G is d_1-open $\Rightarrow \exists$ an open sphere $S(y, r(y))$ such that $S(y, r(y)) \subset G$

Thus, $S^*(y, r(y)) = Y \cap S(y, r(y))$

$$\subset Y \cap G = B$$

Therefore to each $y \in B$, there exists a d_2-open sphere centered at y and contained in B. Hence B is d_2-open subset of Y.

THEOREM 4. **Let (X, d_1) be a metric space and (Y, d_2) be any subspace of $[X, d_1]$. Then every subset B of Y (which is open in Y) be open in X iff Y is open in X.**

PROOF. Let us first suppose every subset B of Y, which is open in Y be also open in X. To show Y is open in X.

Since Y is open in Y and $Y \subset X$, therefore Y is open in X.

Conversely, let B be a subset of Y which is open in Y and Y be open in X. We have to show that B is open in X, therefore, there exists a subset G, open in X such that $B = G \cap Y$.

Now, since Y is open in X, B is open in X

$$(\because \text{ it is the intersection of two open sets}).$$

Hence, B is open in Y.

THEOREM 5. **Let (Y, d_2) be a subspace of a metric space (X, d) and A be a subset of Y. Then**

(i) A is closed in Y if and only if $\exists$ a closed set H in X such that
$A = H \cap Y$

(ii) $\bar{A}^* = \bar{A} \cap Y$, where $\bar{A}^*$ and $\bar{A}$ respectively denote d_2-closure and d_1-closure of A.

(iii) A subset N^* of Y is a d_2-nbd of a point $y \in Y$ if and only if $N^* = N \cap Y$ for some d_1-nbd N of y.

(iv) A point $y \in Y$ is a d_2 limit point of a subset A of Y if and only if it is a d_1-limit point of A.
Also, $D^*(A) = D(A) \cap Y$
where $D^*(A)$ and $D(A)$ respectively denote d_2-derived and d_1-derived set of A.

PROOF.

(i) Let A be a subset of Y.

Then, A is closed in Y

$\Leftrightarrow$ $Y - A$ is open in Y.

$\Leftrightarrow$ $Y - A = G \cap Y$, G is open in X.

$\Leftrightarrow$ $A = Y - (G \cap Y)'$

$\Leftrightarrow$ $(Y - G) \cap (Y - Y)$

$\Leftrightarrow$ $A = Y - G$

$\Leftrightarrow$ $A = Y \cap G'$

$\Leftrightarrow$ $A = Y \cap H$, where $H = G'$ is closed in X ($\because G$ is open in X)

(ii) By definition of closure, we can write

$$\bar{A}^* = \cap \{ H^* : H^* \text{ is closed in } Y \text{ and } A \subset H^* \}$$
$$= \cap \{ H \cap Y : H \text{ is closed in } X \text{ and } A \subset H \cap Y \}$$
$$= \cap \{ H \cap Y : H \text{ is closed in } X \text{ and } A \subset H \}$$
$$= \{ \cap [H : H \text{ is closed in } X \text{ and } A \subset H] \} \cap Y$$
$$= \bar{A} \cap Y$$

(iii) By definition of nbd, we can write

$y \in A^* \subset N^* \Rightarrow \exists$ a d_1-open set A such that $y \in A^* = A \cap Y \subset N^*$

Now, let $N = N^* \cup A$, then N is a d_1-nbd of y, since A is a d_1-open set such that $y \in A \subset N$.

Now, $N \cap Y = (N^* \cup A) \cap Y = (N^* \cap Y) \cup (A \cap Y)$

$$= N^* \cup (A \cap Y) \qquad\qquad\qquad (\because N^* \subset Y)$$
$$= N^* \qquad\qquad\qquad\qquad\qquad (\because A \cap Y \subset N^*)$$

Conversely, let us suppose that $N^* = N \cap Y$, then there exists a d-open set A such that $y \in A \subset N$

$\Rightarrow y \in A \cap Y \subset N \cap Y = N^*$

Since $A \cap Y$ is a d_2-open set, therefore, N^* is a d_2-nbd of y.

(iv) We have y is a d_2-limit point of A.

$\Leftrightarrow (N^* - \{y\}) \cap A \neq \phi \quad \forall d_2$-nbds N^* of y

$\Leftrightarrow [N \cap Y - \{y\}] \subset A \neq \phi \quad \forall d_1$-nbds N of y

$\Leftrightarrow [N - \{y\}] \cap A \neq \phi \quad \forall d_1$-nbds N of y

$\Leftrightarrow y$ is the d_1 limit point of A and hence $D^*(A) = D(A) \cap Y$.

THEOREM 6.   **Let (Y, d_2) be a subspace of a metric space (X, d_1) and A be a subset of Y. Then**

(i) $A^o \subset A^{o*}$ **(ii) $Fr^*(A) \subset Fr(A)$**

where A^{o*} and A^o denote respectively d_2-interior and d_1-interior of A and $Fr^*(A)$ and $Fr(A)$ denote respectively d_2-frontier and d_1-frontier of A.

PROOF.

(i) Let $x \in A^o$.

$\Rightarrow x$ is a d_1-interior point of A

$\Rightarrow A$ is a d_1-nbd of x.

$\Rightarrow A \cap Y$ is a d_1-nbd of x.

$\Rightarrow A$ is a d_2-nbd of x.   ($\because A \subset Y \Rightarrow A \cap Y = A$)

$\Rightarrow x$ is d_2-interior point of A.

$\Rightarrow x \in A^{o^*}$

Since x is arbitrary.

Hence, $A^o \subset A^{o^*}$.

(ii) Let $y \in Fr^*(A)$

$\Rightarrow$ y is d_2-frontier point of A.

$\Rightarrow$ Every d_1-nbd of y intersect both A and $Y - A$

$\Rightarrow$ $N \cap Y$ intersects both A and $X - A$ for every d_1-nbd N of y.

$\Rightarrow$ Every d_1-nbd N of y intersects both A and $X - A$

$\Rightarrow$ y is d_1-frontier point of A

$\Rightarrow$ $y \in Fr(A)$

Since y is arbitrary. Hence, $Fr^*(A) \subset Fr(A)$.

THEOREM 7. *Every subspace of a second countable space is second countable, i.e., the property of being second countable space is hereditary.*

PROOF. Let (X, d_1) be a second countable space.

Let us define a countable base $\mathfrak{B} = [B_n : n \in N]$ for the family G of all d_1-open subset of X. Now let (Y, d_2) be a subspace of (X, d_1). Then clearly the family $\mathfrak{B}^* = [B_n \cap Y : n \in N]$ is a base for the family G^* of d_2-open subset of Y. Also, $\mathfrak{B}^*$ is countable.

Hence, we conclude that (Y, d_2) is second countable.

THEOREM 8. *In a metric space, separability is a hereditary property.*

(or)

Every subspace of a separable metric space is separable.

PROOF. Let (X, d_1) be a separable space and (Y, d_2) be a subspace of (X, d_1). We have to show that (Y, d_2) be separable. Using theorem-2 which states that every separable space is second countable, we can say that (Y, d_2) is second countable. Again using Theorem 2, we can say that (Y, d_2) is separable.

☞ **REMARK**

⇒ If (Y, d_2) be a subspace of a metric space (X, d_1), then every subset B of Y which is open (closed) in Y be open (closed) in Y, it is necessary and sufficient that Y be open (closed) in X.

✎ ILLUSTRATIONS

* **The real line R is separable.**

Since the set Q of rational numbers is a countable dense subset of R (Using denseness property of real numbers). Thus, R is separable.

* **The complex number space C is separable.**

Let us write $C_0 = \{x + iy : x \in Q, y \in Q\}$

Since Q is countable $\Rightarrow$ $Q \times Q$ is also countable.

Further, since $(x, y) \to x + iy$ is a one-to-one correspondence between $Q \times Q$ and C_0

$\Rightarrow$ C_0 is countable.

Now, to show that C_0 is a dense subset of C

⚓ FACTS : TO THE POINT

▶ A is dense in X if and only if each non-empty open set in X contains a point of A.

▶ $]-\infty, -1[\cup]-1, 1[\cup]1, \infty[$ is dense in R.

▶ A non-empty subspace S of a seperable metric space X is seperable.

▶ Union of countable family of seperable spaces of X is seperable.

▶ A non-empty family of pairwise disjoint non-empty open subsets of a seperable metric space is countable.

Let $\qquad\qquad p+iq \in \boldsymbol{C} \sim \boldsymbol{C}_0$

For given $\varepsilon > 0$, since $\boldsymbol{Q}$ is a dense subset of $\boldsymbol{R}$ $\exists$ rational numbers a and b such that

$$|p - a| < \varepsilon / \sqrt{2}$$

and $\qquad\qquad |q - b| < \varepsilon / \sqrt{2}$

$\Rightarrow \qquad\quad |(p+iq) - (a+ib)| < \varepsilon$

$\Rightarrow$ $\boldsymbol{C}_0$ is dense in $\boldsymbol{C}$.

i.e., $\boldsymbol{C}_0$ is a countable dense subset of $\boldsymbol{C}$. Hence $\boldsymbol{C}$ is separable.

* **The Euclidean plane R^2 is separable.**

Clearly $\boldsymbol{Q} \times \boldsymbol{Q}$ is a countable subset of $\boldsymbol{R}^2$. We have to show that $\boldsymbol{Q} \times \boldsymbol{Q}$ is a dense subset of $\boldsymbol{R}^2$. For this, let us consider a point $(p,q) \in \boldsymbol{R}^2 \sim \boldsymbol{Q} \times \boldsymbol{Q}$ and $\varepsilon > 0$. Since $\boldsymbol{Q}$ is a dense subset of $\boldsymbol{R}$, so we can find rational numbers q_1 and q_2 such that $(p - q_1) < \varepsilon / \sqrt{2}$ and $|q - q_2| < \varepsilon / \sqrt{2}$

$\Rightarrow \sqrt{(p - q_1)^2 + (q - q_2)^2} < \varepsilon$

$\Rightarrow d[(p,q),(q_1,q_2)] < \varepsilon$, where d is the usual metric on $\boldsymbol{R}^2$.

And hence $\boldsymbol{Q} \times \boldsymbol{Q}$ is a dense subset of $\boldsymbol{R}^2$. Finally, since there exists a countable dense subset of $\boldsymbol{R}^2$, therefore $\boldsymbol{R}^2$ is separable.

* **The n-dimensional Euclidean space is separable.**

Assume $\qquad S = \boldsymbol{Q}^n = [(x_1, x_2, ..., x_n) : x_1, x_2, ..., x_n \in \boldsymbol{Q}]$

Since $\boldsymbol{Q}$ is countable, thus S is countable.

We have to show that S is a dense subset of $\boldsymbol{R}^n$. For this take any point $(p_1, p_2, ..., p_n) \in \boldsymbol{R}^n \sim S$ and for any $\varepsilon > 0$.

Now, since $\boldsymbol{Q}$ is a dense subset of $\boldsymbol{R}$, therefore for each $i : 1 \le i \le n$, we can find $a_i \in \boldsymbol{Q}$ such that $|p_i - a_i| < \varepsilon / \sqrt{n}$ ...(1)

$\Rightarrow \qquad\quad \sqrt{\sum_{i=1}^{n}(p_i - a_i)^2} < \varepsilon$

$\Rightarrow \qquad\qquad\quad d(p, a) < \varepsilon$

where, $\qquad\qquad \boldsymbol{p} = (p_1, p_2, ..., p_n)$

$\qquad\qquad\qquad \boldsymbol{a} = (a_1, a_2, ..., a_n)$

Since every open sphere centered at $\boldsymbol{p}$ contains a point of S

$\Rightarrow \qquad\qquad\qquad \boldsymbol{p} \in \bar{S}$

Therefore, $\qquad\qquad \boldsymbol{p} \in \boldsymbol{R}^n \sim S \Rightarrow \boldsymbol{p} \in \bar{S}$

$\Rightarrow \qquad\qquad\qquad \bar{S} = \boldsymbol{R}^n$

$\Rightarrow \qquad\qquad S$ is a dense subset of $\boldsymbol{R}^n$.

Since S is a countable dense subset of $\boldsymbol{R}^n$. Hence, $\boldsymbol{R}^n$ is separable.

* **The n-dimensional unitary space C^n is separable.**

Let us write $S = \{(z_1, z_2, ..., z_n) : z_k = x_k + iy_k, \ 1 \le k \le n, x_k, y_k \in \boldsymbol{Q}\}$

Then S is a countable subset of $\boldsymbol{C}^n$.

Let us take any point $(w_1, w_2, ..., w_n) \in \boldsymbol{C}^n \sim S$

Then for any $\varepsilon > 0$ and for each k, $1 \le k \le n$, we can find

$$c_k = a_k + ib_k, a_k, b_k \in \boldsymbol{Q}$$

Such that $|w_k - c_k| < \varepsilon / \sqrt{n}$

Then, we have

$d((w_1, w_2, ..., w_n), (c_1, c_2, ..., c_n)) < \varepsilon$

$\Rightarrow \qquad\qquad\qquad d(\mathbf{w}, \mathbf{c}) < \varepsilon$

where, $\qquad\qquad \mathbf{w} = (w_1, w_2, ..., w_n) \in \mathbf{C}^n \sim S, \quad \mathbf{c} = (c_1, c_2, ..., c_n) \in S$

and d is a usual metric on $\mathbf{C}^n$.

Thus, for each $w \in \mathbf{C}^n \sim S$ and for each $\varepsilon > 0$, we can find $a, c \in S$ such that $d(w, c) < \varepsilon$

$\Rightarrow$ Each point of $\mathbf{C}^n \sim S$ is a limit point of S.

$\Rightarrow$ S is a dense subset of $\mathbf{C}^n$.

Hence, $\mathbf{C}^n$ is separable.

* **Let X be an uncountable set and let d be the discrete metric on X. Then metric space (X, d) is not separable.**

For each $p \in X$, $S(p, 1)$ is an open set.

But $\qquad\qquad\qquad\qquad S(p, 1) = \{p\}$

Thus each singleton is open. Since an arbitrary union of open sets is open, therefore every subset of X is open and consequently every subset of X is closed.

Let S be any countable subset of X. Since S is closed, therefore,

$$\bar{S} = S \neq X \qquad\qquad (\because S \text{ is countable and } X \text{ is uncountable})$$

$\Rightarrow$ S is not dense in X.

Since no countable subset of X is dense in X, thus (X, d) is not separable.

EXERCISE 4.2

1. Let (X, d) be a metric space and A be any subset of X. Then show that following statements are equivalent :
   (a) A is closed
   (b) A contains all its limit points
   (c) $\bar{A} = A$

2. Let (X, d) be a metric space and G be any open set in X. Show that G is disjoint from A iff G is disjoint from $\bar{A}$. (MEERUT 1993,95)

3. Let (X, d) be a metric space and $A \subset X$. Then show that A° equals the set of all those points of A, which are not limit point of A'.

4. Show that the frontier of a subset of a metric space is closed.

5. Let (X, d) be a metric space and $A \subset X$. Find
   (i) A°,
   (ii) $\text{ext}(A)$,
   (iii) $F_r(A)$,
   (iv) $b(A)$.

6. Are the following subsets of $\mathbf{R}$, d-nbds of 3 when d denotes the usual metric defined by $d(x, y) = |x - y|$ for $\mathbf{R}$?
   (i) $]2, 4[$
   (ii) $[1, 3]$
   (iii) $[3, 4[$
   (iv) $\mathbf{N}$

7. Show that every subspace of a discrete metric space is discrete.

8. Consider the following subset of $\mathbf{R}$. Find their closures relative to usual metric $d(x, y) = |x - y|$.
   (i) $A = [1, 2, 3, 4]$
   (ii) $B = [1]$
   (iii) $C =]2, \infty[$
   (iv) $D =]1, 2[\cup]3, 4[$
   (v) $E = \left\{ \dfrac{n+1}{n} : n \in \mathbf{N} \right\}$.

9. Show that the diameter of a subset of a metric space is equal to the diameter of its closure.

10. Show that a closed set is nowhere dense if and only if its complement is everywhere open.

11. Let A be subset of a metric space (X, d).

Prove that A is non-dense in X if and only if $X - \bar{A}$ is dense in X.

12. Let (X, d) be any metric space and let A be any subset of X. Prove that

(i) $\overline{X - A} = X - A^\circ$ (MEERUT 2004)

(ii) $X - \bar{A} = (X - A)^\circ$

13. In the metric space $(\boldsymbol{R}, d)$, where d is the usual metric on $\boldsymbol{R}$, find the boundary of the set of integers $\boldsymbol{Z}$.

14. Given an example of two subsets A and B of $\boldsymbol{R}$ such that $D(A \cap B) \neq D(A) \cap D(B)$. The metric on $\boldsymbol{R}$ is the usual metric.

15. In metric space prove that $(\bar{A})' = (A')^\circ$.

HINTS TO SELECTED PROBLEMS

9. Define $f(A) = \sup\{d(x, y) : x, y \in A\}$.

By definition of closure, A is the smallest closed set containing A. Also A is closed

$\Rightarrow \bar{A} = A \Rightarrow d(\bar{A}) = d(A)$.

10. Since A is dense in X if $\bar{A} = X$ and A is nowhere dense in X if int $(\bar{A}) = \phi$.

$\therefore$ A is non-dense in X.

$\Leftrightarrow$ int $(\bar{A}) = \phi$

$\Leftrightarrow \exists$ no open neighbourhood of any point of $\bar{A}$ such that $N \subset \bar{A}$.

$\Leftrightarrow \bar{A}$ contains no nbd.

$\Leftrightarrow S(r, x) \not\subset \bar{A}, r > 0 \ \forall x \in X$

$\Leftrightarrow S(r, x) \cap (\bar{A})' \neq \phi$

$\Leftrightarrow x$ is adherent point of $(\bar{A})' \ \forall x \in X$

$\Leftrightarrow ((\bar{A}')) = X \Leftrightarrow (\bar{A})'$ is dense in X.

ANSWERS

5. (i) A (ii) A' (iii) ϕ (iv) ϕ

6. (i) Yes   (ii) No   (iii) No   (iv) No

8. (i) A (ii) B (iii) $[2, \infty[$ (iv) $[1, 2] \cup [3, 4]$ (v) $E \cup [1]$

REVIEW QUESTIONS AND ARCHIVE

1. Define the following :

(i) neighbourhood and open set (DELHI-2006, 12, MEERUT-2000, 03, 05, 07, 11, 12, 15, 18)

(ii) Adherent point

(iii) Limit point

(iv) Isolated point

(v) Open set   (MEERUT-1993, 94, 2004, 05, GARHWAL-2004, 06, 08, 10, 12, 15)

(vi) Closed set   (MEERUT-2006, RAJASTHAN-2005, 07, 08, GARHWAL-2009)

(vii) Derived set   (viii) Discrete set

(ix) Perfect set

(x) Closure of a set

(KANPUR-2004, 05, GARHWAL-2000)

2. Differentiate between limit and limit point.

3. Find the closures of the following subset of $\boldsymbol{R}$ relative to usual metric space

(i) $A = \{1, 2, 3, 4, 5\}$

(ii) $B = \{2\}$

(iii) $C =]3, \infty[$

(iv) $D =]1, 2[\cup]3, 4[$

4. Consider the usual metric $d(x, y) = |x - y|$ for $\boldsymbol{R}$ and let $Y = (0, 1)$.

Find whether or not each of the following subsets of Y are open relative to Y :

(i) $\left]\dfrac{1}{2}, 1\right[$ (ii) $\left]\dfrac{1}{3}, \dfrac{2}{3}\right[$

(iii) $\left]0, \dfrac{1}{2}\right[$

5. Show that the diameter of a subset of a metric space is equal to the diameter of its closure.

6. In a metric space, show that arbitrary intersection of open sets is open.

7. Show that a closed set is nowhere dense if and only if its compliment is everywhere dense.

8. Show that every subspace of a discrete metric space is discrete.

9. Show that a subset A of a metric space (X, d) is nowhere dense if and only if every open sphere in X contains a sphere containing no points of A.

10. Prove that in a metric space, following are equivalent :

   (i) A is everywhere dense in X.

   (ii) The only closed set which contains A is X.

   (iii) The only open set disjoint from A is empty.

   (iv) A intersect every non-empty open set.

   (v) A intersect everywhere.

ANSWERS

3. (i) A (ii) B (iii) $[3, \infty[$ (iv) $[1, 2] \cup [3, 4]$

4. (i) Yes   (ii) Yes   (iii) No

OBJECTIVE EVALUATION

▶ FILL IN THE BLANKS

1. In a metric space the union of an arbitrary collection of open set is _____.

2. In a metric space intersection of finite number of open set is _____.

3. In a metric space intersection of infinite number of open set is _____.

4. In a metric space, a subset in it is open iff it is _____ of each of its points.

5. A subset A of metric space is closed iff $D(A)$ _____.

6. Let A be a subset of metric space then $\overline{A} = A \cup$ _____.

7. Let A and B are two closed sets, then:

   (i) $A \cap B$ is _____.

   (ii) $A \cup B$ is _____.

8. Let A and B both are open sets, then:

   (i) $A \cap B$ is _____.

   (ii) $A \cup B$ is _____.

9. A subset A of X is said to be dense set if $\overline{A} =$ _____

10. A subset A of X is said to be nowhere dense set if $(\overline{A})°$ _____.

11. The open interval $]a, b[$ is a nbd of _____ of its point.

12. The set of real number R is a nbd of _____ of its point.

13. The set of integers Z is _____ a nbd of any of its points.

14. The empty set of ϕ is a nbd of _____ of its points.

15. Every open interval is an _____ set.

16. Every closed interval is a _____ set.

17. If A and B are two open sets in a metric space (X, d), then $A \cap B$ is _____.

18. The set $[1, 2] \cup [3, 4]$ is a _____ set.

19. In a metric space, every finite set is _____ set.

20. On the real line, the set Z of integers has _____ limit points.

21. On the real line, the set $S = \left\{ \dfrac{1}{n} : n \in N \right\}$ has _____ limit point.

22. A subset of a metric space is _____ iff it contains all its limit point.

23. In a metric space, the derived set of a metric space is _____.

24. If $S \subset T$, then $S' \subset$ _____.

25. In a metric space, the set of all adherent point is called _____.

26. On a real line, the closure of each of the sets $]0,1[$, $]0,1[$ and $[0,1]$ is _____.

27. A set A is closed iff $\bar{A} =$ _____.

28. A set A is open iff $A° =$ _____.

29. In a metric space (X, d), a point $x \in X$ is an _____ point of a set $S \subset X$ iff every open set containing x contain a point of S.

30. The _____ of a set A is the intersection of all closed supersets of A.

▶ **TRUE OR FALSE**

1. In case of real line, the open sphere $S(x,r)$ is the open interval $]x-r, x+r[$. **(T/F)**

2. For the metric space (X, d), the open sphere centred at z_0 and having radius r is the set $[z : |z - z_0| < r]$. **(T/F)**

3. The closed interval $[a,b]$ is a nbd of each of its points. **(T/F)**

4. In a metric space (X, d), the intersection of the family of all nbds of a point $p \in X$ is $\{p\}$. **(T/F)**

5. A finite set in a metric space can be open. **(T/F)**

6. A finite non-empty set in a metric space can be open. **(T/F)**

7. The intersection of any arbitrary family of open sets in a metric space is always open. **(T/F)**

8. Union of an arbitrary family of open sets in a metric space is always open. **(T/F)**

9. Every infinite set in a metric space is open. **(T/F)**

10. In a discrete metric space, every subset is open. **(T/F)**

11. Every closed interval is a closed set. **(T/F)**

▶ **MULTIPLE CHOICE QUESTIONS** (CHOOSE THE MOST APPROPRIATE ONE)

1. In a metric space, every open sphere is a/an:
   (a) open
   (b) not open
   (c) closed
   (d) none of these

2. In a metric space (X, d) arbitrary union of open sets is:
   (a) necessarily open
   (b) may be open
   (c) not open
   (d) none of these

3. The intersection of finite number of open sets is:
   (a) necessarily open
   (b) may be open
   (c) not open
   (d) none of these

4. The finite union of closed sets is:
   (a) necessarily closed
   (b) may be closed
   (c) not closed
   (d) none of these

5. If $S = \{x \in X : d(x, x_0) = r\}$, $r > 0$ and $x_0 \in X$. Then S is:
   (a) open
   (b) closed
   (c) can't say
   (d) none of these

6. The intesection of an infinite number of open sets is:
   (a) necessarily open
   (b) may or may not be open
   (c) closed
   (d) none of these

7. Which of the following is not true?
   (a) R is a nbd of each of its points
   (b) Z is a nbd of each of its points
   (c) Q is not a nbd of any of its points
   (d) All are true

8. A point $x \in X$ is said to be adherent point of $A \subseteq X$ if $d(x, A)$:
   (a) > 0
   (b) < 0
   (c) $= 0$
   (d) none of these

9. Which of the following is not true?
   (a) $A \subset B \Rightarrow D(A) \subset D(B)$
   (b) $D(A \cap B) \subset D(A) \cap D(B)$
   (c) $D(A \cup B) = D(A) \cup D(B)$
   (d) none of these

10. The closure of the set of integers Z is:
   (a) Z
   (b) N
   (c) Q
   (d) none of these

11. Which of the following is not true?
   (a) $D(Q) = Q$
   (b) $D(R) = R$
   (c) both (a) and (b)
   (d) none of these

12. Which of the following is not true?
   (a) $A \subset B \Rightarrow A° \subset B°$
   (b) $(A \cap B)° = A° \cap B°$

(c) $A° \cup B° \subset (A \cup B)°$

(d) all are true

13. Which of the following is not true?

   (a) $\text{ext}(A) = \text{ext}(\text{ext}(A))$

   (b) $A \subset B \Rightarrow \text{ext}(B) \subset \text{ext}(A)$

   (c) $A \subset B \Rightarrow \text{ext}(A) \subset \text{ext}(B)$

(d) none of these

14. Which of the following is not true?

   (a) $A° = A = \text{Fr}(A)$

   (b) $\text{Fr}(A°) \subset \text{Fr}(A)$

   (c) $\text{Fr}(A \cup B) \subset \text{Fr}(A) \cup \text{Fr}(B)$

   (d) $\text{Fr}(A \cap B) = \text{Fr}(A) \cup \text{Fr}(B)$

ANSWERS

▶ FILL IN THE BLANKS

1. T	**2.** F	**3.** T	**4.** T	**5.** F	**6.** T	**7.** F	**8.** T	**9.** T
10. T	**11.** F	**12.** T	**13.** T	**14.** F	**15.** T	**16.** F	**17.** T	**18.** T
19. T	**20.** F	**21.** T	**22.** T	**23.** F	**24.** T	**25.** F	**26.** T	**27.** T
28. T	**29.** F	**30.** T						

▶ TRUE OR FALSE

1. T	**2.** F	**3.** T	**4.** T	**5.** F	**6.** T	**7.** F	**8.** T	**9.** T
10. T	**11.** F							

▶ MULTIPLE CHOICE QUESTIONS

1. (a)	**2.** (b)	**3.** (a)	**4.** (c)	**5.** (a)	**6.** (b)	**7.** (a)	**8.** (c)	**9.** (a)
10. (b)	**11.** (a)	**12.** (c)	**13.** (a)	**14.** (b)				

CHAPTER SUMMARY

This chapter deals with the concepts of open set and closed set in metric space. The limit point, derived set, adherent point closure and interior of the subset of a metric space have been discussed also in details. Important points discussed in this chapter include:

➲ Sphere, open spheres or open ball or open cell or open disc are same thing.

➲ Open sphere is defined as spherical neighbourhood of a point.

➲ Every open sphere is always non-empty since it contains its centre at least.

➲ In case of real line

   (i) $S(p,r)$ is the open interval $]p-r, p+r[$.

   (ii) The open interval $]a,b[$ is the open ball with centre at the point $p = \frac{1}{2}(a+b)$ and radius $r = \frac{1}{2}(b-a)$.

   (iii) With the same notation as above $S^*(p,r)$ is the closed interval $(p-r, p+r)$.

   (iv) The closed interval $[a,b]$ is the closed sphere $S^*(p,r)$ with $p = \frac{1}{2}(a+b)$ and $r = \frac{1}{2}(b-a)$.

➲ If $X \neq \phi$ and d be the metric on X then open ball centered at a point $p \in X$ and radius r is $[p]$ if $r \leq 1$ and X if $r > 1$.

➲ Every open interval is a nbd of each of its points.

➲ R is a nbd of each of its points.

➲ The closed interval is a nbd of each of its points except the end points.

➲ The set of integers, set of rational numbers is not a nbd of any of its points.

➲ Finite set can not be a nbd of any of its points.

➲ Every open interval is an open set.

➲ R is an open set.

➲ ϕ is an open set.

➲ In a discrete metric space, every singleton is an open set.

➲ A subset A of a X is open iff it is a union of open balls.

➲ In a metric space, finite intersection and arbitrary union of open sets is open.

➲ Hausdorff property of a metric space states that for every pair of distinct points $x, y \in X$, there exists disjoint open sets U and V such that $x \in U$, $y \in V$.

⊃ In a discrete metric space, every subset is open.

⊃ The Cantor set C is not an open set.

⊃ In a metric space, the intersection of two open spheres need not be an open sphere but it will always contain another open sphere.

⊃ Every non-empty set on the real line is the union of a countable collection of pairwise disjoint open intervals.

⊃ In a metric space, every closed sphere is a closed set.

⊃ In a metric space (X, d), the empty set ϕ and whole set X are closed.

⊃ Finite union and arbitrary intersection of closed sets is closed.

⊃ Union of an infinite number of closed sets may or may not be closed.

⊃ The set of all limit points is called derived set.

⊃ On the real line, every real number is a limit point of the set of rational numbers.

⊃ On the real line, every point of $[0, 1]$ is a limit point of open interval $]0, 1[$.

⊃ Set of integers, set of rational numbers has no limit points.

⊃ On the real line, the set $S = \left[\dfrac{1}{n} : n \in \mathbf{N} \right]$ has only one limit point namely 0.

⊃ A finite set has no limit point.

⊃ Derived set of every set is closed.

⊃ In order to show that a point p is not an adherent point of a set S, it is enough to find a nbd N of p such that $N \cap S = \phi$.

⊃ An adherent point of S need not be a limit point of S.

⊃ Every point of S is an adherent point of S, i.e., $S \subset \bar{S}$.

⊃ Every real number is an adherent point of the set of irrational numbers.

⊃ The closure of S is the smallest closed set containing S.

⊃ A space is said to be separable if it possesses a countable dense subset.

⊃ Following are separable spaces :

(i) The real line $\mathbf{R}$

(ii) The complex number space $\mathbf{C}$

(iii) The Euclidean plane $\mathbf{R}^2$

(iv) The n-dimensional Euclidean space

(v) The n-dimensional unitary space

⊃ The interior of a set A is the largest open set contained in A.

FOR ADVANCED LEARNERS

➧ In order to show that a point p is not a limit point of the set A, it is enough to find a nbd N of p such that either $N \cap A = \{p\}$ or $N \cap A = \phi$.

➧ On the real line, every real number is a limit point of the set $\mathbf{Q}$ of all rational numbers.

➧ On the real line, the set $\mathbf{Z}$ of integers has no limit point.

➧ A point $p \in X$ is a limit point of a set $A \subset X$ if any one of the following conditions is satisfied.

(i) For each $n \in \mathbf{Z}^+$, $S\left(p, \dfrac{1}{n}\right) \cap A - \{p\} \neq \phi$

(ii) For each nbd N of p, $N \cap (A - \{p\}) \neq \phi$

(iii) For each nbd N of p, $(N - \{p\}) \cap A \neq \phi$

➧ A limit point of A is also an adherent point of A, while an adherent point of A need not be a limit point of A.

➧ On the real lines, every real number is an adherent point of the set $\mathbf{R} - \mathbf{Q}$ of all irrational numbers.

➧ The closure of the set $\mathbf{Z}$ of integers is $\mathbf{Z}$ itself.

➧ In a metric space (X, d) a point $p \in X$ is in the closure of A iff for each positive rational number r

$$S(p, r) \cap A \neq \phi$$

➧ A family $\mathcal{F}$ of sets in a metric space is said to be locally finite if each point of the space has a nbd which intersects at the most the finite number of members of S.

➧ In a metric space, for an infinite family of sets, the interior of the intersection of the family may be different from the intersection of the interiors of the members of the family.

➧ Cantor's set is a perfect set with empty interior.

●●●●

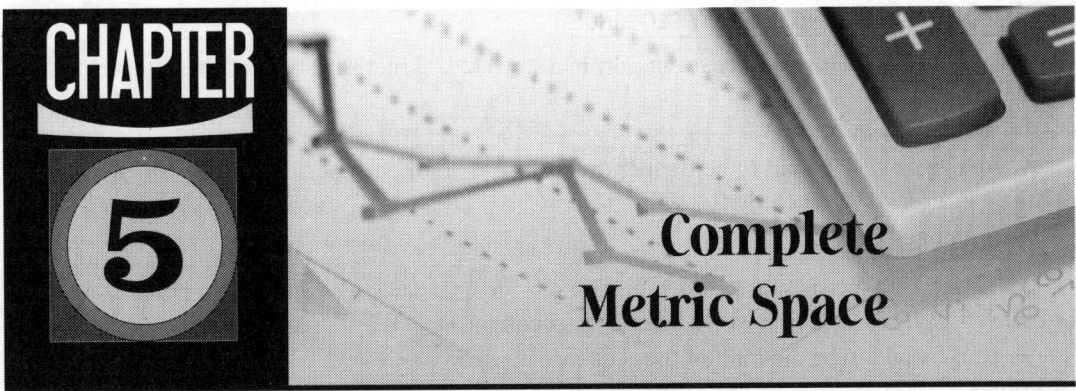

CHAPTER 5

Complete Metric Space

5.1 INTRODUCTION

Recall that a real sequence converges iff it satisfies the Cauchy convergent criterion. In a metric space (X, d), the situation is slightly different. In a metric space, every convergent sequences is a Cauchy sequence but converse is not necessarily true. Here we want to discuss those metric spaces in which every Cauchy sequence converges. Such metric spaces are going to be our main concern in the present chapter.

5.2 SEQUENCE

Let X be a non-empty set. A sequence in X is a function from N into X. It is denoted by $<x_n>$.

☞ REMARKS

➡ In a sequence $<x_n>$, x_n denote the nth term.

➡ The terms of a sequence need not be distinct.

➡ The range set of sequence may be finite or infinite.

5.2.1 BOUNDED SEQUENCE

A sequence $<s_n>$ is said to be bounded if there exists a number $m > 0$ such that

$$|s_n| \leq m \ \forall n \in N$$

For example : The sequence $<s_n>$ defined by $s_n = \dfrac{1}{n}$ is bounded by 1.

5.2.2 CONSTANT SEQUENCE

A sequence $<s_n>$ is said to be constant if

$$s_n = C \ \forall n \in N, \ C \in R.$$

5.2.3 SUBSEQUENCE

Let $<s_n>$ and $<t_n>$ be two sequences in a set X, then $<t_n>$ is said to be a subsequence of $<s_n>$ if there exists a mapping

$$f : N \to N$$

such that

(i) $t = s \circ f$

(ii) for each $n \in N \ \exists m \in N$ such that $f(i) \geq n \ \forall i \geq m$ in N

(or)

If $<s_n>$ is a sequence in X and $<i_n>$ is a sequence of N such that $i_1 < i_2 < ... < i_n < ...$, then $<s_{i_n}>$ is called subsequence of $<s_n>$.

5.2.4 EVENTUALLY AND FREQUENTLY SEQUENCE

A sequence $<s_n>$ is said to be eventually in a set $A \subset X$ if there exists a positive integer m such that $n \geq m$

$$\Rightarrow \qquad s_n \in A$$

A sequence $<s_n>$ is said to be frequently in a set A if for each positive integer m there exists a positive integer $n > m$ such that $s_n \in A$.

5.2.5 CONVERGENT SEQUENCE

Let (X, d) be a metric space. A sequence $<s_n>$ in X is said to converge to a point s_0 in X if for each $\varepsilon > 0$, there exists a positive integer m such that $d(s_n, s_0) < \varepsilon \quad \forall n \geq m$.

Here, s_0 is said to be the limit of the sequence $s_n \to s_0$ as $n \to \infty$.

☛ REMARKS

⟹ From above definition it is clear that
$$\lim_{n \to \infty} d(s_n, s_0) = 0$$

⟹ The limit of a sequence, if exists is unique.

⟹ A sequence $<s_n>$ is said to converge to s_0 in X if it is eventually in every nbd of s_0.

⟹ A sequence which is not convergent, is said to be divergent.

⟹ If a sequence $<s_n>$ converges to s_0, then any subsequence of $<s_n>$ also converges to s_0.

5.2.6 CLUSTER POINT

Let (X, d) be a metric space. A point $s_0 \in X$ is said to be cluster point of a sequence $<s_n>$ in X if $<s_n>$ is frequently in every nbd of s_0.

(or)

A point s_0 is said to be cluster point of the sequence $<s_n>$ if and only if for any $\varepsilon > 0$ and for any positive integer m, there exists an integer $n \geq m$ such that
$$d(s_0, s_n) < \varepsilon$$

☛ REMARK

⟹ From above definition, it is clear that s_0 will be a cluster point of $<s_n>$ iff every open sphere centered at s_0 contains infinitely many terms of the sequence.

5.2.7 CAUCHY SEQUENCE

Let (X, d) be a metric space and let $<s_n>$ be the given sequence in X. Then $<s_n>$ is said to be Cauchy sequence in X if for every $\varepsilon > 0$, there exists a positive integer $n(\varepsilon)$ such that $d(s_m, s_n) < \varepsilon, \forall m, n \geq n(\varepsilon)$ (MEERUT 1991; ROHTAK 2005, 07; LUCKNOW 2006, 07, KURUKSHETRA 2004, 05)

__THEOREM 1.__   ***In a metric space, every convergent sequence has a unique limit.***

__PROOF.__   Let (X, d) be a metric space and $<s_n>$ be a convergent sequence converges to a point say s_0 in X.

We have to show that s_0 is the unique limit of $<s_n>$.

Let if possible, $<s_n>$ converges to another point say t_0 distinct from s_0

i.e., $s_0 \neq t_0$

Let $d(s_0, t_0) = \varepsilon$

Then clearly $S(s_0, \varepsilon/4)$ and $S(t_0, \varepsilon/4)$ are disjoint.

Since $<s_n>$ converges to s_0, therefore, there exists a positive integer n_0 such that
$$s_n \in S(s_0, \varepsilon/4) \ \forall n \geq n_0$$

Also, since $<s_n>$ converges to t_0, therefore there exists a positive integer m_0 such that $s_n \in S(t_0, \varepsilon/4)\ \forall\, n \geq m_0$

Let us define $\qquad m = \max\{n_0, m_0\}$

Then, $<s_m>$ must be a member of both the spheres $S(x_0, \varepsilon/4)$ and $S(y_0, \varepsilon/4)$, which is a contradiction, because both spheres are disjoint. Hence, our assumption is wrong which implies that *every convergent sequence has a unique limit.*

THEOREM 2. *Let (X, d) be a metric space. If s_0 is a limit point of $S \subset X$, then there exists a sequence $<s_n>$ of points of S all distinct from s_0, which converges to s_0.* (GARHWAL 2005, 08; KANPUR 2002, 09)

PROOF. It is given that s_0 is a limit point of S, therefore, by definition of limit point, we can say that every open sphere with centre s_0 must contain a point of S distinct from s_0. Let us take any point s_1 in S and set $r_1 = \min\{1, d(s_0, s_1)\}$.

Then the sphere $S[s_0, s_1]$ contains a point s_2 of S distinct from s_0. Now, setting $r_2 = \min\left\{\dfrac{1}{2}, d(s_0, s_2)\right\}$. Then the sphere $S(s_0, r_2)$ contains a point s_3 of S and so on.

Proceeding in the same manner, we may construct a sequence $<s_n>$ of distinct points all distinct from s_0 such that if

$$r_n = \min\left\{\frac{1}{n},\ d(s_0, s_n)\right\}$$

Then the sphere $S(s_0, r_n)$ contains the points s_{n+1} of S.

Therefore, $d(s_0, s_n) < r_{n-1} \leq \dfrac{1}{n-1}$

Clearly $\dfrac{1}{n-1} \to 0$ as $n \to \infty$

$\Rightarrow \qquad d(s_0, s_n) \to 0$ as $n \to \infty$.

Hence, $<s_n>$ converges to s_0.

THEOREM 3. *If the range set of a convergent sequence in a metric space (X, d) consist of infinitely many distinct points, then the limit of the sequence is an accumulation point of the range set of the sequence.*

PROOF. Let $<s_n>$ be a convergent sequence in a metric space such that $s = <s_n> \to s_0$.

We have to show that s_0 is an accumulation point of the range set $R(S)$ of S. Let if possible, s_0 is not an accumulation point of $R(S)$. By definition of accumulation point, there exists an open sphere $S[s_0, r]$ centered at s_0 which contains no point of $R(S)$ distinct from s_0. Since s_0 is the limit of the sequence, there exists a positive integer m such that $s_n \in S[s_0, r]\ \forall\, n \geq m$ and hence $s_n = s_0\ \forall\, n \geq m$.

$\Rightarrow$ $R(S)$ consist of finitely many distinct points.

which is a contradiction. Hence, s_0 must be an accumulation point.

THEOREM 4. *Let (X, d) be a metric space and s_0, t_0 be two points of X. If $<t_n>$ is a sequence converges to t_0, then $<d(s_0, t_n)>$ converges to $d(s_0, t_0)$.*

PROOF. Consider three points x, y, z of X, then we have

$$|d(x, y) - d(x, z)| \leq d(y, z) \qquad \text{...(1)}$$

Let us set $x = s_0,\ y = t_0,\ z = t_n$. Then (1) becomes

$$|d(s_0, t_0) - d(s_0, t_n)| \leq d(t_0, t_n) \qquad \text{...(2)}$$

Now, since $<t_n>$ converges to t_0, then for given $\varepsilon > 0$, there exists a positive integer p such that

$$d(t_0, t_n) < \varepsilon \quad \forall\, n \geq p \qquad \qquad \ldots(3)$$

Using (3) in (2), we get

$$|d(s_0, t_0) - d(s_0, t_n)| < \in \quad \forall\, n \geq p$$

Hence, we can say that the sequence $<d(s_0, t_n)>$ of real numbers converges to $d(s_0, t_0)$.

THEOREM 5. **In a metric space (X, d), if the sequences $<s_n>$ and $<t_n)$ converge respectively to the points s_0 and t_0 in X, then the sequence $<d(s_n, t_n)>$ of real numbers converges to $d(s_0, t_0)$.** (MEERUT 2000, 01,11; KANPUR 1996)

PROOF. Let (X, d) be a metric space and $<s_n>$ and $<t_n>$ be two sequences which converge to s_0 and t_0 respectively.

By definition of convergent sequence, we can say for given $\varepsilon > 0$, there exists positive integers and m_2 such that

$$d(s_0, s_n) < \varepsilon/2 \qquad \forall\, n \geq m_1 \qquad \ldots(1)$$

and

$$d(t_0, t_n) < \varepsilon/2 \qquad \forall\, n \geq m_2 \qquad \ldots(2)$$

Take $m = \max\{m_1, m_2\}$

Then, from (1) and (2), we have

$$d(s_0, s_n) < \varepsilon/2 \text{ and } d(t_0, t_n) < \varepsilon/2 \quad \forall\, n \geq m \qquad \ldots(3)$$

Consider

$$|d(s_0, t_0) - d(s_n, t_n)| = |d(s_0, t_0) - d(s_n, t_0) + d(s_n, t_0) - d(s_n, t_n)|$$

$$\leq |d(s_0, t_0) - d(s_n, t_0)| + |d(s_n, t_0) - d(s_n, t_n)|$$

(Using triangle inequality)

$$\leq d(s_0, s_n) + d(t_0, t_n)$$

$$< \varepsilon/2 + \varepsilon/2 = \varepsilon \qquad \qquad \text{(Using (3)}$$

$$\Rightarrow \quad |d(s_0, t_0) - d(s_n, t_n)| < \varepsilon$$

Hence, the sequence $<d(s_n, t_n)>$ converges to $d(s_0, t_0)$.

THEOREM 6. **In a metric space, every convergent sequence is Cauchy.**

(KANPUR 2007; GARHWAL 2007, 10; LUCKNOW 2006; KURUKSHETRA 2004, 05, 07; RAJASTHAN 2004, 07; BHOPAL 2009)

PROOF. Let (X, d) be a metric space and $<s_n>$ be a convergent sequence such that $s_n \to s_0$. Then by definition, for given $\varepsilon > 0$ $\exists$ a positive integer p such that

$$d(s_0, s_n) < \varepsilon/2 \,\, \forall\, n \geq p \qquad \ldots(1)$$

Thus, if $m \geq p$ and $n \geq p$, then

$$d(s_0, s_m) < \varepsilon/2 \,\, \forall\, m \geq p \qquad \ldots(2)$$

Then we have

$$d(s_m, s_n) \leq d(s_m, s_0) + d(s_0, s_n)$$

$$< \varepsilon/2 + \varepsilon/2 = \varepsilon \qquad \qquad \text{(Using (1) and (2))}$$

$$\Rightarrow \quad d(s_m, s_n) < \varepsilon \quad \forall\, n, m \geq p$$

Hence, $<s_n>$ is a Cauchy sequence.

☛ **REMARK**

➡ The converse of the above theorem is not true, *i.e.*, "Every Cauchy sequence is not necessarily convergent".

For proof of this, consider the following example

Let (X, d) be a usual metric space and $<s_n>$ be a Cauchy sequence such that $s_n = \dfrac{1}{n} : n \in N$

Clearly $s_n \to 0$, but 0 is not a point of X.

THEOREM 7. *If a Cauchy sequence in a metric space has a convergent subsequence, then the sequence is convergent.*

(MEERUT 2001, 11; GARHWAL 2000, 02, 04, 09; RAJASTHAN 2001, 05, 09)

PROOF. Let (X, d) be a metric space and $< s_n >$ be a Cauchy sequence in X. Further, let $< s_{k_m} >$ be a subsequence of $< s_n >$ converging to s_0 in X, i.e., $s_{k_m} \to s_0$ as $n \to \infty$.

We have already proved that every convergent sequence is Cauchy, so by definition, for given $\varepsilon > 0$, there exists a positive integer p such that $d(s_{k_m}, s_n) < \varepsilon \quad \forall m, n \geq p$.

Since $< k_m >$ is strictly increasing sequence of positive integer (By definition of subsequence), letting $m \to \infty$, we get $d(s_0, s_n) < \varepsilon \quad \forall n \geq p$.

$\Rightarrow \quad < s_n > \to s_0$

i.e., the sequence $< s_n >$ converges to s_0.

THEOREM 8. *If $< s_n >$ is a Cauchy sequence in a metric space, then any cluster point of $< s_n >$ is limit of $< s_n >$.* (MEERUT 2004)

PROOF. Let (X, d) be a metric space and $< s_n >$ be a Cauchy sequence such that s_0 is a cluster point of $< s_n >$.

By definition of Cauchy sequence, for given $\varepsilon > 0$ $\exists$ a positive integer p such that

$$n, m \geq p \Rightarrow d(s_m, s_n) < \varepsilon / 2 \qquad ...(1)$$

Since s_0 is a cluster point of $< s_n >$, therefore there exists a positive integer $n_0 \geq p$ such that

$$d(s_0, s_{n_0}) < \varepsilon / 2 \qquad ...(2)$$

Using triangle inequality, we can write

$$d(s_0, s_n) \leq d(s_0, s_{n_0}) + d(s_{n_0}, s_n)$$

$$< \varepsilon / 2 + \varepsilon / 2$$

$$= \varepsilon \quad \forall n \geq p$$

$\Rightarrow \quad$ Limit of the sequence $< s_n >$ is s_0.

$\Rightarrow \quad$ The sequence $< s_n >$ converges to s_0.

THEOREM 9. *Let $< s_n >$ be a Cauchy sequence in a metric space (X, d) and let $< s_{i_n} >$ be a subsequence of $< s_n >$, then $\lim\limits_{n \to \infty} d(s_n, s_{i_n}) = 0$.*

PROOF. Since $< s_n >$ be a Cauchy sequence. Then for given $\varepsilon > 0$ $\exists$ a positive integer p such that $d(s_n, s_m) < \varepsilon \quad \forall n, m \geq p$

Further, since $i_p \geq p$, therefore

$$d(s_{n_0}, s_{i_{n_0}}) < \varepsilon$$

$\Rightarrow \lim\limits_{n \to \infty} d(s_n, s_{i_n}) = 0$.

THEOREM 10. *Let $< s_n >$ be a Cauchy sequence in a metric space (X, d) and let $< s_{i_n} >$ be a subsequence of $< s_n >$ converging to $s \in X$, then $< s_n >$ also converges to s_0.*

PROOF. Using triangle inequality

$$d(s_0, s_n) \leq d(s_0, s_{i_n}) + d(s_{i_n}, s_n)$$

we get

$$\lim\limits_{n \to \infty} d(s_0, s_n) \leq \lim\limits_{n \to \infty} d(s_0, s_{i_n}) + \lim\limits_{n \to \infty} d(s_{i_n}, s_n) \qquad ...(1)$$

Since $s_{i_n} \to s_0$, therefore

$$\lim\limits_{n \to \infty} d(s_0, s_{i_n}) = 0 \qquad ...(2)$$

Also using previous theorem, we can write

$$\lim_{n \to \infty} d(s_{i_n}, s_n) = 0 \qquad \qquad ...(3)$$

Using (2) and (3) in (1), we get

$$\lim_{n \to \infty} d(s_0, s_n) \leq 0$$

$$\Rightarrow \quad \lim_{n \to \infty} d(s_0, s_n) = 0 \qquad \qquad (\because \text{Distance can not be negative})$$

$$\Rightarrow \quad \text{Limit of } <s_n> \text{ is } s_0.$$

Hence, sequence $<s_n>$ converges to s_0.

THEOREM II. *Let (X, d) be a metric space and $<s_n>$ be a Cauchy sequence in X. If $<t_n>$ be a sequence in X such that $d(s_n, t_n) < \dfrac{1}{n}$ for every positive integer n. Then,*

(i) *$<t_n>$ is also a Cauchy sequence in X.*

(ii) *$<t_n>$ converges to a point $t_0 \in X$ if and only if $<s_n>$ converges to t_0.*

PROOF. (i) Using triangle inequality, we have

$$d(t_m, t_n) \leq d(t_m, s_m) + d(s_m, s_n) + d(s_n, t_n) \qquad ...(1)$$

As per given $<s_n>$ be a Cauchy sequence, then by definition, for a given $\varepsilon > 0$ $\exists$ a positive integer p_1 such that

$$d(s_n, s_m) < \varepsilon / 3 \quad \forall \, n, m \geq p_1 \qquad \qquad ...(2)$$

Further, we can choose a positive integer p_2 such that $\dfrac{1}{p_2} < \varepsilon / 3$.

Then by hypothesis

$$d(s_m, t_m) < \varepsilon / 3 \text{ and } d(s_n, t_n) < \varepsilon / 3 \quad \forall \, n, m \geq p_2 \qquad ...(3)$$

Let us take $p = \max \{p_1, p_2\}$

Then using (2) and (3) in (1) we get

$$d(t_m, t_n) < \varepsilon / 3 + \varepsilon / 3 + \varepsilon / 3 \quad \forall \, n, m \geq p$$

$$\Rightarrow \quad d(t_m, t_n) < \varepsilon \quad \forall \, n, m \geq p$$

Hence, $<t_n>$ is a Cauchy sequence.

(ii) Let us suppose, sequence $<t_n>$ converges to t_0, so that

$$\lim_{n \to \infty} d(t_n, t_0) = 0 \qquad \qquad ...(4)$$

Using triangle inequality, we have

$$d(s_n, t_0) \leq d(s_n, t_n) + d(t_n, t_0)$$

Taking limit of both the sides, we get

$$\lim_{n \to \infty} d(s_n, t_0) \leq \lim_{n \to \infty} d(s_n, t_n) + \lim_{n \to \infty} d(t_n, t_0) \qquad ...(5)$$

But we have

$$\lim_{n \to \infty} d(s_n, t_n) \leq \lim_{n \to \infty} \frac{1}{n} = 0 \qquad \qquad ...(6)$$

Using (4) and (6) in (5), we get

$$\lim_{n \to \infty} d(s_n, t_0) \leq 0 + 0 = 0 \quad \Rightarrow \quad \lim_{n \to \infty} d(s_n, t_0) = 0$$

Hence, we conclude that if $<t_n>$ converges to t_0, then $<s_n>$ also converges to t_0.

Similarly we may show that if $s_n \to t_0$, then $t_n \to t_0$.

5.3 COMPLETE METRIC SPACE

Definition 1. *A metric space* (X, d) *is said to be complete if every Cauchy sequence in* X *converges.*

For example:

(i) R with usual metric is complete

(ii) The metric space $]0, 1]$ with usual metric is not complete.

☛ **REMARK**

➠ A metric space, which is not complete is said to be incomplete.

SOLVED EXAMPLES

EXAMPLE 1. **The real line is a complete metric space.**

SOLUTION. Let $< s_n >$ be a Cauchy sequence in R. Then by definition given any $\varepsilon > 0$ there exist a positive integer k such that

$$|x_n - x_k| < \varepsilon, \ \forall \, n \geq k$$

Let $\varepsilon = 1$.

Then, $|x_n - x_k| < 1, \ \forall \, n \geq k_1$ for some k_1

$\Rightarrow \qquad x_{k_1} - 1 < x_n < x_{k_1} + 1, \ \forall \, n \geq k_1.$

Let us define

$$p = \min \{x_1, x_2, ..., x_{k_1} - 1, \ x_{k_1} + 1\}$$

and $\qquad q = \max \{x_1, x_2, ..., x_{k_1} - 1, x_{k_1} + 1\}$

Then, $\qquad p \leq x_n \leq q, \ \forall n \in N$

$\Rightarrow \quad < x_n >$ is bounded sequence of real numbers.

Let $x = \lim \inf < x_n >$ and $y = \lim \sup < x_n >$.

Let if possible $x \neq y$ and let $y - x = s$ then $s \geq 0$.

Now since $< x_n >$ is a Cauchy sequence, therefore, there exists a positive integer n such that

$$|x_l - x_m| < \varepsilon / 2, \ \forall \, l, m \geq n \qquad \qquad ...(1)$$

Since x is the limit interior of $< x_n >$, therefore $x \leq x_n < x + \dfrac{s}{4}$ for infinitely many values of n. In particular, we can find a positive integer $u > n$ such that

$$x \leq x_n < x + s / 4 \qquad \qquad ...(2)$$

Again since y is the limit superior of $< x_n >$, therefore $y - \dfrac{s}{4} < x_n \leq y$ for infinitely many values of n.

In particular, we can find a positive integer $v > n$ such that

$$y - s / 4 < x_v \leq y \qquad \qquad ...(3)$$

From (2) and (3), we conclude that

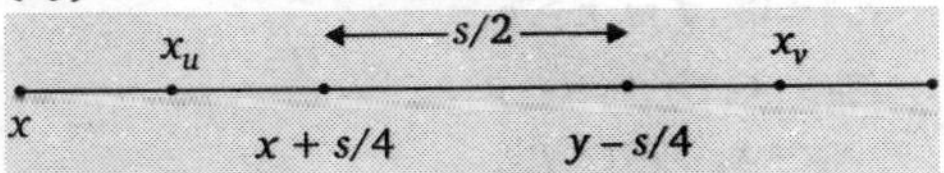

Fig. 1

There exist positive integers $u, v > n$ such that

$$|x_u - x_v| > s / 2 \qquad \qquad ...(4)$$

This contradicts (1) which says that for all positive integers $l, m \geq n$

$$|x_l - x_m| < s / 2 \qquad \qquad ...(5)$$

In view of the above contradiction, we must have $s \not> 0$. Thus $s = 0$ and therefore $x = y$ which gives

$$\liminf < x_n > \ = \ \limsup < x_n >$$

Therefore $\lim_{n \to \infty} < x_n >$ exists.

Hence, the real line is complete.

EXAMPLE 2.   ***Show that C with usual metric is complete.***

SOLUTION.   Let $<z_n>$ be any Cauchy sequence in C such that $z_n = x_n + iy_n$, $x_n, y_n \in R$

We have to prove that $<x_n>$ and $<y_n>$ are Cauchy in R. Let $\varepsilon > 0$ be given. Since, $<z_n>$ is Cauchy, so by definition there exists a positive integer n_0 such that

$$|z_n - z_m| < \varepsilon \;\; \forall \, n, m \geq n_0$$

Now, $|x_n - x_m| \leq |z_n - z_m|$ and $|y_n - y_m| < |z_n - z_m|$

$\Rightarrow$ $|x_n - x_m| < \varepsilon$ and $|y_n - y_m| < \varepsilon$ $(\because |z_n - z_m| < \varepsilon)$

$\Rightarrow$ $<x_n>$ and $<y_n>$ are Cauchy sequence in R.

Since, R is complete, therefore there exists $x, y \in R$ such that $x_n \to x$ and $y_n \to y$

                                               (By definition of complete space)

Let $z = x + iy$. We claim that $z_n \to z$

Consider $|z_n - z| = |(x_n + iy_n) - (x + iy)| = |(x_n - x + i(y_n - y)|$

                    $\leq |x_n - x| + |y_n - y|$ ...(1)

Since, $x_n \to x$ therefore $|x_n - x| < \varepsilon/2$. Similarly, $|y_n - y| < \varepsilon/2$.

Using these values in (1), we get

$$|z_n - z| < \varepsilon/2 + \varepsilon/2 = \varepsilon$$

$\Rightarrow$ $|z_n - z| < \varepsilon$

$\Rightarrow$ $z_n \to z$

$\Rightarrow$ Every Cauchy sequence in convergent.

Hence, C is complete.

EXAMPLE 3.   ***The set Z of integers with the usual metric $(d(x, y) = |x - y| \;\; \forall x, y \in Z)$ is a complete metric space.***

SOLUTION.   Let $< x_n >$ be any Cauchy sequence in Z. Take $\varepsilon = \dfrac{1}{2}$.

Then, we can find a positive integer p such that

$$|x_m - x_n| < \frac{1}{2} \;\; \forall \, m, n \geq p$$

Since $x_n \in Z \;\; \forall \, n \in N$, therefore $|x_m - x_n|$ must be a non-negative integer $\forall m, n \in N$.

Therefore, we have

(i) $|x_m - x_n| < \dfrac{1}{2} \; \forall m, n \geq p$

(ii) $|x_m - x_n|$ is a non-negative integer $\forall \, m, n \in N$.

Hence, we find that

$$|x_m - x_n| = 0 \;\; \forall \, m, n \geq p$$

i.e., $x_m = x_n \; \forall \, m, n \geq p$

Consequently $x_n = x_p \; \forall \, n \geq p$

$\Rightarrow$ $< x_n >$ is eventually a constant sequence.

$\Rightarrow$ $< x_n >$ must converge to x_p.

Since, an arbitrary Cauchy sequence in Z converges.

Therefore, the set Z of integers equipped with the usual metric is a complete metric space.

EXAMPLE 4.   *Let* $X =]0,1[$ *and let* $d(x, y) = |x - y|$ *for all* $x, y \in X$. *Then show that* (X, d) *is not complete.*   (MEERUT 1994)

SOLUTION.   Let $<x_n>$ be a sequence in X defined by setting $x_n = \dfrac{1}{n}$ for all $n \in N$. Let ε be any arbitrary positive number.

Now $d(x_m, x_n) = |x_m - x_n|$

$$= \left| \frac{1}{m} - \frac{1}{n} \right| \leq \frac{1}{m} + \frac{1}{n} \qquad \qquad \ldots (1)$$

Let p be any positive integer greater than $\dfrac{2}{\varepsilon}$, then

$$\frac{1}{m} + \frac{1}{n} < \frac{\varepsilon}{2} + \frac{\varepsilon}{2} \quad \forall\, m, n \geq p$$

Therefore, from (1), we can find that

$$d(x_m, x_n) < \varepsilon \quad \forall\, m, n \geq p$$

$\Rightarrow$ $<x_n>$ is a Cauchy sequence in X.

Now, let $x \in X$. Then by the Archimedean property of real numbers, we can find a positive integer n such that

$$n \leq \frac{1}{n} < n + 1 \Rightarrow \frac{1}{n} \geq x > \frac{1}{n+1}$$

Now there are two different cases :

CASE (I) If $x = \dfrac{1}{n}$, then $\left] \dfrac{1}{(n-1)}, \dfrac{1}{(n+1)} \right[$ is a nbd of x in $]0,1[$ which contains only one element of the sequence $\left\langle \dfrac{1}{n} \right\rangle$.

CASE (II) If $x \neq \dfrac{1}{n}$, then $\left] \dfrac{1}{(n+1)}, \dfrac{1}{n} \right[$ is a nbd of x in $]0,1[$ which does not contain any element of the sequence.

In either case, the sequence can not converge to x.

Since x is any point of X, therefore the sequence $<x_n>$ can not converge to any point of X.

Hence (X, d) is an incomplete metric space.

5.4 PROPERTIES OF COMPLETE METRIC SPACE

THEOREM I.   *Let* (X, d) *be a complete metric space and* Y *be a subspace of* X. *Then* Y *is complete if and only if* Y *is closed.*

(MEERUT 1990, 92, 94, 96, 98, 2000, 01, 02, 03, 05, 06, 08, 18; GARHWAL 2004, 05, 06, 08, 10;
KANPUR 2008, 09; KURUKSHETRA 2000, 02, 03, 05, 07, 09)

PROOF.   Let (X, d) be a complete metric space. Let us first suppose Y be a complete subspace of X. To show Y is closed in X.

Let $p \in X$ be any limit point of . Then by definition of limit point, for every positive integer n, the open sphere $s\left(p, \dfrac{1}{n}\right)$ must contain a point q_n of Y such that the sequence $<q_n>$ converges to p.

$\Rightarrow$ $<q_n>$ is Cauchy in Y ($\because$ Every convergent sequence is Cauchy)

Since Y is complete therefore $p \in Y$.

Now since $p \in Y$ is arbitrary, therefore we can say Y contains all its limit points and hence Y is closed.

Conversely, let Y is closed. To show Y is complete.

Let $< s_n >$ be any Cauchy sequence in Y.

$\Rightarrow \quad < s_n >$ is a Cauchy sequence in X. $\hspace{3cm}$ ($\because Y$ is a subspace of X)

Also, since X is complete, therefore $< s_n >$ must converge to a point $s_0 \in X$. We want to show $s_0 \in Y$.

If the range set of $< s_n >$ consist of finite number of distinct points, then $< s_n >$ must be of the form $< s_1, s_2, ..., s_n, s_0, s_0, s_0, ... >$ where n is finite and hence $s_0 \in Y$. If the range set of $< s_n >$ contains infinitely many points, then s_0 is a limit of the range set of $< s_n' >$.

($\because$ If the range set of a convergent sequence in a metric space consists of infinitely many distinct points, then the limit of the sequence is a limit point of the range set of the sequence).

$\Rightarrow \quad s_0$ is also a limit point of Y.

$\Rightarrow \quad s_0 \in Y$. $\hspace{6cm}$ ($\because Y$ is closed)

$\Rightarrow \quad$ Every Cauchy sequence in Y converges in Y. Hence, Y is complete.

☛ **REMARK**

⟹ Using the above theorem we conclude that $[0, 1]$ with usual metric space is complete.

THEOREM 2. *(Cantor's Intersection Theorem). Let (X, d) be a metric space and let $< F_n >$ be a nested sequence of non-empty closed subset of X such that $\delta(F_n) \to 0$ as $n \to \infty$. Then X is complete if and only if $\overset{\infty}{\underset{n=1}{\cap}} F_n$ consists of exactly one point.* (MEERUT-1994,96,97,2000,01,02,03,14,16; GARHWAL-2007, 10; BANGLURU-2002,03,04,07,08; ROHTAK-2001,03,04,07; KANPUR-2006,07; LUCKNOW-2002,06,09)

PROOF. (i) **Necessary condition.** Let (X, d) be a complete metric space and let $< F_n >$ be the sequence of non-empty closed subsets of X such that $F_{n+1} \subset F_n \ \forall \, n \in \mathbf{N}$ (by definition of nested sequence) and $\delta(F_n) \to 0$ as $n \to \infty$. We have to show that

$\underset{n \in N}{\cap} F_n$ consist of exactly one point.

STEP 1. For each $n \in \mathbf{N}$, choose a point $x_n \in F_n$. It can be easily seen that $< x_n >$ is a Cauchy sequence of points of X. In fact, given $\varepsilon > 0$, we can find a positive integer m such that

$$\delta(F_n) < \varepsilon \quad \forall \, n \geq m \hspace{4cm} ...(1)$$

Since the sequence $< F_n >$ is nested, therefore $F_n \subset F_m \ \forall \, n \geq m$ and consequently

$$x_n \in F_m \quad \forall \, n \geq m \hspace{4cm} ...(2)$$

Since we know that for any non-empty bounded set S in a metric space (X, d)

$$\delta(S) = \sup \{ d(x, y) : x, y \in S \}$$

so that $d(x, y) \leq \delta(S) \ \forall \, x, y \in S \hspace{4cm} ...(3)$

From (2) and (3), we conclude that

$$d(x_n, x_m) \leq \delta(F_m) \quad \forall \, n \geq m \hspace{4cm} ...(4)$$

Now (1) and (4) gives

$$d(x_n, x_m) < \varepsilon \quad \forall \, n \geq m$$

$\Rightarrow \quad < x_n >$ is a Cauchy sequence in X.

STEP 2. Since the metric space (X, d) is complete, therefore the Cauchy sequence $<x_n>$ converges, *i.e.*, there exists $x_0 \in X$ such that $x_n \to x_0$

We shall show that $x_0 \in F_n \; \forall \, n \in \boldsymbol{N}$ and consequently $\underset{n \in N}{\cap} \, F_n$ is non-empty. Let

$k \in \boldsymbol{Z}$ be fixed.

Now from (2)

$$x_n \in F_k \;\; \forall \, n \geq k$$

i.e., $<x_k, x_{k+1}, x_{k+2}, \dots>$ is a sequence in F_k.

The sequence $<x_k, x_{k+1}, \dots>$ converges to x_0 and therefore $x_0 \in F_k$.

Since F_k is closed $\Rightarrow x_0 \in F_k$.

Now, $x_0 \in F_k$ and k is arbitrary, therefore

$$x_0 \in F_n \;\; \forall \, n \in \boldsymbol{N}$$

$$\Rightarrow \qquad x_0 \in \underset{n \in N}{\cap} \, F_n$$

STEP 3 (UNIQUENESS). Let if possible x_0, y_0 be two points in the intersection of

F_n's. To show $x_0 = y_0$.

Suppose if possible $x_0 \neq y_0$

And let $d(x_0, y_0) = \varepsilon > 0$.

Since $\delta(F_n) \to 0$, therefore we can find a positive integer m such that $\delta(F_n) < \varepsilon / 2 \;\; \forall \, n \geq m$

Since $x_0, y_0 \in F_n \;\; \forall \, n \in \boldsymbol{N}$, therefore in particular $x_0, y_0 \in F_m$

$\Rightarrow \quad d(x_0, y_0) \leq \delta(F_m) < \varepsilon / 2$.

Since $d(x_0, y_0) = \varepsilon$, we have

$$\varepsilon = d(x_0, y_0) < \varepsilon / 2$$

which is a contradiction and therefore $x_0 = y_0$ and so $\underset{n \in \boldsymbol{N}}{\cap} \, F_n$ consist of exactly

one point.

(ii) **Condition is sufficient.** Let us suppose that every nested sequence of closed sets with diameter tending to zero has non-empty intersection.

To show X is complete.

Let $<x_n>$ be a Cauchy sequence in X.

Correspond to $\varepsilon = 1/2$, we can find $n_1 \in \boldsymbol{N}$ such that $n_2 > n_1$ and

$$d(x_m, x_n) < \frac{1}{2^n} \;\; \forall \, n \geq n_2.$$

Proceeding in the same manner, we can construct a strictly increasing sequence $<n_1, n_2, n_3, \dots>$ of positive integers such that

$$d(x_{n_1}, x_n) < \frac{1}{2}, \;\; \forall \, n \geq n_1$$

$$d(x_{n_2}, x_n) < \frac{1}{2^2}, \;\; \forall \, n \geq n_2$$

$$\dots \quad \dots \quad \dots \quad \dots \quad \dots \quad \dots \quad \dots$$

$$d(x_{n_k}, x_n) < \frac{1}{2^k}, \;\; \forall \, n \geq n_k$$

$$\dots \quad \dots \quad \dots \quad \dots \quad \dots \quad \dots \quad \dots$$

Let us write $F_k = S^*(x_{n_k}, 2^{-k+1})$, $k = 1, 2, 3, \ldots$

Then $<F_k>$ is a sequence of closed sets with diameters tending to zero.

To see that $<x_k>$ is nested, i.e., $F_{k+1} \subset F_k \ \forall k \geq 1$, take a point $y \in F_{k+1}$.

Then, $\quad d(x_{n_k}, y) \leq d(x_{n_k}, x_{n_{k+1}}) + d(x_{n_{k+1}}, y)$ $\qquad$ (By triangle inequality)

$$\leq 2^{-k} + 2^{-k} = 2^{-k+1}$$

$\Rightarrow \qquad\qquad y \in S(x_{n_k}, 2^{-k+1}) \subset F_k$

$\Rightarrow \qquad\qquad F_{k+1} \subset F_k.$

Thus, we find that $<F_n>$ is a nested sequence of non-empty closed sets with diameters tending to zero, therefore there exists $x_0 \in X$ such that $x_0 \in F_k \ \forall k \in N$.

Consider the sequence $<x_{n_1}, x_{n_2}, x_{n_3}, \ldots>$.

Since $n_1 < n_2 < n_3 \ldots$ therefore the above sequence is a subsequence of $<x_n>$.

Also, $\qquad d(x_{n_k}, x_0) < 2^{-k} \ \forall k \in N$

$$<x_{n_k}> \to x_0$$

Since $<x_n>$ is a Cauchy sequence in x and $<x_{n_k}>$ is a subsequence of $<x_n>$ converging to x_0, therefore $<x_n>$ converges to x_0.

Hence (X, d) is complete.

☛ **REMARKS**

⇒ In the above theorem $\cap F_n$ may be empty if each F_n is not closed.

⇒ In the above theorem $\cap F_n$ may be empty if the hypothesis $\delta(F_n) \to 0$ is ommited.

5.5 METRIC SPACES OF FIRST AND SECOND CATEGORY

Let (X, d) be a metric space. A subset of a metric space is said to be of first category if and only if it can be written as the union of a countable family of nowhere dense sets, otherwise it is said to be the second category.

(MEERUT 1992, 93, 2001)

For example. Set of rational number Q is of second category.

5.5.1 CONTRACTING MAPPING

Let (X, d) be a complete metric space. A mapping $f : X \to X$ is called a contracting mapping (or contraction on X) if there exist a real number α with $0 \leq \alpha < 1$ such that

$$d[f(x), f(y)] \leq \alpha \, d(x, y) < d(x, y) \ \forall x, y \in X$$

(MEERUT 1992, 94, 96, 2002, 03)

5.5.2 FIXED POINT

Let X be a non-empty set and let $f : X \to X$ be a mapping. A point $x \in X$ is said to be a fixed point of f if $f(x) = x$.

For example. Let $f : X \to X$ be the identity mapping on X since $f(x) = x \ \forall x \in X$. Therefore every point of X is a fixed point.

THEOREM I. **Let A be a subset of a metric space X, then the following statements are equivalent:**

(i) A is non-dense in X.

(ii) $\bar{A}$ contains no neighbourhoods.

(iii) $(\bar{A})'$ is dense in X.

PROOF. Let (X, d) be a metric space and A be any subset of X.

Here, we first prove (i) $\Leftrightarrow$ (ii)

A is non-dense in $X \qquad \Leftrightarrow \quad (\bar{A})^o = \phi$

$\Leftrightarrow$ No point of X is interior point of $\bar{A}$

$\Leftrightarrow$ $\bar{A}$ is not a nbd of any of its points

$\Leftrightarrow$ $\bar{A}$ contains no nbd.

We now prove (ii) $\quad \Leftrightarrow \quad$ (iii)

$\bar{A}$ contains no nbd $\quad \Leftrightarrow \quad$ For every $x \in X$, $S(x_0, r) \not\subset \bar{A}$, $r > 0$

$\Leftrightarrow$ $S(x,r) \cap (\bar{A})' \neq \phi$ for every $x \in X$ and every $r > 0$

$\Leftrightarrow$ every nbd of x contains a point of $(\bar{A})'$ for every $x \in X$

$\Leftrightarrow$ x is an adherent point of $(\bar{A})'$ for every $x \in X$

$\Leftrightarrow$ $[(\bar{A}')] = X$ $\qquad$ [By definition of adherent point]

$\Leftrightarrow$ $(\bar{A})'$ is dense in X.

☞ **REMARKS**

➡ Since $(\bar{A})' = ext(A)$, it follows from the above theorem that A is non-dense if and only if $ext(A)$ is everywhere dense.

➡ If A is non-dense in X, then A' is dense in X.

➡ If A is nowhere dense, then $\bar{A}$ is not the entire space X.

THEOREM 2. ***The union of a finite number of nowhere dense sets is no where dense.***

PROOF. Let A and B be two nowhere dense subsets of a metric space (X, d).

Let us write $\quad G = (\overline{A \cup B})^*$

so that $\qquad G \subset \overline{A \cup B} = \bar{A} \cup \bar{B}$.

It follows that

$$G \cap (\bar{B})' \subset (\bar{A} \cup \bar{B}) \cap (\bar{B})'$$
$$= [\bar{A} \cap (\bar{B})'] \cup [\bar{B} \cap (\bar{B})'] \qquad \text{(By distributivity)}$$
$$= \bar{A} \cap (\bar{B})' \qquad [\because \bar{B} \cap (\bar{B})' = \phi]$$
$$\subset \bar{A}$$

$\Rightarrow \quad [G \cap (\bar{B})']^* \subset (\bar{A})^* = \phi \qquad \qquad \ldots(1)$

$[\because A$ is non-dense, $i.e.,$ $(\bar{A})^* = \phi]$

But $\quad [G \cap (\bar{B})']^* = G \cap (\bar{B})' \qquad \qquad \ldots(2)$

$[\because G \cap (\bar{B})'$ is open so use $A^* = A]$

From (1) and (2), we conclude that

$$G \cap (\bar{B})' = \phi$$

$\Rightarrow \qquad G \subset \bar{B} \Rightarrow G^* = (\bar{B})^* = \phi \qquad [\because B$ is non-dense in X so $(\bar{B})^* = \phi]$

But $\qquad G^* = (\overline{A \cup B})^{**} = (\overline{A \cup B})^*$

So that $(\overline{A \cup B})^* = \phi$.

Hence, $A \cup B$ is a non-dense.

In general, the union of a finite number of no where dense set is no where dense.

THEOREM 3. **(Bair's Category Theorem)** ***Every complete metric space is of second category.*** (MEERUT 1990, 92, 93, 96, 2000, 01, 03, 05,13,14,18; BANGLURU-2005, 07; KANPUR-2001, 07)

PROOF. Let (X, d) be a complete metric space. To show X is of second category.

Let if possible, X is not of second category, i.e., X is of first category.

$\Rightarrow \quad X$ can be expressed as a countable union of nowhere dense sets arranged in a sequence $< A_n >$.

Now, since A_1 is non-dense and so there exists a closed sphere K_1 of radius $r_1 < \dfrac{1}{2}$ s.t. $K_1 \cap A_1 = \phi$.

Let the open sphere, with same centre and radius as r_1, be denoted by S_1.

In S_1, we can find a closed sphere K_2 of radius $r_2 < \left(\frac{1}{2}\right)^2$ such that $K_2 \quad A_2 = \phi$ and so $K_2 \cap A_1 = \phi$.

Continuing this process, we constitute a nested sequence $< K_n >$ of closed spheres which have the following properties:

For each positive integer, n, K_n does not intersect $A_1, A_2, ..., A_n$.

The radius of K_n tends to zero as $n \to \infty$. (For $\frac{1}{2^n} \to 0$ as $n \to \infty$)

Since (X, d) is complete, therefore by Cantor's intersection theorem $\bigcap_n K_n$ contains a single point x_0, so that $x_0 \in K_n$, $\forall n$.

Clearly, $x_0 \notin A_n \ \forall n$. By assumption, it is not possible, because X is the union of A_n's. Thus, X is not of first category, which is a contradiction.

Hence, X is of second category.

☞ **REMARK**

⟹ The Bair's category theorem can also be stated as follows :

(i)   "If $< A_n >$ is a sequence of nowhere dense sets in a complete metric space $(X, d) \ \exists$ a point in X, which is not in A_n's."

(ii) "If a complete metric space is the union of a sequence of its subsets, then the closure of at least one set in the sequence must have non-empty interior."

THEOREM 4.   *Every contracting mapping is continuous.*   (MEERUT 1993, 2005)

PROOF.   Let f be a contracting mapping on a metric space (X, d), therefore there exists a positive real number $\alpha < 1$ such that

$$f : (X, d) \to (X, d), \ \forall \, x, y \in X$$

and $d\,[f(x), f(y)] \le \alpha \, d\,(x, y) < d(x, y)$ ...(1)

Taking $d(x, y) < \varepsilon$, we get $d\,[f(x), f(y)] < \varepsilon$.

Given $\varepsilon > 0 \ \exists \ \delta > 0$ such that $d\,[f(x), f(y)] < \varepsilon$, whenever $d(x, y) < \delta$.

Here $\varepsilon = \delta$

Hence, f is a continuous mapping.

THEOREM 5.   **(Banach Fixed Point Theorem).** *If f is a contracting mapping on a complete metric space (X, d), then there exists a unique point p in X such that $f(p) = p$.*

(MEERUT-1992,93,94,96,2000,01,02,03,13; KANPUR-2005,07; RAJASTHAN-2006,08; NET-2007)

PROOF.   Let (X, d) be a complete metric space. Define a contracting mapping $f : (X, d) \to (X, d)$. Then by definition of contracting mapping, there exists a positive real number $\alpha < 1$ such that

$$d\,[f(x), f(y)] \le \alpha \, d(x, y) \ \forall \, x, y \in X \qquad ...(1)$$

Then $d\,[f^2(x), \ f^2(y)] = d[f(f(x)), f(f(y))]$

$$\le \alpha \, d\,[f(x), f(y)] \qquad \text{[using (1)]}$$

$$\le \alpha \cdot \alpha \cdot d(x, y) \qquad \text{[using (1)]}$$

$\Rightarrow \quad d\,[f^2(x), f^2(y)] \le \alpha^2 d(x, y)$.

In general, we get

$$d\,[f^n(x), \ f^n(y)] \le \alpha^n \cdot d(x, y), \ n \in \mathbf{N} \qquad ...(2)$$

Next, we suppose that $x_0 \in X$ and let

$$x_1 = f(x_0), \ x_2 = f(x_1) = ff(x_0) = f^2(x_0)$$
$$x_n = f(x_{n-1}) = ... = f^n(x_0).$$

Thus, $x_n = f^n(x_0)$,

$$x_{n+1} = f^{n+1}(x_0) = f^n[f(x_0)] = f^n(x_1)$$

$$\Rightarrow \quad x_n = f^n(x_0), \ x_{n+1} = f^n(x_1). \quad ...(3)$$

We claim that $< x_n >$ is a Cauchy sequence.

Let $m, n \in N$ be arbitrary such that $m > n$ and $m = n + p$, p being a positive integer ≥ 1.

Now, $d(x_n, x_m) = d(x_n, x_{n+p})$
$$\leq d(x_n, x_{n+1}) + d(x_{n+1}, x_{n+2})$$
$$+ ... + d(x_{n+p-1}, x_{n+p})$$

(By triangle inequality)

Now using (3)

$$d(x_n, x_n) \leq d[f^n(x_0), f^n(x_1)]$$
$$+ d[f^{n+1}(x_0), f^{n+1}(x_1)] + ...$$
$$+ d[f^{n+p-1}(x_0), f^{n+p-1}(x_1)]$$

On using (2)

$$d(x_n, x_m) \leq \alpha^n \, d(x_0, x_1)$$
$$+ \alpha^{n+1} d(x_0, x_1) + ... + \alpha^{n+p-1} d(x_0, x_1)$$
$$= \alpha^n \, d(x_0, x_1)[1 + \alpha + \alpha^2 + ... + \alpha^{p-1}]$$
$$= d(x_0, x_1) . \alpha^n \frac{(1 - \alpha^p)}{(1 - \alpha)}$$

[being the sum of G.P.]

$$\leq d(x_0, x_1) . \frac{\alpha^n}{1 - \alpha} \text{ for } 0 < \alpha < 1.$$

Thus, $d(x_n, x_m) \leq \dfrac{\alpha^n}{1 - \alpha} d(x_0, x_1)$ ...(4)

Therefore, $0 < \alpha < 1 \Rightarrow \lim\limits_{n \to \infty} \alpha^n = 0$.

Now (4) gives

$$d(x_n, x_m) < \varepsilon$$

$\Rightarrow \ < x_n >$ is a Cauchy sequence.

Since (X, d) is complete, therefore by definition, every Cauchy sequence in X converges to some point in X.

$\Rightarrow \quad \exists \, p \in X$ s.t. $x_n \to p$ as $n \to \infty$.

Now, $\lim x_n = p$.

Also, f is contracting mapping

$\Rightarrow \quad f$ is continuous.

($\because$ Every contracting mapping is continuous)

FACTS : TO THE POINT

▶ The open sphere $S(x, \varepsilon)$ contains all but a finite number of terms of the sequence.

▶ $< x_n > \to x$ if and only if the sequence of real numbers $d(x_n, x) \to 0$.

▶ Any discrete metric space is complete.

▶ A subspace of a complete metric space need not be complete.

▶ Q is not complete.

▶ In Cantor's intersection theorem $\cap F_n$ may be empty in any of the following conditions:
  (i) if each F_n is not closed.
  (ii) if the condition $\delta(F_n) \to 0$ is omitted.

▶ Let X be a metric space $A \subseteq X$ then following conditions are equivalent:
  (i) A is nowhere dense in X.
  (ii) $\bar{A}$ does not contain any non-empty open set.
  (iii) Each non-empty open set has a non-empty open subset disjoint from $\bar{A}$.
  (iv) Each non-empty open set has a non-empty open subset disjoint from A.
  (v) Each non-empty open set containing an open sphere disjoint from A.

▶ A subset of a nowhere dense set is nowhere dense set.

▶ The union of a countable number of sets which are of the first category is again a first category.

▶ If a complete metric space X is the union of a sequence of its subsets, then the closure of at least one set in the sequence must have non-empty interior.

▶ Any complete metric space in which every finite subset is nowhere dense is uncountable and hence R is uncountable.

▶ If X is countable then X is of first category.

▶ Any discrete metric space is of second category.

▶ $[a, b], \ ]a, b[$ and $[a, b[$ are of second category.

▶ R with usual metric is complete and of second category.

Therefore $x_n \to p \Rightarrow f(x_n) \to f(p) = \lim f(x_n) = f(p)$.

But $\qquad\qquad x_{n+1} = f(x_n)$.

$\therefore \qquad\qquad f(p) = \lim f(x_n) = \lim x_{n+1} = \lim x_n = p$

or $\qquad\qquad f(p) = p$.

$\Rightarrow \quad p$ is a fixed point.

Now to show the uniqueness of the point p .

Let if possible $\exists$ another fixed point $q \in X$ such that $p \neq q$

and $\qquad\qquad f(q) = q$.

Consider $d[f(p), f(q)] \leq \alpha\, d(p, q)$

$\Rightarrow \qquad\qquad d(p, q) \leq \alpha\, d(p, q)$. But $d(p, q) \neq 0$.

$\Rightarrow \quad \alpha \geq 1$, which is a contradiction. $\qquad\qquad (\because 0 < \alpha < 1)$

Hence, p is unique.

5.6 COMPLETION OF A METRIC SPACES

5.6.1 ISOMETRIC SPACES

A metric space (X, d) is said to be isometric to a space (Y, d_1) if there exist one-one mapping $f : X \to Y$ which preserves the distance, i.e., $d(x, y) = d_1[f(x), f(y)] \quad \forall\, x, y \in X$

5.6.2 COMPLETION OF A METRIC SPACE

If (X, d) is not complete metric space, then we extend the metric space (X, d) to (X_1, d_1) which is complete and (X, d) is a dense subspace.

A metric space (X_1, d_1) is called completion of a metric space (X, d) if X_1 is complete and X is isometric to a dense subset of X .

THEOREM I. *The metric space **(R, d)** is complete where **d** denotes usual metric for R.*

(ROHILKHAND 2001, 07)

PROOF. Let $< s_n >$ be a Cauchy sequence of real numbers. Now, we define a sequence $< n_k >$ of positive integers as n_{k+1} is the smallest integer greater than n_k such that

$$n, m \geq n_k \Rightarrow |s_n - s_m| < \frac{1}{2^{k+1}} \qquad\qquad ...(1)$$

(Since $< s_n >$ is Cauchy sequence)

Now, let I_k be the closed interval $[s_{n_k} - 2^{-k},\ s_{n_k} + 2^{-k}]$.

Here, it is easy to see that $I_{k+1} \subset I_k$, we have $|s_{n_k} - s_{n_{k+1}}| < \frac{1}{2^{k+1}}$.

Also, the length of $I_k \to 0$ as $k \to \infty$. So by nested interval theorem, $\bigcap\limits_{k=1}^{\infty} I_k$ consists of exactly one point, say $a \in R$.

Thus, $a \in I_k \ \forall\, k \in N$ so that

$$|a - s_{n_k}| < \frac{1}{2^k} \text{ for all } k \in N . \qquad\qquad ...(2)$$

Now for $n \geq n_k$, we have from (1)

$$|s_{n_k} - s_n| < \frac{1}{2^{k+1}} < \frac{1}{2^k} \qquad\qquad ...(3)$$

Hence, for all $n \geq n_k$

$$|a - s_n| = |a - s_{n_k} + s_{n_k} - s_n|$$

$$\leq |a - s_{n_k}| + |s_{n_k} - s_n|$$

$$< \frac{1}{2^k} + \frac{1}{2^k} = \frac{1}{2^{k+1}} \qquad \text{[by (2) and (3)]}$$

It follows that $\lim_{n \to \infty} s_n = a$.

Thus every Cauchy sequence in **R** converges to a point in **R**.

Hence, **R** is complete.

THEOREM 2. *The set R^n of all n-tuples $x = (x_1, x_2, ..., x_n)$ of real numbers is a complete metric space with respect to the usual metric*

$$d(x, y) = \left[\sum_{i=1}^{n} (x_i - y_i)^2 \right]^{1/2} \qquad \text{(PATNA 2004)}$$

PROOF. Here, we shall denote the element of R^n by a functional notation. Thus, an element of R^n will be a real function defined on the set $\{1, 2, 3, ..., n\}$. Let $f(m)$ stand for the element $[f_m(1).f_m(2)...f_m(n)]$ of R^n. Now, we shall show that R^n is complete.

Let $< f_m >$ be a Cauchy sequence in R^n, then for a given $\varepsilon > 0$, $\exists$ a positive integer $n(\varepsilon)$ such that

$$p, q \geq n(\varepsilon) \implies d(f_p, f_q) < \varepsilon$$

$$\implies d^2(f_p, f_q) < \varepsilon^2$$

$$\implies \sum_{i=1}^{n} [f_p(i) - f_q(i)]^2 < \varepsilon^2$$

$$\implies [f_p(i) - f_q(i)]^2 < \varepsilon^2$$

$$\implies [f_p(i) - f_q(i)]^2 < \varepsilon^2 \qquad (i = 1, 2, 3, ..., n)$$

$$\implies [f_p(i) - f_q(i)] < \varepsilon \qquad (i = 1, 2, 3, ..., n)$$

So, $< f_m(i) >$ is a Cauchy sequence of real numbers. Since **R** is complete, so every Cauchy sequence in **R** converge to a point in **R** *i.e.*, the sequence $< f_m >$ converges pointwise to a limit function f defined by $\lim f_m(i) = f(i)$.

Since the set $\{1, 2, ..., n\}$ is finite, this convergence is uniform. Hence there exists a positive integer n_0 such that $|f_m(i) - f(i)| < \frac{\varepsilon}{\sqrt{n}}$, for all $m \geq n_0$ and $\forall i$.

Now, squaring and adding the above result for $i = 1, 2, ..., n$, we get

$$\sum_{i=1}^{n} |f_m(i) - f(i)|^2 < \frac{\varepsilon^2}{n} . n = \varepsilon^2$$

or $d^2(f_m . f) < \varepsilon^2 \ \forall m \geq n_0$ thus $d(f_m . f) < \varepsilon, \ \forall m \geq n_0$

which shows that the Cauchy sequence $< f_m >$ converges to the limit f.

Hence, R^n is complete.

THEOREM 3. *The Hilbert space (l_2, d) is complete.* (KANPUR 2006, 08; TAMIL NADU 2005)

PROOF. Let $< s_n >$ be a Cauchy sequence in l_2. Since each s_n is a sequence, so we have

$$s_n = < s_1^n, s_2^n, s_3^3, ... >$$

So that s_k^n denotes the kth term of the sequence s_n. Now

$$d(s_m, s_n) = \left[\sum_{i=1}^{n} (s_i^m - s_i^n)^2 \right]^{1/2}$$

Now, since $< s_n >$ in a Cauchy sequence, so for a given $\varepsilon > 0 \ \exists$ a positive integer m_0 such that

$$m, n \geq m_0 \ \Rightarrow \ d(s_m - s_n) < \varepsilon \ \Rightarrow \ d^2(s_m - s_n) < \varepsilon^2$$

$$\Rightarrow \ \sum_{i=1}^{n} (s_i^m - s_i^n)^2 < \varepsilon^2 \ \Rightarrow \ (s_i^m - s_i^n) < \varepsilon^2, \quad \forall i \in \boldsymbol{N}$$

$$\Rightarrow \ |s_i^m - s_i^n| < \varepsilon \quad \forall i \in \boldsymbol{N}.$$

So, $\{< s_i^n > : i = 1, ..., n\}$ is a Cauchy sequence of real numbers for all $i \in \boldsymbol{N}$, and it must converge to some real number x_i, let $X = < x_1, x_2, ... >$.

Now, we shall show that the sequence $< s_n >$ converges to X, and that $X \in l_2$.

Now, for a fixed integer M, we have $\sum_{i=1}^{m} (s_i^m - s_i^n) < \varepsilon^2$, $\forall m, n \geq m_0$.

If we fix n and let $m \to \infty$, we get

$$\sum_{i=1}^{m} (s_i - s_i^n) < \varepsilon^2 \qquad \qquad \text{...(1)}$$

Since (1) hold for all m, we have for $n \geq m_0$

$$\sum_{i=1}^{\infty} (s_i - x_i^n)^2 < \varepsilon^2. \qquad \qquad \text{...(2)}$$

i.e., $\qquad d^2(X, s_n) < \varepsilon^2 \ \Rightarrow \ d(X, s_n) < \varepsilon$.

Hence, $\qquad \lim_{n \to \infty} s_n = X$.

Now, we shall show that $X \in l_2$, we have

$$x_i^2 = (x_i - s_i^{m_0} + s_i^{m_0})^2$$

$$= (x_i - s_i^{m_0})^2 + (s_i^{m_0})^2 + 2(x_i - s_i^{m_0})(s_i^{m_0})$$

$$\geq 2(x_i - s_i^{m_0})^2 + 2(s_i^{m_0})^2. \qquad \qquad (\because 2ab \leq a^2 + b^2)$$

So, $\qquad \sum_{i=1}^{\infty} x_i^2 \leq 2 \sum_{i=1}^{\infty} (x_i - s_i^{m_0})^2 + 2 \sum_{i=1}^{\infty} (s_i^{m_0})^2$

$$< 2\varepsilon^2 + 2 \sum_{i=1}^{\infty} (s_i^{m_0})^2 \qquad \text{[by (2)]} \qquad \qquad \text{...(3)}$$

Since $s_{m_0} \in l_2$ the series $\sum_{i=1}^{\infty} (s_i^{m_0})^2$ converges so that for some $B > 0$, we have

$$\sum_{i=1}^{\infty} (s_i^{m_0})^2 < B \qquad \qquad \text{...(4)}$$

Now from (2) and (4), we have $\sum_{i=1}^{\infty} x_i^2 < 2\varepsilon^2 + 2B$.

So, the series $\sum_{i=1}^{\infty} x_i^2$ is convergent and consequently $x \in l_2$.

Thus every Cauchy sequence $< s_n >$ in l_2 converges to a point x in l_2.

Hence, (l_2, d) is complete.

Completeness of $C[a, b]$. Let $C[a, b]$ is the set of all real valued continuous functions defined on I $(I = [a, b])$. For $f . g \in C[a, b]$, we define $d(f, g) = \sup \{[f(x) - g(x) : x \in I\}$. We have already seen that d is metric for $C[a, b]$. This space is known as the space of continuous functions on I.

THEOREM 4. *The space $C[a, b]$ is complete.* (DELHI 2003, 07, 10; OSMANIA 2007, 10)

PROOF. Let $I = [a, b]$ and let $< f_m >$ be a Cauchy sequence in $C[a, b]$.

Let for a given $\varepsilon > 0$, there exists a positive integers m_0 such that

$$m, n \geq m_0 \Rightarrow d(f_m, f_n) < \varepsilon$$

$$\Rightarrow \sup \{| f_m(x) - f_n(x)| : x \in I\} < \varepsilon.$$

So $| f_m(x) - f_n(x)| < \varepsilon, \forall m, n \geq m_0$ and for all $x \in I$.

This is the condition for uniform convergence. Thus the sequence $< >$ is a uniform convergence sequence of continuous functions and it must converge to a continuous function f on I.

Thus every Cauchy sequence $< f_m >$ in $C[a, b]$ converges to a point f in $C[a, b]$. Hence, $C[a, b]$ is complete.

THEOREM 5. *Let (X, d) be a metric space and S be the set of all Cauchy sequence in (X, d). Then $\cong$ is an equivalence relation on S.*

PROOF. We have to show that $\cong$ is an equivalence relation. For this we shall prove that is reflexive, symmetric and transitive.

(i) Reflexive : We have

$$d(s_n, s_n) = 0 \text{ for every } n$$

$$\Rightarrow \qquad s \cong s \ \forall s \in S$$

$$\Rightarrow \ \cong \text{ is reflexive.}$$

(ii) Symmetry : Let $s \cong t$

Then, $\lim_{n \to \infty} d(s_n, t_n) = 0$

Also, we have $d(s_n, t_n) = d(t_n, s_n)$

$$\Rightarrow \lim_{n \to \infty} d(t_n, s_n) = 0$$

$$\Rightarrow \qquad t \cong s$$

$$\Rightarrow \ \cong \text{ is symmetric.}$$

(iii) Transitivity : Let us suppose $s \cong t$ and $t \cong u$. We have to show that $s \cong u$.

Since $\qquad s \cong t \Rightarrow \lim_{n \to \infty} d(s_n, t_n) = 0$

Also, $\qquad t \cong u \Rightarrow \lim_{n \to \infty} d(t_n, u_n) = 0$

We have to prove $\lim_{n \to \infty} d(s_n, t_n) = 0$

Consider $\lim_{n \to \infty} d(s_n, t_n) = 0 \leq \lim_{n \to \infty} [d(s_n, t_n) + d(t_n, u_n)]$

$$= \lim_{n \to \infty} d(s_n, t_n) + \lim_{n \to \infty} d(t_n, u_n)$$

$$= 0 + 0$$

$$\Rightarrow \qquad \lim_{n \to \infty} d(s_n, u_n) \leq 0$$

$$\Rightarrow \qquad d(s_n, u_n) \leq 0$$

But $\qquad d(s_n, u_n) \geq 0$

Thus we conclude that $d(s_n, u_n) = 0$

$$= \lim_{n \to \infty} d(s_n, u_n) = 0 \quad \forall \in N$$

$$\Rightarrow \qquad s \cong u$$

Hence $\cong$ is transitive.

THEOREM 6. **(Completion Theorem).** *Let (X, d) be a metric space. There exists a complex metric space (X^*, d^*) in which (X, d) can be isometrically embedded in such a way that X is dense in X^*, i.e., (X, d) is symmetric to a dense subspace of $[X^*, d^*]$.*

PROOF. **STEP 1.** Let S be the set of all Cauchy sequence in (X, d) where X is not complete. Define a relation $\cong$ in S such that $s \cong t$ where $s = < s_n >$ and $t = < t_n >$ are two members of S if and only if $\lim_{n \to \infty} d(s_n, t_n) = 0$.

We know that $\cong$ is an equivalence relation (see previous theorem). Let us denote the set of all equivalence classes by X^*. To show X^* can be turned into a metric space by a suitable choice of distance function defined on X^*. Clearly X^* have two types of members. First type of element is an equivalence class of Cauchy sequence which converge to points in X and an element of the second type is an equivalence class which consist of equivalent Cauchy sequence not converging to any point of X. We shall denote an equivalence class contemporary to a sequence $S = < s_n >$ by $[S]$. Thus a typical number x^* or X^* is of the form $[s]$ where $[s]$ is an equivalence class represented by a Cauchy sequence $s = < s_n >$.

STEP 2. Define a mapping

$$d^* : X^* \times X^* \to R$$

By $d^*([s], [t]) = \lim_{n \to \infty} d(s_n, t_n)$ and show that d^* is a metric for X^*. We first show that $d^*([s], [t])$ actually exists for every $s \in S$ and every $t \in S$. We have

$$|d(s_m, t_m) - d(s_n, t_n)| = |d(s_m, t_m) - d(s_m, t_n) + d(s_m, t_n) - d(s_n, t_n)|$$

$$\leq |d(s_m, t_m) - d(s_m, t_n)| + |d(s_m, t_n) - d(s_n, t_n)| \qquad \dots(1)$$

(Using triangular inequality)

Also, we know that

$$|d(x, y) - d(y, z)| \leq d(x, z) \qquad \dots(2)$$

Applying (2) to (1), we have

$$|d(s_m, t_m) - d(s_n, t_n)| \leq d(t_m, t_n) + d(s_m, s_n) \qquad \dots(3)$$

Since $s = < s_n >$ and $t = < t_n >$ are both Cauchy sequences in (X, d) for a given $\varepsilon > 0$, there exists a positive integer $n(\varepsilon)$ such that

$$m, n \geq n(\varepsilon) \Rightarrow d(t_m, t_n) < \varepsilon / 2$$

and $\qquad d(s_m, s_n) < \varepsilon / 2$ $\qquad \qquad \dots(4)$

From (3) and (4), we obtain

$$|d(s_m, t_m) - d(s_n, t_n)| < \frac{\varepsilon}{2} + \frac{\varepsilon}{2} = \varepsilon$$

Whenever $m, n \geq n(t)$, it follows that $< d(s_n, t_n) >$ is a Cauchy sequence in R. Since R is complete $d^*([s], [t]) = \lim_{n \to \infty} d(s_n, t_n)$ exists.

We now show that d^* is well defined. In other words, we shall show that if $[s] = [u]$ and $[t] = [v]$, then $d^*([s], [t]) = d^*([u], [v])$.

Now for every positive integer n , we have

$$d(s_n, t_n) \le d(s_n, u_n) + d(u_n, t_n)$$
$$\le d(s_n, u_n) + d(u_n, v_n) + d(v_n, t_n)$$
$$\lim d(s_n, t_n) \le \lim d(s_n, u_n) + \lim d(u_n, v_n) + \lim d(v_n, t_n).$$

Hence, $d^*([s], [t]) \le \lim d(s_n, u_n) + d^*([u], [v]) + \lim d(v_n, t_n)$ ...(5)

Since $[s] = [u]$. We have $s \cong u$ so that by definition of $\cong \lim d(s_n, u_n) = 0$.
Similarly, $\lim d(v_n, t_n) = 0$. Then (5) gives

$$d^*([s], [t]) \in d^*([u], [v]).$$ ...(6)

On the other hand

$$d(u_n, v_n) \le d(u_n, s_n) + d(s_n, t_n) + d(t_n, v_n).$$

Arguing exactly in the same manner as above, we shall obtain

$$d^*([u], [v]) \le d^*([s], [t])$$ ...(7)

From (6) and (7), we get

$$d^*([s], [t]) = d^*([u], [v])$$

Thus d^* is well defined. It will now be shown that d^* is a metric on X^* .
Since $d(s_n, t_n) \ge 0$ for every positive integer n , it follows

$$d^*([s], [t]) = \lim_{n \to \infty} d(s_n, t_n) \ge 0$$

so that $[m_1]$ is satisfied. To prove $[m_2]$, we have

$$[s] = [t] \Leftrightarrow s \cong t \Leftrightarrow \lim d(s_n, t_n) = 0 \Leftrightarrow d^*([s], [t]) = 0$$

Thus $[s] = [t] \Leftrightarrow$ iff $d^*([s], [t]) = 0$. Again

$$d^*([s], [t]) = \lim d(s_n, t_n) = \lim d(t_n, s_n)$$
$$= d^*([t], [s])$$

so that $[m_3]$ is also satisfied. To prove $[m_4]$, we first observed that

$$d(s_n, t_n) \le d(s_n, u_n) + d(u_n, t_n)$$

For every integer n

$$\lim d(s_n, t_n) \le \lim (s_n, u_n) + \lim d(u_n, t_n)$$

or $\quad d^*([s], [t]) \le d^*([s], [u]) + d^*([u], [t])$.

It follows that d^* is a metric on X^* .

STEP 3. Let $s^x = <s_n^x>$ denote the Cauchy sequence in X defined by $S_n^x = x$ for every positive integer n so that $[s^x]$ is the equivalence class of all Cauchy sequences which converge to x . Let Y^* be the subset of X^* defined by $Y^* = \{[s^x] : x \in X\}$. We shall show that (Y^*, d^*) and (X, d) are isometric. Define a mapping $f: X \to Y^*$ by $f(x) = [s^x] \; \forall \, x \in X$. If $y \in Y^*$, then $y = [s^x]$ for some $x \in X$ so that $y = f(x)$. Hence f is an onto mapping. Now let x, y be any two distinct members of X . Then, $x \ne y \Rightarrow d(x, y) = c > 0$.

$\Rightarrow \lim d(s_n^x, s_n^y) = c > 0 \Rightarrow s^x$ is not equivalent to s^y :

$[s^x]$ and $[s^y]$ are disjoint $\Rightarrow f(x) \ne f(y)$.

Therefore, f is one-one mapping. Finally, if x and y are any two members of X, then

$$d^*(f(x),\ f(y)) = d^*([s^x],[s^y]) = \lim d(s_n^x, s_n^y) = d(x, y) \qquad \ldots(8)$$

Hence, (X, d) and (Y^*, d^*) are isometric.

STEP 4. It will now be shown that (Y^*, d) is a dense subspace of (X^*, d^*). We shall show that every point of $X^* \sim Y^*$ is a limit point of Y^*. Let $[s]$ be any arbitrary member of $X^* \sim Y^*$. Since $s =< s_n >$ is a Cauchy sequence in X, for a given $\varepsilon > 0$, there exists an integer k such that $m, n \ge k \Rightarrow d(s_m, s_n) < \varepsilon / 2$. $\ldots(9)$

Define a sequence $y^\varepsilon =< y_n^\varepsilon >$ in X by $y^\varepsilon = s_k$ for every positive integer n. Evidently y^ε is a Cauchy sequence converging to $s_k \in X$ and so $[y^\varepsilon] \in Y^*$. Moreover $d^*([s],[y^\varepsilon]) = \lim\limits_{n \to \infty} d(s_n, y_n^\varepsilon) = \lim\limits_{n \to \infty} d(s_n, s_k)$.

Since $d(s_n, s_k) < \varepsilon / 2$ for every $n \ge k$, we have

$$\lim_{n \to \infty} d(s_n, s_k) \le \varepsilon / 2 < \varepsilon$$

or $d^*([s],[y^\varepsilon]) \le \varepsilon$

$\Rightarrow$ $[y^\varepsilon] \in S([s],\ \varepsilon)$

Thus, every open ε-sphere about $[s]$ contains a point y^ε of Y^*. Since $[s] \notin Y^*$, it follows that $[s]$ is limit point of Y^*. It follows that $X^* \sim Y^* \subset D(Y^*)$

Hence, $\overline{Y}^* = Y * \cup D(Y^*) \supset Y^* \cup (X^* - Y^*) = X^*$

But $X^* \supset (\overline{Y}^*)$ always. Therefore $(Y^*) = X^*$. Consequently Y^* is dense in X^*.

STEP 5. Finally we show that (X^*, d^*) is complete.

We first prove that every Cauchy sequence of point of Y^* converges to a point on X^* and complete the proof by using the fact that Y^* is dense in X^*.

We shall denote a Cauchy sequence in X which converge to the point x_n in X. Then $< x_n >$ is a sequence in X such that $d^*(s_m^x, s_n^x) = d(x_m, x_n)$ by (8) $\ldots(10)$

Since $< \lceil x_n \rceil >$ is a Cauchy sequence in Y^*, it follows from (10) that $< x_n >$ is a Cauchy sequence in X. Let us denote the equivalence class containing the sequence $< x_n >$ by $[x]$ so that $[x] \in X^*$.

Let $< y_n >$ denote a Cauchy sequence in the equivalence class $[y]$. Then $d^*([x],[y]) = \lim\limits_{m \to \infty} d(x_n, y_n)$.

If we take $[y]$ to be the point $[s_n^x]$ of Y^* which contains the constant sequence all of whose members are x_n, then

$$d^*([x],[s_n^x]) = \lim_{n \to \infty} (d(x_m, x_n))$$

But since $< x_n >$ is a Cauchy sequence in X for every positive number ε, there exists n_1 such that $m > n \ge n_1 \Rightarrow d(x_m, x_n) > \varepsilon / 2$

It follows that $d^*([x],[s_n^*]) = \lim\limits_{n \to \infty} d(x_m, x_n) \le \varepsilon / 2 < \varepsilon$

Whenever $n \ge m$.

Hence, the Cauchy sequence $< s_n^x >$ of points of Y^* converge to the point $[x]$ of X^*.

Finally let $< x_n^* >$ be any Cauchy sequence in X^*.

Then for every $\varepsilon > 0$, there exists an integer $n_2 \geq 0$ such that

$$m > n \geq n_2 \Rightarrow d^*(x_m^*, x_n^*) < \varepsilon / 3 \qquad \text{...(11)}$$

Since Y^* is dense in Y^*, for every positive integer n, there exists a point y_n^* of Y^* such that

$$d^*(x_n^*, y_n^*) < \varepsilon / 3 \qquad \text{...(12)}$$

Now using triangle inequality, we get

$$d^*(y_n^*, y_n^*) \leq d^*(y_m^*, x_n^*) + d(x_n^*, y_n^*)$$
$$\leq d^*(y_m^*, x_m^*) + d^*(x_m^*, x_n^*) + d^*(x_n^*, y_n^*)$$
$$< \varepsilon / 3 + \varepsilon / 3 + \varepsilon / 3 = \varepsilon \text{ by (11) and (12)}$$

It follows that $< y_n^* >$ is a Cauchy sequence of points in Y^*.

By what has been proved above $< y_n^* >$ must converge to a point, say $y^* \in X^*$ but $d^*(y_m^*, x_n^*)$

$$\leq d(y_m^*, x_n^*) + d^*(x_m^*, x_n^*)$$
$$< \varepsilon / 3 + \varepsilon / 3 = 2\varepsilon / 3$$

where $m > m \leq n_2$. Letting $m \to \infty$, we obtain $d^*(y^*, x_n^*) \leq 2\varepsilon / 3 < \varepsilon$.

Whenever $n \geq n_2$. Hence $< x_n^* >$ converges to $y^* \in X^*$.

Accordingly (X^*, d^*) is complete.

THEOREM 7. **All completions of a metric space are isometric.**

PROOF. Let (X, d) be a metric space and let (X^*, d^*) and $(\overline{X}, \overline{d})$ being two completions of it. We can define an isometry between X^* and $\overline{X}$ as follows :

Since X is dense in X^*, to each $x^* \in X^*$, there exists some sequence $< x_n >$ of points of X converging to x^*. But we may also consider $< x_n >$ as a Cauchy sequence in X and since $\overline{X}$ is complete $< x_n >$ must converge to some point $\overline{x} \in \overline{X}$.

The above construction enables us to define a mapping of X^* onto $\overline{X}$ which turns out to be an isometry. We define the mapping $f : X^* \to X$ by setting $f(x^*) = \overline{x} \ \forall x^* \in X^*$. It can be easily shown that this construction is independent of the particular sequence $< x_n >$ converging to x^* so that f is well defined. We leave it to the reader to show that f is one-one mapping of X^* onto $\overline{X}$. Evidently $f(x) = x \ \forall \ x \in X$.

Now let $x_n \to x^*$ in X and $x_n \to \overline{x}$ in $\overline{X}$. Also, let $y_n \to y^*$ in X and $y_n \to \overline{y}$ in $\overline{X}$.

Then, $\qquad d^*(x^*, y^*) = \lim_{n \to \infty} d(x_n, y_n)$ and $d(\overline{x}, \overline{y}) = \lim_{n \to \infty} d(x_n, y_n)$

$\therefore \qquad\qquad \overline{d}(\overline{x}, \overline{y}) = d^*(x^*, y^*)$

$$\overline{x} = f(x^*) \text{ and } \overline{y} = f(y^*)$$

$$\overline{d} f((x^*), f(y^*)) = d^*(x^*, y^*)$$

It follows that f is isometry. Hence all completions of X are isometric.

SOLVED EXAMPLES

EXAMPLE 1. *Show that the set C of complex number with usual metric is a complete metric space.*

SOLUTION. Let C is the set of complex number and $z_1, z_2 \in C$ such that $z_1 = x_1 + i y_1$ and $z_2 = x_2 + i y_2$ where $x_1, y_1, x_2, y_2 \in R$.

Let us define a metric on C as $d(z_1, z_2) = |z_1 - z_2|$.

Further, let $< z_n >$ be a Cauchy sequence in C.

To show that C is a complete metric space, we have to show that $< z_n >$ converges to a point $z \in C$. Let for given $\varepsilon > 0$, there exist $\in$ such·that

$$|z_n - z_m| < \varepsilon, \quad \forall n \geq m$$

$$\Rightarrow \quad |(x_n + i y_n) - (x_m + i y_m)| < \varepsilon$$

$$\Rightarrow \quad |(x_n - x_m) + i(y_n - y_m)| < \varepsilon$$

$$\Rightarrow \quad |(x_n - x_m) + i(y_n - y_m)|^2 < \varepsilon^2$$

$$\Rightarrow \quad |x_n - x_m|^2 + |y_n - y_m|^2 < \varepsilon^2$$

$$\Rightarrow \quad |x_n - x_m| < \varepsilon \text{ and } |y_n - y_m| < \varepsilon, \forall n \geq m$$

$$\Rightarrow \quad < x_n > \text{ and } < y_n > \text{ are Cauchy sequence in } R.$$

And we know that every Cauchy sequence converges to a point that is every Cauchy sequence is convergent

$$\Rightarrow \quad x_n \to x \text{ and } y_n \to y \text{ in } R$$

$$\Rightarrow \quad z_n \to x + iy = z \text{ in } C$$

$$\Rightarrow \quad < z_n > \text{ converges to a point } z \text{ in } C.$$

$$\Rightarrow \quad C \text{ is convergent.}$$

$$\Rightarrow \quad (C, d) \text{ is a complete metric space.}$$

THEOREM 8. *Let C [a, b] be the set of all continuous functions on [a, b]. For $f, g \in C[a, b]$, define*

$$\rho(f, g) = \left\{ \int_a^b |f(x) - g(x)|^2 \, dx \right\}^{1/2}$$

then ρ is a metric for $C[a, b]$, which is not complete.

(GARHWAL 2005; HIMACHAL 2009)

PROOF. It is easy to see that ρ is a metric for $C[a, b]$.

Now we shall show that the space is not complete. For this, consider the sequence of continuous functions $< f_n >$ defined on $[-1, 1]$ by

$$f_n(x) = \begin{cases} 0 & \text{if } -1 \leq x \leq 0 \\ nx & \text{if } 0 \leq x \leq \frac{1}{n} \\ 1 & \text{if } \frac{1}{n} \leq x \leq 1 \end{cases}$$

Then for $m > n$, we have

$$d^2(f_m, f_n) = (m - n)^2 \int_0^{1/m} x^2 \, dx + \int_{1/m}^{1/n} (1 - nx)^2 \, dx$$

$$= \frac{(m - n)^2}{3m^3} + \left[\frac{1}{3n} \left(1 - \frac{n}{m} \right)^3 \right]$$

$$= \frac{(m - n)^2}{3m^3} + \frac{(m - n)^3}{3nm^3} = \frac{(m - n)^2}{3nm^2}$$

$$< \frac{1}{3n} < \varepsilon \quad \text{if} \quad n > \frac{1}{3\varepsilon}.$$

Therefore, the sequence is a Cauchy sequence.

Suppose, if possible, this Cauchy sequence converges to a continuous function f so that

$$d(f_n, f) = \int_{-1}^{1} |f_n(x) - f(x)|^2 \, dx \to 0$$

This implies that the integral with any limits between ± 1 also tends to 0. Thus $\int_{-1}^{0} |f_n(x) - f(x)|^2 \, dx \to 0$.

But $f_n(x) = 0$ when $x \le 0$ and hence this interval is independent of n. So the continuous function f is such that $\int_{-1}^{0} |f(x)|^2 \, dx = 0$.

It follows that $f(x) = 0$ when $x \le 0$. Again, if $c > 0$, then $\int_{c}^{1} |f_n(x) - f(x)|^2 \, dx \to 0$ as $n \to \infty$.

If we choose $n \to \frac{1}{c}$, we have $\int_{c}^{1} |1 - f(x)|^2 \, dx \to 0$ as $n \to \infty$.

As the interval is independent of n, vanishes, and since $f(x)$ is continuous, we have $f(x) = 1$ for $x \ge c$. But we can choose as near to zero as we want, thus there exists a continuous function which vanishes when $x \le 0$ which is equal to 1 when $x > 0$. So the Cauchy sequence does not converge to a point of $C[a,b]$. Hence the space is not complete.

THEOREM 9. *Let (X, d) and (Y, e) be two complete metric spaces then the product space $Z = X \times Y$ with metric*

$$\rho(z_1, z_2) = \sqrt{[d^2(x_1, x_2) + e^2(y_1, y_2)]}$$

is complete where $z_1 = (x_1, y_1)$ and $z_2 = (x_2, y_2)$.

PROOF. We know that ρ is a metric for Z. Now we show that the product space (Z, ρ) is complete. Let $< z_n >$ be a Cauchy sequence in Z. Then for a given $\varepsilon > 0$, there exists a positive integer m_0 such that

$$m, n \ge m_0 \Rightarrow \rho(z_m, z_n) < \varepsilon$$

$$\Rightarrow \rho^2(z_m, z_n) < \varepsilon^2$$

$$\Rightarrow d^2(x_m, x_n) + e^2(y_m, y_n) < \varepsilon^2$$

$$\Rightarrow d^2(x_m, x_n) < \varepsilon^2 \quad \text{and} \quad e^2(y_m, y_n) < \varepsilon^2$$

$$\Rightarrow d(x_m, x_n) < \varepsilon \quad \text{and} \quad e(y_m, y_n) < \varepsilon.$$

It follows that $< x_n >$ and $< y_n >$ are Cauchy sequences in the space X and Y respectively. Since these spaces are complete, the sequences $< x_n >$ and $< y_n >$ converges respectively to points $x \in X$ and $y \in Y$. It follows that the sequence $< z_n >$ converges to $z = (x, y) \in Z$ and consequently the product space $Z = X \times Y$ is complete.

☛ **REMARK**

⟹ The space defined in the above theorem is known as product metric space.

EXERCISE 5.1

1. Show that the set of real numbers is complete.

2. Show that the Hilbert space is complete.

3. Define a complete metric space. Give an example of a complete metric space.

4. Define an incomplete metric space. Give an example of an incomplete metric space.

5. Show the set C^n of all ordered n-tuples $z = (z_1, z_2, ..., z_n)$ of complex number is a complete metric space with respect to the usual metric d defined by

$$d(z, u) = \left[\sum_{i=1}^{n} |z_i - u_i|^2 \right]^{1/2},$$

where $(z = z_1, z_2, ..., z_n)$ and $u = (u_1, u_2, ..., u_n)$

6. Show that the metric space of rational numbers with the usual metric is incomplete.

7. Let $D(a, b)$ denote the set of all functions f on $[a, b]$ which have continuous derivatives at all points of $I = [a, b]$. For $f, g \in D[a, b]$, define
$$d(f, g) = |f(a) - g(b)|$$
$$+ \sup\{ |f'(x) - g'(x)| : x \in \mathbf{Z} \}$$

Show that d is a metric for $D(a, b)$ and that the space $(D[a, b], d)$ is complete.

8. Show that the completeness is preserved under isometrics.

9. Let X consist of all ordered pairs $\mathbf{X} = (x_1, x_2)$ be real numbers with metric

$$d(x, y) = \max\{ |x_1 - y_1|, |x_2 - y_2| \}.$$

Prove that X is complete.

10. Let X consist of all bounded sequences $X = \langle x_n \rangle$ in $\mathbf{R}$. Prove that

$$d(x, y) = \sup\{ |x_i - y_i| : i \in \mathbf{N} \}$$

is a metric on X and X is complete.

11. Which of the following are Cauchy sequence in $\mathbf{R}$ with respect to the usual metric d

   (i) $\left\langle \dfrac{1}{n} \right\rangle$ (ii) $\left\langle \dfrac{n-1}{n+1} \right\rangle$

  (iii) $\left\langle \dfrac{2^{n+1}}{2^n} + 1 \right\rangle$ (iv) $\left\langle 1 + \dfrac{(-1)^n}{n} \right\rangle$

12. Let X consist of all sequence $X = \langle x_n \rangle$ in $\mathbf{R}$. Show that $\dfrac{2}{\varepsilon}$ is a metric on

$$d(s_n, s_m) \le \frac{1}{n} + \frac{1}{m} < \frac{\varepsilon}{2} + \frac{\varepsilon}{2} \quad \forall m, n \ge p$$

and X is complete.

HINTS TO SELECTED PROBLEMS

10. Let $\langle x_n \rangle$ be a Cauchy sequence in the given space X so that given $\varepsilon > 0, \exists n_0 \in \mathbf{N}$ such that

$m, n \ge n_0 \Rightarrow d(x_m - x_n) < \varepsilon$

$\Rightarrow \sup\{ |x_m - x_n| : n \in \mathbf{N} \} < \varepsilon$

$\Rightarrow |x_m - x_n| < \varepsilon$

$\Rightarrow \langle x_n \rangle$ is uniformly convergent.

$\Rightarrow \langle x_n \rangle$ is convergent.

$\Rightarrow X$ is complete.

11. Let $s_n = \dfrac{1}{n}$

$$d(s_n, s_m) = |s_n - s_m| = \left| \frac{1}{n} - \frac{1}{m} \right|$$

$$\le \frac{1}{n} + \frac{1}{m}$$

Now there may exist a positive integer p greater than $\dfrac{2}{\varepsilon}$, then

$$\frac{1}{n} + \frac{1}{m} < \frac{\varepsilon}{2} + \frac{\varepsilon}{2} \quad \forall m, n \ge p$$

$\therefore \quad d(s_n, s_m) \le \dfrac{1}{n} + \dfrac{1}{m}$

$$< \frac{\varepsilon}{2} + \frac{\varepsilon}{2} \quad \forall m, n \ge p$$

$\Rightarrow \quad d(s_n, s_m) < \varepsilon$

$\Rightarrow \quad \langle s_n \rangle$ is a **Cauchy sequence**

REVIEW QUESTIONS AND ARCHIVE

1. Define the following:
   (i) Convergent sequence (MEERUT–2018)
   (ii) Cauchy sequence (MEERUT–2001, 05, 10, 12, 18
       ROHTAK–2005, 07, LUCKNOW–2006, 07, 12,
       KURUKSHETRA–2004, 05, 14)
   (iii) Complete metric space
       (MEERUT–2015, 16, 17, 18)
   (iv) Contracting mapping
       (MEERUT–2002, 03, 12, PATNA–2010)
   (v) First and second category of metric spaces
       (MEERUT–2014, ROHILKHAND–2012)
   (vi) Isometric spaces (ROHTAK–2012)

2. Prove that in a metric space, every convergent sequence has a limit point.

3. Prove that in a metric space, every convergent sequence is Cauchy.

4. Prove that if a Cauchy sequence in a metric space has a convergent subsequence, then the sequence is convergent.

5. Prove that the real line is a complete metric space.

6. Prove that the union of a finite number of nowhere dense sets is nowhere dense.

7. Prove that every contracting mapping is continuous.

8. The metric space $(\mathbf{R}, d)$ is complete where d denotes usual metric for $\mathbf{R}$.

9. The space $C[a, b]$ is complete.

10. Show that the set $\mathbf{C}$ of complex number with usual metric is a complete metric space.

11. Let (X, d) be a complete metric space and Y be a subspace of X. Then prove that Y is complete if and only if Y is closed.

12. Let (X, d) be a metric space and let $<f_n>$ be a nested sequence of non-empty closed subset of X such that $\delta(f_n) \to 0$ as $n \to \infty$. Then prove that X is complete if and only if

$$\bigcap_{n=1}^{\infty} f_n$$ consist of exactly one point.

13. Prove that every complete metric space is of second category.

14. State and prove Banach fixed point theorem.

15. Prove that the Hilbert space (l_2, d) is complete.

OBJECTIVE EVALUATION

▶ FILL IN THE BLANKS

1. A sequence $<s_n>$ is said to be _____ if $\exists$ a number $M > 0$ such that $|s_n| < M \; \forall n \in \mathbf{N}$.

2. A sequence which is not convergent is said to be _____.

3. A sequence $<s_n>$ in X is said to converge to a point $s_0 \in X$ if for each $\varepsilon > 0$ of a positive integer m such that $d(s_n, s_0) < \varepsilon$ for all _____.

4. In a metric space the convergent sequence has a _____ limit.

5. Every convergent sequence in a metric space is _____ sequence.

6. A metric space (X, d) is said to be _____ iff every Cauchy sequence in X converges to a point in X.

7. The open interval is _____.

8. A subspace of a metric space is complete iff it is _____.

▶ **TRUE OR FALSE**

1. In a metric space every convergent sequence is Cauchy. **(T/F)**

2. In a metric space, every Cauchy sequence is convergent. **(T/F)**

3. In a complete metric space, every Cauchy sequence is convergent. **(T/F)**

4. A sequence in a metric space can converge to at most one point of that space. **(T/F)**

5. If a Cauchy sequence in a metric space has a convergent subsequences, then the sequence is convergent. **(T/F)**

6. Completeness is not preserved under isometrics. **(T/F)**

7. The real line is complete. **(T/F)**

8. The image of a Cauchy sequence under a uniformly continuous mapping is a Cauchy sequence. **(T/F)**

▶ **MULTIPLE CHOICE QUESTIONS (CHOOSE THE MOST APPROPRIATE ONE)**

1. A complete subspace of a metric space is:
   (a) closed
   (b) open
   (c) neither open nor closed
   (d) none of these

2. The set of integers with the usual metric is:
   (a) incomplete
   (b) complete
   (c) both are same
   (d) none of these

3. Set of rational numbers with the usual metric is:
   (a) always complete
   (b) may or may not be complete

   (c) not complete
   (d) none of these

4. The set of irrational with the usual metric is:
   (a) always complete
   (b) may or may not be complete
   (c) incomplete
   (d) none of these

5. Which space is complete, with the usual metric:
   (a) $X =]0, 1[$
   (b) $X = [0, 1[$
   (c) $X =]0, \infty[$
   (d) none of these

ANSWERS

▶ **FILL IN THE BLANKS**

1. bounded	2. divergent	3. $n \geq m$	4. unique	5. Cauchy
6. complete	7. open set	8. closed		

▶ **TRUE OR FALSE**

1. T	2. F	3. T	4. T	5. T	6. T	7. T	8. T

▶ **MULTIPLE CHOICE QUESTIONS**

1. (b)	2. (b)	3. (c)	4. (c)	5. (a)

CHAPTER SUMMARY

This chapter deals with the concept of complete metric space. Concept of Cauchy sequence and convergent sequences are main idea discussed in this chapter. Important points discussed in this chapter are given below :

➲Let X be a non-empty set. A sequence in X is a function from N into X. It is denoted by $<x_n>$.

➲A sequence $<s_n>$ is said to be bounded if there exists a number $m > 0$ such that $|s_n| \le m \ \forall n \in N$

➲In a metric space every convergent sequence has a unique limit.

➲Let (X, d) be a metric space. If s_0 is a limit point of $S \subset X$, then there exists a sequence $<s_n>$ of points of S all distinct from s_0, which converges to s_0.

➲If the range set of a convergent sequence in a metric space (X, d) consist of infinitely many distinct points, then the limit of the sequence is an accumulation point of the range set of the sequence.

➲Let (X, d) be a metric space and s_0, t_0 be two points of X. If $<t_n>$ is a sequence converges to t_0, then $<d(s_n, t_n)>$ converges to $d(s_0, t_0)$.

➲In a metric space (X, d), if the sequences $<s_n>$ and $<t_n>$ converge respectively to the points s_0 and t_0 in X, then the sequence $<d(s_n, t_n)>$ of real numbers converges to $d(s_0, t_0)$.

➲In a metric space, every convergent sequence is Cauchy.

➲If a Cauchy sequence in a metric space has a convergent subsequence, then the sequence is convergent.

➲If $<s_n>$ is a Cauchy sequence in a metric space, then any cluster point of $<s_n>$ is limit of $<s_n>$.

➲Let $<s_n>$ be a Cauchy sequence in a metric space (X, d) and let $<s_{i_n}>$ be a subsequence of $<s_n>$ converging to $s_0 \in X$, then $<s_n>$ also converges to s_0.

➲A metric space (X, d) is said to be complete if every Cauchy sequence in X converges.

➲A metric space, which is not complete is said to be incomplete.

➲The real line is a complete metric space.

➲Let (X, d) be a complete metric space and Y be a subspace of X. Then Y is complete if and only if Y is closed.

➲**Cantor's Intersection Theorem.** Let (X, d) be a metric space and let $<F_n>$ be a nested sequence of non-empty closed subset of X such that $\delta(f_n) \to 0$ as $n \to \infty$. Then

X is complete if and only if $\bigcap\limits_{n=1}^{\infty} F_n$ consists of exactly one point.

➲Let (X, d) be a metric space. A subset of a metric space is said to be of first category if and only if it can be written as the union of a countable family of nowhere dense sets : otherwise it is said to be the second category.

➲The union of a finite number of nowhere dense sets is no where dense.

➲**Bair's Category Theorem.** Every complete metric space is of second category.

➲Every contracting mapping is continuous.

➲**Banach Fixed Point Theorem.** If f is a contracting mapping on a complete metric space (X, d), then there exists a unique point p in X such that $f(p) = p$.

➲The metric space (R, d) is complete where d denotes usual metric for R. (MEERUT-2010)

➲The Hilbert space (l_2, d) is complete.

➲The space $C[a, b]$ is complete.

➲Let (X, d) be a metric space and S be the set of all Cauchy sequence in (X, d). Then $\cong$ is an equivalence relation on S.

➲**Completion Theorem.** Let (X, d) be a metric space. There exists a complex metric space (X^*, d^*) in which (X, d) can be isometrically embedded in such a way that X is dense in X^*, i.e., (X, d) is symmetric to a dense subspace of (X^*, d^*).

➲All completions of a metric space are isometric.

FOR ADVANCED LEARNERS

➠ Completeness is preserved under isometrices but not preserved under homeomorphism.

➠ A uniformly continuous image of a Cauchy sequence is Cauchy.

➠ The set Z of integers with the usual metric is complete.

➠ The set Q of rationals with the usual metric is not complete.

➠ The set I of irrationals with the usual metric is not complete.

➠ If the diameters of the given sequence of sets does not tends to zero, the intersection of the sequence may be empty.

➠ Given a metric space (X, d) there exists a complete metric space (Y, D) such that X is a dense subset of Y.

➠ The completion of each of the spaces $]0, 1]$ and $[0, 1[$ is $[0, 1]$.

➠ The completion of a metric space is seperable iff the space itself is seperable.

➠ The completion of the product of two metric spaces is isometric to the product of their completion.

➠ If X is not complete, then there exists a uniformly continuous mapping of X into R^+ with infimum 0.

➠ If X is not complete, then there exists an unbounded continuous mapping of X into R.

➠ If every continuous mapping of X into R is uniformly continuous then X is complete.

CHAPTER 6

Continuous Functions

6.1 CONTINUITY AT A POINT (LOCAL CONTINUITY)

Let (X, d) and (Y, e) be two metric spaces. A function $f : (X, d) \to (Y, e)$ is said to be continuous at a point $x \in X$ if given $\varepsilon > 0, \exists a\ \delta > 0$ such that

$$d(x, x_0) < \delta \Rightarrow e[f(x), f(x_0)] < \varepsilon.$$

Or, in other words f is continuous at x_0 iff to each open sphere $S(f(x_0), \varepsilon)$, there exists an open sphere $S(x_0, \delta)$ such that

$$f[S(x_0, \delta)] \subset S(f(x_0), \varepsilon).$$

☞ REMARKS

- ⇒ The above statement simply says that the f-images of the open sphere $S(x_0, \delta)$ is contained in the open sphere $S(f(x_0), \varepsilon)$.
- ⇒ f is said to be continuous on X iff it is continuous at every point of X. This is known as Global continuity.

Definition. *A function $f : X \to Y$ is said to be continuous at a point $x_0 \in X$ if given a nbd N of $f(x_0), \exists a$ nbd M of x_0 such that*

$$f(M) \subset N.$$

✎ ILLUSTRATIONS

- **✱ Let $f : X_1 \to X_2$ be given such that $f(x) = a, a \in X_2$ then f is continuous.**
  For, let $x \in X_1$ and $\varepsilon > 0$ be given then for and $\delta > 0, f[S(x, \delta)] = \{a\} \subseteq S(a, \varepsilon) \Rightarrow f$ is continuous at a.
  Since a is arbitrary $\Rightarrow f$ is continuous.
- ✱ A constant function is always continuous.
- ✱ Every inclusion function is continuous.
- ✱ Every identity function is continuous.

6.2 HOMEOMORPHISM

Definition 1. *A map $f : (X, d) \to (Y, e)$ is said to be homeomorphism or topological mapping if*
(i) *f is one-one*
(ii) *f and f^{-1} are continuous.*

☞ REMARK

- ⇒ In the above case, the spaces X and Y are said to be homeomorphic or topological equivalent to one another and Y is called the homeomorphic image of X.

✍ ILLUSTRATIONS

* The identity mapping on a metric space to itself is a homeomorphism.
* The mapping $x \to x^2$ from R to itself is not a homeomorphism.
* The mapping $x \to 2x + 3$ from R to itself is a homeomorphism.

Definition 2. *A topological property, which remains invariant under a homeomorphism is said to be topological invariant property or intrinsic quality property.*

For Example: Bases, closed sets, open sets, axioms of countability, etc.

Definition 3. *A mapping $f : X \to Y$ is called a bicontinuous mapping if f and f^{-1} both are continuous mapping.*

☛ **REMARK**

➠ A mapping which is one-one and onto is called bijective map.

Definition 4. *Let (X, d) and (Y, e) be two metric spaces. A mapping $f : X \to Y$ is called an open map or interior map if for any d-open set G, $f(G)$ is e-open.*

Also, the map f is called closed map if any d-closed set F implies $f(F)$ is e-closed set.

6.3 UNIFORM CONTINUITY IN A METRIC SPACE

Let (X, d) and (Y, e) be any two metric spaces. Let $f : X \to Y$ be a mapping. Then f is said to be continuous at $x_0 \in X$ if given $\varepsilon > 0$, $\exists \delta > 0$ such that

$$d(x, x_0) < \delta \;\; \forall \, x \in X \implies e(f(x), f(x_0)) < \varepsilon$$

Here, we observe that

(i) If ε is made smaller, then there will be corresponding decrement in the value of δ. Hence δ depend upon ε.

(ii) If ε is kept fixed, then it is not necessary that

$$d(x, x_0) < \delta \;\; \forall \, x \in X \implies d(x_1, x_2) < \delta \;\; \forall \, x_1, x_2 \in X.$$

Therefore δ depends upon the point x_0.

Definition. *If given $\varepsilon > 0$, we can find a $\delta > 0$ which works throughout the entire space X, then f is said to be uniformly continuous, i.e., f is said to be uniformly continuous if for given $\varepsilon > 0$ $\exists \delta > 0$ such that*

$$d(x_1, x_2) < \delta \implies e[f(x_1), f(x_2)] < \varepsilon, \; \forall \, x_1, x_2 \in X$$

☛ **REMARKS**

➠ If f is uniformly continuous, then it is also continuous but converse is not necessarily true.

➠ The uniform continuity is a property of function on the entire set whereas continuity can be defined at a single point. Uniform continuity at a single point has no sense.

6.4 ISOMORPHISM

Let (X, d) and (Y, e) be two metric spaces. A map $f : X \to Y$ is called isomorphism iff f is bijective and both f and f^{-1} are uniformly continuous.

6.4.1 ISOMETRY

Let (X, d) and (Y, e) be two metric spaces. Then a one-one map $f : X \to Y$ is called an isometry if

$$d(x, y) = e(f(x), f(y)), \; \forall \, x, y \in X.$$

In this case, the **metric spaces are called isometric spaces.**

☛ **REMARK**

➠ **Isometry is a uniformly continuous map.**

✎ ILLUSTRATIONS

* The identity mapping on a metric space to itself is an isometry.
* The mapping $x \to 2x + 3$ from R to itself is not an isometry.
* The inverse of an isometry is an isometry.
* The composition of two isometries is an isometry.

6.4.2 RESTRICTION MAP

Let $f : X \to Y$ be a map. Let $X_1 \subset X$ be arbitrary.

Then f induces a function $f' : X_1 \to Y$ defined by $f'(x) = f(x)$, $\forall x \in X_1$.

The map f' is called a restriction of f to X_1 and is denoted by f / X_1.

Now, assume that X_2 is any set such that $X_2 \supset X$.

Then f induces a function $F : X_2 \to Y$ defined by $F(x) = f(x)$, $\forall x \in X_2$

This map F is called an extension of f.

6.4.3 LINEAR SPACE

Let L be a non-empty set and R be the set of real numbers, the elements of L are called scalars. Let $x, y, z \in L$ and $a, b \in R$. Then L is called linear space over R if the following conditions holds :

(1) $(L, +)$ must be abelian, *i.e.*

   (i) For $x, y \in L$, $x + y \in L$

   (ii) $\exists\, 0 \in L : x + 0 = 0 + x = x$

   (iii) For every $x \in L$ $\exists -x \in L : x + (-x) = 0$

   (iv) $x + y = y + x$ $\forall x, y \in L$

   (v) $(x + y) + z = x + (y + z)$ $\forall x, y, z \in L$

(2) Scalar multiplication law, *i.e.,* $a \in R, x \in L$
$\Rightarrow ax \in L$ satisfies the following conditions:

   (i) $a(x + y) = ax + ay$

   (ii) $(a + b)x = ax + bx$

   (iii) $1 . x = x$

☛ REMARK

➠ The linear space defined above is called vector space and its elements are called vectors.

6.4.4 LINEAR SUBSPACE

A non-empty subset M of linear space L over R is called linear subspace of L if:

(i) For all $x \in M$, $a \in R \Rightarrow ax \in M$

(ii) For all $x, y \in M \Rightarrow x + y \in M$

6.4.5 NORMED LINEAR SPACE

A linear space L together with norm is called normed linear space if the real number $|x|$ corresponding to every $x \in L$ satisfying the following conditions:

(i) $||x|| \geq 0$

FACTS : TO THE POINT

▶ Under a continuous map, the image of an open set need not be an open set.

▶ If f is a continuous bijection, then f^{-1} need not be continuous.

▶ The map $f : R^n \to R$ defined by $f(x_1, x_2, ..., x_n) = x_1$ is continuous.

▶ Let $f : X_1 \to X_2$ be a bijection then f^{-1} is continuous iff f is an open map.

▶ Let $f : X_1 \to X_2$ be a bijection then f^{-1} is continuous iff f is a closed map.

▶ Completeness of metric space is not preserved under homeomorphism.

▶ Homeomorphism is an equivalence relation among metric spaces.

▶ Any two closed intervals are homeomorphic.

▶ Any two open intervals are homeomorphic.

▶ Isometry is an equivalence relation among metric spaces.

▶ Any homeomorphism from one metric space to another is an isometry.

▶ Any isometry from one metric space to another is homeomorphism.

▶ The set of points of discontinuities of a monotonic function is countable.

▶ If $f : X_1 \to X_2$ is continuous at every point of X_1, then f is uniformly continuous on X_1.

▶ If $f : X_1 \to X_2$ is uniformly continuous, then f is continuous at every point of X_1.

  (ii) $||x|| = 0 \iff x = 0$

  (iii) $||x + y|| \leq ||x|| + ||y||$

  (iv) $||ax|| = |a| \cdot ||x||$

where $x, y \in L$ and $a \in \mathbf{R}$.

☛ REMARK

➠ A normed linear space is a metric space with respect to induced metric defined by
$$d(x, y) = ||x - y||$$

6.4.6 BANACH SPACE

A Banach space is a normed linear space which is complete as a metric.

6.4.7 ALGEBRA

A linear space is said to be an algebra if vector multiplication satisfies the following conditions :

  (i) $x(yz) = (xy)z$

  (ii) $x(y + z) = xy + xz$

  (iii) $(x + y)z = xz + yz$

  (iv) $\alpha(xy) = x(\alpha y), \ \alpha \in \mathbf{R}$.

☛ REMARKS

➠ An algebra is called real algebra or complex algebra according as the scalars are real or complex numbers.

➠ An algebra in which vector multiplication is commutative is said to be commutative algebra.

6.4.8 THE METRIC FUNCTION

Let (X, d) be a metric space and $(X \times X, D)$ the metric product of the metric space (X, d) with itself and $d : X \times X \to \mathbf{R}$, the function
$$(x, y) \to d(x, y) \ \forall \ x, y \in X \times X$$
is said to be metric function.

☛ REMARK

➠ The function d is a continuous function from $(X \times X, d)$ into the real line.

The squaring function. Let $f : \mathbf{R} \to \mathbf{R}$ be a function defined by setting $f(x) - x^2 \ \forall x \in \mathbf{R}$. Then f is a continuous function from the real line to itself.

THEOREM 1.   *If f and g are continuous map on a metric space (X, d). Then $f + g$ and $f.g$ are continuous map.*

                                                              (MEERUT 1990, 2005)

PROOF.   Let f and g be continuous mapping on a metric space (X, d).

         Let $x_0 \in X$ be arbitrary.

         By definition of continuity (local), we have

         Give $\varepsilon > 0, \exists \delta_1 > 0$ such that
$$d(x, x_0) < \delta_1 \Rightarrow |f(x) - f(x_0)| < \varepsilon$$

         Also for given $\varepsilon > 0, \exists \delta_2 > 0$ such that
$$d(x, x_0) < \delta_2 \Rightarrow |g(x) - g(x_0)| < \varepsilon.$$

         Define $\delta = \min \{\delta_1, \delta_2\}$.

         Then, for all x such that
$$d(x, x_0) < \delta \Rightarrow |f(x) - f(x_0)| < \varepsilon \text{ and } |g(x) - g(x_0)| < \varepsilon.$$

Now to show $f + g$ is continuous at x_0. Let $d(x, x_0) < \delta$.

Consider

$$|(f + g)(x) - (f + g)(x_0)| = |f(x) - f(x_0) + g(x) - g(x_0)|$$
$$\leq |f(x) - f(x_0)| + |g(x) - g(x_0)|$$
$$< \varepsilon + \varepsilon = 2\varepsilon = \varepsilon_1 \quad \text{(say)}$$

$\Rightarrow \quad |(f + g)(x) - (f + g)(x_0| < \varepsilon_1 \quad \forall \ x \text{ such that } d(x, x_0) < \delta$

$\Rightarrow \quad (f + g)$ is continuous at x_0.

Now to show $f \cdot g$ is continuous at x_0.

Consider

$$|(fg)(x) - (fg)(x_0)| = |f(x)\, g(x) - f(x_0)\, g(x_0) + f(x)g(x_0) - f(x)\, g(x_0)|$$
$$= |f(x)\{g(x) - g(x_0)\} + g(x_0)\{f(x) - f(x_0)\}|$$
$$\leq |f(x)||g(x) - g(x_0)| + |g(x_0)||f(x) - f(x_0)|$$
$$\leq \varepsilon\{|f(x)| + |g(x_0)|\}$$
$$< \varepsilon\{|f(x) - f(x_0)| + |f(x_0) + g(x_0)|\}$$
$$< \varepsilon[\varepsilon + |f(x_0) + g(x_0)|] = \varepsilon_2 \quad \text{(say)}$$

$\Rightarrow \quad |fg(x) - fg(x_0)| < \varepsilon_2$ with any x such that $d(x, x_0) < \delta$.

$\Rightarrow \quad fg$ is continuous at x_0.

☞ **REMARK**

⟹ In a similar way, we can prove that if f is continuous at x_0 and $\alpha \in R$. Then αf is continuous.

THEOREM 2. *If $f : (X_1, d_1) \to (X_2, d_2)$ and $g : (X_2, d_2) \to (X_3, d_3)$ be continuous mapping on the metric spaces, then the composite map $g \circ f : (X_1, d_1) \to (X_3, d_3)$ is continuous.*
(MEERUT 1995, 2005)

PROOF. Let $x_1 \in X_1$ and $x_2 = f(x_1) \in X_2$.

Now since $g : X_2 \to X_3$ is continuous map at $x_2 = f(x_1)$.

Therefore given any open sphere $S((gf)(x_1), \varepsilon) \in X_3$ there exists an open sphere $S(g(x_1), \delta) \in X_2$ such that

$$g\,[S\,(f(x_1), \delta)] \subset S\,[(gf)(x_1), \varepsilon] \qquad \qquad \text{...(1)}$$

Also, $f : X_1 \to X_2$ is continuous at x_1, therefore given any open sphere $S(f(x_1), \delta) \in X_2$ there exists an open sphere.

$$S(x_1, \eta) \in X_1 \text{ such that } f\,[S(x_1, \eta)] \subset S(f(x_1), \delta). \qquad \qquad \text{...(2)}$$

Combining (1) and (2), we get

$$(gf)\,[S(x_1, \eta)] = g\,\{f(S(x_1, \eta))\}$$
$$\subset g\,[S(f(x_1), \delta)] \qquad \qquad \text{[using (2)]}$$
$$= g\,[S(f(x_1), \delta)] \subset S((gf)(x_1), \varepsilon) \qquad \qquad \text{[using (1)]}$$

or $\quad (g \circ f)\,[S(x_1, \eta)] \subset S((gf)(x_1), \varepsilon)$

Hence, $g \circ f$ is continuous mapping.

THEOREM 3. *Let $f : (X, d) \to (Y, e)$ be a mapping in a metric space. Then f is continuous iff the inverse $f^{-1}(V)$ of each e-open set V of Y is d-open subset of X.*
(MEERUT 1990, 93, 96, 2013, 15, 18)

PROOF. Let us first suppose $f : X \to Y$ be a continuous mapping.

Let $V \subset Y$ be any e-open set.

To show $f^{-1}(V)$ in X is d-open.

Let $x \in f^{-1}(V)$ be arbitrary, then $f(x) \in V$.

Since, every open set is a nbd of each of its points and so V is a e-nbd of $f(x)$.

$\Rightarrow$ $f^{-1}(V)$ is d-nbd of x.

$\Rightarrow$ $f^{-1}(V)$ is d-open.

Conversely, let $f^{-1}(V)$ is d-open set. To show f is continuous.

Let $x \in X$ be the arbitrary and N be e-nbd of $f(x)$ in Y.

Then by definition of nbd of an open sphere at $f(x)$, we have $S(f(x), \varepsilon) \subset N$ for some $\varepsilon > 0$.

Now, we can select e-open set V with $f(x) \in V \subset N$.

By assumption, $f^{-1}(V) = M$, say, is d-open.

Therefore, M is a d-nbd of x with $f(M) = V \subset N$ or $f(M) \subset N$.

This implies f is continuous at x.

Hence, f is a continuous mapping.

THEOREM 4. **Let (X, d) and (Y, e) be two metric spaces. A map $f : X \to Y$ is continuous if and only if $f^{-1}(C)$ is closed in X for every closed set $C \subset Y$.**

(MEERUT 2000, 15)

PROOF. Let $f : X \to Y$ be a continuous mapping.

To show that $f^{-1}(C)$ is closed in X for each closed set $C \subset Y$.

Let $C \subset Y$ be an arbitrary closed set.

$\Rightarrow$ $Y - C$ is open.   ($\because$ Compliment of a closed set is open)

$\Rightarrow$ $f^{-1}(Y - C)$ is open.   ($\because$ f is continuous)

$\Rightarrow$ $f^{-1}(Y) - f^{-1}(C)$ is open in X.

$\Rightarrow$ $X - f^{-1}(C)$ is open in X.   ($\because$ $X = f^{-1}(Y)$)

$\Rightarrow$ $f^{-1}(C)$ is closed in X.

Conversely, let $f^{-1}(C)$ is closed. To show f is continuous.

Let $G \subset Y$ be an arbitrary open set.

$\Rightarrow$ $Y - G$ is closed in Y.

$\Rightarrow$ $f^{-1}(Y - G)$ is closed in X.   (By hypothesis)

$\Rightarrow$ $f^{-1}(Y) - f^{-1}(G)$ is closed in X.

$\Rightarrow$ $X - f^{-1}(G)$ is closed in X.

$\Rightarrow$ $f^{-1}(G)$ is open in X.

$\Rightarrow$ Inverse image of an open set under f is open.

Hence, f is continuous.

THEOREM 5. **A mapping $f : X \to Y$ is continuous if and only if**

$$\overline{f^{-1}(B)} \subset f^{-1}(\overline{B})$$

for any set $B \subset Y$, where (X, d) and (Y, e) be two metric spaces.

PROOF. Let us first suppose $f : X \to Y$ be a continuous mapping. Let $B \subset Y$ be arbitrary.

To show $\overline{f^{-1}(B)} \subset f^{-1}(\overline{B})$

Since $B \subset \overline{B}$

$\Rightarrow$ $\overline{f^{-1}(B)} \subset \overline{f^{-1}(\overline{B})}$

...(1)

Since $\bar{B}$ is closed in Y and f is continuous, therefore by above theorem $f^{-1}(\bar{B})$ is closed in X.

$\Rightarrow$ $\qquad \overline{f^{-1}(\bar{B})} \subset f^{-1}(\bar{B})$ $\qquad$ ($\because A$ is closed iff $\bar{A} = A$) $\qquad$ ...(2)

From (1) and (2), we conclude that

$$\overline{f^{-1}(B)} \subset f^{-1}(\bar{B}) \qquad\qquad ...(3)$$

Conversely, let $\overline{f^{-1}(B)} \subset f^{-1}(\bar{B})$. To show f is continuous.

Let $C \subset Y$ be an arbitrary closed set.

Then, $\qquad\qquad \bar{C} = C$ $\qquad$ ($\because A$ is closed iff $\bar{A} = A$) $\qquad$ ...(4)

$\Rightarrow$ $\qquad\qquad \overline{f^{-1}(C)} \subset f^{-1}(\bar{C})$ $\qquad\qquad$ (By hypothesis)

Using (4), we get

$$\overline{f^{-1}(C)} \subset f^{-1}(C) \qquad\qquad ...(5)$$

But $\qquad\qquad f^{-1}(C) \subset \overline{f^{-1}(C)}$ $\qquad$ (for any $C \subset Y$) $\qquad$ ...(6)

From (5) and (6), we conclude that

$$f^{-1}(C) = \overline{f^{-1}(C)}$$

$\Rightarrow$ $\quad f^{-1}(C)$ is closed in X.

$\Rightarrow$ $\quad$ Inverse image of a closed set under f is closed.

Hence, f is continuous.

THEOREM 6. **Let $f : (X, d) \to (Y, e)$ be a map. Then f is continuous if and only if $f(\bar{A}) \subset \overline{f(A)}$ for any set $A \subset X$.** $\qquad$ (MEERUT 1992, 95, 2016, 18)

PROOF. Let us first suppose $f : (X, d) \to (Y, e)$ be a continuous mapping. Let $A \subset X$ be arbitrary.

To show $f(\bar{A}) \subset \overline{f(A)}$.

Now, let $f(A) = B$. Clearly $B \subset Y$.

Since f is continuous $\Rightarrow$ $\overline{f^{-1}(f(A))} \subset f^{-1}(\bar{B})$

$\Rightarrow$ $\overline{f^{-1}(f(A))} \subset f^{-1}(\overline{f(A)})$

$\Rightarrow$ $f^{-1}(f(\bar{A})) \subset f^{-1}(\overline{f(A)})$

$\Rightarrow$ $f(\bar{A}) \subset \overline{f(A)}$

Conversely, suppose that $f : (X, d) \to (Y, e)$ be a map such that $f(\bar{A}) \subset \overline{f(A)}$ for any $A \subset X$.

To show f is a continuous map.

Let $B \subset Y$ be an arbitrary closed set, then $f^{-1}(B) \subset X$.

By hypothesis

$$f[\overline{f^{-1}(B)}] \subset \overline{f[f^{-1}(B)]}$$

$\Rightarrow$ $\qquad f[\overline{f^{-1}(B)}] \subset f[f^{-1}(\bar{B})]$

$\Rightarrow$ $\qquad\qquad \overline{f^{-1}(B)} \subset f^{-1}(\bar{B})$ $\qquad\qquad$ ...(1)

But $\qquad\qquad f^{-1}(B) \subset \overline{f^{-1}(B)}$ $\qquad\qquad$ ...(2)

$\Rightarrow$ $\quad B$ is closed.

$\Rightarrow$ $\qquad\qquad \bar{B} = B$ $\qquad\qquad$ ...(3)

From (1) and (3), we conclude that

$$\overline{f^{-1}(B)} \subset f^{-1}(B) \qquad\qquad ...(4)$$

Combining (2) and (4), we get

$$f^{-1}(B) = \overline{f^{-1}(B)}$$

$\Rightarrow \quad f^{-1}(B)$ is closed.

$\Rightarrow \quad$ Inverse image $f^{-1}(B)$ of a closed set B under f is closed.

Hence, f is continuous.

THEOREM 7. *A one-one onto mapping $f : (X, d) \to (Y, e)$ is homeomorphism if and only if $f(\overline{A}) = \overline{f(A)}$ for any $A \subset X$.*

(MEERUT 1978, 79)

PROOF. Let $f : (X, d) \to (Y, e)$ be a one-one onto mapping.

Let us first suppose $f(\overline{A}) = \overline{f(A)}$ for any set $A \subset X$.

To show f is a homeomorphism.

For this we shall show that

(i) f is one-one onto mapping.

(ii) f is continuous.

(iii) f^{-1} is continuous.

Result (i) follows by the initial assumption.

Let $A \subset X$ be arbitrary.

Now, $\qquad f(\overline{A}) = \overline{f(A)}$ (By hypothesis)

which implies

$$f(\overline{A}) \subset \overline{f(A)} \qquad \qquad \ldots(1)$$

and $\qquad \overline{f(A)} \subset f(\overline{A}) \qquad \qquad \ldots(2)$

Then from (2), it is clear that f is continuous (By previous theorem).

Now, let $B = f(A)$.

$\Rightarrow \qquad f^{-1}(B) = f^{-1}[f(A)] = A$

$\Rightarrow \qquad f^{-1}(B) = A$.

Equation (2) can also be written in as

$$\overline{f(f^{-1}(B))} \subset f[\overline{f^{-1}(B)}]$$

$\Rightarrow \qquad f[\overline{f^{-1}(\overline{B})}] \subset f[\overline{f^{-1}(B)}]$

$\Rightarrow \qquad \overline{f^{-1}(\overline{B})} \subset \overline{f^{-1}(R)}$

$\Rightarrow \quad f^{-1}$ is continuous. (By previous theorem)

Hence, f is a homeomorphism.

Conversely, let us suppose $f : (X, d) \to (Y, e)$ be one-one onto mapping and f is a homeomorphism.

To show that $f(\overline{A}) = \overline{f(A)}$ for any $A \subset X$.

Let $A \subset X$ be arbitrary and B is its image under f, i.e., $B = f(A)$.

Now, $\qquad B = f(A) \Rightarrow f^{-1}(B) = f^{-1}(f(A)) = A$

$\qquad \qquad \qquad \Rightarrow f^{-1}(B) = A$

Since f is continuous, $f(\overline{A}) \subset \overline{f(A)}$. $\qquad \ldots(3)$

Since f^{-1} is also continuous.

$\Rightarrow \qquad f^{-1}(\overline{B}) \subset \overline{f^{-1}(B)}$

$\Rightarrow \qquad f^{-1}(\overline{f(A)}) \subset \overline{f^{-1}(f(A))}$

$\Rightarrow \qquad f^{-1}[\overline{f(A)}] \subset f^{-1}[f(\overline{A})]$

$$\Rightarrow \qquad\qquad \overline{f(A)} \subset f(\bar{A}). \qquad\qquad\qquad ...(4)$$

From (3) and (4), we conclude that $f(\bar{A}) = \overline{f(A)}$.

THEOREM 8. *A function $f : (X, d) \to (Y, e)$ is continuous if and only if $[f^{-1}(B)]^o \supset f^{-1}(B^o)$ for $B \subset Y$.*

(MEERUT 1996, 2016)

PROOF. Let $f : (X, d) \to (Y, e)$ be a mapping and $B \subset Y$.

Let us first suppose, f is continuous.

To show $\quad [f^{-1}(B)]^o \supset f^{-1}(B^o)$.

Now, $B \subset Y \Rightarrow B^o$ is open in Y.

$\Rightarrow f^{-1}(B^o)$ is open in X. $\qquad\qquad$ ($\because f$ is continuous)

$\Rightarrow [f^{-1}(B^o)]^o = f^{-1}(B^o) \qquad\qquad\qquad ...(1)$

Since $B^o \subset B$ $\qquad\qquad\qquad$ (By definition of interior)

Therefore $\quad f^{-1}(B^o) \subset f^{-1}(B)$

$\Rightarrow \qquad f^{-1}(B) \supset f^{-1}(B^o)$

$\Rightarrow \qquad [f^{-1}(B)]^o \supset [f^{-1}(B^o)]^o = f^{-1}(B^o) \qquad\qquad$ [using (1)]

$\Rightarrow \qquad [f^{-1}(B)]^o \supset f^{-1}(B^o)$.

Conversely, let us suppose that $[f^{-1}(B)]^o \supset f^{-1}(B^o)$.

To show f is continuous.

Let G be any open set in Y.

Since G is open, therefore $G^o = G$ $\qquad\qquad$ ($\because A$ is open iff $A^o = A$)

$\Rightarrow \qquad [f^{-1}(G)]^o \supset f^{-1}[G^o]$ $\qquad\qquad$ (By hypothesis)

$\qquad\qquad\qquad = f^{-1}(G)$

$\therefore \qquad [f^{-1}(G)]^o \supset f^{-1}(G)$.

But $\qquad [f^{-1}(G)]^o \subset f^{-1}(G)$

$\therefore \qquad [f^{-1}(G)]^o = f^{-1}(G)$.

$\Rightarrow \quad f^{-1}(G)$ is open in X.

$\Rightarrow \quad$ Inverse image of an open set is open.

Hence, f is continuous.

THEOREM 9. *A homeomorphic image of a second countable space is second countable.*

(MEERUT 2005)

PROOF. Let (X, d) be a second countable space. Then by definition there exists a countable base $\mathbf{B} = \{B_n : n \in \mathbf{N}\}$ for the system of open subset of X.

Let $f : (X, d) \to (Y, e)$ be a homeomorphism.

To show (Y, e) is second countable.

Since B is countable $\Rightarrow \quad \{B_n : n \in \mathbf{N}\}$ is countable.

$\Rightarrow \quad \{f(B_n) : n \in \mathbf{N}\}$ is countable.

Also, since B_n is d-open $\Rightarrow f(B)$ is e-open. $\qquad\qquad$ ($\because f$ is continuous)

Let G be a e-open set.

f is a homeomorphism $\qquad \Rightarrow \quad f$ is continuous.

$\Rightarrow \quad f^{-1}(G)$ is d-open sets.

$\Rightarrow \quad f^{-1}(G) = \bigcup [B_n : n \in \mathbf{N}]$

$\Rightarrow \quad G = f[\bigcup \{B_n : n \in \mathbf{N}\}]$

$\qquad\qquad = \bigcup \{f(B_n) : n \in \mathbf{N}\}$

= a union of e-open set.

$\Rightarrow$ $\{f(B_n) : n \in N\}$ is a countable base for the system of open subset of Y

$\Rightarrow$ Y is second countable.

SOLVED EXAMPLES

TYPE 1. BASED ON CONTINUITY

EXAMPLE I. *Let f be a continuous real valued function defined on a metric space X and $A = \{x \in X : f(x) \geq 0\}$. Prove that A is colsed.*

SOLUTION. We have, $A = \{x \in X : f(x) \geq 0\} = \{x \in X : f(x) \in [0, \infty[\}$

$= f^{-1}([0, \infty[)$

Since $[0, \infty[$ is a closed subset of $\mathbf{R}$ and f is continuous, there $f^{-1}([0, \infty[)$ is closed in X. Hence, A is closed.

EXAMPLE 2. *Let f, g be continuous real valued function on a metric space X and let $A = \{x : x \in X \text{ and } f(x) < g(x)\}$. Prove that A is open.*

SOLUTION. Let f, g be continuous real valued function on X. Then we can easily prove that $f - g$ is also continuous.

Now, consider $A = \{x \in X : f(x) < g(x)\}$

$= \{x \in X : f(x) - g(x) < 0\}$

$= \{x \in X : (f - g)x < 0\}$

$= \{x \in X : (f - g)x \in]-\infty, 0[\}$

$= (f - g)^{-1}\{]-\infty, 0[\}$

Since, $]-\infty, 0[$ is open in $\mathbf{R}$ and $f - g$ is continuous, therefore, $(f - g)^{-1}\{]-\infty, 0[\}$ is open in X. Hence, A is open in X.

EXAMPLE 3. *Let f be a function from R^2 onto R defined by $f(x, y) = x \; \forall \; (x, y) \in R^2$. Show that f is continuous in R^2.*

SOLUTION. Let $(x, y) \in \mathbf{R}^2$. Suppose that $<(x_n, y_n)>$ be a sequence in $\mathbf{R}^2$ converging to (x, y). Then clearly $<x_n> \to x$ and $<y_n> \to y$.

$\Rightarrow$ $f < (x_n, y_n) > \to f(x, y)$

Hence, f is continuous.

EXAMPLE 4. *Let (X, d) be a metric space and $a \in X$. Show that the function $f : X \to R$ defined by $f(x) = d(x, a)$ is continuous.*

SOLUTION. Let $x \in X$ and $<x_n>$ be a sequence in X such that $<x_n> \to x$. We want to prove that $<f(x_n)> \to f(x)$.

Let $\varepsilon > 0$ be given. **Consider**

$$|f(x_n) - f(x)| = |d(x_n, a) - d(x, a)| \leq d(x_n, x) \qquad \dots(1)$$

Since $<x_n> \to x \Rightarrow d(x_n, x) < \varepsilon \; \forall \; n \geq n_1$. Using this result in (1), we get

$$|f(x_n) - f(x)| < \varepsilon \; \forall \; n \geq n_1$$

$\Rightarrow$ $<f(x_n)> \to f(x)$

Hence, f is continuous.

EXAMPLE 5. *If $f : R \to R$ and $g : R \to R$ are continuous function on R and if $h : R^2 \to R^2$ is defined by $h(x, y) = (f(x), g(y))$. Prove that h is continuous on R^2.*

SOLUTION. Let $<x_n, y_n>$ be a sequence in R^2 converging to (x, y).

We want to prove that $<h(x_n, y_n)>$ converges to $h(x, y)$

Since $<(x_n, y_n)> \to (x, y)$ in R^2 so $<x_n> \to x, <y_n> \to y$ in R.

Also, f and g are continuous, therefore, $<f(x_n)> \to f(x)$ and $<g(y_n)> \to g(y)$

$\Rightarrow$ $<f(x_n), g(y_n)> \to <f(x), g(y)>$

$\Rightarrow$ $<h(x_n, y_n)> \to h(x, y)$

Hence, h is continuous on R^2.

EXAMPLE 6. *Let G be an open subset of R. Prove that the characteristic function on G defined by* $\chi_G^{(x)} = \begin{cases} 1 & \text{if } x \in G \\ 0 & \text{if } x \notin G \end{cases}$ *is continuous at every point of G.*

SOLUTION. Let $x \in G$ so that $\chi_G(x) = 1$

Now, let $\varepsilon > 0$ be given. Since G is open and $x \in G$, we can find a $\delta > 0$ such that $S(x, \delta) \subseteq G$.

So, $\chi_G(S(x, \delta)) \subseteq \chi_G(G) = \{1\}$

$\subseteq S(1, \varepsilon)$

$\Rightarrow$ $\chi_G(S(x, \delta)) \subseteq S(\chi_G(x), \varepsilon)$

$\Rightarrow$ χ_G is continuous at x.

Now, since $x \in G$ is arbitrary, therefore, χ_G is continuous on G.

EXAMPLE 7. *Let d be the usual metric on R such that*

$$f(x) = \begin{cases} x, & x < 1 \\ 1, & 1 \le x \le 2 \\ x^2 / 4, & x > 2 \end{cases}$$

Show that f(x) is continuous but not open. (MEERUT 1998)

SOLUTION. Let $G =]a, b[$ be d-open set, then $a, b \in R$ s.t. $a < b$.

Now, $f^{-1}(G) = \{x \in R : f(x) \in G =]a, b[\}$

$= \{x \in R : a < f(x) < b\}$

Therefore,

$$f^{-1}(G) = \begin{cases}]a, b[, & a < b < 1 \\ ]a, 2\sqrt{b}[, & a < 1 < b \le 2 \\ ]2\sqrt{a}, 2\sqrt{b}[, & 2 < a < b \end{cases}$$

In every case f^{-1} is d-open set.

$\Rightarrow$ f is continuous.

Now image of an open set $\left(\dfrac{5}{4}, \dfrac{3}{2}\right)$ is $[1]$, which is not open.

Hence, f is not open.

TYPE 2. BASED ON UNIFORM CONTINUITY

EXAMPLE 8. *Let R be the metric space of real numbers and R^* be the set of function of the form $f(x) = \alpha x, \forall \alpha \in R$.*

If $f(x) = \alpha(x), \quad g(x) = \beta(x)$

Define $d(f, g) = |\alpha - \beta|$

Show that (R^*, d) is a metric space. Again define a map $T : R^* \to R$ by

$T(f) = \alpha$, *where* $f(x) = \alpha x$.

Show that the map T is isometry and uniformly continuous. (MEERUT 1994)

SOLUTION. Let $f, g, h \in R^*$, where R^* is the set of function of the form

$$f(x) = \alpha x, \ \alpha \in \mathbf{R}$$

such that $\quad f(x) = \alpha, \ g(x) = \beta x, \ h(w) = \gamma(x)$

$$d(f, g) = |\alpha - \beta|.$$

To show d is a metric on R^*.

(i) $\quad d(f, g) \geq 0$.

Because the modulus of any number is always greater than equal to zero.

(ii) $\quad d(f, g) = 0 \Leftrightarrow f = g$.

For $\quad d(f, g) = 0 \Leftrightarrow f = g$

So $\quad d(f, g) = 0 \Leftrightarrow |\alpha - \beta| = 0 \Leftrightarrow \alpha = \beta$

$$\Leftrightarrow \alpha x = \beta x \Leftrightarrow f(x) = g(x) \qquad \left(\because \begin{array}{l} f(x) = \alpha k \\ g(x) = \beta k \end{array} \right)$$

$$\Leftrightarrow f = g$$

(iii) $\quad d(f, g) = d(g, f)$

Because $|\alpha - \beta| = |\beta - \alpha|$

(iv) $\quad d(f, h) \leq d(f, g) + d(g, h)$

$$d(f, h) = |\alpha - \gamma| = |(\alpha - \beta) + (\beta - \gamma)|$$

$$\leq |\alpha - \beta| + |\beta - \gamma|$$

$$= d(f, g) + d(g, h)$$

Hence, d is a metric on R^*.

EXAMPLE 9. **Prove that the function $f : [0, b] \to R$ defined by $f(x) = x^2$ is uniformly continuous, where $b > 0$.** (MEERUT 1993, 95)

SOLUTION. Let $x, y \in [0, b]$, then

$$|f(x) - f(y)| = |x^2 - y^2| = |x - y| \cdot |x + y|$$

$$\leq [|x| + |y|] \cdot |x - y| \leq (b + b) |x - y| \qquad (\text{Since} |x| = b \text{ and} |y| = b)$$

or $\quad |f(x) - f(y)| < 2b |x - y|$

Now take $|x - y| < \dfrac{\varepsilon}{2b}$, $\delta = \dfrac{\varepsilon}{2b}$, then we get

$|f(x) - f(y)| < 2\varepsilon$ where $|x - y| < \delta \ \forall \ x, y \in [0, b]$

Hence, f is uniformly continuous.

EXAMPLE 10. **Show that the function $f : [0, 1] \to R$ such that $f(x) = \dfrac{1}{x}$ is not uniformly continuous.**

SOLUTION. To show $f(x)$ is not uniformly continuous.

Let $x, y \in [0, 1]$. Consider

$$|f(x) - f(y)| = \left| \frac{1}{x} - \frac{1}{y} \right| = \left| \frac{y - x}{yx} \right| = \left| \frac{x - y}{xy} \right|$$

or $\quad |f(x) - f(y)| = \dfrac{|x - y|}{|xy|}$

Since $x, y \in [0, 1]$, then $x, y \leq 1 \Rightarrow \dfrac{1}{x}, \dfrac{1}{y} \geq 1$.

Take $x = \varepsilon > 0$, $y = x + \varepsilon$, then we have

$$|f(x) - f(y)| = \frac{|x - y|}{|xy|} = \frac{\varepsilon}{|x| \cdot |y|} \geq \varepsilon$$

or $\quad |f(x) - f(y)| \geq \varepsilon$ for $\delta = \varepsilon$.

Hence, f is not uniform continuous function.

EXAMPLE 11. *Prove that the function $f : R \to R$ defined by $f(x) = \sin x$ is uniformly continuous on R.*

SOLUTION. Let $x, y \in R$ and $x > y$

Now, $\sin x - \sin y = (x - y) \cos z$, where $x > z > y$

(By Lagrange's mean value theorem)

Therefore, $|\sin x - \sin y| = |x - y| |\cos z|$

$\qquad\qquad\qquad\qquad \leq |x - y| \qquad\qquad\qquad\qquad (\because |\cos z| \leq 1)$

$\Rightarrow \quad$ For a given $\varepsilon > 0$, if we choose $\delta = \varepsilon$, we have

$$|x - y| < \delta \Rightarrow |f(x) - f(y)| = |\sin x - \sin y| < \delta$$

Hence, $f(x) = \sin x$ is uniformly continuous on R.

TYPE 3. BASED ON HOMEOMORPHISM AND ISOMETRY

EXAMPLE 12. *Show that the metric space [0, 1] and [0, 2] with usual metric are homeomorphism.*

SOLUTION. Let us define $f : [0, 1] \to [0, 2]$ by $f(x) = 2x$

Clearly, f is one-one and onto.

Further, $f^{-1}(x) = \dfrac{1}{2} x$

Also, we may easily verify that f and f^{-1}, both are continuous.

Hence, f is a homeomorphism.

EXAMPLE 13. *Show that R^2 with usual metric and C with usual metric are isometric and $f : R^2 \to C$ defined by $f(x, y) = x + iy$ is the required isometry.*

SOLUTION. Let d_1 be the usual metric on R^2 and d_2, the usual metric on C.

Also, let $a = (x_1, y_1)$, $b = (x_2, y_2)$ be any two elements of R^2.

Then we have

$$d_1(a, b) = \sqrt{(x_1 - x_2)^2 + (y_1 - y_2)^2}$$
$$= |(x_1 - x_2) + i(y_1 - y_2)| = |(x_1 + iy_1) - (x_2 + iy_2)|$$
$$= d_2(f(a), f(b))$$

Hence, f is an isometry.

EXAMPLE 14. *Show that R with usual metric is not homeomorphic to R with discrete metric.*

SOLUTION. Let $X_1 = R$ with usual metric and $X_2 = R$ with discrete metric.

Also, let $f : X_1 \to X_2$ be any bijection.

Now, $\{a\}$ is open in X_2. But $f^{-1}(\{a\}) = \{f^{-1}(a)\}$ is not open in X_1.

$\Rightarrow \quad f$ is not continuous.

Therefore, any projection $f : X_1 \to X_2$ is not a homeomorphism.

Hence, X_1 is not homeomorphic to X_2.

EXAMPLE 15. *Show that the metric space]0, 1[and]0, ∞[with usual metrices are homeomorphic.*

SOLUTION. Let us define a map $f :]0, 1[\to]0, \infty[$ by $f(x) = \dfrac{x}{1 - x}$.

We have to prove that f is one-one and onto.

Now $f(x) = f(y)$ $\Leftrightarrow$ $\dfrac{x}{1-x} = \dfrac{y}{1-y}$

$\Leftrightarrow$ $x(1-y) = y(1-x)$

$\Leftrightarrow$ $x - xy = y - xy$

$\Leftrightarrow$ $x = y$

Thus, f is one-one.

Now, let $y \in]0, \infty[$ then $f(x) = y$ $\Rightarrow$ $\dfrac{x}{1-x} = y$

$\Rightarrow$ $y - xy = x$

$\Rightarrow$ $x + xy = y$

$\Rightarrow$ $x(1+y) = y$

$\Rightarrow$ $x = \dfrac{y}{1+y}$

Therefore, $\dfrac{y}{1+y} \in]0,1[$ is the preimage of y under f.

Also, f and f^{-1} are continuous.

Hence, f is a homeomorphism.

EXAMPLE 16. **Let d_1 be the usual metric on [0, 1] and d_2 be the usual metric on [0, 2]. Show that the map $f : [0,1] \to [0,2]$ defined by $f(x) = 2x$ is not an isometry.**

SOLUTION. Let $x, y \in [0,1]$. Then consider

$$d_2[f(x), f(y)] = |f(x) - f(y)|$$

$$= |2x - 2y| = 2|x - y|$$

$$= 2d_1(x, y)$$

which shows that $d_1(x, y) \neq d_2[f(x), f(y)]$

Hence, f is not an isometry.

EXAMPLE 17. **Prove that the mapping $f : R \to]-1,1[$ defined by $f(x) = \dfrac{x}{1+|x|}$ is homeomorphism, but is not an isometry.**

SOLUTION. (i) Since $1 + |x| \neq 0$ for any $x \in R$ $\Rightarrow$ f is well defined on R.

(ii) We have $|f(x)| - \dfrac{|x|}{1+|x|} < 1$ $\forall x \in R$, therefore $-1 < f(x) < 1$ $\forall x \in R$

We shall show that $f(R) =]-1, 1[$ i.e. f is onto $]-1, 1[$.

If $y \in]-1,1[$ then

$f(x) = y$ $\Leftrightarrow$ $\dfrac{x}{1+|x|} = y$

$\Leftrightarrow$ $\dfrac{|x|}{1+|x|} = |y|$ $\qquad (\because x > 0 \Leftrightarrow y > 0)$

$\Leftrightarrow$ $1 - \dfrac{1}{1+|x|} = |y|$

$\Leftrightarrow$ $1 + |x| = \dfrac{1}{1-|y|}$

$\Leftrightarrow$ $x(1 - |y|) = y$

$\Leftrightarrow$ $x = \dfrac{y}{1-|y|}$

showing that for each $y \in]-1,1[$, there exists an $x \in R$ such that $f(x) = y$. Thus,

$f : \mathbf{R} \rightarrow\,]-1, 1[$ is onto.

(iii) Now, $f(x_1) = f(x_2)$ $\Leftrightarrow$ $\dfrac{x_1}{1+|x_1|} = \dfrac{x_2}{1+|x_2|}$

$\Leftrightarrow$ $\dfrac{|x_1|}{1+|x_1|} = \dfrac{|x_2|}{1+|x_2|}$

$\Leftrightarrow$ $|x_1| = |x_2|$ $\qquad$ (By using $\dfrac{x_1}{1+|x_1|} = \dfrac{x_2}{1+|x_2|}$)

$\Leftrightarrow$ $x_1 = x_2$

Therefore, f is one-one.

(iv) From the above, we find that f is a bijection from $\mathbf{R}$ onto $]-1, 1[$. Now, since every bijective mapping is invertible. Therefore, f is invertible.

Now, from (ii) above, we get

$$y = \frac{x}{1+|x|} \qquad \Leftrightarrow \qquad x = \frac{y}{1-|y|}$$

$\Rightarrow$ Inverse of f is given by $g(y) = \dfrac{y}{1-|y|}$

(v) Since the mapping $x \rightarrow |x|$ and $x \rightarrow 1$ are continuous and the sum of two continuous mappings is again a continuous mapping, thus the mapping $x \rightarrow x$ is continuous and the quotient of two continuous mappings is continuous. So, the mapping $x \rightarrow \dfrac{x}{1+|x|}$ is continuous *i.e.* f is continuous.

(vi) By the same argument as in (v), the mapping $g :]-1, 1] \rightarrow \mathbf{R}$ defined by

$$g(y) = \frac{y}{1-|y|} \quad \forall y \in\,]-1, 1[$$

is continuous. Since, f is a bijective mapping from $\mathbf{R}$ onto $]-1, 1[$ such that both f and $f^{-1} (= g)$ are continuous. Hence, f is a homeomorphism.

Now, it remains to prove that it is not an isometry. For this it is sufficient to find two points x_1 and x_2 in $\mathbf{R}$ such that the distance between them is not equal to the distance between $f(x_1)$ and $f(x_2)$.

Consider the point $x_1 = 0$ and $x_2 = 1$

Then $f(x_1) = 0$ and $f(x_2) = \dfrac{x_2}{1+|x_2|} = \dfrac{1}{2}$

Since, $|x_1 - x_2| = 1$ but $|f(x_1) - f(x_2)| = \dfrac{1}{2}$, therefore,

$$|x_1 - x_2| \neq |f(x_1) - f(x_2)|$$

$\Rightarrow$ f does not preserve distance.

Hence, f is not an isometry.

EXERCISE 6.1

1. Show that the image of a Cauchy sequence under a uniform continuous mapping is Cauchy sequence.

2. Let (X, d) and (Y, ρ) be two metric spaces and Y is complete. Let $A \subset X$ be dense. If $f : A \rightarrow Y$ be uniformly continuous. Show that f can be extended uniquely to a uniform continuous map $g : X \rightarrow Y$.

3. Let (X, d) and (Y, ρ) be the metric spaces and let $f : X \rightarrow Y$ be a map and $x \in X$. Show that the following statements are equivalent :

(i) f is (d, ρ) continuous at x.

(ii) For each ρ nbd N of $f(x)$ in Y, $\exists$ some d-nbd M of x in X such that $f(M) \subset N$.

(iii) The inverse image $f^{-1}(N)$ of each ρ-nbd N of $f(x)$ in Y is a d-nbd of $x \in X$.

4. If (X, d) is the product metric space of the metric spaces (X_1, d_1) and (X_2, d_2), show that the projection maps π_1 and π_2 are continuous and open. (MEERUT 1984)

5. Prove that A map $f : (X, d) \to (Y, e)$ is closed if and only if $\overline{f(A)} \subset f(\overline{A})$ for every $A \subset X$.

6. If $f : R \to R$ is a constant mapping prove that f is continuous.

7. Show that the function $f : (R, d) \to (R, d)$ s.t. $f(x) = 1 \ \forall x \in R$, where d is a usual metric on R is continuous and closed but not open.

8. Let $f : R \to R$ be a continuous mapping also let $f(x) = 0$, $\forall x \in R$ such that x is rational. Then show that $f(x) = 0$, $\forall x \in R$.

9. Let (X, d) and (Y, e) be two metric spaces. Let $A \subset X$, if f, g are continuous maps from $X \to Y$ s.t. $f(x) = g(x), \forall x \in A$, then show that $f(x) = g(x) \ \forall x \in A$.

10. Let (X, d) be a metric space and $A \subset X$. For any $x \in X$, let $f(x) = d(x, A) = \inf\{d(\varepsilon, a) : x \in A\}$. Show that $f(x)$ is a continuous map.

REVIEW QUESTIONS AND ARCHIVE

1. Define the following:
  (i) Continuity at a point (MEERUT–2012, 14, LUCKNOW–2011, HIMACHAL–2010, 12)
  (ii) Uniform continuity (BHOPAL–2012, DELHI–2010)
  (iii) Isometry (MEERUT–2011, 12)
  (iv) Homeomorphism (GARHWAL–2012, MEERUT–2011)

2. Show that the homeomorphic image of a second countable space is second countable. (MEERUT–2005, 11)

3. Show that a mapping $f : (R, d) \to (R, d)$ such that $f(x) = x^2 \ \forall x \in R$ is a continuous mapping but not uniformly continuous

mapping where d is a usual metric on R.

4. In a discrete metric space X, prove the following statements :
  (i) Every map f from X to a metric space Y is continuous.
  (ii) If $f(x_n)$ is a sequence of distinct terms in X, then it is divergent.

5. Let f be a continuous mapping on a metric space X. Let $Z(f)$ be the set of all $p \in X$ at which $f(p) = 0$. Show that $Z(f)$ is closed.

6. Let X be a infinite discrete space. Show that every map from X to a metric space y is continuous.

OBJECTIVE EVALUATION

▶ FILL IN THE BLANKS

1. The image of a Cauchy sequence under a uniform continuous mapping is a _____ sequence.

2. If f and g are continuous, then $f + g$ and fg _____.

3. _____ continuity is a property on the entire set.

4. A Banach space is a normed linear space which is _____.

▶ TRUE OR FALSE

1. The image of a Cauchy sequence under a uniform continuous map is Cauchy sequence. **(T/F)**

2. A mapping $f : (X, d) \to (Y, e)$ is closed if $\overline{f(A)} \subset f(\overline{A}) \ \forall A \subset X$. **(T/F)**

3. The projection map is always open. **(T/F)**

4. A homeomorphic image of a second countable is space is second countable. **(T/F)**

► **MULTIPLE CHOICE QUESTIONS** (CHOOSE THE MOST APPROPRIATE ONE)

1. Which of the following statement is not true?
   Continuous image of _____
   (a) a connected metric space is connected
   (b) a complete metric space is complete
   (c) a compact metric space is complete
   (d) none of these

2. In usual metric space, there exists a continuous function from:
   (a)]0, 1[onto R (b)]0, 1[onto [0, 1]
   (c) [0,1] onto]0,1[(d) none of these

3. If $f : R \to R$ is continuous then f is:
   (a) one-one
   (b) onto
   (c) uniformly continuous
   (d) none of these

4. Let $f : R \to R$ be a continuous function and let $A = \{x \in R, f(x) = 0\}$. Then A is:
   (a) open   (b) closed
   (c) bounded   (d) none of these

5. Let $f : R \to R$ and $g : R \to R$ be uniformly continuous function. Which of the following statement is not true?
   (a) $(f + g)$ is uniformly continuous
   (b) $(f - g)$ is uniformly continuous
   (c) fg is uniformly continuous
   (d) none of these

6. Which of the following pairs of metric spaces (all with usual metric) are homeomorphic?
   (a)]0, 1[and]0,2[(b)]0, 4[and [0, 4]
   (c)]0,∞[and [1,∞[(d) none of these

7. Which of the following pairs of metric spaces (all with usual metric) are isometric?
   (a)]0, 1[and]0, 2[(b)]0, 4[and [0, 4]
   (c)]0, 4[and]2, 6[(d) none of these

8. Let $f : R \to R$ be defined by
   $$f(x) = \begin{cases} 1; & \text{if } x \text{ is rational} \\ 0; & \text{if } x \text{ is irrational} \end{cases}.$$ Then set of discontinuities of f is:
   (a) Q (b) R
   (c) $R - Q$ (d) none of these

CHAPTER SUMMARY

The important points discussed in this chapter are as follows :

➲ Let (X, d) and (Y, e) be two metric spaces. A function $f : (X, d) \to (Y, e)$ is said to be continuous at a point $x \in X$ if given $\varepsilon > 0, \exists a \, \delta > 0$ such that
$$d(x, x_0) < \delta \Rightarrow e[f(x), f(x_0)] < \varepsilon$$

➲ A topological property, which remains invariant under a homeomorphism is said to be topological invariant property or intrinsic quality property.

➲ A Banach space is a normed linear space which is complete as a metric.

➲ A homeomorphic image of a second countable space is second countable.

➲ A function $f : X \to Y$ is said to be continuous at a point $x_0 \in X$ if given a nbd N of $f(x_0), \exists a$ nbd M of x_0 such that $f(M) \subset N$.

➲ A topological property, which remains invariant under a homeomorphism is said to be topological invariant property or intrinsic quality property.

➲ A mapping which is one-one and onto is called bijective map.

⊃ Let (X, d) and (Y, e) be two metric spaces. A mapping $f : X \to Y$ is called an open map or interior map if for any d-open set G, $f(G)$ is e-open.

⊃ If f is uniformly continuous, then it is also continuous but converse is not necessarily true.

⊃ The uniform continuity is a property of function on the entire set whereas continuity can be defined by at a single point. Uniform continuity at a single point has no sense.

⊃ Let (X, d) and (Y, e) be two metric spaces. A map $f : X \to Y$ is called isomorphism iff f is bijective and both f and f^{-1} are uniformly continuous.

⊃ Let (X, d) and (Y, e) be two metric spaces. Then a one-one map $f : X \to Y$ is called an isometry   if $d(x, y) = e(f(x), f(y))$, $\forall \, x, y \in X$.

⊃ Isometry is a uniformly continuous map.

⊃ The linear space defined above is called vector space and its elements are called vectors.

⊃ An algebra is called real algebra or complex algebra according as the scalars are real or complex numbers.

⊃ An algebra in which vector multiplication is commutative is said to be commutative algebra.

⊃ If f and g are continuous map on a metric space (X, d). Then $f + g$ and $f.g$ are continuous map.

⊃ If $\qquad f : (X_1, d_1) \to (X_2, d_2) \qquad$ and $g : (X_2, d_2) \to (X_3, d_3)$ be continuous mapping on the metric spaces, then the composite map $g \circ f : (X_1, d_1) \to (X_3, d_3)$ is continuous.

⊃ Let $f : (X, d) \to (Y, e)$ be a mapping in a metric space. Then f is continuous iff the inverse $f^{-1}(V)$ of each e-open set V of Y is d-open subset of X.

⊃ Let (X, d) and (Y, e) be two metric spaces. A map $f : X \to Y$ is continuous if and only if $f^{-1}(C)$ is **closed in** X for every closed set $C \subset Y$.

⊃ Let $f : (X, d) \to (Y, e)$ be a map. Then f is continuous if and only if $f(\bar{A}) \subset \overline{f(A)}$ for any set $A \subset X$.

⊃ A one-one onto mapping $f : (X, d) \to (Y, e)$ is homeomorphism if and only if $f(\bar{A}) = \overline{f(A)}$ for any $A \subset X$.

⊃ A function $f : (X, d) \to (Y, e)$ is continuous if and only if $[f^{-1}(B)]^o \supset f^{-1}(B^o)$ for $B \subset Y$.

◼◻◻ ⟩**FOR ADVANCED LEARNERS**/ ◼◻◻

➡ If f and g are continuous real valued functions defined on a metric space X and let $A = \{x \in X : f(x) < g(x)\}$, then A is open.

➡ An isometry is an isomorphism.

➡ A mapping f between metric spaces X, Y is uniformly continuous if and only if $d(f(A), f(B)) = 0$ whenever $A, B \subset X$ and $d(A, B) = 0$.

➡ If every continuous mapping of X into $\textbf{\textit{R}}$ is uniformly continuous then X is complete.

➡ For each pair A, B of non-empty disjoint closed subsets of X, there exists a uniformly continuous mapping $f : X \to [0, 1]$ such that $f(A) = \{0\}$ and $f(B) = \{1\}$, then X is complete.

➡ In a metric space (X, d) following conditions are equivalent:

(i) Every continuous function $f : X \to \textbf{\textit{R}}$ is uniformly continuous.

(ii) $d(A, B) > 0$ for all non-empty disjoint closed subset A, B of X.

➡ Every real valued function defined and continuous on a closed and bounded interval is uniformly continuous.

➡ Every continuous function on a compact metric space is uniformly continuous.

❋❋❋❋

CHAPTER 7

Connectedness in Metric Spaces

7.1 INTRODUCTION

Connectedness is one of the most important property of metric spaces. By 'connected', we mean 'in one piece'. In this chapter we shall discuss the concept of 'connectedness'. In order to have a precise mathematical formulation of the concept of connectedness we must have to know the concept of separatedness.

7.2 SEPARATED SETS

Definition 1. *Let* (X, d) *be a metric space. Two non empty subsets* A *and* B *of* X *are said to be separated if*

$$A \cap \bar{B} = \phi \quad and \quad \bar{A} \cap B = \phi \qquad \text{(MEERUT 1990)}$$

☞ **REMARKS**

⟹ The above definition states that no point of A should be in the closure of B and no point of B should be in the closure of A.

⟹ The above two conditions are equivalent to a single condition

$$(A \cap \bar{B}) \cup (\bar{A} \cap B) = \phi$$

Thus, if $(A \cap \bar{B}) \cup (\bar{A} \cap B) = \phi$, then we say that A and B are separated.

⟹ A and B are separated iff both A and B are non-empty and disjoint. As we see, that these conditions are not enough for any two sets to be separated. For example, the subsets $A = [2, 3[$ and $B = [3, 4[$ are non-empty and disjoint but are not separated sets of the real line.

Since, $A = [2, 3[$ and $B = [3, 4[$

We see $\bar{A} = [2, 3], \bar{B} = [3, 4]$

so that $A \cap \bar{B} = [2, 3[\cap [3, 4] = \phi$

$$\bar{A} \cap B = [2, 3] \cap [3, 4[= \{3\} \neq \phi$$

Thus, A and B are not separated subsets of real line.

⟹ Every two separated sets are disjoint, but the converse is not necessarily true, *i.e.*, every two separated sets are disjoint, but two disjoint sets are not necessarily true.

Thus, A and B are separated iff **A and B are** disjoint and neither of them contains the limit point of the other. As **we have**

$$A \cap \bar{B} = \phi \Rightarrow A \cap (B \cup D(B)) = \phi$$

$$\Rightarrow (A \cap B) \cup (A \cap D(B)) = \phi$$

$$\Rightarrow A \cap D(B) = \phi \quad \text{(Since } A \text{ and } B \text{ are disjoint, so, } A \cap B = \phi \text{)}$$

$$\Rightarrow A \text{ does not contrain any limit point of } B.$$

Similarly, we can prove that B has no limit point of A.

✎ ILLUSTRATIONS

* Consider the following subsets of R (with usual metric)

$$A =]-\infty, 0[\quad \text{and} \quad B = [0, \infty[$$

Writing $\quad A =]-\infty, 0[\quad \text{and} \quad B = [0, \infty[$

We find that $\quad \bar{A} =]-\infty, 0]; \quad \bar{B} = [0, \infty[$

So that $\quad A \cap \bar{B} =]-\infty, 0[\cap [0, \infty[$

$$= \phi$$

$$\bar{A} \cap B =]-\infty, 0] \cap [0, \infty[$$

$$= \{0\} \neq \phi$$

$\Rightarrow$ A and B are not separated.

* Consider the space (R, d) and let

$$A =]-4, -3[, \quad B =]1, 2[\quad \text{and} \quad C =]5, 6[$$

We have $\quad A =]-4, -3[, \quad B =]1, 2[, \quad C =]5, 6[$

$$\bar{A} = [-4, -3], \bar{B} = [1, 2], \bar{C} = [5, 6]$$

So that $\quad A \cap \bar{B} =]-4, -3[\cap [1, 2]$

$$= \phi$$

$$\bar{A} \cap B = [-4, -3] \cap]1, 2[$$

$$= \phi$$

$\Rightarrow$ Sets A and B are separated.

Similarly, we can show, sets A and C are also separated.

* Consider the following subsets of R with usual metric

$$A =]0, 1[, \quad B =]1, 2[\quad \text{and} \quad C = [1, 2]$$

We find that $\bar{A} = [0, 1], \quad \bar{B} = [1, 2] \quad \text{and} \quad \bar{C} = [1, 2]$

So that $A \cap \bar{B} =]0, 1[\cap [1, 2] = \phi$

$$\bar{A} \cap B = [0, 1] \cap]1, 2[= \phi$$

$\Rightarrow$ A and B are separated sets.

But, A and C are not separated sets

Since $\quad \bar{A} \cap C = [0, 1] \cap [1, 2] = \{1\} \neq \phi$

THEOREM 1. *Let (Y, d^*) be a subspace of a metric space (X, d) and let A, B be two subsets of Y. Then A, B are d-separated if and only if d^* separated.*

PROOF. We are given A and B be two subsets of Y. Let $\bar{A}^*, \bar{A}$ be the d^* closure and d closure of A, respectively. Also,

$$\bar{A}^* = \bar{A} \cap Y$$

$$\bar{B}^* = \bar{B} \cap Y$$

Now, $(\bar{A}^* \cap B) \cup (A \cap \bar{B}^*) = (\bar{A} \cap Y \cap B) \cup (A \cap \bar{B} \cap Y)$

$$= (\bar{A} \cap B) \cup (A \cap \bar{B}) \qquad\qquad [\because A, B \subset Y]$$

Therefore, $(\bar{A} \cap B) \cup (A \cap \bar{B}) = \phi$

iff $\qquad (\bar{A}^* \cap B) \cup (A \cap \bar{B}^*) = \phi$

Hence, A, B are d-separated, iff they are d^*-separated.

THEOREM 2. *(X, d) be a metric space and let A and B be two separated subsets of X and C, D be two non-empty sets such that $C \subset A$ and $D \subset B$. Then the sets C and D are also separated.*

PROOF. We are given

$$C \subset A \Rightarrow \bar{C} \subset \bar{A} \qquad \qquad \text{...(1)}$$

and

$$D \subset B \Rightarrow \bar{D} \subset \bar{B} \qquad \qquad \text{...(2)}$$

If A and B are separated subsets of (X, d), then

$$A \cap \bar{B} = \phi \text{ and } \bar{A} \cap B = \phi \qquad \qquad \text{...(3)}$$

Collectively, $(A \cap \bar{B}) \cup (\bar{A} \cap B) = \phi \qquad \qquad \text{...(4)}$

From (1), (2) and (3), we conclude that

$$(C \cap \bar{D}) \cup (\bar{C} \cap D) = \phi$$

Hence, the set C and D are also separated.

THEOREM 3. *Let (X, d) be a metric space and let A, B and C be any three non-empty subsets of X such that A and B are separated, and also the sets A and C are separated. Then the sets A and $B \cup C$ are separated.*

PROOF. We are given A and B are separated subsets of (X, d)

$$\Rightarrow \qquad (A \cap \bar{B}) \cup (\bar{A} \cap B) = \phi \qquad \qquad \text{...(1)}$$

Also, A and C are separated subsets of (X, d)

$$\Rightarrow \qquad (A \cap \bar{C}) \cup (\bar{A} \cap C) = \phi \qquad \qquad \text{...(2)}$$

From (1) and (2), we see that

$$(\bar{A} \cap B) = \phi \text{ and } (\bar{A} \cap C) = \phi$$

Similarly, $(A \cap \bar{B}) = \phi$ and $(A \cap \bar{C}) = \phi$

$$\text{i.e.,} \qquad \bar{A} \cap (B \cup C) = \phi \qquad \qquad \text{...(3)}$$

and

$$A \cap (\overline{B \cup C}) = \phi \qquad \qquad \text{...(4)}$$

$$[\because A \cap (\bar{B} \cup \bar{C}) = \phi]$$

On combining (3) and (4), we get

$$[\bar{A} \cap (B \cup C)] \cup (A \cap (\overline{B \cup C})) = \phi$$

Also, A and $B \cup C$ are non-empty.

Hence, A and $B \cup C$ are separated.

THEOREM 4. **(i)** *(X, d) be a metric space and let A and B be two closed subsets of X, then A and B are separated iff they are disjoint.*

(ii) *(X, d) be a metric space, and let A and B be two open subsets of X, then A and B are separated iff they are disjoint.*

PROOF. (i) (X, d) is a metric space and let A and B be two closed subsets of X.

Since, any two separated sets are disjoint, therefore, we need only to prove that two disjoint closed subsets of a metric space are separated.

Thus, we need only to prove that A and B are separated.

$$\text{i.e.,} \qquad (A \cap \bar{B}) \cup (\bar{A} \cap B) = \phi$$

Since A and B are closed sets, then $\bar{A} = A$ and $\bar{B} = B$.

Also, A and B are disjoint $\Rightarrow A \cap B = \phi$

It follows that

$$\bar{A} \cap B = A \cap B = \phi$$

and

$$A \cap \bar{B} = A \cap B = \phi$$

Hence, A and B are separated sets.

(ii) Since, any two separated sets are disjoint, we need only to prove that two disjoint open sets are separated. Let A and B be two disjoint open subsets of (X, d) so that

$$A \cap B = \phi \qquad \qquad \text{...(1)}$$

Suppose, A and B are not separated.

$\Rightarrow$ either $A \cap \bar{B}$ or $\bar{A} \cap B$ is non-empty.

Let $A \cap \bar{B} \neq \phi$, then there exists a point $x \in X$ s.t. $x \in A$ and $x \in \bar{B}$

Since A is open, it is a neighbourhood of x.

Again, since $x \in \bar{B}$

$\Rightarrow$ x is an adherent point of B

$\Rightarrow$ A must contain a point of B

$\Rightarrow$ $\qquad A \cap B \neq \phi$

which contradicts (1).

$\Rightarrow$ Our assumption is wrong.

i.e., A and B must be separated.

THEOREM 5. **(X, d) be a metric space and let A and B be two separated subsets of X**
(i) If $A \cup B$ is closed, then A and B are closed sets.
(ii) If $A \cup B$ is open, then A and B are open sets.

PROOF. (X, d) is a metric space and A and B are two separated subsets of X.

Since A and B are separated sets, we have

$$A \cap \bar{B} = \phi \text{ and } \bar{A} \cap B = \phi \qquad \qquad ...(1)$$

(i) Let $A \cup B$ be closed, then we have to prove A and B are closed sets.

Since $A \cup B$ is closed.

$\Rightarrow \qquad \overline{A \cup B} = A \cup B = \bar{A} \cup \bar{B} \qquad \qquad ...(2)$

Hence, $\bar{A} = \bar{A} \cap (\bar{A} \cup \bar{B}) \qquad \qquad [\because \bar{A} \subset \bar{A} \cup \bar{B}]$

$\qquad = \bar{A} \cap (A \cup B) \qquad \qquad \text{[by (2)]}$

$\qquad = (\bar{A} \cap A) \cup (\bar{A} \cap B) \qquad \qquad \text{[By distributive law]}$

$\qquad = (\bar{A} \cap A) \cup \phi \qquad \qquad \text{[by (1)]}$

$\qquad = \bar{A} \cap A$

$\qquad = A$

$\qquad \qquad \text{[Since } \bar{A} \text{ is the smallest closed set containing } A, \text{ i.e., } A \subset \bar{A}]$

Since $\bar{A}$ is a closed set $\Rightarrow A$ is also closed.

Similarly, we can show, B is also closed.

Thus, if $A \cup B$ is closed, then A and B are also closed sets.

(ii) Let $A \cup B$ be open, then we have to prove that A and B both are open sets.

Since $\bar{B}$ is closed

$\Rightarrow X - \bar{B}$ is open.

Also, $A \cup B$ is open $\Rightarrow (A \cup B) \cap (X - \bar{B})$ is open.

But, $(A \cup B) \cap (X - \bar{B})$

$\qquad = [A \cap (X - \bar{B})] \cup [B \cap (X - \bar{B})]$

$\qquad = [A \cap (X - \bar{B})] \cup \phi$

$\qquad \qquad [\because B \cap (X - B) = \phi \text{ and } B \subset \bar{B} \Rightarrow B \cap (X - \bar{B}) = \phi]$

$\qquad = A \cap (X - \bar{B}) = A \qquad \qquad [\because A \cap \bar{B} = \phi \Rightarrow A \subset X - \bar{B}]$

Since $(A \cup B) \cap (X - \bar{B})$ is open.

$\Rightarrow$ **A is open.**

Similarly, $(A \cup B) \cap (X - \bar{A}) = B$ is open.

Hence, if $A \cup B$ is open, then A and B are open sets.

THEOREM 6. **(X, d) be a metric space and let A and B be two non-empty disjoint subsets of X. Then**

(i) A and B are separated iff each is closed in the subspace $A \cup B$.

(ii) A and B are separated iff each is open in the subspace $A \cup B$.

(iii) A and B are separated iff A is both open and closed in the subspace $A \cap B$.

PROOF. A and B are two non-empty disjoint subsets of X.

i.e., $\qquad A \cap B = \phi$ ...(1)

(i) Now, $A \cap \bar{B} = \phi$ iff A contains no accumulation point of B, i.e., iff B contains all of its accumulation points which are in $A \cup B$.

$\Rightarrow A \cap \bar{B} = \phi$ iff B is closed in subspace $A \cup B$.

Similarly, $\bar{A} \cap B = \phi$ iff A is closed in subspace $A \cup B$.

$\Rightarrow A$ and B are separated iff each is closed in subspace $A \cup B$.

(ii) Let us denote $A \cup B$ by E, i.e., $A \cup B = E$

Since $B \subset \bar{B}$, we have $A \cap \bar{B} = \phi \Rightarrow A \cap B = \phi$

Then, $\qquad A = E - B$ ...(1)

and $\qquad B = E - A$ ...(2)

Thus, keeping (1) and (2) in view, it follows from above part (i) that A and B are separated if and only if each is open in the subspace $A \cup B$.

(iii) From (i) and (ii), we get A and B are separated iff A is both open and closed in the subspace $A \cup B$.

Hence, the theorem.

THEOREM 7. **(X, d) be a metric space and let A, B be two subsets of X such that $A \cup B = X$. If A', B' (compliments of A, B respectively) are separated sets, then for every subset M of X, we have $\bar{M} = (\overline{M \cap A} \cap A) \cup (\overline{M \cap B} \cap B)$.**

PROOF. Given (X, d) is a metric space and A, B be two subsets of X s.t. $A \cup B = X$.

Now, since M is a subset of X then

$$M \cap A \subset M \Rightarrow \overline{M \cap A} \subset \bar{M}$$

$$\Rightarrow \overline{M \cap A} \cap A \subset \bar{M}.$$

Similarly, $\overline{M \cap B} \cap B \subset \bar{M}$.

Hence, $(\overline{M \cap A} \cap A) \cup (\overline{M \cap B} \cap B) \subset \bar{M}$ ...(1)

Now, it remains to prove that

$$(\overline{M \cap A} \cap A) \cup (\overline{M \cap B} \cap B) \supset \bar{M}$$

To prove it, we begin with the relation

$$M = (M \cap A) \cup (M \cap A')$$

$$\Rightarrow \bar{M} = \overline{(M \cap A)} \cup \overline{(M \cap A')}$$

$$\Rightarrow \bar{M} \cap A \subset \overline{(M \cap A} \cap A) \cup \overline{(M \cap A')}$$ ...(2)

Since, it is given that A' and B' are separated, we have

$$\bar{A}' \cap B' = \phi \text{ and } A' \cap \bar{B}' = \phi$$

$$\Rightarrow \bar{A}' \cap B = \phi$$

$$\Rightarrow \bar{A}' \subset B$$ ...(3)

Also, $\qquad \bar{A}' \subset B \Rightarrow A' \subset B$

$$\Rightarrow M \cap A' \subset M \cap B$$

$$\Rightarrow \overline{(M \cap A')} \subset \overline{(M \cap B)}$$

and $M \cap A' \subset A' \Rightarrow (\overline{M \cap A'}) \subset \bar{A}' \subset B$ [By (3)]

Hence, $\overline{M \cap A'} \subset \overline{M \cap B} \cap B$ ...(4)

Then by (2) and (4), we have

$$\bar{M} \cap A \subset (\overline{M \cap A} \cap A) \cup (\overline{M \cap B} \cap B)$$

Also by symmetry

$$\bar{M} \cap B \subset (\overline{M \cap A} \cap A) \cup (\overline{M \cap B} \cap B)$$

Therefore

$$(\bar{M} \cap A) \cup (\bar{M} \cap B) \subset (\overline{M \cap A} \cap A) \cup (\overline{M \cap B} \cap B)$$

But $(\bar{M} \cap A) \cup (\bar{M} \cap B) = \bar{M} \cap (A \cup B) = \bar{M} \cap X = \bar{M}$

Hence, $\bar{M} \subset (\overline{M \cap A} \cap A) \cup (\overline{M \cap B} \cap B)$ ...(5)

From (1) and (5), we have

$$\bar{M} = (\overline{M \cap A} \cap A) \cup (\overline{M \cap B} \cap B)$$

THEOREM 8.   **(X, d) be a metric space and let A, B be any two non-empty subsets of X if $d(A, B) > 0$, then A and B are separated sets.**

PROOF.   Let $d(A, B) = \lambda > 0$, then $\lambda = \inf \{d(x, y) : x \in A, \ y \in B\} > 0$

By the definition of infimum

$$d(x, y) \geq \lambda \ \forall \ x \in A, \ y \in B$$

If $x \in A$, then x can not be a limit point of B since $S(\lambda/2, x)$ contains no point of B.

$\Rightarrow$ A does not have an accumulation point of B.

$\Rightarrow$ $A \cap \bar{B} = \phi$ ...(1)

Similarly, we can show,

B does not have an accumulation point of A

i.e., $\bar{A} \cap B = \phi$ ...(2)

Hence, from (1) and (2), we conclude A and B are separated sets.

☛ **REMARKS**

⇒ The converse of above theorem is not true.

⇒ In usual metric space $\boldsymbol{R}_n$, the sets

$A = \{x : x < 0\}$ and $B = \{x : x > 0\}$

are separated while $d(A, B) = 0$.

7.3 CONNECTED AND DISCONNECTED SETS

Connected metric space is one which consists of a single piece.

Definition 1. *Let (X, d) be a metric space and let A be a subset of X. Then A is said to be disconnected if A is the union of two non-empty separated sets, i.e., if there exists two sets C and D such that $A = C \cup D$ and $C \cap \bar{D} = \phi$ and $\bar{C} \cap D = \phi$. If these conditions be satisfied, then the pair (C, D) is called a separation (or a disconnection) of A.*

In particular, the space X is disconnected if there exists two non-empty sets A and B such that

$$X = A \cup B, \ A \cap \bar{B} = \phi \text{ and } \bar{A} \cap B = \phi$$ (MEERUT 1994, 95, 2011)

A set (or space) is said to be connected if it is not disconnected.   (MEERUT 1990, KANPUR 1992)

Definition 2. *Let (X, d) be a metric space and let a, b be any two points of the metric space X, then a and b are said to be connected if they are contained in a connected subset of X.*

☛ **REMARKS**

➠ The set R with usual metric is connected.

➠ The empty set ϕ and the singleton sets $\{x\}$ are assumed to be connected.

✍ ILLUSTRATIONS

* The space of rationals with the usual distance function is disconnected.

* R^n is connected.

* The real line is connected.

* Every discrete space (X, d) where X contains more than one point is disconnected.

* Every indiscrete space is connected.

* Let X be any infinite set, and let d be a metric on X defined by setting

$$d(x, y) = 0 \quad \text{if} \quad x = y$$
$$d(x, y) = 1 \quad \text{if} \quad x \neq y$$

The metric space (X, d) is disconnected.

* Every line segment in R^2 is connected.

* The closure of connected set is connected.

* Union of two non-empty separated subsets of a metric space is disconnected.

* Every sub-space E of the real line R is connected iff it is an interval.

🏛 FACTS : TO THE POINT

▶ Let (X, d) be a metric space. Then following are equivalent:

   (i) X is connected

   (ii) X can not be written as the union of two disjoint closed sets.

   (iii) X can't be written as the union of two non-empty sets A and B such that $A \cap \bar{B} = \phi$ and $\bar{A} \cap B = \phi$.

   (iv) X and ϕ are the only sets which are both open and closed.

▶ Any discrete metric space with more than one point is disconnected.

▶ A metric space X is connected if and only if every continuous function $f : X \to \{0, 1\}$ is not onto.

▶ Any connected subset of R containing more than one point is uncountable.

▶ A subset of a discrete space is connected if and only if it is a singleton set.

▶ A unit circle is a connected subset of R^2.

▶ The open unit disc $\{(x,y) : x^2+y^2 < 1\}$ is connected subset of Euclidean plane.

▶ The closed unit disc $\{(x,y) : x^2+y^2 \leq 1\}$ is a connected subset of Euclidean space.

▶ Every line segment in $R^?$ is connected.

THEOREM 1. *Let (X, d) be a metric space, and let $A \subset X$. The following statements are equivalent :*

   (i) A is disconnected.

   (ii) A is expressible as the union of two separated subsets of X.

   (iii) There exists closed sets G and H such that

$$A \subset G \cup H, A \cap G \neq \phi, A \cap H \neq \phi,$$
$$A \cap G \cap H = \phi$$

   (iv) There exists d-open sets P and Q such that

$$A \subset P \cup Q, \ A \cap P \neq \phi, \ A \cap Q \neq \phi, \ A \cap P \cap Q = \phi$$

PROOF.   (i) **(i) $\Rightarrow$ (ii)**

A is disconnected.

$\Rightarrow$ There exists two disjoint, non-empty, relatively closed subsets P and Q of A s.t. $A = P \cup Q$

Since P and Q are relatively closed

$\Rightarrow \qquad P = \bar{P} \cap A \quad Q = \bar{Q} \cap A$

$\Rightarrow \qquad P \cap \bar{Q} = \bar{P} \cap Q = P \cap Q = \bar{P} \cap \bar{Q} \cap A$

Since P and Q are disjoint.

$\therefore \qquad P \cap Q = \phi$

$\Rightarrow \qquad P \cap \bar{Q} = \bar{P} \cap Q = \phi$

$\Rightarrow \quad A$ is expressible as the union of two separated sets P and Q.

(ii) $\Rightarrow$ (iii)

Since, A is expressible as the union of two separated sets P and Q, we can write $A = P \cup Q$ and $P \cap \bar{Q} = \bar{P} \cap Q = \phi$.

We can write

$$P = P \cup (\bar{P} \cap Q)$$
$$= \bar{P} \cap (P \cup Q)$$
$$= \bar{P} \cap A$$

Similarly, we have

$$Q = \bar{Q} \cap A$$

Setting $\bar{P} = G$ and $\bar{Q} = H$, we find that G and H are d-closed subsets of X such that

$$G \cap A \neq \phi, \ H \cap A \neq \phi, \ A \subset G \cup H, \ G \cap H \cap A = \phi$$

(iii) $\Rightarrow$ (iv)

Since G and H be d-closed sets satisfying the conditions

$$A \subset G \cup H, \ G \cap A \neq \phi, \ H \cap A \neq \phi, \ G \cap H \cap A = \phi$$

Then $G_1 = G \cap A$ and $H_1 = H \cap A$ are disjoint, non-empty, relatively closed sets whose union is A.

$\Rightarrow \quad G_1$ and H_1 are both relatively open as well as closed.

$\therefore \ \exists \ d$-open sets P and Q s.t.

$$G_1 = P \cap A, \ H_1 = Q \cap A$$

Since the sets G_1 and H_1 are disjoint, non-empty and $G_1 \cup H_1 = A$

$\Rightarrow \ P \cap Q \cap A = \phi, \ P \cap A \neq \phi, \ Q \cap A \neq \phi$ and $A \subset P \cup Q$

(iv) $\Rightarrow$ (i)

Since P and Q be d-open sets satisfying conditions

$$A \subset P \cup Q, \ P \cap A \neq \phi, \ Q \cap A \neq \phi, \ P \cap Q \cap A = \phi$$

Writing $P \cap A = G$ and $Q \cap A = H$, we find G and H are non-empty, relatively open, disjoint sets having A as union.

$\Rightarrow \quad A$ is disconnected.

THEOREM 2. *Let (Y, d^*) be a subspace of a metric space (X, d). Then Y as a subset of X is disconnected iff the subspace (Y, d) is disconnected as a space in its own right.*

PROOF. We have already proved that, two empty subsets of Y are d-separated iff they are d^*-separated.

$\therefore \ Y$ is the union of two d-separated sets iff it is the union of two d^*-separated sets. Hence the theorem.

THEOREM 3. *A metric space (X, d) is disconnected iff there exists a non-empty proper subset of X which is both d-open and d-closed.* [MEERUT 2010,16]

PROOF. Let A be a non-empty proper subset of X which is both open and closed. Then, we wish to show (X, d) is disconnected.

Let $B = X - A$. Then B is non-empty. $[\because A$ is proper subset of X]

Also, $A \cup B = X$ and $A \cap B = \phi$ ...(1)

Since, A is open

$\Rightarrow \quad B = X - A$ is closed, $i.e.$, $\bar{B} = B$.

$\Rightarrow \qquad\qquad A \cap \bar{B} = \phi$ [By (1)]

Again, since A is closed.

$\Rightarrow \qquad\qquad \bar{A} = A$

$\Rightarrow \qquad\qquad \bar{A} \cap B = \phi$ [by (1)]

Thus, X has been expressed as the union of two non-empty subsets A and B of X s.t. $A \cap \bar{B} = \phi$ and $\bar{A} \cap B = \phi$

$\Rightarrow \quad (X, d)$ is disconnected.

Conversely, let (X, d) be disconnected.

$\Rightarrow \quad$ There exists non-empty subsets A, B of X s.t. $A \cup B = X$, $A \cap \bar{B} = \phi$ and $\bar{A} \cap B = \phi$.

Also, since $A \subset \bar{A}$

Then, $\qquad \bar{A} \cap B = \phi \Rightarrow A \cap B = \phi$

Thus, $A \cup B = X$ and $A \cap B = \phi$

$\therefore \ A = X - B$ and since B is non-empty, it follows that A is a proper subset of X.

Also, by hypothesis, A is non-empty.

Now, $\qquad A \cup \bar{B} = X$

$[\because A \cup B = X$ and $\bar{B} \supset B \Rightarrow A \cup \bar{B} \supset X$ but $X \supset A \cup \bar{B}$ always]

Also, $\qquad A \cap \bar{B} = \phi$

Implies that $\qquad A = X - \bar{B}$ ...(2)

Similarly, $\qquad B = X - \bar{A}$ ...(3)

$\bar{A}$ and $\bar{B}$ are d-closed.

From (2), it follows that A is open and from (3), it follows that B is open.

$\Rightarrow \quad A$ and B are open.

Also we have proved above that $A = X - B$

$\Rightarrow \quad A$ is closed. [as B is open]

$\Rightarrow \quad A$ is a non-empty, proper subset of X which is both open and closed.

☞ REMARKS

➡ A metric space X is connected iff the only non-empty subset of X which is both open and closed is X itself.

➡ Let (X, d) be a metric space and $Y \subset X$. Then Y is disconnected iff there exists a non-empty proper subset A of Y which is both open and closed in Y, $i.e.$, iff $\exists$ a open set G and a closed set F such that $A = G \cap Y$ and $A = F \cap Y$.

THEOREM 4. **A metric space X is disconnected if and only if any one of the following statements hold :**

    (i) X is the union of two non-empty disjoint open sets.

    (ii) X is the union of two non-empty disjoint closed sets.

PROOF. **If part**

Let X be disconnected.

$\Rightarrow \quad \exists$ a non-empty proper subset A of X which is both open and closed.

$\Rightarrow$ A' is also both open and closed.

Also, $A \cup A' = X$ and A' is non-empty since A is a proper subset of X.

Thus, we get two non-empty sets A and A' which satisfy the requirement of (i) and (ii).

Only if part

Let $X = A \cup B$ and $A \cap B = \phi$ where A and B are two non-empty d-open sets.

Since, $A \cup B = X$ and $A \cap B = \phi$

$\Rightarrow$ $A = X - B$

$\Rightarrow$ A is closed.   (Since B is open)

Also, A is non-empty.   ($\because$ B is non-empty)

$\Rightarrow$ A is a proper closed subset of X.

Given that A is open.

$\Rightarrow$ A is a proper subset of X which is both open and closed.

$\Rightarrow$ X is disconnected.   [By theorem 3]

Again, let X C D and $C \cap D = \phi$, where C and D are non-empty closed sets.

$\Rightarrow$ $C = X - D$

$\Rightarrow$ C is open.   ($\because$ D is closed)

Also, C is non-empty.   ($\because$ D is non-empty)

$\Rightarrow$ C is a proper subset **of** X which is both open and closed.

$\Rightarrow$ X is disconnected.   [By Theorem 3]

Thus, if any one of the condition (i) and (ii) hold, X is disconnected.

Hence, the theorem.

☛ **REMARKS**

⟼ A subset Y of a metric space X is disconnected iff Y is the union of two non-empty disjoint sets both open (closed) in Y.

⟼ A subset Y of a metric space X is disconnected iff there exists non-empty open (closed) sets A and B in X s.t. $A \cap Y \neq \phi$, $B \cap Y \neq \phi$, $Y \subset A \cup B$ and $A \cap B \subset X - Y$.

<u>**THEOREM 5.**</u>   **Let (X, d) be a metric space and let P be a connected subset of X such that it contained in the union of two separated sets, then it must be contained in one of them.**   (MEERUT 2014)

Or

Let (X, d) be a metric space and let P be a connected subset of X s.t. $P \subset A \cup B$ where A and B are separated sets. Then $P \subset A$ or $P \subset B$.

<u>**PROOF.**</u>   Let P be connected and let A, B be two separated sets s.t. $P \subset A \cup B$.

Since, A and B are separated sets

$\Rightarrow$ $A \cap \bar{B} = \phi$, $\bar{A} \cap B = \phi$

Now, $P \subset A \cup B \Rightarrow P = P \cap (A \cup B)$

                    $= (P \cap A) \cup (P \cap B)$ ...(1)

We claim that at least one of the sets $P \cap A$ and $P \cap B$ is empty. If possible, suppose none of these sets is empty.

i.e., $P \cap A \neq \phi$ and $P \cap B \neq \phi$

Then $(P \cap A) \cap \overline{(P \cap B)} = (P \cap A) \cap (\bar{P} \cap \bar{B})$

                                $= (P \cap \bar{P}) \cap (A \cap \bar{B})$

                                $= (P \cap \bar{P}) \cap \phi$

                                $= \phi$

Similarly,
$$\overline{(P \cap A)} \cap (P \cap B) = \phi$$
$\Rightarrow$ $P \cap A$ and $P \cap B$ are separated sets.

$\Rightarrow$ P has been expressed as the union of two non-empty separated sets.

$\Rightarrow$ P is disconnected.

But, this is a contradiction, hence our assumption is wrong, i.e., at least one of the sets $P \cap A$ and $P \cap B$ is empty.

If $P \cap A = \phi$, then from (1) we get
$$P = \phi \cup (P \cap B) = P \cap B$$
$\Rightarrow$ $P \subset B$

Similarly, if $P \cap B = \phi$, then $P \subset A$

Hence, $P \subset A$ or $P \subset B$.

THEOREM 6. *Let (X, d) be a metric space and let P be a connected subset of X. If Q is a subset of X s.t. $P \subset Q \subset \overline{P}$, then Q is connected.*   (MEERUT 2014, 18)

PROOF.   Let Q is disconnected.

Then there exists non-empty separated sets whose union is Q

i.e., $\exists$ A and B s.t.
$$A \cap \overline{B} = \phi, \ \overline{A} \cap B = \phi \text{ and } A \cup B = Q$$

Since $P \subset Q$, we have $P \subset A \cup B$

$\Rightarrow$ the connected set P is contained in the union of two separated sets A and B.

$\Rightarrow$ either $P \subset A$ or $P \subset B$ [by previous theorem]

Let $P \subset A \Rightarrow \overline{P} \subset \overline{A}$

$\Rightarrow$ $\overline{P} \cap B \subset \overline{A} \cap B$

$\Rightarrow$ $\overline{P} \cap B \subset \phi$ ...(1)

Since ϕ is the subset of every set

$\Rightarrow$ $\phi \subset \overline{P} \cap B$ ...(2)

From (1) and (2)
$$\overline{P} \cap B = \phi$$ ...(3)

Again, $Q = A \cup B$ and $Q \subset \overline{P} \Rightarrow B \subset Q \subset \overline{P}$

Hence, $\overline{P} \cap B = B$ ...(4)

From (3) and (4), we conclude that $B = \phi$

which is a contradiction, since B is non-empty (by hypothesis).

$\Rightarrow$ Q is connected.

COROLLARY I. *Let (X, d) be a metric space and let Q be a connected subset of X. If Q is a subset of X s.t. $P \subset Q \subset \overline{P}$, then $\overline{P}$ is connected.*

THEOREM 7. *The closure of connected set is connected.*

Or

Let (X, d) be a metric space and let Y be any subset of X then $\overline{Y}$ is connected if Y is connected.

PROOF.   Let Y be any subset of (X, d), then we can write $Y \subset Y \subset \overline{Y}$ and since Y is connected, by using above theorem, we get $\overline{Y}$ is connected.

Thus, this theorem is a particular case of the preceding theorem. But because of its importance, an independent proof is given here.

Suppose $\overline{Y}$ is disconnected.

$\Rightarrow$ there exists non-empty sets A and B s.t.
$$A \cap \overline{B} = \phi, \ \overline{A} \cap B = \phi \text{ and } A \cup B = \overline{Y}$$ ...(1)

Since $\qquad Y \subset \bar{Y}$

$\Rightarrow \qquad Y \subset A \cup B$

$\Rightarrow$ Connected set Y is contained in the union of two separated sets A and B.

$\Rightarrow$ Either $Y \subset A$ or $Y \subset B$ \hfill [Using Theorem 5]

Let $\qquad Y \subset A$

Since $Y \subset A \Rightarrow \bar{Y} \subset \bar{A} \Rightarrow \bar{Y} \cap B \subset \bar{A} \cap B$

$\Rightarrow \qquad \bar{Y} \cap B \subset \phi$ \hfill [From (1)]

But, since ϕ is the subset of every set

$\Rightarrow \qquad \bar{Y} \cap B = \phi$ \hfill ...(2)

Again $\qquad \bar{Y} = A \cup B$

$\Rightarrow \qquad B \subset \bar{Y}$

$\Rightarrow \qquad \bar{Y} \cap B = B$ \hfill ...(3)

From (2) and (3), we get

$\qquad B = \phi$

which is a contradiction, since B is non-empty [By our hypothesis]

$\Rightarrow$ Our assumption is wrong.

$\Rightarrow \quad \bar{}$ is connected.

✎ ILLUSTRATIONS

* The intervals $[a, b]$, $]a, b[$, $[a, b[$ are connected subsets of the real line.
* For each real number a, the ray $\{x : x \geq a\}$ is the closure of the connected set $\{x : x > a\}$ and is therefore connected.
* For each real number a, the ray $\{x : x \leq a\}$ is the closure of the connected set $\{x : x < a\}$ and is therefore connected.
* The set $P = \left\{ \left(x, \sin \dfrac{1}{x} \right) : x > 0 \right\} \cup \{(0, y) : -1 < y < 1\}$ is a connected subset of R^2.

  Since, if $E = \{(x, \sin \frac{1}{x}) : x > 0\}$. Then, $\bar{E} = \{(x, \sin \frac{1}{x}) : x > 0\} \cup \{(0, y) : -1 \leq y \leq 1\}$

  Thus, we get $E \subset P \subset \bar{E}$

  $\Rightarrow P$ is connected (since E is connected)

* Let $f : (X, d) \rightarrow (Y, d^*)$ be a continuous mapping and let (X, d) be connected. Then the graph of f is connected subset of the product space $(X, d) \times (Y, d^*)$.

📥 SOLVED EXAMPLES

EXAMPLE 1. *Show that every singleton set is connected.*

SOLUTION. We know that ϕ is regarded as a connected subset of X. Thus, every singleton set can not be expressed as the union of two non-empty separated sets.

Hence, every singleton set is connected.

EXAMPLE 2. *Show that every discrete space (X, d) where X contains more than one point is disconnected.*

SOLUTION. X contains more than one point. (Given)

Every singleton subset of X is connected. (Proved above)

i.e., every singleton subset of X is a non-empty proper subset of X which is both open and closed.

$\Rightarrow \quad (X, d)$ is disconnected.

EXAMPLE 3. *Prove that the real line R is connected.*

(MEERUT 2016)

SOLUTION.   We have to prove that the real line **R** is connected. If possible, let real line **R** be disconnected.

Thus, by the definition of disconnectedness, there exists two non-empty sets A and B s.t. $R = A \cup B$, $\bar{A} \cap B = \phi$ and $A \cap \bar{B} = \phi$.

Let there exists points $x \in A$ and $y \in B$, also assume that $x > y$.

Again, $[x, y] \subset R = A \cup B$

Hence, every point of $[x, y]$ is either in A or in B.

Let $a = \sup([x, y] \cap A)$. Evidently $x \le a \le y$.

Since a is the sup $([x, y] \cap A)$, for each $\in > 0$

$\exists$ some $b \in [x, y] \cap A$ s.t. $a - \in < b \le a$. This shows that every nbd of a contains a point of $[x, y] \cap A$ and hence a point of A.

$\Rightarrow$ either $a \in A$ or a is a limit point of A.

Since A is closed.   [By Theorem 4]

$\Rightarrow$ Every limit point of A belongs to A.

$\Rightarrow$ In either case $a \in A$. Since $A \cap B = \phi$, $a \notin B$

Again since $y \in B$, we have $y \ne a$

$\Rightarrow$ $a < y$

Moreover, by the definition of a, $a + \in \in B$ for every $\in > 0$ s.t. $a + \in \le y$

$\Rightarrow$ every nbd of a contains a point of B other than a

$\Rightarrow$ a is the limit point of B.

But, since B is also closed and every limit point of a closed set belongs to it.

$\Rightarrow$ $a \in B$

$\Rightarrow$ $a \in A \cap B$

which is a contradiction, since $A \cap B = \phi$

$\Rightarrow$ Our assumption is wrong.

Thus, **R** is connected.

EXAMPLE 4.   *If A and B are connected sets which are not separated then $A \cup B$ is connected set.*   (MEERUT–2013)

SOLUTION.   Let A and B be separated subsets of metric space (X, d). We have to prove that $A \cup B$ is disconnected.

Let $X - \bar{A} = G$ and $X - \bar{B} = H$

$\therefore$ $\bar{A}$ and $\bar{B}$ are non-empty closed subsets of X.

$\Rightarrow$ G and H are non-empty open subsets of X.

Also, $G \cup H = (X - \bar{A}) \cup (X - \bar{B}) = X - (\bar{A} \cap \bar{B})$

Now, $(A \cup B) \cap G = (A \cup B) \cap (X - \bar{A})$

$= A \cap (X - \bar{A}) \cup B \cap (X - \bar{A})$

$= \phi \cup B = B$

$\Rightarrow$ $(A \cup B) \cap G = B$.

Similarly, it can be easily shown that

$(A \cup B) \cap H = A$

Also, since A and B are separated subsets of (X, d)

$\Rightarrow$ $\bar{A} \cap B = \phi \Rightarrow B \subset X - \bar{A}$

$\Rightarrow$ $B \cap (X - \bar{A}) = B$

and $A \subset \bar{A} \Rightarrow (X - \bar{A}) \subset X - A$

$\Rightarrow$ $A \cap (X - \bar{A}) \subset A \cap (X - A)$

$\Rightarrow$ $A \cap (X - \bar{A}) \subset \phi$

$$\Rightarrow \quad A \cap (X - \bar{A}) = \phi$$

[Since ϕ is the subset of every set]

Now, $A \neq \phi$ and $B \neq \phi$

$$\Rightarrow \qquad (A \cup B) \cap G \neq \phi \text{ and } (A \cup B) \cap H \neq \phi$$

$$\because \quad [(A \cup B) \cap G] \cap [(A \cup B) \cap H] = A \cap B = \phi$$

$$[(A \cup B) \cap G] \cup [(A \cup B) \cap H] = A \cup B$$

Thus, we have shown that $\exists \ G, H \subset X$ s.t.

$$[(A \cup B) \cap G] \neq \phi, \quad [(A \cup B) \cap H] \neq \phi$$

$$[(A \cup B) \cap G] \cap [(A \cup B) \cap H] = \phi$$

$$[(A \cup B) \cap G] \cup [(A \cup B) \cap H] = A \cup B$$

$\Rightarrow \quad G \cap H$ is a disconnection of $A \cup B$ and hence $A \cup B$ is disconnected.

EXAMPLE 5. **Let A and B be closed subsets of X such that $A \cup B$ and $A \cap B$ are connected. Show that each of the sets A and B is connected.**

SOLUTION. Let (X, d) be a metric space.

Given. A and B are closed subsets of X and $A \cup B$ and $A \cap B$ are connected.

To prove. A and B both are connected.

Proof. If $A \subset B$ or $B \subset A$, the result is trivially true (since if $A \subset B$, then $A \cap B = A$ and $A \cup B = B$).

Now, let neither A is a subset of B nor B subset of A, then each of the set $A \sim B$ and $B \sim A$ is non-empty. Also, $A \sim B$ and $B \sim A$ are disjoint sets.

$$A \sim B = (A \cup B) \cap (X \sim B)$$

$$B \sim A = (A \cup B) \cap (X \sim A)$$

$\Rightarrow \quad A \sim B$ and $B \sim A$ are relatively open subsets of $A \cup B$.

$\Rightarrow \quad A \sim B$ and $B \sim A$ are non-empty, disjoint, relatively open subsets of $A \cup B$.

$\Rightarrow \quad A \sim B$ and $B \sim A$ are relatively separated.

Also, $(A \sim B) \cup (A \cap B)$ and $(B \sim A) \cup (A \cap B)$ are connected.

$\Rightarrow \quad$ Sets A and B are connected.

$[(A \cup B) \sim (A \cap B) = (A \sim B) \cup (B \sim A)$, $A \cup B$ and $A \cap B$ are connected]

THEOREM 8. **Let (X, d) be a connected metric space, and let C be a connected subset of X. If $X \sim C = A \cup B$, where A and $\quad$ are separated sets, then the sets $A \cup C$ and $B \cup C$ is connected.**

Given : C is a connected subset of (X, d) s.t.

$$X \sim C = A \cup B$$

To prove : $A \cup C$ and $B \cup C$ is connected.

PROOF. Let, if possible, $A \cup C$ is disconnected.

Then, there exists two non-empty sets P and Q such that

$$A \cup C = P \cup Q, \ P \cap \bar{Q} = \phi, \ \bar{P} \cap Q = \phi$$

Since $C \subset P \cup Q$ and C is connected, we have

Either $C \subset P$ or $C \subset Q$

Suppose, $C \subset P$. Then $C \cap H = \phi$

Also, $\qquad Q \subset A \cup C \qquad\qquad [\because P \cup Q = A \cup C \Rightarrow Q \subset A \cup C]$

and $\qquad C \cap Q = \phi$

$\Rightarrow \qquad Q \subset A.$

Also, given that A and B are separated sets.

$\Rightarrow \quad A$ and B are separated sets and $Q \subset A$ and $B \subset B$

$\Rightarrow \quad Q$ and B are separated sets.

$\Rightarrow \quad Q$ is separated from P as well as from B

$\Rightarrow$ Q is separated from $P \cup B$.

Again, $X \sim C = A \cup B \Rightarrow X = C \cup A \cup B$

$= (P \cup Q) \cup B$ $[\because A \cup C = P \cup Q]$

$= P \cup Q \cup B$

$\Rightarrow$ $X = Q \cup (P \cup B)$

$\Rightarrow$ X can be expressed as the union of two separated sets Q and $P \cup B$, *i.e.*, X is disconnected.

which is a contradiction (since (X, d) is connected space)

$\Rightarrow$ Our assumption is wrong.

i.e., $A \cup C$ is connected.

Similarly, if we take $C \subset Q$, it will be easily shown that B C is connected.

Hence the theorem.

THEOREM 9.   **Let (X, d) be a metric space and let $\{C_\lambda : \lambda \in \Lambda\}$ be a non-empty collection of connected subsets of X, such that $\cap \{C_\lambda : \lambda \in \Lambda\} \neq \phi$. Then $\cup \{C_\lambda : \lambda \in \Lambda\}$ is a connected set.**   (MEERUT 2001, 03, 07(B.P.), 18)

PROOF.   Let, $G = \cup \{C_\lambda : \lambda \in \Lambda\}$

We have to show that G is a connected set.

Let, if possible, suppose G is non-connected, then there exists two non-empty sets A and B such that $G = A \cup B$, $\bar{A} \cap B = \phi$ and $A \cap \bar{B} = \phi$.

Since, $\cap \{C_\lambda : \lambda \in \Lambda\} \neq \phi$, then there exists some point

$a \in \cap \{C_\lambda : \lambda \in \Lambda\}$

then surely a must belong to G.

i.e., $a \in A \cup B$ (since $a \in G$ and $G = A \cup B$)

$\Rightarrow$ either $a \in A$ or $a \in B$.

Suppose that $a \in A$

Since $a \in \cap \{C_\lambda : \lambda \in \Lambda\}$

$\Rightarrow$ $a \in C_\lambda$ $\forall$ $\lambda \in \Lambda$

$\Rightarrow$ $A \cap C_\lambda \neq \phi$ $\forall \lambda \in \Lambda$ $[$Since $a \in A \cap C_\lambda]$

Also, it is given that each C_λ is a connected subset of X, therefore

$C_\lambda \subset A \cup B$ $[A \cup B = G = \cup \{C_\lambda : \lambda \in \Lambda\}]$

Now, using Theorem 5 of Section 7.3

$C_\lambda \subset A$ or $C_\lambda \subset B$

$\Rightarrow$ $C_\lambda \subset A$ $\forall$ $\lambda \in \Lambda$

$[\because A$ and B are disjoint sets and $C_\lambda \cap A \neq \phi$ $\forall \lambda \in \Lambda]$

$\Rightarrow$ $\cup \{C_\lambda : \lambda \in \Lambda\} \subset A$

$\Rightarrow$ $G \subset A$

$\Rightarrow$ $A \cup B \subset A$

$\Rightarrow$ $B = \phi$

which is a contradiction. $[$Since B is non-empty$]$

Thus, our hypothesis is wrong.

Hence, G is connected.

$\Rightarrow$ $\cup \{C_\lambda : \lambda \in \Lambda\}$ is connected set.

Hence the theorem.

THEOREM 10.   **Let $\{C_\lambda : \lambda \in \Lambda\}$ be a family of connected subsets of X such that one of the members of this family intersects every other member. Then $\cup \{C_\lambda : \lambda \in \Lambda\}$ is connected.**   (DELHI 2005, 08; BILASPUR 2003; ALLAHABAD 2007)

Or

Prove that in a metric space (X, d) the union of the family is non-empty class of connected subspace is connected if the intersection of the family is non-empty.

PROOF.
Let, $G = \bigcup \{C_\lambda : \lambda \in \Lambda\}$.

We have to prove that G is connected.

Given that the intersection of the family is non-empty, *i.e.*, there exists a fixed member C_{λ_1} of the given family s.t.

$$C_{\lambda_1} \cap C_\lambda \neq \phi \quad \forall \ \lambda \in \Lambda$$

Let us denote $C_{\lambda_1} \cup C_\lambda$ by A_λ, *i.e.*, $A_\lambda = C_{\lambda_1} \cup C_\lambda \quad \forall \ \lambda \in \Lambda$

Then A_λ is connected. [Using previous theorem]

Now, $\bigcup \{A_\lambda : \lambda \in \Lambda\} = \bigcup \{C_{\lambda_1} \cup C_\lambda : \lambda \in \Lambda\}$

$$= C_{\lambda_1} \cup \{\bigcup \{C_\lambda : \lambda \in \Lambda\}$$

$$= \bigcup \{C_\lambda : \lambda \in \Lambda\} \quad \text{[Since } C_{\lambda_1} \text{ is any one member of the } C_\lambda \text{;s]}$$

and $\bigcap \{A_\lambda : \lambda \in \Lambda\} = \bigcap \{C_{\lambda_1} \cup C_\lambda : \lambda \in \Lambda\}$

$$= C_{\lambda_1} \cup \{\bigcap \{C_\lambda : \lambda \in \Lambda\} \neq \phi \quad \text{[Given]}$$

$\Rightarrow \quad \{A_\lambda : \lambda \in \Lambda\}$ is the collection of connected subsets of X having a non-zero intersection.

By the previous theorem, if A_λ is the union of connected subsets of X having non-zero intersection. Then $\bigcup \{A_\lambda : \lambda \in \Lambda\}$ is connected, we have

$\bigcup \{A_\lambda : \lambda \in \Lambda\}$ is connected.

i.e., $\bigcup \{C_\lambda : \lambda \in \Lambda\}$ is connected. $\qquad [\because \bigcup \{A_\lambda : \lambda \in \Lambda\} = \bigcup \{C_\lambda : \lambda \in \Lambda\}]$

 SOLVED EXAMPLES

EXAMPLE 1. **Show that every line segment in $\mathbf{R}^2$ is connected.**

SOLUTION.
Let PQ be a line segment in $\mathbf{R}^2$. Choose P as the origin, PQ produced as the axis of X, and the perpendicular from P on PX as the axis of y.

Let Q be the point $(x, 0)$.

Consider the mapping f from $[0, x[$ to $\mathbf{R}^2$ defined by $f(x) = (x, 0)$

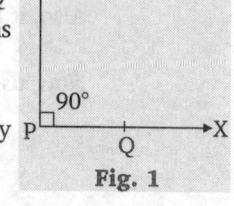

Fig. 1

$\Rightarrow \quad f$ is continuous.

Let x_0 be any point of $[0, x[$ and $<x_n>$ be a sequence converging to x_0, then

$$f(x_n) = (x_n, 0) \to (x_0, 0)$$

$$\Rightarrow \qquad f(x_n) \to f(x_0)$$

$\Rightarrow \quad f$ is continuous at x_0.

Since x_0 is any point of $[0, x[$, therefore it follows that f is continuous on $[0, x[$.

$\because$ Every interval on a real line is connected

$\Rightarrow \quad [0, x[$ is connected.

And f is continuous, therefore $f(0, x)$ is connected subset of $\mathbf{R}^2$.

Also, $f[0, x] = PQ$.

$\Rightarrow \quad PQ$ is connected.

Since PQ is any arbitrary line segment

$\Rightarrow$ Every line segment in $\mathbf{R}^2$ is connected.

EXAMPLE 2. **Prove that every circle is a connected subset of R^2.**

SOLUTION. In parametric form, the equation of circle of radius r can be written as $x = r\cos\theta;\ y = r\sin\theta$, where $0 \le \theta \le 2\pi$.

Let us define the mapping $f : \mathbf{R} \to \mathbf{R}^2$ s.t.

$$f(\theta) = [r\cos\theta,\ r\sin\theta] \qquad \ldots(1)$$

First, we shall prove that f is continuous.

Let $\theta_0 \in \mathbf{R}$ and d be the usual metric on $\mathbf{R}^2$, then

$$d(f(\theta), f(\theta_0)) = \sqrt{(r\cos\theta - r\cos\theta_0)^2 + (r\sin\theta - r\sin\theta_0)^2}$$

$$= 2r \left| \sin\frac{(\theta - \theta_0)}{2} \right| \qquad \ldots(2)$$

We know that the sin function is continuous, therefore for given $\in > 0\ \exists\ a\ \delta > 0$ s.t.

$$\left| \sin\frac{(\theta - \theta_0)}{2} \right| < \in/2r \text{, whenever } \left| \sin\frac{(\theta - \theta_0)}{2} \right| < \frac{\delta}{2}$$

i.e., $d(f(\theta), f(\theta_0)) < \in$, whenever $|\theta - \theta_0| < \delta$

$\Rightarrow$ f is continuous at θ_0.

Since θ_0 is any arbitrary point of $\mathbf{R}$.

$\Rightarrow$ f is continuous on $\mathbf{R}$.

Also, from equation (1), it is clear that the given circle is the continuous image of interval $[0, 2\pi[$ under f.

Since every continuous image of a connected set is connected and $]0, 2\pi[$ is connected.

[∵ every interval on real line is connected]

$\Rightarrow$ Every circle is a connected subset of $\mathbf{R}^2$.

7.4 CONTINUITY AND CONNECTEDNESS

THEOREM I. **Continuous image of a connected space is connected.**

[MEERUT 2005, 05(B.P.), 08, 10, 11, 14, 16, 18; ROHILKHAND 2000, 01, 08; BHOPAL 2003, 08; KANPUR 2002, 04, 2005, 09; HIMACHAL 2007; GARHWAL 2000]

PROOF. Let $f : X \to Y$ be a continuous mapping of a connected space X into an arbitrary metric space Y.

We wish to show that $f[X]$ is connected as a subspace of Y.

Let, if possible, $f[X]$ is disconnected.

Then there exists sets A and B both open in Y such that

$$A \cap f[X] \ne \phi,\ B \cap f[X] \ne \phi$$

$$(A \cap f[X]) \cap (B \cap f[X]) = \phi$$

and $$(A \cap f[X]) \cup [B \cap f[X]) = f[X]$$

Now, $\phi = f^{-1}[\phi] = f^{-1}[(A \cap f[X]) \cap (B \cap f[X])]$

$= f^{-1}[[A \cap B] \cap f[X]]$

$= f^{-1}[A] \cap f^{-1}[B\} \cap f^{-1}[f[X]]$

$= f^{-1}[A] \cap f^{-1}[B] \cap X$

$= f^{-1}[A] \cap f^{-1}[B]$

and $X = f^{-1}[f[X]] = f^{-1}[(A \cap f[X]) \cup [B \cap f[X]]$

$= f^{-1}[(A \cup B) \cap f[X]]$

$$= f^{-1}[A \cup B] \cap f^{-1}(f[X])$$
$$= \{f^{-1}[A] \cup f^{-1}[B]\} \cap X$$
$$= f^{-1}[A] \cup f^{-1}[B]$$

Since f is continuous and A, B are open in Y both intersecting $f[X]$, it follows that $f^{-1}[A]$ and $f^{-1}[B]$ are non-empty open subsets of X.

$\Rightarrow$ X is expressible as a union of two disjoint non-empty open subsets of X.

$\Rightarrow$ X is disconnected, which is a contradiction.

Hence, our assumption is wrong.

i.e., $f[X]$ is connected.

$\Rightarrow$ continuous image of a connected space is connected.

THEOREM 2. *If f is a continuous mapping of a connected space onto an arbitrary metric space Y, then Y is connected.*
(AGRA 2004, 06, 07)

PROOF. Let (X, d) be a connected metric space and let $f : X \to Y$ be a continuous mapping from X onto Y.

We have to show that Y is connected.

Let, if possible, Y is disconnected.

Then, by definition of disconnected space there exists a non-empty proper subset A of Y which is both open and closed in Y.

Since f is continuous and onto Y.

$\Rightarrow$ $f^{-1}(A)$ is a non-empty proper subset of X which is both open and closed in X.

$\Rightarrow$ X is disconnected, which is a contradiction.

Hence, Y must be connected.

THEOREM 3. *If f is a continuous mapping of a connected space X into R, then $f[X]$ is an interval.*

Or

The range of a continuous real valued map defined on connected space is an interval.
(HIMACHAL 2003, 2007, 2009)

PROOF. The proof of this corollary is a direct consequence of the two theorems, namely continuous image of a connected space is connected and any subset of **R** is connected iff it is an interval.

It also leads to the following generalization of Weirstrass intermediate value theorem.

If $f : X \to \mathbf{R}$ is a continuous map of a connected space X into **R**, then f takes all values between any two of its values.

> **FACTS : TO THE POINT**
>
> ▶ Continuous image of a connected metric space is connected.
> ▶ Connectedness is preserved under homeomorphisms.
> ▶ Connectedness is preserved under isomorphisms.
> ▶ If the product of two metric spaces be a connected metric space, then each factor space is connected.

7.5 CHARACTERISATION OF CONNECTED SUBSETS OF THE REAL LINE

Definition 1. *Let A be any subset of **R**. Then $m \in \mathbf{R}$ is called a lower bound for A iff $m \le x$ $\forall x \in A$. Also m' is called a greatest lower bound or infimum for A iff m' is lower bound for A and if m is any lower bound for A then $m \le m'$.*

Definition 2. *Let A be any subset of **R**. Then $M \in \mathbf{R}$ is called a upper bound for A iff $x \le M$ $\forall x \in A$. Also M' is called a least upper bound or supremum for A iff M' is upper bound for A and if M is any upper bound for A then $M' \le M$.*

THEOREM 1.   ***A subset of the real line is connected if it is an interval.***   (MEERUT 2015)

PROOF.   The 'Only if part'

Let P be a connected subset of $\mathbf{R}$. Let if possible, P is not an interval. Then there exists real numbers x, a, y with $x < a < y$ s.t. $x, y \in P$ but $a \notin P$.

Let us define $G_1 =]-\infty, a[$ and $H_1 =]a, \infty[$

Then $x \in G_1$ and $y \in H_1$ so that G_1 and H_1 are non-empty disjoint open sets.

Let $A = P \cap G_1$ and $B = P \cap H_1$. Then A and B are non-empty since $x \in A$ and $y \in B$. Also, both A and B are open in P.

Now, $$A \cap B = (P \cap G_1) \cap (P \cap H_1)$$
$$= P \cap (G_1 \cap H_1)$$
$$= P \cap \phi \qquad\qquad [\because G_1 \text{ and } H_1 \text{ are disjoint sets}]$$
$$= \phi$$

and $$A \cup B = (P \cap G_1) \cup (P \cap H_1)$$
$$= P \cap (G_1 \cup H_1)$$
$$= P \cap (]-\infty, a[\cup]a, \infty[)$$
$$= P \cap (R - \{a\})$$
$$= P \qquad\qquad [\because a \notin P \text{ and } P \subset R - \{a\}]$$

$\Rightarrow$ P has been expressed as a union of two non-empty disjoint sets both open in P and so P is disconnected, which is a contradiction.

$\Rightarrow$ Our assumption is wrong and P must be an interval.

The '**if part**'

Let P be an interval. Then we have to show that P is connected.

Let, if possible, suppose that P is disconnected.

Then there exists two non-empty disjoint sets A and B both closed in P such that $P = A \cup B$.

$$\bar{A} \cap B = \phi, \quad A \cap \bar{B} = \phi.$$

Since A and B both are non-empty sets.

Then, we may choose a point $x \in A$ and $y \in B$ since $A \cap B = \phi \Rightarrow x \neq y$ then either $x > y$ or $x < y$ (by Archimedian property of real nos.)

Without loss of generality, we may assume $x < y$.

Since P is an interval and $x, y \in P$.

$\Rightarrow$ $[x, y] \subset P$ $(\because P = A \cup B)$

$\Rightarrow$ Each point of $[x, y]$ is either in A or in B.

Let us define two sets G and H by setting $G = [x, y] \cap A$, $H = [x, y] \cap B$.

Firstly, we shall show that the sets G and H are separated since $x \in A$ and $y \in B$.

$\Rightarrow$ $x \in G$ and $y \in H$.

Thus, the sets G and H are non-empty.

Since $G \subset A$ and $H \subset B$ and A and B are separated, therefore, the sets G and H are also separated.

Now, we shall show that G has a supremum and that the sup $G \in [x, y]$.

The set G is non empty (since $x \in G$)

Also, G is bounded above by y $(\because G \subset [x, y])$

$\Rightarrow$ G has a supremum, say M.

Since M is the supremum of G and y is an upper bound of H.

$\therefore$ $M \leq y$.

Also, since M is the supremum of G and $x \in G$.

$$\Rightarrow \qquad\qquad x \le M$$

Thus, we get $x \le M \le y$

We shall show now that $M \notin G$.

If $M \in G$, then M must be the greatest element of G.

$$\therefore \qquad\qquad]M, y] \subset H$$

$$\Rightarrow \qquad\qquad [M, y] \subset \bar{H}$$

$$\Rightarrow \qquad\qquad M \notin G \qquad\qquad\qquad \text{(since } G \cap H = \phi \text{ and } M \in H\text{)}$$

Now, we shall show that $M \notin H$.

Since M is the supremum of G, therefore for every $\in > 0$, there exists some $a \in G$ s.t. $a > M - \in$.

Also, $a \le M$. Since $M \notin G$ $\therefore a \ne M$.

$\Rightarrow$ There exists a different from M s.t.

$$]M - \in, M + \in[\cap G \ne \phi$$

$\Rightarrow \qquad M$ is a limit point of G

$$\Rightarrow \qquad\qquad M \in \bar{G} \qquad\qquad [\because \text{ every closed set has all its limit points}]$$

$$\Rightarrow \qquad\qquad M \notin H \qquad\qquad\qquad\qquad \text{(since } \bar{G} \cap H = \phi\text{)}$$

$$\Rightarrow \qquad\qquad M \notin G \text{ and } M \notin H$$

$$\Rightarrow \qquad M \notin G \cup H \Rightarrow M \notin [x, y]$$

which contradicts the fact that $x \le M \le y$.

$\Rightarrow \quad P$ must be connected.

✎ ILLUSTRATIONS

* **R** is connected.
* All the rays $]a, \infty[, [a, \infty[,]-\infty, a[,]-\infty, a]$ are connected.
* All the bounded intervals $[a, b],]a, b], [a, b[$ and $]a, b[$ are connected.

7.6 CONNECTEDNESS OF UNIONS

We shall discuss the theorem which gives sufficient condition for the union of family of connected sets to be connected, in this section and after it, we shall discuss some of the simple consequences of unions of connectedness.

THEOREM 1.   *Let $G = \{C_\lambda : \lambda \in \Lambda\}$ be a collection of connected subsets of a metric space (X, d) having the following property :*

If A and B are any two sets in G, then there exists a finite chain $C_0, C_1, ..., C_n$ of members of G s.t. $C_0 = A$ and $C_n = B$ for each m $(0 \le m \le n - 1)$, the sets C_m and C_{m+1} are not separated. Then, $\cup \{C_\lambda : \lambda \in \Lambda\}$ is connected.

PROOF.   We have $G = \{C_\lambda : \lambda \in \Lambda\}$ be a family of connected subsets of X. Let, if possible, suppose that the set $\cup \{C_\lambda : \lambda \in \Lambda\}$ is disconnected.

Then there exists a pair of non-empty disjoint set P and Q s.t.

$$P \cap \bar{Q} = \phi; \ \bar{P} \cap Q = \phi \text{ and } P \cup Q = \cup \{C_\lambda : \lambda \in \Lambda\}$$

Since each C_λ is connected and the sets P and Q are separated

$\therefore$ Each C_λ must be contained either in P or in Q.

Also, $C_p \subset P$, $C_q \subset Q$ for some $C_{p, q} \in \lambda, (p \ne q)$ (Since P and Q are non-empty)

Again, by our hypothesis, there exists a finite chain $C_0, C_1, ..., C_n$ of G s.t. $C_0 = C_p$, $C_n = C_q$ and for each m s.t. $0 \le m < n - 1$, the sets C_m and C_{m+1} are not separated. Each C_m belongs either to P or Q.

Let M be the largest index $(M = 0, 1, ..., n-1)$ s.t. $C_M \subset P$. Then $C_{M+1} \subset Q$.

$\Rightarrow$ C_M and C_{M+1} must also be separated

(since P and Q are separated)

which contradicts our hypothesis.

$\Rightarrow$ $\cup \{C_\lambda : \lambda \in \Lambda\}$ must be connected.

DEDUCTION I. *If G be the collection of connected subsets of a metric X and if one of them, say C_0 be not separated from any of the remaining members of G, then $\cup \{C_\lambda : \lambda \in \Lambda\}$ is connected.*

PROOF. Let C_p and C_q be any two members of G. Then C_p, C_0, C_q is a finite chain of members of G s.t. no member of this chain is separated from the preceeding one.

$\Rightarrow$ $\cup \{C_\lambda : \lambda \in \Lambda\}$ is connected.

DEDUCTION 2. *Let G be a collection of connected subsets of a metric space (X, d) s.t. each member of G intersects a fixed member, say C_0 of G. Then $\cup \{C_\lambda : \lambda \in \Lambda\}$ is connected.*

DEDUCTION 3. *If $< C_n >$ be a sequence of connected subsets of a metric space (X, d) s.t. for each $n \in Z^+$, C_n is not separated from C_{n+1} then $\cup \{C_n : n \in Z^+\}$ is connected.*

PROOF. If C_p, C_q (say $p < q$) be any two members of the given sequence, then $C_p, C_{p+1}, C_{p+2}, ..., C_q$ is a finite chain of members of $< C_n >$ s.t. no member of the chain is separated from the preceeding one

$\Rightarrow$ $\cup \{C_n : Z \in Z^+\}$ is connected (using above theorem).

DEDUCTION 4. *If $G = \{C_\lambda : \lambda \in \Lambda\}$ be a collection of connected subsets of a metric space (X, d) such that no two members of G are separated then $\cup \{C_\lambda : \lambda \in \Lambda\}$ is connected.*

PROOF. Let C_p and C_q $(p \neq q)$ be any two members of G, then C_p, C_q is a finite chain of members of G which are not separated.

Hence, using Theorem 1 of Section 7.6 $\cup \{C_\lambda : \lambda \in \Lambda\}$ is connected.

DEDUCTION 5. *If G be the collection of connected subsets of a metric space s.t. for each pair C_p, C_q of members of G s.t. $p \neq q$ if $C_p \cap C_q \neq \phi$ then $< C_n >$ be a sequence of connected subsets of a metric space (X, d) s.t. for each $n \in I^+$, C_n is not separated from C_{n+1} then $\cup \{C_\lambda : \lambda \in \Lambda\}$ is connected.*

☛ REMARK

⟹ The above Deduction 5 can also be stated in a general form as "If $G = \{C_\lambda : \lambda \in \Lambda\}$ be a collection of connected subsets of metric space (X, d) s.t. $\cap \{C_\lambda : \lambda \in \Lambda\} \neq \phi$, then $\cup \{C_\lambda : \lambda \in \Lambda\}$ is connected."

SOLVED EXAMPLES

EXAMPLE 1. *Let AP and PB be any two line segments in the plane. Show that the broken line segment APB is a connected set.*

SOLUTION. AP and PB be any two line segments in the plane. Since every line segment in $\mathbf{R}^2$ plane is connected and every connected subset of $\mathbf{R}$ is an interval

$\Rightarrow$ AP and PB is the interval in $\mathbf{R}$.

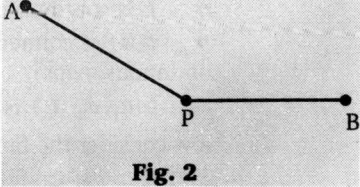

Fig. 2

$\Rightarrow$ *AP* and *PB* are connected sets having a common point *P*, therefore their union *APB* is connected.

EXAMPLE 2.   ***Show that the set obtained by removing one point from the Euclidean plane is connected.***

SOLUTION.   Take the point that has been removed (say *A*) as the origin and a pair of perpendicular lines through *A* as the axes of co-ordinates.

We have to show that $\mathbf{R}^2 - \{A\}$ is a connected subset of $\mathbf{R}^2$.

Let $S_o \equiv \{(a, 0) : a \in \mathbf{R}^+\}$

and for each real number $r > 0$; let $S_r \equiv \{(a, b) : a^2 + b^2 = r^2\}$

The set S_o is a ray in $\mathbf{R}^2$ and since each ray is connected in $\mathbf{R}^2 \Rightarrow S_o$ is connected.

The set S_r is a circle also since every circle is a connected subset of $\mathbf{R}^2 \Rightarrow S_r$ is connected.

The family $G = \{S_\lambda : \lambda \in \mathbf{R}^+ \cup \{\mathbf{Q}\}\}$ is a family of connected subsets in $\mathbf{R}^2$ and every connected subset S_λ of G intersects S_o

$\therefore$ $\cup \{S_\lambda : \lambda \in \mathbf{R}^+ \cup \{\mathbf{Q}\}\}$ is connected.

$\Rightarrow$ $\mathbf{R}^2 \sim \{A\}$ is connected subset of the plane.

Thus, the set obtained by removing one point from the Euclidean plane is connected.

EXAMPLE 3.   ***Show that the Euclidean space $\mathbf{R}^3$ is connected.***

SOLUTION.   Take an arbitrary but fixed point $P(a_1, a_2, a_3)$ in $\mathbf{R}^3$ and consider the line joining it to the origin.

The equation of this line *OA* is given by

$$\frac{x_1}{a_1} = \frac{x_2}{a_2} = \frac{x_3}{a_3} = c \text{ (say)}$$

The set of all points on this line is given by

$$L(a_1, a_2, a_3) = \{(ca_1, ca_2, ca_3) : c \in \mathbf{R}\}$$

Let us define a continuous map

$f : c \to (ca_1, ca_2, ca_3) \ \forall c \in \mathbf{R}$, then $L(a_1, a_2, a_3)$ can be proved a connected set if we shall show that

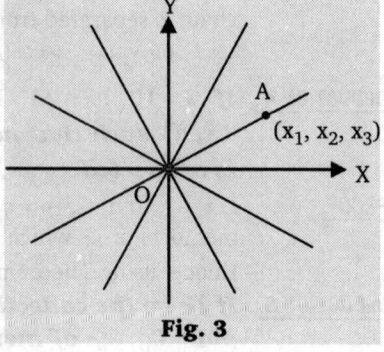

Fig. 3

$L(a_1, a_2, a_3)$ is the continuous image of $\mathbf{R}$ under f.

Let us take a point $c_0 \in \mathbf{R}$ any sequence $<c_n>$ converging to c_0.

The sequence $< f(c_n) > = < c_n a_1, c_n a_2, c_n a_3 >$

$\Rightarrow$ $c_n a_1 \to c_0 a_1; \ c_n a_2 \to c_0 a_2; \ c_n a_3 \to c_0 a_3$.

$\Rightarrow$ $f(c_n) \to (c_0 a_1, c_0 a_2, c_0 a_3)$

i.e., $f(c_n) \to f(c_0)$

$\Rightarrow$ f is continuous.

$\Rightarrow$ $f(\mathbf{R})$ is connected (since $\mathbf{R}$ is connected and connectedness is preserved under continuous maps).

$\Rightarrow$ $L(a_1, a_2, a_3)$ is connected.

Now consider the family ζ of such lines $L(a_1, a_2, a_3)$

i.e., $\zeta = \{L(a_1, a_2, a_3) : (a_1, a_2, a_3) \in R^3\}$

ζ is a family of connected sets.

∴ The union of the family ζ is connected.

⇒ $\mathbf{R}^3$ is connected.

☞ **REMARK**

➠ R^n is connected. It can be easily shown by replacing (a_1, a_2, a_3) by $(a_1, a_2, ..., a_n)$ and $\mathbf{R}^3$ by $\mathbf{R}^n$ everywhere.

EXAMPLE 4. *If* $A_0, A_1, ..., A_n$ *are* $n+1$ *distinct points in the plane. Show that the polygon line* $A_0A_1A_2...A_n$ *is connected.*

SOLUTION. We know that every line segment in $\mathbf{R}^2$ plane is connected. By joining these points, we get the line segments $A_0A_1, A_1A_2, ..., A_{n-1}A_n$.

Thus, $A_0A_1, A_1A_2, ..., A_{n-1}A_n$ are connected sets having common points. Since none of them is separated as A_0A_1 is not separated from A_1A_2 and so on.

∴ The union of all these line segments is connected.

⇒ The polygon line $A_0A_1A_2...A_n$ is connected.

EXAMPLE 5. *Show that the set that remains after removing countably many points from the plane is connected.*

SOLUTION. Let $\{A_1, A_2, ..., A_n, ...\}$ be a countable set of points to be removed. Let G denote the set that remains after removing all the points $A_1, A_2, ..., A_n, ...$ from the plane, *i.e.*,

$G = \mathbf{R}^2 - \{A_1, A_2, ..., A_n, ...\}$

We wish to show G is connected.

Take two points P and Q in G. We shall use the following theorem, that 'A subset A of a metric space (X, d) is connected iff for every pair of points $a, b \in A$ ∃ a connected set containing a and b and contained in A.

Thus to prove G a connected set, we shall find a connected set containing P and Q and contained in G.

Let l denote the perpendicular bisector of the line segment PQ. Let the lines PA_1 and QA_1 meet l in the point B_1 and G respectively.

For each of the points A_i, we get two points B_i and C_i on l. Thus resulting in a countable set of points $B_1, C_1, B_2, C_2, B_3, C_3, ...$ on l.

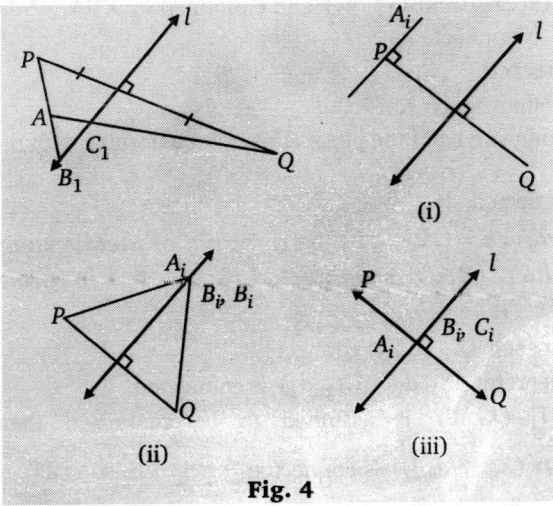

Fig. 4

The above three cases arise, but these special situations do not affect the proof.

Thus, choose a point R on l different from all the points $B_1, C_1, B_2, C_2, \ldots$ Since there are uncountably many points on l and $\{B_1, C_1, B_2, C_2, \ldots\}$ is countable set.

$\therefore$ It is always possible to choose R on l.

Draw the line segments PR and QR. Both these segments are in G.

Also, both these segments are connected (having a point R in common)

$\Rightarrow$ The broken line PRQ is connected, which also lies in G

Since none of the points $A_1, A_2, A_3, \ldots, A_n, \ldots$ lies on PRQ. Thus we have A connected set containing the points P, Q in G since P and Q are any two points of G.

$\therefore$ We have for any pair of points $P, Q \in G$, a connected set containing P, Q and contained in G.

$\Rightarrow$ G is connected set.

7.7 CONNECTEDNESS OF PRODUCTS

THEOREM 1. *The product of two metric spaces is connected iff each of them is connected.*

PROOF. Consider two metric spaces (X, d) and (Y, e).

Let (x_1, x_2) and (y_1, y_2) be any two elements of $X \times Y$.

To prove $X \times Y$ a connected metric space.

We shall show that there exists a connected subset of $X \times Y$ which contains (x_1, x_2) and (y_1, y_2).

Let us define a map $f : (Y, e) \to (\{x_1\} \times Y, e)$

then $\{x_1\} \times Y$ is a copy of the connected metric space (Y, e) and is therefore connected.

Also, $\{x_1\} \times Y$ contains the points (x_1, x_2) and (x_1, y_2).

Similarly, $X \times \{y_2\}$ is a copy of the connected metric space (X, d) and is therefore connected and contains the points (x_1, y_2) and (x_2, y_2).

Now, the connected set $\{x_1\} \times Y$ and $X \times \{y_2\}$ have non-empty intersection. [since $(x_1, y_2) \in$ to both the sets]

$\therefore$ The set $G = (\{x_1\} \times Y) \cup (X \times \{y_2\})$ is connected.

Obviously, G contains the points (x_1, x_2) and (y_1, y_2).

$\Rightarrow$ For each pair of points (x_1, x_2) and (y_1, y_2) of $X \times Y$ $\exists$ a connected set G containing (x_1, x_2) and (y_1, y_2) and contained in $X \times Y$.

$\Rightarrow$ $X \times Y$ is connected.

COROLLARY 1. R^2 *is connected.*

PROOF. R^2 is the product space $R \times R$.

Since R is connected and the plane R^2 is homeomorphic with the product of two copies of real line.

$\therefore$ R^2 is connected. [By above theorem]

THEOREM 2. *Let $\{(X_k, d_k) : k = 1, 2, \ldots, n\}$ be a family of metric spaces and let (X, d) be their product. (X, d) is connected iff (X_k, d_k) is connected for each index k s.t. $1 \le x \le n$.*

PROOF. Assume that each (X_k, d_k) is connected.

By above theorem, $(X_1, d_1) \times (X_2, d_2)$ is connected.

Also, if $\prod_{1 \le i \le m} (X_i, d_i)$ is assumed to be connected then the product space

$\prod_{1 \le i \le m} (X_i, d_i) \times (X_{m+1}, d_{m+1})$ is connected.

$\Rightarrow$ Product $\prod\limits_{1 \le i \le m+1} (X_i, d_i)$ being   a   homeomorphism   with   the   mapping

$((x_1, ..., x_p), x_{p+1}) \to (x_1, x_2, ..., x_{p+1})$ is connected.

$\Rightarrow$ The product (X, d) of the family of metric space $\{(X_k, d_k) : k = 1, 2, ..., n\}$ is connected.

☛ **REMARK**

⇒ R^n is connected, being homeomorphic to the product of n copies of the real line.

7.8 COMPONENTS

If a metric space is not itself connected, then it can be decomposed into a disjoint class of maximal connected subspaces. We shall show that this is always possible.

Definition 1. *A maximally connected subspace of a metric space (X, d) is called a component of the space.*

Definition 2. *A subspace of the metric space (X, d) is called a component of X iff it is connected and is not properly contained in any larger connected subspace of X.*

Definition 3. *Let Y be any subset of the metric space X, then a component of Y with respect to its induced metric is a maximal connected subset of Y.*

THEOREM 1.   ***Every component of a metric space X is closed.***   (MEERUT 2018, AGRA 2001, 04, 07, 08)

Or

Show that the connected components of a metric space X are closed.

(MEERUT 2005 (B.P.); ROHTAK 2003; HIMACHAL 2002, 04)

PROOF.   Let C be a component of metric space X.

Being a component, C is connected.

We have to show that C is closed.

Since, C is connected and the closure of connected set is connected.

$\Rightarrow \bar{C}$ is connected.

🗔 FACTS : TO THE POINT
▶ Components are non-empty, since singleton sets of a metric space are connected.
▶ Let (X, d) be a connected metric then the only component of X is X itself.
▶ If (X, d) is a discrete metric then the only connected subsets of X are singleton sets. Hence, each singleton set is a component for a discrete space.

By the definition of component, we know that the component is the maximal connected set, thus it is not contained in any larger connected subspace.

$\Rightarrow$ $\bar{C} \subset C$ ...(1)

But $C \subset \bar{C}$

$[\because \bar{C}$ is the smallest closed set containing $C]$

                                                  ...(2)

From (1) and (2)

$\Rightarrow$ $C = \bar{C}$

Hence, C is closed.

THEOREM 2.   ***If (X, d) is a metric space, then***

     (i) Each point in X is contained in exactly one component of X.

     (ii) The components of X form a partition of X, i.e., any two components are either disjoint or identical and the union of all the components is X.

     (iii) Each connected subset of X is contained in a component of X.

(iv) Each connected subset of X which is both open and closed is a component of X.

(i) Let x be any arbitrary point of X. Let $\{C_\lambda\}$ be the collection of all connected subsets of X which contain k. Since every singleton set $\{\ \}$ is connected

$\therefore$ $\qquad \{C_\lambda\} \neq \phi$

Also, $\qquad \cup\{C_\lambda\} \neq \phi$ $\qquad\qquad$ (Since x is a point of each C_λ)

Hence, $C_x = \cup C_\lambda$ is connected.

Also, $x \in C_x \Rightarrow C_x$ is evidently maximal.

Let A be any connected set s.t. $C_x \subset A$

Then $\qquad x \in A$ $\qquad\qquad$ [$\because A$ is one of the members of $\{C_\lambda\}$]

$\Rightarrow \qquad\qquad A \subset C_x$

$\Rightarrow \qquad\qquad C_x = A$

Hence, C_x is a component of X containing x.

$\Rightarrow$ We have shown that C_x is the only component of X containing x. Let C_x^* be any other component of X containing x.

$\because$ C_x^* is one of the C_λ's and so $C_x^* \subset C_x$.

But since C_x^* is maximal connected subset of X.

$\Rightarrow \qquad\qquad C_x \subset C_x^*$

i.e., $\qquad\qquad C_x = C_x^*$.

(ii) Let $C = \{C_x : x \in X\}$, where C_x is defined as in (i)

Let C contains all the components of X.

Let A be any other component of X.

Being a component A is non-empty contains some point, say x_0.

$\Rightarrow \qquad\qquad A = C_{x_0} \in C$

We shall show that C forms a partition of X.

Evidently, $X = \cup\{C_x : x \in X\}$

Let C_p and C_q be any two distinct components such that $C_p \cap C_q \neq \phi$

We have to show that $A_p = A_q$

Let $a \in C_p \cap C_q$

$\Rightarrow a \in C_p$ and $a \in C_q$

Both C_p and C_q are connected subsets containing a and C_a is a component containing a, we have $C_p \subset C_a$ and $C_q \subset C_a$.

But C_p, C_q are components so that we must have $C_p - C_a = C_q$

Thus, components of X form a partition of X.

(iii) Let C be any connected subset of X.

If $C = \phi$, then C is contained in every component.

If $C \neq \phi$, then C contains a point $x_0 \in X$, then $C \subset C_{x_0}$ (by i)

(iv) Let C be any connected subset of X which is both open and closed. By (iii) $C \subset A$, where A is a component of .

We claim that $C = A$.

Let, if possible, C be a proper subset of A,

Then $A \neq \phi$ and $C' \cap A \neq \phi$.

Since C is both open and closed.

$\Rightarrow$ C' is both open and closed.

Also, $(C \cap A) \cap (C' \cap A) = (C \cap C') \cap A$

$$= \phi \cap A = \phi$$

and $(C \cap A) \cup (C' \cap A) = (C \cup C') \cap A$

$$= X \cap A = A$$

$\Rightarrow$ There exists two non-empty disjoint sets whose union is A.

$\Rightarrow$ A is disconnected.

which is a contradiction (since being a component A is connected).

$\Rightarrow$ Our assumption is wrong.

and hence $A = C$

and consequently C is a component of X.

☛ **REMARK**

➠ The component of a metric space is not necessarily open. In other words, the component of a metric space is need not to open.

7.9 TOTALLY DISCONNECTED SPACES

Definition 1. *Let (X, d) be a metric space. Then (X, d) is said to be totally disconnected iff for each pair of distinct element $a, b \in X$ there exists a disconnection $A \cup B$ of X s.t. $a \in A$ and $b \in B$.*

(KANPUR 1992, 97; ROHILKHAND 2005; AGRA 2009)

☛ **REMARKS**

➠ Every singleton set is both connected and totally disconnected.

➠ Every totally disconnected space is Hausdroff.

➠ Every discrete space is totally disconnected.

SOLVED EXAMPLES

EXAMPLE 1.   ***Show that every discrete space is totally disconnected.***

SOLUTION.   Let a, b be any two points of the discrete space (X, d). Then $A = \{a\}$ and $B = X - \{a\}$ both are non-empty, disjoint sets whose union is X s.t. $a \in A$, $b \in B$.

$\Rightarrow$ $A \cup B$ is a disconnection of X with $a \in A$, $b \in B$.

$\Rightarrow$ (X, d) is totally disconnected.

EXAMPLE 2.   ***Show that the set I of all irrationul numbers w.r.t. the induced usual metric is totally disconnected.***

(MEERUT 2006)

SOLUTION.   Let a, b be any two distinct irrational numbers. Let $a < b$, then there exists a rotational number c s.t. $a < c < b$. Let $A = I \cap]-\infty, c[$ and $B = I \cap]c, \infty[$.

Then $A \cap B = \phi$

and $A \cup B = (I \cap]-\infty, c[) \cup (I \cap]c, \infty[)$

$$= I \cup []-\infty, c[\cap]c, \infty[]$$

$$= I \cup \phi = I$$

$\Rightarrow$ $A \cup B$ is a disconnection of I with $a \in A$ and $b \in B$

$\Rightarrow$ I is totally disconnected.

☞ REMARK

⇒ The set **Q** of all rational numbers w.r.t. the induced usual metric is totally disconnected. (Proof lies on the same line as that of Ex. 2 since between any two rationals there is a irrational number).

THEOREM I. *The components of a totally disconnected space (X, d) are the singleton subsets of X.*

(MEERUT 1995, 2016; AGRA 1995, 2005; ROHILKHAND 2003)

PROOF. Let (X, d) be a totally disconnected space. To prove that the components of X are the singleton subsets of X, it would be sufficient to prove every subset of X containing more than one point disconnected. Let E be any subset of X which contains more than one point. Let a, b be any two points s.t. $a \neq b$ belongs to E.

Since X is totally disconnected.

⇒ $A \cup B$ is a disconnection of X with $a \in A$ and $b \in B$.

Let us define two sets G_1 and H_1 s.t. $G_1 = A \cap E$ and $H_1 = B \cap E$.

Then both the sets G_1 and H_1 are non-empty since

$$a \in G_1 \text{ and } b \in H_1$$

Also, $G_1 \cap H_1 = (A \cap E) \cap (B \cap E)$

$$= (A \cap B) \cap E$$

$$= \phi \cap E \qquad\qquad \text{(Since } A \text{ and } B \text{ are disjoint)}$$

$$= \phi$$

and $G_1 \cup H_1 = (A \cap E) \cup (B \cap E)$

$$= (A \cup B) \cap E$$

$$= X \cap E \qquad\qquad (A \cup B \text{ being disconnection of } X)$$

$$= E$$

⇒ $G_1 \cup H_1$ is a disconnection of E.

⇒ E is disconnected.

⇒   Every subset of X containing more than one point is disconnected.

(Since E is arbitrary)

⇒   The singleton subsets are the only connected sets, i.e., components of X.
Hence the theorem.

7.10 LOCALLY CONNECTED SPACES

Definition 1. *Let (X, d) be a metric space, then it is said to be locally connected at $a \in X$ iff every open nbd of a contains a connected open nbd of a, i.e., iff the collection of all connected open nbds of a forms a local base at a.*

(MEERUT 2005, 05(B.P.), 2008, 09)

Definition 2. *The space (X, d) is said to be locally connected iff it is locally connected at each of its points.*

THEOREM I. *Every component of a locally connected space is open.*

PROOF. Given that (X, d) is a locally connected space and let c be a component of X. Let $a \in C$

⇒ a must belong to at least one connected open set G_a s.t.

$$a \in G_a \subset C \qquad\qquad [\because X \text{ is locally connected space}]$$

⇒ $C = \cup \{G_a : a \in c\}$

⇒   Being the arbitrary union of open sets, C is open.

Thus, every component of a locally connected space is open.

THEOREM 2. *A metric space (X, d) is locally connected iff the components of every open subspace of X are open.*

PROOF. Let (Y, e) be an open subspace of (X, d), then any set A is said to be open in Y iff it is open in X.

Necessary Condition.

Suppose that the components of every open subspace Y of X are open.

Let be any point of Y, then a must belong to some component of Y, say

By the hypothesis, C_a is open. Also, being a component C_a is connected.

$\Rightarrow$ C_a is open and connected in X.

$\Rightarrow$ X is locally connected (by definition).

Sufficient Condition.

Let X be locally connected. Let Y be an open subspace of X and let C be any component of Y so that C is open in Y

(∵ Every component of a locally connected space is open)

For every $a \in C$, there exists an open set G_a in Y s.t.

$$a \in G_a \subset C \qquad \text{[By definition of local connectedness]}$$

$\Rightarrow$ $G_a \cup C$ is connected in Y.

Since C is maximal connected set (being a component)

$\Rightarrow \qquad G_a \cup C \subset C \qquad \Rightarrow \qquad G_a \subset C$

$\Rightarrow$ C contains a nbd of each of its points.

$\Rightarrow$ C is open in Y.

$\Rightarrow$ Every open subspace of X has open components.

THEOREM 3. *Local connectedness neither implies nor is implied by connectedness.*

PROOF. It can be easily verified by giving example of the sets which are locally connected but not connected.

Thus, the union of two disjoint open intervals on real line forms a space which is locally connected but not connected.

Let X be any subspace of Euclidean space defined by $X = A \cup B$, where

$$A = \{(x, y) : x = 0, \ y \in [-1, 1]\} \qquad 0 \le x \le 1, \ y = \sin 1/x$$

Consider the map $f : (0, 1) \to \mathbf{R}^2$

$$f(x) = (x, \sin 1/x),$$

Then f is continuous and $f(0, 1) = B$ = continuous image of connected set $(0, 1)$

$\Rightarrow$ B is connected $\Rightarrow$ $X = \overline{B} \Rightarrow X$ is connected.

$\Rightarrow$ B is connected but not locally connected.

$\Rightarrow$ X is not locally connected.

Thus, local connectedness neither implies nor is implied by connectedness.

SOLVED EXAMPLES

EXAMPLE 1. *Give an example of a space which is connected but not locally connected.*

(MEERUT 2008 (B.P.); RAJASTHAN 2001, 03, 05, 10)

SOLUTION. Consider the subspace Y of the Euclidean plane R^2 defined by $Y = A \cup B$ where $A = \{(0, y) : y \in [-1, 1]\}$ and $B = \{(x, y) : x \in]0, 1]$ and $y = \frac{1}{x}\}$

Consider the continuous mapping $f : [0, 1] \to R^2$ s.t.

$$f(x) = x \sin \frac{1}{x}$$

$\Rightarrow$ B is the image of the interval $]0, 1]$ under f.

$\Rightarrow$ B is connected. $\qquad (\because]0, 1]$ is connected)

But it is not locally connected.

EXAMPLE 2. *Show that every discrete space is locally connected.*

(HIMACHAL 2005, 08; GARHWAL 2010)

SOLUTION. Let (X, d) be a discrete metric space and x be any arbitrary point of X.

Since every singleton set is connected and every subset of a discrete space is open.

$\Rightarrow$ $\{x\}$ is connected open nbd of x.

$\Rightarrow$ Every open nbd of x must contain $\{x\}$

$\Rightarrow$ (X, d) is locally connected.

EXAMPLE 3. *Give two examples of locally connected spaces which are not connected.*

SOLUTION. (1) Let (X, d) be a discrete metric space having more than one points.

$\qquad \Rightarrow$ X is disconnected.

$\qquad \Rightarrow$ X has a non-empty proper subset of X both open and closed in X.

$\qquad \Rightarrow$ X is locally disconnected.

(2) Let Y be the union of two disjoint open sets on real line (*i.e.*, open intervals on R)

$\qquad \Rightarrow$ $Y =]a, b[\cup]c, d[$, where $a \leq b \subset c \leq d$

Then, Y is locally connected but not connected as since $]a, b[$ and $]c, d[$ are open in R $\Rightarrow$ also open in Y

$\qquad \Rightarrow$ $]a, b[\cup]c, d[= Y$, *i.e.*, Y is the union of two non-empty disjoint sets both open in Y.

$\qquad \Rightarrow$ Y is disconnected.

Now, it remains to prove that Y is locally connected.

Let x be any point of $\Rightarrow$, then $\exists G_x$ any open nbd of x in Y, then there exists $\in > 0$ s.t. $\qquad]x - \in, x + \in[\subset G_x$

Since, being an interval $]x - \in, x + \in[$ is connected in R

$\qquad \Rightarrow$ Connected in Y

$\qquad \Rightarrow$ Every open nbd of x in Y contains an open connected nbd of x in Y.

$\qquad \Rightarrow$ Y is locally connected (by definition).

EXAMPLE 4. *Give an example of a locally connected space which is totally connected.*

SOLUTION. Every discrete space is locally connected as well as totally connected.

THEOREM 4. *The image of a locally connected space under a mapping which is both continuous and open is locally connected.*

(HIMACHAL 2003, 07, 2010; MEERUT 2009, 09(B.P.))

PROOF. Let (X, d) be a locally connected space and f be a continuous and open mapping $f : X \to Y$, where Y is an arbitrary space.

We have to show that the image $f[X]$ of X is locally connected.

For any $x \in X \ \exists \ y \in Y$ s.t. $y = f(x)$.

Let y be any arbitrary point of $f[X]$ and V be any open nbd of y in Y.

Since f is continuous, then $f^{-1}[V]$ is open in X and contains x

$$(\because V \text{ contains } y = f(x)).$$

$\Rightarrow \quad f^{-1}[V]$ is an open nbd of x.

Since X is locally connected, there exists a connected open nbd N of x s.t.

$$x \in N \subset f^{-1}[V]$$

$\Rightarrow \qquad\qquad y = f(x) \in f[N] \subset V$

$\Rightarrow \quad f(N)$ is open subset of $f[X]$ $[\because f$ is open mapping$]$

$\Rightarrow \quad f(N)$ is connected in $f[X]$ $[\because f$ is continuous and N is connected$]$

$\Rightarrow \quad$ to each open nbd V of y, there exists connected open nbd $f[N]$ of y s.t.

$y \in f[N] \subset V$

$\Rightarrow \quad f[X]$ is locally connected (by definition)

REVIEW QUESTIONS AND ARCHIVE

1. Define the following :
   (i) Separated sets   (MEERUT–2001, 11, 12, 15)
   (ii) Connected and disconnected sets
          (MEERUT–2000, 11, 12, 14, KANPUR–2016)
   (iii) Component of a metric space
   (iv) Totally disconnected space

2. Let (X, d) be a metric space and let
   $$A =]0,1[, \ B =]1,2[, \ C = [1,2]$$
   then show that A and B are separated but A and C are not separated.

3. Define locally connected space with an example.

4. Give an example of a space which is connected but not locally connected.
            (MEERUT 2008)

5. If A and B are any two subsets of a metric space X, both A and B are closed or open, then show that $A \sim B$ is separated from $B \sim A$.

6. Let (X, d) be a metric space and let Y be a subset of X. Then   is disconnected iff there exists non-empty open (closed) sets G and H in X s.t. $G \cap Y \neq \phi$, $H \cap Y \neq \phi$, $Y \subset G \cup H$ and $G \cap H \subset X - Y$.

7. Prove that every connected space has at least two components.

8. Prove that a set A is connected iff A is not the union of two separated sets.

9. Give an example to show that components of a connected space need not be open.

10. Prove that connectedness is preserved under continuous map.

11. Prove that the product of any non-empty class of connected space is connected.

12. Let X be an infinite set, and let d be the discrete metric on X. Show that every subset of X having more than one point is disconnected.

13. Let A and B be two separated subsets of X. Show that there exists two disjoint open sets containing A and B.   (MDU(ROHTAK) 2005)

14. Define disconnected sets and show that every singleton set is connected.

15. Prove that any continuous image of a connected set is connected.

16. (X, d) is a connected metric space iff every continuous function from (X, d) into (Y, e) is constant, where $Y = \{0, 1\}$ and e is the discrete metric on Y.

17. Prove that the unit circle is connected subset of $\mathbf{R}^2$.

18. Prove that every line segment in R^3 is connected.

19. Let X be a locally connected space. If Y is an open subspace of X, then show that each component of Y is open in X.

(BHOPAL 2004)

▰ OBJECTIVE EVALUATION

▶ FILL IN THE BLANKS

1. The empty set ϕ and the singleton sets are _____ .

2. The set R with _____ metric is connected.

3. Metric space X is disconnected if and only if it is the union of two _____ , _____ , _____ sets.

4. The _____ of a connected set is connected.

5. Every line segment in R^2 is _____ .

6. Component of a metric space X is need not to _____ .

7. The set of rational numbers Q with rest to induced usual metric is _____ .

8. Component of a _____ is open.

▶ TRUE OR FALSE

1. Q a connected subset of R with respect to metric $d(x,y) = \dfrac{|x-y|}{1+|x-y|}$. **(T/F)**

2. $\{1\}$ and $\{2,3,4\}$ are separated subsets of R. **(T/F)**

3. Every discrete space is connected. **(T/F)**

4. Connectedness is preserved under continuous map. **(T/F)**

5. Every connected space has at least two components. **(T/F)**

6. Each component of locally connected space is closed. **(T/F)**

7. Local connectedness implies connectedness. **(T/F)**

8. Every indiscrete space is connected. **(T/F)**

9. The components of a metric space X is closed. **(T/F)**

10. Discrete space is totally disconnected as well as locally disconnected. **(T/F)**

▶ MULTIPLE CHOICE QUESTIONS (CHOOSE THE MOST APPROPRIATE ONE)

1. Which of the following pairs of subsets of R are separated?

(a) Q and $R \sim Q$

(b) $]1,2[$ and $]7,8[$

(c) $]-3,-2]$ and $]-2,4[$

(d) $(]-2,0] \cup]5,\infty[)$ and $]4,6[$

2. Every discrete space is: (DELHI 2007)

(a) Connected

(b) Totally disconnected

(c) Disconnected

(d) None of these

3. Every indiscrete space is: (NAGPUR 2004)

(a) Connected

(b) Totally disconnected

(c) Disconnected

(d) None of these

4. Which of the following subset is connected:

(KANPUR 2001)

(a) $\{(x,y) : (x^2 - y^2) \geq 4\}$ of R^2

(b) Real line R with usual metric

(c) $A = \{x \in R : |x| > 2\}$ as a subset of real line with usual metric

(d) None of these

5. The closure of connected set is: (IAS 2007)

(a) Connected

(b) Disconnected

(c) Totally disconnected

(d) Locally disconnected

6. The set R of real numbers with usual metric is:

(a) Connected

(b) Disconnected

(c) Totally disconnected

(d) None of these

7. If (X,d) is a metric space, which of the following is incorrect: (KANPUR 2002)

(a) The closure of a connected space is connected

(b) The components of totallty disconnected space are its points

(c) Each component of locally connected space is closed

(d) None of these

8. If (X, d) is a metric space, which of the following is correct:

(a) Every component of a metric space X is open.

(b) Every discrete space is connected.

(c) Local connectedness implies connectedness.

(d) The components of a totally disconnected space are its points.

ANSWERS

▶ **FILL IN THE BLANKS**

1. Connected   **2.** Usual   **3.** Non-empty, disjoint, open   **4.** Closure

5. Connected   **6.** Open   **7.** Totally disconnected   **8.** Locally connected space

▶ **TRUE OR FALSE**

1. F   **2.** T   **3.** F   **4.** T   **5.** T   **6.** F   **7.** F   **8.** T   **9.** T

10. T

▶ **MULTIPLE CHOICE QUESTIONS**

1. (b)   **2.** (b)   **3.** (a)   **4.** (b)   **5.** (c)   **6.** (a)   **7.** (c)   **8.** (d)

CHAPTER SUMMARY

Connectedness is the most important property of metric space. The chapter deals with the concept of connectedness and separated sets. The concepts of locally connected space and totally disconnected sets have also been discussed in detail. The important points discussed in this chapter are as follows :

➲ Two non-empty subsets A and B of X are said to be separated if $A \cap \bar{B} = \phi$ and $\bar{A} \cap B = \phi$.

➲ Two disjoint sets are not necessarily separated.

➲ A set is said to be connected, if it is not disconnected.

➲ A metric space is disconnected iff $\exists$ a proper non-empty subset of X which both open and closed.

➲ The closure of connected set is connected.

➲ Union of two non-empty separated subsets of a metric space is disconnected.

➲ A subspace of the real line R is connected iff it is an interval.

➲ R is connected.

➲ A maximal connected subspace of a metric space is called its component.

➲ Every component of metric space X is closed.

➲ Every discrete space is totally disconnected.

➲ The components of a totally disconnected space are its points.

➲ A metric space is said to be locally connected iff it is locally connected at each of its points.

➲ Every discrete space is locally connected.

➲ Every component of a locally connected space is open.

➲ Local connectedness neither implies nor implied by connectedness.

FOR ADVANCED LEARNERS

⇒ If A, B be non-empty closed subsets of a metric space X such that $A \cup B$ and $A \cap B$ are connected. Then A and B are connected.

⇒ If A, B be connected subsets of a metric space X such that $\overline{A} \cap B$ is non-empty. Then $A \cup B$ is connected.

⇒ Let $<S_n>$ be a sequence of connected subsets of a metric space X such that $S_n \cap S_{n+1} \neq \phi \; \forall n$. Then $\cup S_n$ is connected.

⇒ A metric space X is said to be chain connected if for each pair a, b of X and

each $\varepsilon > 0$ there exists finitely many points $a = x_0, x_1, ..., x_n = b$ such that $d(x_i, x_{i+1}) < \varepsilon$ for $i = 0, 1, ..., n - 1$.

⇒ The range of a continuous mapping from a connected metric space into a metric space is connected.

⇒ Let X be an unbounded connected metric space. Then for each $x \in X$, $r > 0 \; \exists \; y \in X$ such that $d(x, y) = r$.

⇒ Let S be the connected subset of the Euclidean space $\mathbf{R}^n$ then for each $r > 0$ the set $\{x \in \mathbf{R} : d(x, S) \leq r\}$ is also connected.

❋❋❋❋

CHAPTER 8

Compactness in Metric Spaces

8.1 INTRODUCTION

In this chapter, we shall discuss an important property of metric space namely compactness. The motivation for the study of compactness in metric spaces stems from the structure of real line. Such type of property is motivated by well known Heine-Borel theorem which states that every open cover of a closed and bounded interval in $\mathbf{R}$ has a finite subcover.

8.2 COMPACT METRIC SPACE

Definition 1. *Let (X, d) be a metric space and A be any subset of X. The open cover of A is a collection $G = \{G_\lambda : \lambda \in \Lambda\}$ of open subsets of X such that*

$$A \subset \cup \{G_\lambda : \lambda \in \Lambda\}$$

Further, G is said to be open cover of X if

$$X = \cup \{G_\lambda : \lambda \in \Lambda\}$$

Definition 2. *Let G be an open cover of A. Then a subcollection G' of G is said to be subcover of A if G' covers A.*

Definition 3. *A metric space (X, d) is said to be compact if every open cover of X has a finite subcover.*

✎ ILLUSTRATIONS

* Let $X = [0, 1]$ and $d(x, y) = |x - y| \; \forall \, x, y \in X$ (Usual metric space), then (X, d) is compact, i.e., usual metric space is a compact metric space.
* Let X be a non-empty finite set and d be any metric space on X, then it is compact.
* Discrete metric space defined on an infinite set X is not compact.
* Hilbert space is not compact.
* Hilbert cube is compact.

THEOREM 1. **Let (X, d) be a metric space and Y be a subspace of X such that $A \subset Y$. Then A is compact relative to X if and only if A is compact relative to Y.**

PROOF. Let us first suppose that A is compact relative to X. We have to show that A is compact relative to Y. Since A is compact relative to X. Suppose that $\{V_\lambda : \lambda \in \Lambda\}$ be a collection of sets, open relative to Y which covers A such that $A \subset \cap \{V_\lambda : \lambda \in \Lambda\}$.

$\Rightarrow$ there exist G_λ, open relative to X such that

$$V_\lambda = Y \cap G_\lambda \;\; \forall \, \lambda \in \Lambda$$

$\Rightarrow \qquad A \subset \cup \{G_\lambda : \lambda \in \Lambda\}$

$\Rightarrow \quad \{G_\lambda : \lambda \in \Lambda\}$ is an open cover of A relative to X, but A is compact relative to X, so there exists finitely many indices $\lambda_1, \lambda_2, ..., \lambda_n$ such that

$$A \cap G_{\lambda_1} \cup ... \cup G_{\lambda_n} \qquad ...(1)$$

Since $A \subset Y$, therefore

$$A \subset Y \cap [G_{\lambda_1} \cup ... \cup G_{\lambda_n}] = [(Y \cap G_{\lambda_1}) \cup ... \cup (Y \cap G_{\lambda_n})]$$

Further, since $Y \cap G_{\lambda_i} = V_{\lambda_i} \ (i = 1, 2, ..., n)$, we get

$$A \subset V_{\lambda_i} \cup ... \cup V_{\lambda_n}$$

$\Rightarrow \quad A$ is compact relative to Y.

Conversely, let us suppose that A is compact relative to Y. We have to show that A is compact relative to X.

Let us suppose that $\{G_\lambda : \lambda \in \Lambda\}$ be a collection of open subsets of X which covers A such that

$$A \subset \cup \{G_\lambda : \lambda \in \Lambda\}$$

Since $\qquad A \subset Y \Rightarrow A \subset Y \cap \{\cup G_\lambda : \lambda \in \Lambda\}$

$$= \cup \{Y \cap G_\lambda : \lambda \in \Lambda\}$$

Clearly, since $Y \cap G_\lambda$ is open, relative to Y, then the collection $\{Y \cap G_\lambda : \lambda \in \Lambda\}$ is an open cover of A relative to Y. Since A is compact relative to Y, so by definition, we have $A \subset [Y \cap G_{\lambda_1}] \cup ... \cup [Y \cap G_{\lambda n}]$

$\Rightarrow \qquad A \subset G_{\lambda_1} \cup ... \cup G_{\lambda_n}$.

$\Rightarrow \quad A$ is compact relative to X.

THEOREM 2. *Let (X, d) be a metric spaces and $A \subseteq X$. Then A is compact iff given a family of open sets $\{G_\alpha\}$ in X such that $\cup G_\alpha \supseteq A$, there exists a subfamily $\{G_{\alpha_1}, G_{\alpha_2}, ..., G_{\alpha_n}\}$ such that $\overset{n}{\underset{i=1}{\cup}} G_{\alpha_i} \supseteq A$.*

PROOF. Let us first suppose that A is a compact subset of X and $\{G_\alpha\}$ be a family of open sets in X such that $\cup G_\alpha \supseteq A$.

Clearly, $(\cup G_\alpha) \cap A = A$

$\Rightarrow \quad \cup (G_\alpha \cap A) = A$

Also, $G_\alpha \cap A$ is open in A. Therefore, the family $\{G_\alpha \cap A\}$ is an open cover for A. But A is compact, so by definition there exists a finite subcover.

$$G_{\alpha_1} \cap A, G_{\alpha_2} \cap A, ..., G_{\alpha_n} \cap A$$

such that $\overset{n}{\underset{i=1}{\cup}} (G_{\alpha_i} \cap A) = A$

$\Rightarrow \quad \left(\overset{n}{\underset{i=1}{\cup}} G_{\alpha_i} \right) \cap A = A$

$\Rightarrow \quad \overset{n}{\underset{i=1}{\cup}} G_{\alpha_i} \supseteq A$

Conversely let $\{H_\alpha\}$ be an open cover for A. Therefore, each H_α is open in A.

$\Rightarrow \quad H_\alpha = G_\alpha \cap A$, where G_α is open in X.

Now, $\cup H_\alpha = A \qquad \Rightarrow \quad \cup \{G_\alpha \cap A\} = A$

$\Rightarrow \quad (\cup G_\alpha \cap A) = A$

$\Rightarrow \quad \cup G_\alpha \supseteq A$

Thus, by our hypothesis, there exists a finite subfamily $\{G_{\alpha_1}, G_{\alpha_2}, ..., G_{\alpha_n}\}$ such that

$\displaystyle\bigcup_{i=1}^{n} G_{\alpha_i} \supseteq A.$ Therefore $\left(\displaystyle\bigcup_{i=1}^{n} G_{\alpha_i}\right) \cap A = A.$

$\Rightarrow \quad \displaystyle\bigcup_{i=1}^{n} (G_{\alpha_i} \cap A) = A$

$\Rightarrow \quad \displaystyle\bigcup_{i=1}^{n} H_{\alpha_i} = A$

$\Rightarrow \quad \{H_{\alpha_1}, H_{\alpha_2}, \ldots, H_{\alpha_n}\}$ is a finite subcover of open cover $\{H_\alpha\}$.

Hence, A is compact.

THEOREM 3. ***Any compact subset of a metric space is bounded.***

PROOF. Let (X, d) be a metric space and A be a compact subset of X. We have to prove that A is closed.

Let $x_0 \in A$.

Consider $\{S(x_0, n) : n \in \mathbf{N}\}$

Clearly, $\displaystyle\bigcup_{n=1}^{\infty} S(x_0, n) = X \qquad \Rightarrow \qquad \displaystyle\bigcup_{n=1}^{\infty} S(x_0, n) \supseteq A$

But A is compact so there exists a finite subfamily $\{S(x_0, n_1), S(x_0, n_2), \ldots, S(x_0, n_k)\}$ such that $\displaystyle\bigcup_{i=1}^{K} S(x_0, n_i) \supseteq A$.

Let $n_0 = \max\{n_1, n_2, \ldots, n_k\}$

Then $\displaystyle\bigcup_{i=1}^{K} S(x_0, n_i) = S(x_0, n_0)$

$\Rightarrow \quad S(x_0, n_0) \supseteq A$

Finally, since $S(x_0, n_0)$ is always bounded, therefore, A being the subset of a bounded set is bounded.

THEOREM 4. ***Every compact subset of a metric space is closed.***

(MEERUT 1994, 2000, 15, 16, 18; AGRA 1994, 98)

PROOF. Let (X, d) be a metric space and A be a subset of X. We have to show that A is closed.

For this, we shall prove that A' is open.

Let $p \in A'$. Using Hausdorff axiom for X, for every $q \in A$, there exists open nbds of p and q denoted by $M(q)$ and $N(q)$, respectively such that $M(q) \cap N(q) = \phi$.

The collection $\{N(q) : q \in A\}$ is an open cover of A. Since A is compact

$\Rightarrow \quad \exists$ finite number of points $q_i : i = 1, 2, \ldots, n$ such that

$$A \subset \bigcup_{i=1}^{n} N(q_i) \qquad\qquad\qquad \ldots(1)$$

Now, let

$$M = \bigcap_{i=1}^{n} M(q_i) \text{ and } N = \bigcup_{i=1}^{n} N(q_i)$$

Then M is an open nbd of p, because it is the intersection of a finite of open nbds of p.

We have to show that $M \cap N = \phi$.

Let $\qquad x \in N \Rightarrow x \in N(q_i)$ for any i

$\qquad\qquad\qquad \Rightarrow x \notin M(q_i) \qquad\qquad\qquad (\because M(q) \cap N(q) = \phi)$

$\qquad\qquad\qquad \Rightarrow x \notin M$

$$\Rightarrow \quad M \cap N = \phi$$

Now, since $A \subset N \Rightarrow A \cap M = \phi \Rightarrow M \subset A'$

$\Rightarrow \quad A'$ contains a nbd of each of its points.

$\Rightarrow \quad A'$ is open.

$\Rightarrow \quad A$ is closed.

☞ **REMARKS**

⇒ The above theorem can be restated as follows : "A compact subset of a Hausdorff space is closed".

⇒ The converse of the above theorem is not true. For example, $[0, \infty[$ is a closed subset of **R**, which is not compact.

THEOREM 5.   *Closed subsets of compact sets are compact.*

(MEERUT 1995, 98, 2000, 03, 06, 15, 18; KANPUR 1993, 95, 98; AGRA 1992, 94, 98)

PROOF.   Let (X, d) be a metric space and Y be a compact subset of X. Let us suppose that F be a subset of Y closed relative to X. We have to show that F is compact.

Let us suppose that

$$G_1 = \{G_\lambda : \lambda \in \Lambda\}$$

be an open cover of F.

Then the collection $\quad G_2 = \{G_\lambda : \lambda \in \Lambda\} \cup [X - F]$

forms an open cover of Y, but since Y is compact, there is a finite subcover G_2' of G_2 which cover Y.

$\Rightarrow \quad G_2$ covers F.

If $X - F \in G_2'$, remove it from G_2'. The remaining set be also an open cover of F.

$\Rightarrow \quad$ A finite subcollection of G_1 covers F.

Hence, F is compact.

THEOREM 6.   *A metric space X is compact if and only if every basic open cover of X has a finite subcover.*

(MEERUT 2015)

PROOF.   Let (X, d) be a metric space which is compact. We have to show that every basic open cover of X has a finite subcover. Since X is compact, so by definition every open cover of X has a finite subcover. In particular, we can say that every basic open cover of X has a finite subcover.

Conversely, let us suppose that every basic open cover of X has a finite subcover. We have to show that X is compact.

Let $G = \{G_\lambda : \lambda \in \Lambda\}$ be an open cover of X.

If $\mathfrak{B} = \{B_\alpha : \alpha \in \Lambda\}$ be an open base for X, then each G_λ is a union of some member of $\mathfrak{B}$ and totality of all such members of $\mathfrak{B}$ is clearly a basic open cover of   .

Then by our hypothesis, the collection of $\mathfrak{B}$ has a finite subcover, say $\{B_{\alpha_i} : i = 1, ..., n\}$.

Now, for each B_{α_i} in this finite subcover, we can select a G_λ from G_1 such that $B_{\alpha_i} \subset G_{\lambda_i}$

$\Rightarrow \quad$ A finite subcollection $\{G_{\lambda_i} : i = 1, 2, ..., n\}$ is a subcover of G. Hence, X is compact.

8.3 EQUIVALENT CHARACTERISATIONS FOR COMPACTNESS

8.3.1 FINITE INTERSECTION PROPERTY

A collection G of sets is said to have the finite intersection property (FIP) if the intersection of members of each finite subcollection of G is non-empty.

8.3.2 BOLZANO-WEIERSTRASS PROPERTY

A metric space (X, d) is said to have Bolzano-Weierstrass property (BWP) if every infinite set in X has a limit point.

☞ REMARK

➡ A space with Bolzano-Weierstrass property is also called Frechet Compact.

THEOREM 1. *A metric space X is compact if and only if every collection of closed subsets of X with F.I.P. has a non-empty intersection.*

PROOF. Let (X, d) be a metric space. Let us first suppose that X is compact.

Suppose that $F = \{F_\lambda : \lambda \in \Lambda\}$ be a collection of closed subsets of X with finite intersection property. We have to show that

$\cap \{F_\lambda : \lambda \in \Lambda\}' = X$

$\Rightarrow \quad \cup \{F_\lambda' : \lambda \in \Lambda\} = X$ (By Demorgan's law)

$\Rightarrow \quad \{F_\lambda' : \lambda \in \Lambda\}$ is an open cover of X, because each closed F_λ implies that F_λ' is open for each λ.

Since X is compact $\Rightarrow$ Every open cover has a finite subcover.

Suppose that $\cup [F_{\lambda_1}' : i = 1, 2, ..., n] = X$

$\Rightarrow \quad [\cap \{F_\lambda : i = 1, 2, ..., n\}] = X$
(By Demorgan's law)

$\Rightarrow \quad \cap [F_{\lambda_i} : i = 1, 2, ..., n] \neq \phi$

Conversely, suppose that every collection of closed subsets of X with finite intersection property have a non-empty intersection and suppose that $G = \{G_\lambda : \lambda \in \Lambda\}$ be an open cover of X such that $X = \cup [G_\lambda : \lambda \in \Lambda]$.

Taking complement, we get

$\phi = (\cup \{G_\lambda : \lambda \in \Lambda\})' = \cap \{G_\lambda' : \lambda \in \Lambda\}$

$\Rightarrow \quad \{G_\lambda' : \lambda \in \Delta\}$ is a collection of closed sets with empty intersection. Thus, by hypothesis, the collection does not have finite intersection property.

$\Rightarrow \quad \exists$ a finite number of sets $G_{\lambda_i} : i \in N$ such that

$$\phi = \cap \{G_{\lambda_i}' : i = 1, 2, ..., n\} = (\cup \{G_{\lambda_i} : i = 1, 2, ..., n\})'$$

Hence, $\quad X = \cup \{G_{\lambda_i} : i = 1, 2, ..., n\}$

$\Rightarrow \quad X$ is compact.

FACTS : TO THE POINT

▶ We can apply the definition of compactness to a topological space X even if the topology is not metrizable.

▶ A subset of the Euclidean space R^n is compact if and only if it is bounded and closed.

▶ Let K be a compact subset of an open set $U \subset X$. Then there exists $r > 0$ such that $d(x, K) \leq r$ then $x \in U$.

▶ Any open cover of a seperable metric space has a countable subcover.

▶ A continuous mapping f of a compact metric space X into R is bounded. Also, f attains its bounds in the sense that there exists points a, b in X such that $f(a) = \inf f$, $f(b) = \sup f$.

▶ A continuous mapping of a compact space X into R^+ has a positive infimum.

▶ If f is continuous one-one mapping of a comapact metric space X onto a metric space Y. Then the inverse mapping $f^{-1} : Y \to X$ is continuous.

THEOREM 2. *Compact sets have Bolzano-Weierstrass property.* (MEERUT 2016)

PROOF. Let (X, d) be a metric space, Y be a compact subset of X and A is an infinite subset of Y.

Let us suppose that A has no limit point in Y. Then for every $y \in Y'$, there exists an open neighbourhood N_y of y which contains no point of A other than y (By definition of limit point).

Then the collection $\{G_y : y \in Y\}$ forms an open cover of Y. Since Y is compact

$\Rightarrow \quad \exists$ finitely many points $y_1, y_2, ..., y_n$ in Y such that

$$Y \subset G_{y_1} \cup ... \cup G_{y_n}$$

and hence $\qquad A \subset G_{y_1} \cup ... \cup G_{y_n}$

Since each G_{y_i} contains at most one point of A, but above equation shows that A has at most n points.

$\Rightarrow \quad A$ is finite, which is a contradiction.

Hence, A must have a limit point in Y.

8.4 COMPACT SUBSETS OF THE REAL LINE

Here we propose to characteristic compact subsets of the real line as closed and bounded subsets.

THEOREM 1.   *The space (R, d) (where d is the usual metric on R) is not compact.*

(MEERUT 2016)

PROOF.   Consider a collection

$$G = \{G_n : n \in N\}$$

where $\qquad G_n =]-n, n[$

Then clearly G is an open cover of $\mathbf{R}$. If $\{G_{n_1}, G_{n_2}, ..., G_{n_k}\}$ is any finite subcollection of G.

Suppose $n_0 = \max \{n_1, n_2, ..., n_k\}$

$\Rightarrow \qquad n_0 \notin G_{n_i} : i = 1, 2, ..., k$

Now, since $n_0 \in \mathbf{R} \Rightarrow$ No finite subcollection of G can cover $\mathbf{R}$.

Hence, $(\mathbf{R}, d)$ is not compact.

THEOREM 2.   *Every closed and bounded interval of real number R is compact, where R has the usual metric d.*

(MEERUT 1997, 2001, 13)

PROOF.   **STEP 1.**   Let $G = \{G_i : i \in \Lambda\}$ be an open cover of $[a, b]$. Set $S = [\{x : a \le x \le b\}$ and $[a, x]$ is covered by a finite subfamily of G].

Then clearly $S \ne \phi$ for $a \in S$.

Further, S is bounded by b, therefore by completeness axioms, S must have a supremum, say c.

Then clearly, $a \le c \le b$

          STEP 2.   Now, assume that $c < b$. Since G is a cover of $[a, b]$, therefore there exists $G_c \in G$ such that $c \in G_c$.

Since G_c is open $\Rightarrow \exists \varepsilon > 0$ such that $]c - \varepsilon, c + \varepsilon[\subset G_c$.

Now, since $c < b$, so without loss of any generality, we may assume that ε to be small such that $c + \varepsilon < b$.

Since c is the supremum of S, so there exist $d \in S$ such that

$$c - \varepsilon < d \le c < c + \varepsilon < b$$

Now, by definition of S, there must exist a finite subfamily

$$F = \{G_1, ..., G_k\} \text{ of } G \text{ that cover } [a, d]$$

So, $F \cup \{G_c\}$ is a finite subfamily of G that covers $[a, c + \varepsilon]$ and therefore $[a, c + \frac{\varepsilon}{2}]$ also.

$$\Rightarrow \quad c + \frac{\varepsilon}{2} \in S$$

which contradicts the fact that c is the supremum of S

$$\Rightarrow \qquad\qquad c \not< b$$

Thus　　　$c = b$.

STEP 3. Since G is a cover of $[a, b]$, there exists $G_b \in G$ such that $b \in G_b$.

Now, for some $\varepsilon > 0$, $]b - \varepsilon, b + \varepsilon[\subset G_b$

Since b is the supremum of S, therefore, there exists a finite subfamily F^* of G that cover $[a, b]$

$\Rightarrow F^* \cup [G_b]$ is a finite subfamily G that covers $[a, b]$.

Hence, $[a, b]$ is compact.

THEOREM 3. ***Every compact subset of real line is closed and bounded.*** (MEERUT–2013)

PROOF. We have already proved that every compact subset of a metric space is closed. So, it is sufficient to prove that every compact subset of the real line is bounded.

Let us suppose that S be a compact subset of **R**. The family $G = \{\,]-n, n[: n \in \mathbf{Z}^+\}$ is an open cover of S.

Since S is compact, so there exists a finite subcollection

$$F = \{\,]-n_1, n_1[, ...,]-n_k, n_k[\,\}$$

of G cover S. If max $\{n_1, n_2, ..., n_k\}$

Then clearly S is contained in $]-s, s[$.

Hence, S is bounded.

8.5 COMPACT SUBSET IN R^n

THEOREM 1. ***If $<I_n>$ is a sequence of closed intervals in R such that $I_n \supset I_{n+1}$, $n = 1, 2, ...$ Then $\cap \{I_n : n \in N\} \neq \phi$.***

PROOF. Let us define　　　$I_n = [a_n, b_n]$

Suppose that A is the set of all a_n. Then, A is non-empty and bounded above by b_1.

$\Rightarrow$　A must have the supremum say u.

If $m, n \in \mathbf{Z}^+$, then

$$a_n \le a_{m+n} \le b_{m+n} \le b_m$$

$$\Rightarrow \qquad\qquad u \le b_m \quad \forall\, m$$

Also,　　　　　$a_m \le u \quad \forall\, m$

$\Rightarrow$　$u \in I_m$ for $m = 1, 2, ...$

Hence, $\cap [I_n : n \in N] \neq \phi$.

THEOREM 2. ***If $< I_n >$ is a sequence of k-cells (k is a positive integer) in R^n such that $I_n \supset I_{n+1}$ $(n \in N)$. Then $\cup [I_n : n \in N] \neq \phi$.***

PROOF. We know that a k-cell I_n consists of all points

$$\boldsymbol{x} = [x_1, x_2, ..., x_k]$$

such that　　　$a_{n,i} \le x_i \le b_{n,i}$　　　$(1 \le i \le k : n = 1, 2, ...)$

Putting $I_{n,i} = [a_{n,i}, b_{n,i}]$

Now for each i, the sequence $< I_{n,i} >$ satisfy the condition of previous theorem.

Therefore, there exists real numbers x_i^* $(1 \le i \le k)$ such that

$$a_{n,i} \le x_i^* \le b_{n,i} \qquad (1 \le i \le k; \; n = 1, 2, 3, \ldots\ldots)$$

Set $\qquad\qquad x^* = (x_i^*, \ldots\ldots, x_k^*)$

Then, $\qquad\qquad x^* \in I_n$

$\Rightarrow \qquad\qquad \cap [I_n : n \in N] \ne \phi$

☛ REMARK

⇒ Similarly, we may prove that every k-cell is compact.

THEOREM 3. **(Generalized Heine-Borel Theorem). *A subset of R^n is compact if and only if it is closed and bounded.***

(MEERUT 2011, 16, 17)

PROOF. Let us first suppose that A be a closed and bounded subset of R^n.

$\Rightarrow \quad A \subset I$, where I be a k-cell.

Then using above remark, I is compact.

And we know that a closed subset of a compact set is compact.

Hence, A is compact.

Conversely, let us suppose that A be a compact subset of R^n.

We have to show that A is closed and bounded.

Then, we have already proved that 'A closed and bounded subset of a compact set is compact $\Rightarrow A$ is compact.

Conversely, let A be a compact subset of R^n. We have to show that A is closed and bounded.

Since we have already proved that every compact subset of a metric space is closed.

$\Rightarrow \quad A$ is closed.

Now, it remains to prove that A is bounded.

Let if possible, A is not bounded, then A contains points x_n with

$$|x_n| > n, \; n = 1, 2, \ldots$$

The set containing of the points x_n is infinite and clearly has no limit point in R^n and hence has none in A, which is a contradiction (Because if A is an infinite subset of a compact set Y, then A has limit point in X).

Hence,   is bounded.

8.6 COUNTABLY COMPACTNESS

In the definition of compactness, we require that every open cover must have a finite subcover, while in the definition of countable compactness, we require that every countable open cover has a finite subcover.

Definition. *A metric space is said to be countably compact if every countable open cover of the space have a finite subcover.*

☛ REMARKS

⇒ Every compact metric space is countably compact.

⇒ We shall later on show that countable compactness is actually equivalent to compactness.

THEOREM I. ***A metric space (X, d) is countably compact if and only if every countable family of non-empty closed sets with F.I.P. has non-empty intersection.***

PROOF. Let us first suppose that (X, d) be countably compact.

Suppose that $F = \{F_n : n \in N\}$ be a countable family of non-empty closed sets with F.I.P.

If $\qquad \cap\{F_n : n \in \boldsymbol{N}\} = \phi$

Then, $\qquad G = \{X - F_n : n \in \boldsymbol{N}\}$

is a countable open cover of X. By definition of countable compactness, there exists a finite subcover, say $\{X \sim F_{n_1}, X \sim F_{n_2}, ..., X \sim F_{n_k}\}$

$\Rightarrow \quad F_{n_1} \cap F_{n_2} \cap ... \cap F_{n_k} = \phi$

which contradict the fact that the family has F.I.P. and hence

$\qquad \cap\{F_n : n \in \boldsymbol{N}\} \neq \phi$

Conversely, suppose that every countable family of non-empty closed sets with F.I.P. has non-empty intersection. Now, let

$$G = \{G_n : n \in \boldsymbol{N}\}$$

be any countable open cover of X. If any $G_n = X$, then our result is obvious. Otherwise $F = \{X - G_n : n \in \boldsymbol{N}\}$ is a countable family of non-empty closed sets with empty intersection and hence there exists a finite subfamily $\{X - G_{n_1}, X - G_{n_2}, ..., X - G_{n_k}\}$ with empty intersection. Because if there were no such subfamily, then F would have F.I.P. and therefore $\cap \{X - G_n : n \in \boldsymbol{N}\}$ would be non-empty.

Hence, $\{G_{n_1}, G_{n_2}, ..., G_{n_k}\}$ is a finite subfamily of G that covers X.

DEDUCTION. A metric space is compact if and only if it is countably compact.

THEOREM 2. *A metric space (X, d) is countably compact if and only if every sequence in X has a cluster point in X.*

PROOF. Let (X, d) be a metric space. Let us first suppose (X, d) is countably compact. Let $<x_n>$ be a sequence in X.

Let $\qquad A_n = \{x_k : k \geq n\}$

$\Rightarrow \quad \{\bar{A}_n : n \in \boldsymbol{N}\}$ is a countable collection of closed subsets of X with finite intersection property.

Using above theorem, we can say that this countable family of closed sets has a non-empty intersection.

Let x_0 be any point of the intersection of this family so that $x_0 \in \bar{A}_n \ \forall n$

$\Rightarrow \quad x_n$ is a cluster point of the sequence $<x_n>$.

For, let N_{x_0} be the nbd of x_0 then for any integer m.

Then $x_0 \in \bar{A}_m$ $\qquad\qquad\qquad\qquad\qquad\qquad (\because x_0 \in \bar{A}_n \ \forall n)$

$\Rightarrow \quad N_{x_0} \cap A_m \neq \phi$

$\Rightarrow \quad x_p \in N$ for some $p \geq m$.

$\Rightarrow \quad x_0$ is a cluster point of the sequence.

Conversely, suppose that every sequence in X has a cluster point. We have to show that the space is countably compact.

Consider $\{A_n : n \in \boldsymbol{N}\}$, a countable collection of closed subsets of X with finite intersection property.

For each n, select a point x_n from $\cap \{A_i : i = 1, 2, ..., n\}$

Further, let x_0 be a cluster point of $[x_n]$. Let N_{x_0} be any nbd of x_0 and m be any positive integer.

Then, $x_p \in N_{x_0}$ for some $p \geq m$.

But, $x_p \in \cap \{A_i : i = 1, 2, \ldots\ldots, p\}$

$\Rightarrow \qquad\qquad x_p \in A_m \qquad\qquad\qquad\qquad (\because p \geq m)$

and $\qquad A_m \cap N_{x_0} \neq \phi$

$\Rightarrow \quad x_0$ is an adherent point of $A_m \Rightarrow x_0 \in \bar{A}_m$

But A_m is closed $\Rightarrow \bar{A}_m = A_m$

$\Rightarrow \qquad\qquad x_0 \in A_m$

$\Rightarrow \qquad \cap \{A_n : n \in N\} \neq \phi$

$\Rightarrow \quad$ Every countable collection of closed subsets of $\quad$ has finite intersection property. Hence, using previous theorem, we conclude that space (X, d) is countably compact.

THEOREM 3. ***Every countable compact metric space has Bolzano-Weierstrass Property (BWP).***

(MEERUT 2010, 16)

PROOF. Let (X, d) be a metric space which is countable compact. Let if possible, it does not have BWP.

Then, there exists an infinite set S having no limit point

$\Rightarrow \quad A$ is a closed set.

Also, for each $a_n \in A$, a_n is not a limit point of A

$\Rightarrow \quad \exists$ an open set G_n, such that $a_n \in G_n$ and $G_n \cap A = [a_n]$.

Then, clearly, the collection $\{G_n : n \in N\} \cup A'$ is a countable open cover of X. This cover has no finite subcover. Because if we remove a single G_n, it will not be a cover of X since then a_n will not be covered. Hence, X is not countably compact which is a contradiction.

Hence, space (X, d) have a BWP.

8.7 SEQUENTIAL COMPACTNESS

A metric space (X, d) is said to be sequentially compact if every sequence has a convergent subsequence.

ILLUSTRATIONS

- The real line is not sequentially compact. Because the sequence $< 1, 2, 3, 4, \ldots\ldots >$ does not have any convergent subsequence.
- The closed and bounded interval $[0, 1]$ is sequentially compact.
- Let X be an infinite set and d be the discrete metric on X, then (X, d) is not sequentially compact.

THEOREM 1. ***A continuous image of a sequentially compact metric space is sequentially compact.***

PROOF. Let (X, d) and (Y, ρ) be two metric spaces. Consider a continuous mapping $f : X \xrightarrow{\text{onto}} Y$.

Let us first suppose (X, d) be sequentially compact. We have to show that (Y, ρ) be sequentially compact.

Let $< y_n >$ be any sequence in Y. Now, since f is onto, so for every $n \in N$, we can choose an $x_n \in X$ such that $f(x_n) = y_n$.

Now, since $<x_n>$ is a sequence in a sequentially compact metric space $(X, d) \Rightarrow$ It must have a convergent subsequence $<x_{n_k}>$ converging to some $x_0 \in X$.

Since $x_{n_k} \to x_0$ and f is continuous $\Rightarrow \quad f(x_{n_k}) \to f(x_0)$

$\Rightarrow \quad y_{n_k} \to f(x_0)$

So, $<y_{n_k}>$ is a convergent subsequence of $<y_n>$. Since $<y_{n_k}>$ is arbitrary. Hence (Y, ρ) is sequentially compact.

THEOREM 2. ***Every closed subspace of a sequentially compact metric space is sequentially compact.***

PROOF. Let (X, d) be a sequentially compact metric space and Y be a closed subset of X. We have to show that every sequence in Y has a subsequence which convergent to a point of Y with respect to the subspace metric.

Let $<y_n>$ be a sequence in Y. Since $Y \subset X$, therefore, $<y_n>$ is a sequence in X. By definition of sequential compactness of X, there exists a subsequence $<y_{n_k}>$ of $<y_n>$ and a point $y_0 \in X$ such that $y_{n_k} \to y_0$. Now, since Y is closed. Therefore, limit of every convergent sequence in Y must belong to $Y \Rightarrow y_0 \in Y$.

Since $y_{n_k} \to y_0$, therefore given $\varepsilon > 0 \ \exists$ a positive integer m such that $d(y_{n_k}, y_0) < \varepsilon$ whenever $k \geq m$.

Since $y_{n_k} \in Y \ \forall k \in \mathbf{N}$ and $y_0 \in Y$. Thus, $\quad d(y_{n_k}, y_0) = d_Y(y_{n_k}, y_0) \ \forall k \in \mathbf{N}$.

Hence, we find that given $\varepsilon > 0 \ \exists$ a positive integer m such that $d_Y(y_{n_k}, y_0) < \varepsilon$ $\forall \ k \geq m$.

$\Rightarrow \quad <y_{n_k}>$ converges to a point in Y with respect to the subspace metric.

Hence, (Y, d_Y) is sequentially compact.

THEOREM 3. ***Every sequentially compact metric space is countably compact.***

(MEERUT 1990, 92, 97, 2000; AGRA 1990)

PROOF. Let (X, d) be a sequentially compact space.

If X is finite then space is definitely compact.

Now, let X is infinite. We prove this theorem contrapositively. Let $\{G_i : i \in \mathbf{N}\}$ be a countable open cover of X which has no finite subcover. Define the sequence $<x_1, x_2, ..., x_n, ...>$ as follows :

Choose the smallest positive integer n_1 such that $X \cap G_{n_1} \neq \phi$.

Now, choose $x_1 \in X \cap G_{n_1}$. Further, let n_2 be the least positive integer greater than n_1 such that $X \cap G_{n_2} \neq \phi$.

Choose $\quad x_2 \in (X \cap G_{n_2}) - (X \cap G_{n_1})$

Clearly such a point x_2 always exists. If it is not, then G_{n_1} will be a cover of X. Continuing in this way, we get a sequence $<x_1, x_2, ..., x_n, ...>$ with the property that for each $i \in \mathbf{N}$.

$$x_i \in X \cap G_{n_i}, \quad x_i \notin \bigcup \{X \cap G_k : k = 1, 2, ..., n-1\} \text{ and } n_i > n_{i-1}.$$

Now, it can be seen that $<x_n>$ has no convergent subsequence in X. For, let $x \in X$, then there exists $G_{m_0} \in \{G_n : n \in \mathbf{N}\}$ such that $x \in G_{m_0}$

Now, since $X \cap G_{m_0} \neq \phi \Rightarrow \exists k_0 \in N$ such that $G_{n_{k_0}} = G_{m_0}$

By the choice of the sequence $< x_1, x_2, ... >$, we have $i > k_0 \Rightarrow x_i \notin G_{m_0}$.

Clearly, since G_{m_0} is an open set containing x, no subsequence of $< x_n >$ converges to x, since x is arbitrary, so we can show that space X is not sequentially compact.

THEOREM 4. *A metric space (X, d) is sequentially compact if and only if each sequence in X has a cluster point.*

PROOF. For sequentially compactness, we have to show the following:

"A sequence $< x_n >$ in a metric space has a convergent subsequence $< x_{n_k} >$ converging to some $x \in X$ if and only if $< x_{n_k} >$ has x as a cluster point."

Let us first assume that $< x_n >$ has a subsequence $< x_{n_k} >$ converging to some $x \in X$.

If N be the nbd of x, then $\exists p \in N$ such that $x_{n_k} \in N \ \forall k \geq p$. If m be any positive integer, let $q = \max \{n_p, n_m\}$.

Now, since $q \geq n_m \geq m$ and $x_q \in N \Rightarrow x$ is a cluster point of $< x_n >$.

Conversely, assume that $< x_n >$ has a cluster point, say x. Then choose

$$n_1 > 1 \quad \text{such that} \quad x_{n_1} \in S(x, 1)$$

$$n_2 > n_1 \quad \text{such that} \quad x_{n_2} \in S(x, \tfrac{1}{2})$$

$$n_3 > n_1 \quad \text{such that} \quad x_{n_3} \in S(x, \tfrac{1}{3})$$

... ...

and so on.

Thus, we can inductively define a subsequence $< x_{n_k} >$ of $< x_n >$ such that $x_{n_p} \in S(x, \tfrac{1}{k}) \ \forall \ p \geq k$.

Hence, $x_{n_k} \to x$.

8.8 TOTAL BOUNDEDNESS

Definition 1. *Let (X, d) be a metric space and $\varepsilon > 0$ be given. A finite subset A of X is said to be an ε-net for X if and only if for every $x \in X \ \exists$ a point $a \in A$ such that*

$$d(a, x) < \varepsilon$$

or we can say that A is an ε-net for X iff A is finite and $X = \bigcup \{S(a, \varepsilon) : a \in A\}$.

Definition 2. *A metric space (X, d) is said to be totally bounded iff X has an ε-net for every $\varepsilon > 0$.*

Definition 3. *Let (X, d) be a metric space and let*

$$G = \{G_\lambda : \lambda \in \Delta\}$$

be an open cover of X. A real number $l > 0$ is called a Lebesgue number for G if and only if every subset of X with diameter less than l is contained in at least one G_λ.

THEOREM 1. *Every sequentially compact metric space is totally bounded.*

PROOF. Let (X, d) be a metric space. Suppose X is not totally bounded. Then by definition, there exists $\varepsilon > 0$ such that X has no ε-net. Now, let $x \in X$. Then there must exist a point $x_2 \in X$ such that $d(x_1, x_2) \geq \varepsilon$, otherwise $\{x_1\}$ would be an ε-net for X. Again, there must exist a point $x_3 \in X$ such that $d(x_1, x_2) \geq \varepsilon$ and $d(x_1, x_3) \geq \varepsilon$ for otherwise $[x_1, x_2]$ would be an ε-net for X. Continuing in the same way, we get a sequence $\{x_1, x_2, ..., x_n\}$ with the property that $d(x_i, x_j) \geq \varepsilon$ for $i \neq j$.

Hence, the sequence $< x_n >$ can not contain any convergent subsequence.

$\Rightarrow$ X is not sequentially compact.

THEOREM 2. (**Lebesgue Covering Lemma**). *Every open cover of a sequentially compact metric space has a Lebesgue number.*

(MEERUT 1992, 96, 97, 2003, 04, 16; KANPUR 1997; AGRA 1995)

PROOF. Let (X, d) be a metric space with the Bolzano-Weierstrass property. Let $G = \{G_j : j \in \Delta\}$ be an open cover of X.

If G does not have any Lebesgue number, then for each positive integer n, there exists a point $x_n \in X$ such that $S(x_n, \frac{1}{n})$ is not contained in any member of G.

Now, let $S = \{x_1, x_2, ..., x_n, ...\}$

If S is finite, then $x_n = x$ (say) for infinitely many indices, say n. Since G is a cover of X, so we can find some $G_x \in G$ such that $x \in G_x$. Since G_x is open, we can find a positive integer m such that $S(x, \frac{1}{m}) \subset G_x$. Again, since $x_n = x$ for infinitely many indices n, choose a positive integer $p > m$ such that $x_p = x$. Now $S(x_p, \frac{1}{p}) \subset S(x, \frac{1}{m}) \subset G_x$

Which is a contradiction, because the open sphere $S(x, -)$ is not contained in **any** member of G. Hence, S must be infinite.

Since S is infinite, so by Bolzano-Weierstrass property, it has a limit point, say x_0.

Since G is a cover of X, $\exists G_{x_0} \in G$ such that $x_0 \in G_{x_0}$. Now, since G_{x_0} is open, so we can find a positive integer k such that $S(x_0, \frac{1}{k}) \subset G_{x_0}$. Again since x_0 is a limit point of S, therefore the open sphere $S(x_0, \frac{1}{2k})$ contains infinitely many points of S. Particularly, we can choose a positive integer $q > 2k$ such that $x_q \in S(x_0, \frac{1}{2k})$.

Now, we show that $S(x_q, \frac{1}{q}) \subset G_{x_0}$

Let $y \in S(x_q, \frac{1}{q})$

$\Rightarrow d(y, x_0) \leq d(y, x_q) + d(x_q, x_0)$

$$< \frac{1}{q} + \frac{1}{2k}$$

$$< \frac{1}{k}$$

$\Rightarrow \quad y \in S(x_0, \frac{1}{k}) \subset G_{x_0}$

$\Rightarrow S(x_q, \frac{1}{q}) \subset G_{x_0}$

which contradict the fact that $S(x_q, \frac{1}{q})$ can not be contained in any member of the cover and hence the cover G must have a Lebesgue number.

FACTS : TO THE POINT

▶ A bounded interval in R is totally bounded.

▶ If a metric space is either sequentially compact or totally bounded space is seperable.

▶ A totally bounded space is seperable.

▶ Let f be a uniformly continuous mapping of a totally bounded space into a metric space, then the range of f is totally bounded.

▶ Let f be a function of bounded variation on a compact interval $I \subset R$, then $f(I)$ is totally bounded.

▶ In a metric space, following conditions are equivalent:
  (i) X is compact.
  (ii) X is sequentially compact.
  (iii) X is totally bounded and complete

▶ Let X be a metric space, which is not totally bounded then there exists a sequence $<x_n>$ in X and a positive no. α such that $d(x_m, x_n) \geq \alpha$ whenever $m \neq n$.

THEOREM 3. *A metric space X is sequentially compact if and only if it has Bolzano-Weierstrass property (BWP), i.e., every infinite subset of X has a limit point.*
(MEERUT 1992; AGRA 1993)

PROOF. Let us first suppose that metric space (X, d) is sequentially compact. Let A be any infinite set. We have to show that A has a limit point. Let $< x_n >$ be any sequence of distinct points of A. Since X is assumed to be sequentially compact. Therefore, $< x_n >$ contains a convergent sequence $< x_{i_n} >$ with limit x (say).

$\Rightarrow$ x is a limit point of A of X has a Bolzano-Weierstrass property.

Conversely, let X have a Bolzano-Weierstrass property and $< x_n >$ be any sequence in X. Now, if the range set $A = \{x_1, x_2, ..., x_n, ...\}$ is finite, then one of the point, say x_{m_0} is such that $x_i = x_{m_0}$ for infinitely many $i \in N$.

$\Rightarrow$ $< x_{m_0}, x_{m_0}, x_{m_0}, ... >$ is a subsequence of $< x_n >$ which converges to $x_{m_0} \in X$. Let the range set A is infinite. Since X has Bolzano-Weierstrass property. A has a limit point, say x_0. Then we can easily define a subsequence of $< x_n >$ which converges to x_0.

Hence, X is sequentially compact.

THEOREM 4. *A metric space X is compact if and only if it is sequentially compact.*

PROOF. Let us first suppose X is compact. We have to show that X is sequentially compact. Since we know that every compact space X has Bolzano-Weierstrass property $\Rightarrow$ X has BWP.

Then using previous theorem, we can say that X is sequentially compact. Conversely, let us suppose that X be sequentially compact. We have to show that X is compact. For this, we shall prove that every open cover of X has a finite subcover.

Let $G = \{A_\lambda : \lambda \in \Lambda\}$ be an open cover of X.

To prove, G has a finite subcover.

Since X is sequentially compact, the cover G has a Lebesgue number, say $l > 0$. Also, we know that every sequentially compact space is totally bounded. Using the definition of total boundedness for $\varepsilon = \dfrac{1}{3}, \exists$ ε-net $A = \{a_1, a_2, ..., a_n\}$, where a's belong to X such that

$$X = \bigcup \{S(a, \varepsilon) : a_i \subset \Lambda\} \qquad ...(1)$$

$$\Rightarrow \quad \delta [S(a_i, \varepsilon)] \leq 2\varepsilon = \frac{2l}{3} < l \text{ for } i = 1, 2, ..., n.$$

Now, by definition of Lebesgue number, for each i, there exists an $A_{\lambda_i} \in G$ such that

$$S(a_i, \varepsilon) \subset A_{\lambda_i} \qquad ...(2)$$

From (1) and (2), we conclude that

$$X = \bigcup \{A_{\lambda_i} : i = 1, 2, ..., n\}$$

Hence, X is compact.

THEOREM 5. *Every compact metric space is complete.*

PROOF. Let (X, d) be a metric space which is compact. We have to show that X is complete. For this, we shall prove that every Cauchy sequence in X is convergent.

Let $< x_n >$ be a Cauchy sequence in X. We know that a space is compact if and only if it is sequentially compact.

$\Rightarrow$ X is sequentially compact.

$\Rightarrow$ **The Cauchy sequence $< x_n >$ has a subsequence $< x_{n_k} >$ converges to some** $x_0 \in X$.

We have to show that $< x_n >$ also converges to x_0.

By definition of Cauchy sequence, we can write

For given $\varepsilon > 0 \, \exists \, n_0 \in N$ such that

$$d(x_m, x_n) < \varepsilon / 2 \quad \forall \, m, n \geq n_0 \qquad \text{...(1)}$$

Further since $< x_{n_k} >$ converges to $x_0 \, \exists \, k_0 \in N$ such that

$$d(x_{n_k}, x_0) < \varepsilon / 2 \quad \forall \, k \geq k_0 \qquad \text{...(2)}$$

Choose $k_1 \geq k_0$ such that $n_{k_1} \geq n$. Then from (1) and (2), we have

$$n \geq n_{k_1} \Rightarrow d(x_n, x_{n_{k_1}}) < \varepsilon / 2$$

and $\quad d(x_{n_{k_1}}, x_0) < \varepsilon / 2$

Now, using triangle inequality, we can write

$$d(x_n, x_0) \leq d(x_n, x_{n_{k_1}}) + d(x_{n_{k_1}}, x_0) < \varepsilon / 2 + \varepsilon / 2 = \varepsilon \quad \forall \, n \geq n_{k_1}$$

Hence, $\lim\limits_{n \to \infty} x_n = x_0$.

$\Rightarrow \quad X$ is complete.

THEOREM 6.   ***A compact subset of a metric space is closed and bounded.***

PROOF.   Let (X, d) be a metric space and Y be a subset of X such that (Y, d_1) is compact. We have to show that Y is closed and bounded.

CASE I: Y is finite then it is surely closed and bounded.

CASE II: If Y is infinite. Since every compact metric space is sequentially compact $\Rightarrow \, Y$ is sequentially compact. We have to show that Y is bounded. Let if possible, Y is not bounded. Then we can always find a pair of points of Y at arbitrary large distance. Using this property to construct a sequence of points of Y which contains no convergent subsequence, start with $y_1 \in Y$. Choose a point y_2 such that $d(y_1, y_2) > 1$.

Then select a point y_3 such that $d(y_1, y_3) > 1 + d(y_1, y_2)$ and so on.

So, for every $n \in N$, we may choose y_n such that $d(y_1, y_n) > 1 + d(y_1, y_{n-1})$

$\Rightarrow \quad$ If $m > n$

$$d(y_1, y_m) > 1 + d(y_1, y_n).$$

Hence, $\quad d(y_m, y_n) \geq | d(y_1, y_m) - d(y_1, y_n) | > 1$

The sequence $< y_n >$ contains no convergent subsequence, which is a contradiction, because Y is sequentially compact. Hence, Y must be bounded.

Now, we have to show that Y is closed. For this, we shall prove that it contains all its limit points. Let y be any arbitrary limit point of Y.

$\Rightarrow \quad \exists$ a sequence of point $< y_n >$ of Y which converges to y and every subsequence of $< y_n >$ converges to y.

But since Y is a complete subspace, $< y_n >$ has a subsequence which converges to a point of Y.

$\Rightarrow \qquad\qquad y \in Y$.

Since y is arbitrary, so we can say that Y contains all its limit points. Hence, Y is closed.

THEOREM 7.   ***A metric space is compact if and only if it is complete and totally bounded.***

(MEERUT 1993, 95, 98, 2003, 04; AGRA 1995)

PROOF.

Let (X, d) be a metric space.

Let us first suppose that space is compact. We have to show that it is complete and totally bounded.

Since we have already proved the following theorems:

   (i) Every compact metric space is complete.

   (ii) Every compact space is sequentially compact, and

   (iii) Every sequentially compact space is bounded.

Hence, using above result, we conclude that X is complete and totally bounded.

Conversely, let X is complete and totally bounded. We have to show that X is compact. Since we know that a space is compact if and only if it is sequentially compact. So we have to show that X is sequentially compact.

Consider an arbitrary sequence $s = \langle x_1, x_2, ..., x_n, ... \rangle$ in X, since X is totally bounded.

$\Rightarrow$ $\exists$ an ε-net for X for $\varepsilon > 0$.

Let us take $\varepsilon = 1$, then X is contained in the union of a finite collection of open spheres of radius 1.

Let one of these sphere be denoted by $S(a_1, 1)$, $a_1 \in X$, which must contain a subsequence $s_1 = \{x_1^{(1)}, x_2^{(1)}, ..., x_n^{(1)}, ...\}$ of s.

The distance between any two points of s_1 is less than s.

Similarly for $\varepsilon = \dfrac{1}{2}$, we get a subsequence $s_2 = \{x_1^{(2)}, x_2^{(2)}, ...\}$ all of whose points lie in the open sphere $S = (a_2, \tfrac{1}{2})$, $a_2 \in X$ of radius $\tfrac{1}{2}$. Here, clearly the distance between any two points of s_2 is less than 1.

Proceeding in the same way, we may get for $\varepsilon = \tfrac{1}{k}$ a subsequence $s_k = \{x_1^{(k)}, x_2^{(k)}, ...\}$ all of whose points lie in an open sphere $S(a, \tfrac{1}{k})$ such that the distance between any two points of s_k is less than $\dfrac{2}{k}$. We claim that the diagonal sequence $d = \{x_1^{(1)}, x_2^{(1)}, ..., x_n^{(1)}, ...\}$ is a Cauchy sequence of s.

Observe that

$$x_n^{(n)} \in S(a_n, \tfrac{1}{n}) \subset S(u_{n-1}, \tfrac{1}{n-1}) \subset ...... \subset S(a_1, 1) \qquad ...(1)$$

Now for each $\varepsilon > 0$, we can chose $n_0 \in \mathbf{N}$ such that $\dfrac{2}{n_0} > \varepsilon$.

So, $\qquad m, n > n_0 \Rightarrow x_n^{(n)}, x_m^{(m)} \in S(a_{n_0}, \dfrac{1}{n_0})$

$$\Rightarrow d(x_n^{(n)}, x_m^{(m)}) < \dfrac{2}{n_0}$$

$$\Rightarrow d(x_n^{(n)}, x_m^{(m)}) < \varepsilon$$

$\Rightarrow$ d is a Cauchy subsequence of s.

Since X is complete $\Rightarrow$ d is convergent to a point in X.

$\Rightarrow$ Every sequence in X has a subsequence which converges to a point in X.

$\Rightarrow$ X is sequentially compact $\Rightarrow$ X is compact.

THEOREM 8. *A subset of a complete metric space* (X, d) *is compact if and only if A is closed and totally bounded.*

(MEERUT 1990, 91, 98)

PROOF. Let (X, d) be a complete metric space and $A \subseteq X$.

Let us first suppose A is compact. We have to show that A is closed and totally bounded.

Since A is compact $\Rightarrow$ A is sequentially compact.

Also, A is compact subset of a complete metric space.

$\Rightarrow A$ is closed $\qquad$ ($\because$ Compact subset of a complete metric space is closed)

$\Rightarrow A$ is totally bounded.

$\qquad\qquad$ ($\because$ Every sequentially compact metric space is totally bounded)

Conversely, let us suppose that A is closed and totally bounded.

We have to show that A is compact.

Since we know that a closed subset of a complete metric space is complete.

$\Rightarrow \quad A$ is complete.

Also, A is totally bounded.

Using previous theorem, we can say that a metric space is compact if and only if it is complete and totally bounded.

$\therefore \qquad A$ is complete and totally bounded $\Rightarrow$ A is compact.

THEOREM 9. $\quad$ ***A totally bounded metric space is separable.***

PROOF. $\quad$ Let (X, d) be a totally bounded metric space. Using the definition of totally boundedness $\exists$ a ε-net A_n for every positive integer n.

Let $\qquad\qquad\qquad A = \overset{\infty}{\underset{n=1}{\cup}} A_n$

$\Rightarrow A$ is the union of a countable family of finite sets.

$\Rightarrow A$ is countable.

Let p be any point of X and $S(p, \varepsilon)$ be any spherical neighbourhood of p. Choose a positive integer m such that $\frac{1}{m} < \varepsilon$.

$\Rightarrow \quad$ There exists a point a_i in the $\frac{1}{m}$-net A_m

$\Rightarrow \quad a_i \in A$ such that $d(p, a_i) < \frac{1}{m}$

Since $\frac{1}{m} < \varepsilon$, we have $a_i \in S(p, \varepsilon)$

$\Rightarrow \quad$ Every neighbourhood of p contains a point of A.

$\Rightarrow \quad$ Every point of X is an adherent point of A.

$\Rightarrow \quad A$ is dense in X. Hence, X is separable.

☛ **REMARK**

⇒ Since every compact metric space is totally bounded so using above theorem, we conclude that 'A compact metric space is separable'.

THEOREM 10. $\quad$ ***A metric space (X, d) is totally bounded if and only if every sequence in X has a Cauchy subsequence.***

PROOF. $\quad$ Let us first suppose every sequence in X has a Cauchy subsequence.

We have to prove that X is totally bounded.

Let $\varepsilon > 0$, choose $x_1 \in X$, then:

(i) If $S(x_1, \varepsilon) = X$, then obviously X is totally bounded.

(ii) If $S(x_1, \varepsilon) \neq X$. Choose $x_2 \in X - S(x_1, \varepsilon)$ so that $d(x_1, x_2) \geq \varepsilon$

Now, if $S(x_1, \varepsilon) \cup S(x_2, \varepsilon) = X$, then obviously X is totally bounded.

FACTS : TO THE POINT

For a metric space (X, d), the following properties are equivalent :

▶ (X, d) is compact.

▶ Every family of non-empty closed sets in X with finite intersection property has non-empty intersection.

▶ Every cover of X of open spheres has a finite subcover.

▶ Every countable open cover of X has a finite subcover.

▶ Every countable family of non-empty closed sets with finite intersection property has non-empty intersection.

▶ Every sequence in X has a limit point.

▶ Every infinite subset of X has a limit point.

▶ Every infinite open cover of X has a proper subcover.

▶ (X, d) is complete and totally bounded.

If not, choose $x_3 = X - \{S(x_1, \varepsilon) \cup S(x_2, \varepsilon)\}$ and so on.

Suppose this process does not stop at a finite stage. Then we get a sequence $x_1, x_2, ...,$ $x_n, ...$ such that $d(x_n, x_m) \geq \varepsilon$ if $n \neq m$.

Clearly, this sequence $<x_n>$ can't have a Cauchy subsequence, which contradict our hypothesis. Thus the above process stops at a finite stage and we get a finite set of points $\{x_1, x_2, ..., x_n, ...\}$ such that $X = S(x_1, \varepsilon) \cup S(x_2, \varepsilon) \cup ... \cup S(x_n, \varepsilon)$.

Hence, X is totally bounded.

Conversely, suppose that X is totally bounded. Let $S_1 = \{x_{11}, x_{12}, x_{13}, ..., x_{1n}, ...\}$ be a sequence in X. If one term of the sequence is infinitely repeated then S_1 contains a constant subsequence which is obviously a Cauchy subsequence. Therefore, we assume that no term of S_1 is infinitely repeated so that the range of S_1 is infinite. Now, since X is totally bounded therefore, X can be covered by a finite number of open balls of radius $1/2$.

$\Rightarrow$ At least one of these sphere must contain an infinite no. of terms of the sequence S_1.

$\Rightarrow$ S_1 contains a subsequence $S_2 = \{x_{21}, x_{22}, ..., x_{2n}, ...\}$ all terms of which lie within an open sphere of radius $1/2$.

Similarly, S_2 contains a subsequence $S_3 = \{x_{31}, x_{32}, ..., x_{3n}, ...\}$ all terms of which lie within an open sphere of radius $1/3$.

Repeat this process of forming successive subsequences and finally we take the diagonal sequence.

$$S = (x_{11}, x_{22}, ..., x_{nn}, ...)$$

We claim that S is a Cauchy subsequence of S_1.

If $m > n$, both x_{mm} and x_{nn} lie within an open sphere of radius $1/n$.

So, $d(x_{mm}, x_{nn}) < \dfrac{2}{n}$

$\Rightarrow$ $\quad d(x_{mn}, x_{nn}) < \varepsilon$ if $n, m > \dfrac{2}{\varepsilon}$

$\Rightarrow$ $\quad S$ is a Cauchy subsequence of S_1.

Hence, every sequence in X contains a Cauchy subsequence.

THEOREM II. *Let $<x_n>$ be a Cauchy sequence in a metric space X. If $<x_n>$ has a subsequence $<x_{n_k}>$ converging to x then $<x_n>$ converges to x.*

PROOF. Let $\varepsilon > 0$ be given. Since, $<x_n>$ is a Cauchy sequence, there exists a positive integer m_1 such that

$$d(x_n, x_m) < \frac{\varepsilon}{2} \quad \forall n, m \geq m_1 \qquad \qquad ...(1)$$

Further, since $<x_{n_k}> \to x$, therefore, there exists a positive integer m_2 such that

$$d(x_{n_k}, x) < \frac{\varepsilon}{2} \quad \forall n_k \geq m_2 \qquad \qquad ...(2)$$

Let $m_0 = \max\{m_1, m_2\}$ and fix $n_k \geq m_0$

Then $\quad d(x_n, x) \leq d(x_n, x_{n_k}) + d(x_{n_k}, x)$

$\qquad\qquad\quad < \varepsilon/2 + \varepsilon/2 \qquad\qquad\qquad\qquad$ (By (1) and (2))

$\qquad\qquad\quad = \varepsilon \quad \forall\, n \geq m_0$

$\Rightarrow \qquad d(x_n, x) < \varepsilon \quad \forall\, n \geq m_0$

Hence, $<x_n> \to x$.

8.9 COMPACTNESS AND CONTINUITY

THEOREM 1. *Continuous image of a compact metric space is compact.*

PROOF. Let f be a continuous mapping from a compact metric space X_1 to any metric space X_2. We have to prove that $f(X_1)$ is also compact. Without loss of any generality we may assume that $f(X_1) = f(X_2)$ i.e. we have to prove that X_2 is compact.

Let $\{G_n\}$ be a family of open sets in X_2 such that $\cup G_\alpha = X_2$.

Therefore, $\cup G_\alpha = f(X_1)$ $\Rightarrow$ $f^{-1}(\cup G_\alpha) = X_1$

$\Rightarrow$ $\cup f^{-1}(G_\alpha) = X_1$

Also, since f is continuous, $f^{-1}(G_\alpha)$ is open in X_1, for each α.

$\Rightarrow$ $\{f^{-1}(G_\alpha)\}$ is an open cover for X_1, which is compact.

$\Rightarrow$ $\exists$ a finite subcover $f^{-1}(G_{\alpha_1}) \cup ... \cup f^{-1}(G_{\alpha_n})$ such that

$$f^{-1}(G_{\alpha_1}) \cup f^{-1}(G_{\alpha_2}) \cup ... \cup f^{-1}(G_{\alpha_n}) = X_1$$

$\Rightarrow$ $f^{-1}\left(\overset{n}{\underset{i=1}{\cup}} G_{\alpha_i} \right) = X_1$

$\Rightarrow$ $\overset{n}{\underset{i=1}{\cup}} G_{\alpha_i} = f(X_1) = X_2$

$\Rightarrow$ $G_{\alpha_1}, G_{\alpha_2}, ..., G_{\alpha_n}$ is a cover for X_2.

Thus, the given open cover $\{G_\alpha\}$ for X_2 has a finite subcover.

Hence, $X_2 = f(X_1)$ is compact.

THEOREM 2. *Any continuous real valued function f defined on a compact metric space is bounded and attains its bounds.*

PROOF. Let X be a compact metric space and $f : X \to \mathbf{R}$ be a continuous real valued function. Then $f(X)$ is a compact subset of $\mathbf{R}$ and hence, $f(X)$ is closed and bounded subset of $\mathbf{R}$.

Since, $f(X)$ is bounded, so f is a bounded function.

Now, let $a = l.u.b.$ of $f(X)$ and $b = g.l.b.$ of $f(x)$. Then by definition of $l.u.b.$ and $g.l.b.$

$a, b \in \overline{f(X)}$ but $f(X)$ is closed therefore, $f(X) = \overline{f(X)}$.

$\Rightarrow$ $a, b \in f(X)$

$\Rightarrow$ There exists $x, y \in X$ such that $f(x) = a$ and $f(y) = b$.

Hence, f attains its bounds.

☞ **REMARK**

⟼ Above result is not true if and only if X is not compact.

THEOREM 3. *Let f be a one-one continuous function from a compact metric space X_1 onto any metric space X_2. Then f^{-1} is continuous on X_2. Hence f is a homeomorphism from X_1 to X_2.*

PROOF. Here, we have to prove that f^{-1} is continuous. For this, it is sufficient to prove that F is a closed set in X_1.

$\Rightarrow$ $(f^{-1})^{-1}(F) = f(F)$ is compact subset of X_2.

Let F be a closed set in X_1. Since, X_1 is compact, so F is compact.

Also, since f is continuous, $f(F)$ is compact subset of X_2.

$\Rightarrow$ $f(F)$ is a closed subset of X_2.

$\Rightarrow$ f^{-1} is continuous on X_2.

THEOREM 4. *Any continuous mapping f defined on a compact metric space (X_1, d_1) into any other metric space (X_2, d_2) is uniformly continuous on X_1.*

PROOF.   Let $\varepsilon > 0$ be given. Let $x \in X_1$. Now since, f is continuous at x, there exists $\delta_x > 0$ such that

$$d_1(y, x) < \delta_x \Rightarrow d_2(f(y), f(x)) < \frac{\varepsilon}{2} \qquad \qquad ...(1)$$

Now, the family of open spheres $\left\{ S\left(x, \dfrac{\delta_x}{2} \right) : x \in X_1 \right\}$ is an open cover for X_1. But X_1 is

compact, this cover has a finite subcover say $S\left(x_1, \dfrac{\delta_{x_1}}{2} \right), ..., S\left(x_n, \dfrac{\delta_{x_n}}{2} \right)$

Let $\delta = \min \left\{ \dfrac{1}{2} \delta_{x_1}, ..., \dfrac{1}{2} \delta_{x_n} \right\}$

We claim that $d_1(p, q) < \delta \Rightarrow d_2(f(p), f(q)) < \varepsilon$

Let $p \in S\left(x_i, \dfrac{\delta_{x_i}}{2} \right)$ for some i, $1 \leq i \leq n$

Therefore, $d_1(q, x_i) < \dfrac{\delta_{x_i}}{2}$. So, $d_2(f(p), f(x_i)) < \dfrac{\varepsilon}{2}$ (By (1)) $...(2)$

Now, $\qquad d_1(q, x_i) \leq d_1(q, p) + d_1(p, x_i)$

$$\leq \delta + \frac{1}{2} \delta_{x_i}$$

$$\leq \frac{\delta_{x_i}}{2} + \frac{\delta_{x_i}}{2} = \delta_{x_i}$$

$\Rightarrow \qquad d_1(q, x_i) < \delta_{x_i}$

$\Rightarrow \qquad d_2(f(q), f(x_i)) < \dfrac{\varepsilon}{2}$ (Using (1))

Finally, $d_2(f(p), f(q)) \leq d_2(f(p), f(x_i)) + d_2(f(x_i), f(q))$

$$< \varepsilon/2 + \varepsilon/2 = \varepsilon \qquad \qquad \text{(By (2))}$$

Therefore, $\qquad d_1(p, q) < d \Rightarrow d_2(f(p), f(q)) < \varepsilon$

Hence, f is uniformly continuous on X.

☞ **REMARK**

⇒ Above theorem is not true if X_1 is not compact.

THEOREM 5.   *The range of a continuous real valued function f on a compact connected metric space X must be either a single point or a closed and bounded interval.*

PROOF.   Let $f : X \to R$ be a continuous function. If f is a constant function, then the range of f is a single point. Now, let f is not constant, then the range of f contains more than one point. Since, X is connected, so $f(X)$ is connected subset of R. Thus, $f(X)$ is an interval in R. Further, since X is compact and f is continuous, $f(X)$ is a compact subset of R.

$\Rightarrow \quad f(X)$ is a closed and bounded subset of R.

Hence, $f(X)$ is a closed and bounded interval of R.

THEOREM 6.   *Every continuous function on a compact metric space is uniformly continuous.*

PROOF.   Consider a mapping $f : (X, d) \to (Y, \rho)$ which is continuous. Suppose (X, d) is compact.

We have to show that f is uniformly continuous. Let $\varepsilon > 0$ and $x \in X$.

Since f is continuous at x, $\exists a$ $\delta_x > 0$ such that

$$d(x, y) < \delta_x \Rightarrow \rho[f(x), f(y)] < \frac{\varepsilon}{2} \qquad \ldots(1)$$

Keeping ε fixed, let us vary x over X, i.e., for the same ε and for each $x \in X$, find $a, \delta x$ satisfying (1).

Clearly $\{S(x, \frac{1}{2}\delta x) : x \in X\}$ is a cover of X. Since (X, d) is compact, there exists a finite subcover

$$\{S(x_i, \frac{1}{2}\delta x_i) : i = 1, 2, \ldots, n\} \qquad \ldots(2)$$

Let $\quad \delta = \frac{1}{2} \min \{\delta x_1, \delta x_2, \ldots, \delta x_n\}$

We shall prove that for every pair of points p and q in X such that $d(p, q) < \delta$, we shall have $\rho[f(p), f(q)] < \varepsilon$

Let $p, q \in X$ be such that $d(p, q) < \delta$.

Since the open sphere in (2) form a cover of X, so for some $x_i \in X$, $p \in S(x_i, \frac{1}{2}\delta x_i)$.

Now, since $d(p, x_i) < \frac{1}{2}\delta x_i < \delta x_i$, from (1) we find that

$$\rho[f(p), f(q)] < \frac{1}{2}\varepsilon \qquad \ldots(3)$$

Now, $\qquad d(q, x_i) \leq d(q, p) + d(p, x_i)$

$$< \delta + \frac{1}{2}\delta x_i$$

$$< \frac{1}{2}\delta x_i + \frac{1}{2}\delta x_i$$

$$= \delta x_i$$

$\therefore \quad$ From (1)

$$\rho[f(p), f(q)] < \frac{1}{2}\varepsilon \qquad \ldots(4)$$

From (3) and (4), we conclude that

$$\rho[f(p), f(q)] \leq \rho(f(p), f(x_i)) + \rho(f(x_i), f(q))$$

$$< \frac{\varepsilon}{2} + \frac{\varepsilon}{2} = \varepsilon$$

Therefore, $d(p, q) < \delta \Rightarrow \rho(f(p), f(q)) < \varepsilon$

Hence, f is uniformly continuous.

8.10 LOCALLY COMPACT SPACES

A metric space (X, d) is said to be locally compact if and only if every point in X has at least one neighbourhood whose closure is compact.

THEOREM 1. ***Every compact metric space is locally compact.***

PROOF. Let (X, d) be a metric space which is compact. We have to show that X is locally compact.

Since X is both open and closed.

$\rightarrow \quad \overline{X} = X$ and it is a neighbourhood of each of its points.

$\overline{X} = X$ and X is compact $\Rightarrow \overline{X}$ is compact.

Hence, we conclude that X is locally compact.

THEOREM 2. ***Every closed subspace of a locally compact space is locally compact.***

PROOF. Let (X, d) be a metric space which is locally compact. Let us suppose that Y be a closed subset of X.

Let $y \in Y$ be arbitrary. Since $y \in Y \subset X \Rightarrow y \in X$.

Now, since X is locally compact, there exists an open neighbourhood N of y such that $\bar{N}$ is compact.

$\Rightarrow$ $N \cap Y$ is an open neighbourhood of y in Y such that $\overline{N \cap Y} \subset \bar{N}$

$\Rightarrow$ $\overline{N \cap Y}$ is a closed subset of the compact set N and is therefore compact.

Further, Y is closed in X.

$\Rightarrow$ Closure of $N \cap Y$ in X is the same as its closure in Y.

$\Rightarrow$ Every point in Y has a neighbourhood in Y whose closure in Y is compact.

Hence, Y is locally compact.

8.11 LINDELOF SPACES

A metric space (X, d) is said to be Lindelof if every open cover of X has a countable subcover.

☛ **REMARK**

➠ Every compact metric space is Lindelof.

THEOREM I. *Every Lindelof metric space is second countable.*

PROOF. Let (X, d) be a metric space which is Lindelof. We have to show that it is second countable.

For each positive integer m, we have $G = \{S(x, \frac{1}{m}) : x \in X\}$ an open cover of X.

Since the space is Lindelof, G has a countable subcover, say

$$\left\{ S\left(x_n^{(m)}, \frac{1}{m} \right) : n \in N \right\}$$

We claim that the collection

$$G' = \left\{ S\left(x_n^{(m)}, \frac{1}{m} \right) : n \in N, m \in N \right\}$$

is a countable open base for the set G of all open subsets of X. Let x be any point of an open set G. Then $\exists \varepsilon > 0$ such that $S(x, \varepsilon) \subset G$

Now, let us choose an integer $m_0 > \frac{2}{\varepsilon}$. Since $\left\{ S\left(x_n^{(m_0)}, \frac{1}{m_0} \right) : n \in N \right\}$ is a cover of X $\exists$

a positive integer n_0 such that $x \in S\left(x_{n_0}^{(n_0)}, \frac{1}{m_0} \right)$.

Since x is arbitrary, therefore, we want to show that

$$S\left(x_{n_0}^{(m_0)}, \frac{1}{m_0} \right) \subset S(x, \varepsilon)$$

Let $y \in S\left(x_0^{(m_0)}, \frac{1}{m_0} \right)$

Then $d(x, y) \le d(x, x_{n_0}^{m_0}) + d(x_0^{(m_0)}, y)$

$$< \frac{1}{m_0} + \frac{1}{m_0} = \frac{2}{m_0} < \varepsilon$$

But $d(x, y) < \varepsilon \Rightarrow y \in S(x, \varepsilon)$

$\Rightarrow$ $x \in S\left(x_{n_0}^{(m_0)}, \frac{1}{m_0} \right) \subset S(x, \varepsilon) \subset G$

$\Rightarrow$ For every point x of an **open set** G $\exists$ a member of G' contained in G and containing x.

$\Rightarrow$ G' is a countable open base for G. Hence, (X, d) is second countable.

8.12 PRODUCT OF TWO COMPACT METRIC SPACES

THEOREM I. *The product of two metric spaces is compact if and only if each factor space is compact.*

PROOF. Let (X_1, d_1) and (X_2, d_2) be two metric spaces and

$$Z = X_1 \times X_2 \text{ and } d = \sqrt{d_1^2 + d_2^2}$$

i.e., (Z, d) be the product of (X_1, d_1) and (X_2, d_2)

Now, (Z, d) is compact.

$\Leftrightarrow$ (Z, d) is complete and totally bounded.

$\Leftrightarrow$ (X_1, d_1) and (X_2, d_2) are both complete and totally bounded.

$\Leftrightarrow$ (X_1, d_1) and (X_2, d_2) are both compact.

SOLVED EXAMPLES

EXAMPLE 1. *Let A be a compact subset of a metric space (X, d). Show that for every $B \subset X$ $\exists$ a point $x_0 \in A$ such that $d(x_0, B) = d(A, B)$*

SOLUTION. Let us define

$$d(A, B) = p$$

By definition, $d(A, B) = \inf [d(x, y) : x \in A, \ y \in B]$

$\Rightarrow \forall n \in N$, there exists $x_n \in A$ and $y_n \in A$ such that $d(x_n, y_n) < p + \frac{1}{n}$

Since A is compact

$\Rightarrow A$ is sequentially compact.

$\Rightarrow < x_n >$ has a subsequence which converges to a point $x_0 \in A$.

We want to show that

$$d(x_0, B) = d(A, B) = p$$

Let if possible, $d(x_0, B) > p$

i.e., Suppose $d(x_0, B) = p + \varepsilon$, where $\varepsilon > 0$

Now, since a subsequence of $< x_n >$ converges to $x_0 \exists n_0 \in N$, such that

$$d(x_0, x_{n_0}) < \frac{\varepsilon}{2}$$

and $\qquad d(x_{n_0}, y_{n_0}) < p + \frac{1}{n_0} < p + \frac{\varepsilon}{2}$

But then

$$d(x_0, x_{n_0}) + d(x_{n_0}, y_{n_0}) < \frac{\varepsilon}{2} + p + \frac{\varepsilon}{2}$$

$$= p + \varepsilon = d(x_0, B) \leq d(x_0, y_{n_0})$$

which contradict the triangle inequality.

Hence, we conclude that $d(x_0, B) = d(A, B)$

EXAMPLE 2. *Let (X, d) be a metric space and A be a compact subset of X and B be a closed subspace of X such that $A \cap B = \phi$, then show that $d(A, B) > 0$.*

(MEERUT 1990)

SOLUTION. Let if possible, $d(A, B) = 0$, then using previous example $\exists x_0 \in A$, such that

$$d(x_0, B) = d(A, B) = 0.$$

Now, since B is a closed subset of X, it must contain all those points whose distance from B is zero.

$$\Rightarrow \qquad\qquad x_0 \in B$$

Therefore, $x_0 \in A$ and $x_0 \in B \Rightarrow x_0 \in A \cap B$

Which is a contradiction because $A \cap B = \phi$.

Hence, $\qquad d(A, B) > 0$.

EXAMPLE 3. *Give an example of a closed and bounded subset of l_2, which is not compact.*

SOLUTION. Consider, $\mathbf{O} = (0, 0, 0, ...) \in l_2$ and consider the closed sphere $S[\mathbf{0}, 1]$.

Clearly, $S[\mathbf{0}, 1]$ is closed set. $\qquad$ ($\because$ Every closed sphere is a closed set)

We claim that $S[\mathbf{0}, 1]$ is not compact.

Consider $e_1 = (1, 0, 0, ...); e_2 = (0, 1, 0, ...) ... e_n = (0, 0, 0, ..., 1, 0, ...)$.

Now, $d(\mathbf{0}, e_n) = 1$. Therefore, $e_n \in S[\mathbf{0}, 1]$ $\forall n$

$\Rightarrow$ $<e_n>$ is a sequence in $S[\mathbf{0}, 1]$.

Also, $d(e_n, e_m) = \sqrt{2}$ if $n \neq m$.

Hence, the sequence $<e_n>$ does not contain a Cauchy subsequence.

$\Rightarrow$ $S[\mathbf{0}, 1]$ is not totally bounded.

Hence, $S[\mathbf{0}, 1]$ is not compact.

EXAMPLE 4. *Prove that the closure of a totally bounded set is totally bounded.*

SOLUTION. Let A be a totally bounded subset of a metric space X. We claim that $\bar{A}$ is totally bounded. For this, we shall show that every sequence in $\bar{A}$ contains a Cauchy subsequence.

Let $<x_n>$ be a sequence in $\bar{A}$ and $\varepsilon > 0$ be given.
Then since $x_n \in \bar{A}, S\left(x_n, \dfrac{\varepsilon}{3} \right) \cap A \neq \phi$.

Choose $y_n \in S\left(x_n, \dfrac{\varepsilon}{3} \right) \cap A \quad \Rightarrow \quad d(y_n, x_n) < \dfrac{\varepsilon}{3}$ $\qquad$ ...(1)

Now, $<y_n>$ is a sequence in A. Since A is totally bounded, $<y_n>$ contains a Cauchy sequence say $< y_{n_k} >$. Therefore, there exists a natural number m such that

$$d(y_{n_i}, y_{n_j}) < \dfrac{\varepsilon}{3} \quad \forall n_0, n_j \geq m \qquad\qquad\qquad ...(2)$$

Therefore, $\qquad d(x_{n_i}, x_{n_j}) \leq d(x_{n_i}, y_{n_i}) + d(y_{n_i}, y_{n_j}) - d(y_{n_j}, x_{n_j})$

$$< \dfrac{\varepsilon}{3} + \dfrac{\varepsilon}{3} + \dfrac{\varepsilon}{3} \quad \forall n_i, n_j \geq m \qquad\qquad \text{(By (1) and (2))}$$

$$= \varepsilon \quad \forall n_0, n_j \geq m$$

$\Rightarrow$ $<x_{n_k}>$ is a Cauchy subsequence of $<x_n>$.

Hence, $\bar{A}$ is totally bounded.

EXAMPLE 5.   ***Let A be a totally bounded subset of R. Show that $\bar{A}$ is compact.***

SOLUTION.   Since, A is totally bounded, $\bar{A}$ is also totally bounded.

Also, since $\bar{A}$ is a closed subset of **R** and **R** is complete, $\bar{A}$ is complete.

$\Rightarrow$ $\bar{A}$ is totally bounded and complete.

Hence, $\bar{A}$ is compact.

REVIEW QUESTIONS AND ARCHIVE

1. Define the following :
   (i) Open cover   (MEERUT–2012, 14)
   (ii) Finite subcover   (DELHI–2012)
   (iii) Compact set   (MEERUT–2016, GARHWAL–2011)
   (iv) Sequentially compact   (PATNA–2004)
   (v) Countable compactness
   (vi) Bolzano - Weierstrass Property (BWP)
   (vii) Finite Intersection Property (FIP)
   (viii) Total boundedness

2. State Lebesgue covering lemma.

3. Show that every totally bounded metric space is bounded.

4. Show that every compact metric space is locally compact.

5. Show that a compact metric space is separable.

6. Show that every sequentially compact metric space is totally bounded.

7. State and prove Lebesgue covering lemma.

8. Show that intersection and union of two compact subsets is compact.

9. Show that every bounded infinite subset of **R** has at least one limit point with respect to usual metric for **R**.

10. Let A and B be two non-empty subsets of a metric space (X, d) such that B is compact. Show that $d(A, B) = 0$ if and only if $\bar{A} \cap B$ is non-empty.

OBJECTIVE EVALUATION

▶ FILL IN THE BLANKS

1. A uniformly continuous image of a totally bounded metric spaces is _____.

2. The closure of a totally bounded set is itself _____.

3. Every closed subset of a compact metric space is _____.

4. Every compact subset of a metric space is _____.

▶ TRUE OR FALSE

1. Any closed and bounded subset of a metric space is compact.   **(T/F)**

2. Any closed and bounded subset of **R** is compact.   **(T/F)**

3. A closed subspaces of a compact metric space is compact.   **(T/F)**

4. Any compact metric space is totally bounded. **(T/F)**

5. Any totally bounded metric space is compact. **(T/F)**

6. Any compact metric space is complete. **(T/F)**

7. Any infinite subset of a compact metric space has a limit point. **(T/F)**

8. Any infinite subset of **R** has a limit point. **(T/F)**

9. Any sequence in a compact metric space has a convergent subsequence. **(T/F)**

10. A bounded infinite subset of **R** has a limit point. **(T/F)**

▶ **MULTIPLE CHOICE QUESTIONS (CHOOSE THE MOST APPROPRIATE ONE)**

1. The real line is:
(a) compact   (b) not compact
(c) may be compact   (d) None of these

2. Every finite subset of a metric space is:
(a) bounded
(b) totally bounded
(c) both (a) and (b) are true
(d) None of these

3. Which of the following is / are true?
(a) The closed interval [a, b] is compact
(b) Every compact subset of real line is closed
(c) Both (a) and (b) are true
(d) None of these

4. A metric spaces has the BWP if it is:
(a) compact
(b) sequentially compact
(c) both (a) and (b) are true
(d) none of these

5. If a metric space has the BWP then it is:
(a) bounded
(b) totally bounded
(c) both (a) and (b) are true
(d) none of these

6. A metric space is compact if it is:
(a) complete
(b) totally bounded
(c) complete and totally bounded
(d) none of these

ANSWERS

▶ **FILL IN THE BLANKS**

1. totally bounded   **2.** totally bounded   **3.** compact **4.** closed

▶ **TRUE OR FALSE**

1. F   **2.** T   **3.** T   **4.** T   **5.** F   **6.** T   **7.** T   **8.** F   **9.** T
10. T

▶ **MULTIPLE CHOICE QUESTIONS**

1. (b)   **2.** (c)   **3.** (c)   **4.** (b)   **5.** (c)   **6.** (c)

CHAPTER SUMMARY

In this chapter, we have discussed the most important concept of metric space, *i.e.*, compactness. The concept of locally compact and sequentially compact have also been discussed in detail. The important points discussed in this chapter are as follows :

- Let G be an open cover of A. Then a subcollection G' of G is said to be subcover of A if G' covers A.
- A metric space (X, d) is said to be compact if every open cover of X has a finite subcover.
- Every compact subset of a metric space is closed.
- Closed subsets of compact sets are compact.
- A metric space X is compact if and only if every basic open cover of X has a finite subcover.
- A collection G of sets is said to have the finite intersection property (FIP) if the intersection of members of each finite subcollection of G is non-empty.
- A metric space (X, d) is said to have Bolzano-Weierstrass property (BWP) if every infinite set in X has a limit point.
- A metric space X is compact if and only if every collection of closed subsets of with F.I.P. has a non-empty intersection.
- Compact sets have Bolzano-Weierstrass property.
- The space $(\mathbf{R}, d)$ (where d is the usual metric on $\mathbf{R}$) is not compact.
- Every closed and bounded interval on real numbers $\mathbf{R}$ is compact, where $\mathbf{R}$ has the usual metric d.
- Every compact subset of real line is closed and bounded.
- A subset of $\mathbf{R}^n$ is compact if and only if it is closed and bounded.
- A metric space is said to be countably compact if every countable open cover of the space having a finite subcover.
- A metric space is compact if and only if it is countably compact.
- A metric space (X, d) is countably compact if and only if every sequence in X has a cluster point in X.
- Every countable compact metric space has

Bolzano-Weierstrass Property (BWP).

- A metric space (X, d) is said to be sequentially compact if every sequence has a convergent subsequence.
- A continuous image of a sequentially compact metric space is sequentially compact.
- Every closed subspace of a sequentially compact metric space is sequentially compact.
- Every sequentially compact metric space is countably compact.
- A metric space (X, d) is sequentially compact if and only if each sequence in X has a cluster point.
- A metric space (X, d) is said to be totally bounded iff X has an ε-net for every $\varepsilon > 0$.
- Every sequentially compact metric space is totally bounded.
- Every open cover of a sequentially compact metric space has a Lebesgue number.
- A metric space X is compact if and only if it is sequentially compact.
- Every compact metric space is complete.
- A compact subset of a metric space is closed and bounded.
- A metric space is compact if and only if it is complete and totally bounded.
- A subset of a complete metric space (X, d) is compact if and only if A is closed and totally bounded.
- A totally bounded metric space is separable.
- Every continuous function on a compact metric space is uniformly continuous.
- Every compact metric space is locally compact.
- Every closed subspace of a locally compact space is locally compact.
- Every Lindelof metric space is second countable.
- The product of two metric spaces is compact if and only if each factor space is compact.

FOR ADVANCED LEARNERS

➡ If X is compact, then there exists points a, b of X such that $d(a, b) = \text{diam}(X)$.

➡ Let A, B be non-empty disjoint subsets of a metric space X with A closed and B compact then $d(A, B) > 0$.

➡ Let X be a compact space in which each point x isolated then X is finite.

➡ Let (X, d) be a metric space and suppose that X is complete with respect to every metric equivalent to d, then X is compact.

➡ In a metric space (X, d), following conditions are equivalent:

(i) If ρ is a metric equivalent to d and A, B are disjoint closed subset of (X, ρ) then $d(A, B) > 0$

(ii) X is compact

➡ The following are equivalent conditions on a metric space X:

(i) X has the Lebesgue covering property.

(ii) every continuous mapping of X into a metric space is uniformly continuous.

(iii) every continuous mapping of X into $\boldsymbol{R}$ is uniformly continuous.

➡ Let A and B be totally compact subspaces of a locally compact metric space X then $A \cap B$ is locally compact.

➡ Let X be a metric space in which every bounded set is contained in a compact set. Then X is locally compact and seperable.

➡ Let X be a compact metric space and suppose that the closure of any open sphere $S(a, r)$ in X is the closed sphere $\overline{S}(a,r)$, then any open or closed sphere in X is connected.

➡ A subset of the product space $\boldsymbol{R}^2$ or $\boldsymbol{C}^2$ is compact if and only if it is closed and bounded.

➡ If X be an infinite set and d be the discrete metric on X. Then (X, d) is not sequentially compact.

➡ Sequentially compactness $\Leftrightarrow$ Countable compactness.

CHAPTER 9

The Metric Space C[a, b]: Uniform and Pointwise Convergence

9.1 INTRODUCTION

In this chapter, we shall discuss the uniform convergence of sequence and series of functions. Mainly, we want to draw attention on the most important aspects of the probelms of sequence and series of functions.

9.2 POINTWISE CONVERGENCE

Let $\langle f_n \rangle$ be a sequence of real valued functions on a metric space (X, d). Let the function f_n be tends to a definite limit for all values of $x \in X$ as $n \to \infty$. Therefore, to each point $t \in X$, there corresponds a sequence of numbers $\langle f_n(t) \rangle$ with terms $f_1(t), f_2(t), f_3(t), \ldots$

Let this sequence $\langle f_n(t) \rangle$ converge to $f_n(t)$. Then pointwise converges can be defined as follows:

Definition. *Let $X \neq \phi$ and f be a function from X to $\mathbf{R}$. Also, each $n \in \mathbf{N}$ let $f_n : X \to \mathbf{R}$. Then, the sequence of functions $\langle f_n \rangle$ converges pointwise to the function f, if for each $x \in X$, the sequence of real numbers $\langle f_n(x) \rangle$ converges to the real number $f(x)$.*

Therefore, $\langle f_n \rangle$ converges pointwise to f if $\lim_{n \to \infty} f_n(x) = f(x) \forall x \in X$

✎ ILLUSTRATIONS

* For each $n \in \mathbf{N}$. Let us define $f_n : \mathbf{R} \to \mathbf{R}$ by $f_n(x) = \dfrac{x}{n}, \forall x \in \mathbf{R}$

  Then $\langle f_n(x) \rangle$ converges to $f(x) = 0, \forall x \in \mathbf{R}$

* The sequence $\langle f_n(x) \rangle = \langle x^n \rangle$ converges pointwise to the function $f : [0, 1] \to \mathbf{R}$ defined by
  $$f(x) = \begin{cases} 0 & \text{if } x \in]0,1[\\ 1 & \text{if } x = 1 \end{cases}$$

* The sequence $\langle x(1-x)^n \rangle$ converges pointwise to the function f that vanish identically.

* The sequence $\left\langle f_n = 1 + \dfrac{x}{1+nx} \right\rangle$ converges pointwise to the function f defined by $f(x) = 1$,

  $\forall x \in]0, \infty[$.

* The geometric series $1 + x + x^2 + x^3 + \cdots$ converges to $(1 - x)^{-1}, \forall x \in]-1, 1[$.

☛ REMARKS

➠ If the sequence f_n converges to f, then, we can write $\lim_{n \to \infty} f_n(x) = f(x)$ for $x \in X$

➠ Pointwise convergent of a sequence of functions depends on the metric space as well as on the functions.

Definition. *Let* (X, d) *be a metric space and f be a function from X to **R** and* $f_n : X \to \mathbf{R}, \forall n \in \mathbf{N}$. *The sequence of function* $\langle f_n \rangle$ *converges pointwise to f if and only if for each* $x \in X$ *and for each positive real number* ε, $\exists$ *a positive integer m such that* $n \geq m \Rightarrow |f_n(x) - f(x)| < \varepsilon$

SOLVED EXAMPLES

EXAMPLE 1. **Let** $<f_n>$ **be the sequence defined by** $f_n : R \to R$ **such that**

$$f_n(x) = \frac{x}{n} \, \forall x \in R, n \in N$$

Show that the sequence converges pointwise to the zero function.

SOLUTION. Here, we want to show that given sequence converges pointwise to the zero function *i.e.*, $f(x) = 0$, $x \in R$, then we must show that given $\varepsilon > 0$, we can find $m \in \mathbf{N}$ such that

$$\forall n \geq m \Rightarrow \left| \frac{x}{n} - 0 \right| = \frac{|x|}{n} \qquad ...(1)$$

Let us choose $m > \dfrac{|x|}{\varepsilon}$

Then (1) gives

$$\forall n \geq m \Rightarrow \left| \frac{x}{n} - 0 \right| = \frac{|x|}{n} < \varepsilon$$

Here, the given sequence converges pointwise to the zero function.

9.3 UNIFORM CONVERGENCE OF SEQUENCES

Let us suppose the sequence $\langle f_n(x) \rangle$ converges for every point x in X. Therefore, f_n tends to a definite limit as $n \to \infty$ for every $x \in X$. The limit is also a function of x.

Then by definition of limit, we must have for every $\varepsilon > 0$ $\exists$ a positive integer m such that

$$n \geq m \Rightarrow |f_n(x) - f(x)| < \varepsilon.$$

Here, it must be noted that the integer m depends upon x as well as ε.

Definition. *The sequence* $\langle f_n(x) \rangle$ *of functions is said to converge uniformly on X to a function f, if for every* $\varepsilon > 0$, *we can find a positive integer m such that*

$$n \geq m \Rightarrow |f_n(x) - f(x)| < \varepsilon, \forall x \in X$$

The convergence of a sequence $\langle f_n(x) \rangle$ at every point $x \in X$ (*i.e.*, pointwise convergence) does not necessarily ensure its uniform convergence on X. A sequence of functions may be convergent at every point of X, yet it may not be uniformly convergent on X.

Let us consider the following examples :

(1) The sequence of function $\langle f_n \rangle$ defined on **R** such that

$$f_n(x) = \frac{x}{n}, \forall n \in \mathbf{N} \text{ converges pointwise to the zero}$$

FACTS : TO THE POINT

▶ Pointwise convergence (or simple convergence) is a local property (to be satisfied at a point) whereas uniform convergence is essential associated with an interval, *i.e.*, uniform convergence is a global property.

▶ The interval of uniform convergence is always closed, or the functions are bounded above, *i.e.*, $a < x < b$. This is a convenient, though not a necessary assumption.

▶ In case of convergence, a positive integer m depend on both ε and x does not remain bounded, but in case of uniform convergence, m solely dependent on ε but independent on x remains bounded.

▶ Convergence at each point of a closed interval does not mean uniform convergence in that interval.

▶ Each term of the series is the function of x for uniform convergence but it is not necessary for convergence.

▶ Uniform convergence is a far reaching *sufficient* condition for the validity of interchanging certain limits, but it does not provide the complete answer of the general question of whether we can reverse the order of two limit processes. That is to say, the conditions of uniform convergence is merely *sufficient* for the truth of theorems regarding continuity of a sum, term by term integration, and termwise differentiation, but it is by no means a necessary condition (Bromwich).

function (*i.e.*, $f(x) = 0$) while, this sequence does not converges uniformly to this function. We will prove that convergence is not uniform.

Let us suppose the sequence $\left\langle \dfrac{x}{n} \right\rangle$ converges uniformly to the zero function on $\boldsymbol{R}$, then there

is some $m \in \boldsymbol{N}$ (m depending only on $\varepsilon = 1$) such that

$$n \geq m \Rightarrow |f_n(x) - f(x)| = \frac{|x|}{n} < 1, \forall x \in \boldsymbol{R}$$

which is not true for all $x \in \boldsymbol{R}$ for if $n = m$ and $x = m$, then $\dfrac{|x|}{m} = 1$

(2) Let $\langle f(x) \rangle = \langle x^n \rangle$ be the sequence of function defined on [0, 1]. Then we can easily verify that the given sequence $\langle f_n(x) \rangle$ converges pointwise to the limit function f, defined by

$$f(x) = \begin{cases} 0 & \text{if} \quad 0 \leq x < 1 \\ 1 & \text{if} \quad x = 1 \end{cases} \quad \text{for every } x \in [0, 1].$$

To check that the convergence is uniform, we consider the interval [0, 1]. Let $\varepsilon > 1$ be given. Then, we have

$$|f_n(x) - f(x)| < \varepsilon \Rightarrow |x^n - 0| < \varepsilon \Rightarrow x^n < \varepsilon$$

$$\Rightarrow \qquad \frac{1}{x^n} > \frac{1}{\varepsilon} \Rightarrow n \log \frac{1}{x} > \log \frac{1}{\varepsilon}$$

i.e., $\qquad\qquad n > \dfrac{\log(1 / \varepsilon)}{\log(1 / x)}$ \qquad\qquad\qquad ... (1)

Therefore, when $x \neq 1$, $m \in$ N such that $m > \dfrac{\log(1 / \varepsilon)}{\log(1 / x)}$

In particular, when $x = 0$, $m = 1$.

Now as x increases from 0 to 1, it is clear from (1) that $n \to \infty$.

Therefore, it is not possible to find $m \in \boldsymbol{N}$ such that

$$n \geq m \Rightarrow |f_n(x) - f(x)| < \varepsilon \text{ for all } x \in [0,1]. \text{ Hence, the given}$$

sequence is not uniformly convergent on [0, 1].

☞ **REMARK**

➠ If we consider the interval]0, k[, where $0 < k < 1$, then the greatest value of log $(1/\varepsilon)$/ log $(1/x)$ is log $(1/\varepsilon)$/ log $(1/k)$ so that if we take $m > (\log 1/\varepsilon)/\log(1 / k) \in \boldsymbol{N}$, we have

$$n \geq m \Rightarrow |f_n(x) - f(x)| < \varepsilon \ \forall \ x \in [0, k].$$

Therefore, $\langle f_n(x) \rangle$ converges uniformly on [0, k].

(3) The sequence of functions $< 1/(1 + nx) >$ does not converges uniformly on $\boldsymbol{R}$ to the function f defined by

$$f(x) = \begin{cases} 0, & \text{if} \quad x \neq 0 \\ 1, & \text{if} \quad x = 0 \end{cases}$$

(4) Let a be any positive real number and for each $n \in \boldsymbol{N}$.

Define $\qquad f_n(x) = \dfrac{1}{1 + nx^2} \forall x \in [a, \infty]$

The sequence $\langle f_n(x) \rangle$ converges uniformly to the zero function *i.e.*, $f(x) = 0$ on [a, ∞[, because

of $m \in$ N, $m > (1 - \varepsilon)/a^2$, then

$$n \geq m \Rightarrow |f_n(x) - 0| = \frac{1}{1 + nx^2} \leq \frac{1}{1 + mx^2} \leq \frac{1}{1 + ma^2} < \varepsilon \forall x \in [a, \infty[$$

9.3.1 POINT OF NON-UNIFORM CONVERGENCE

A point such that the sequence does not converge uniformly in any neighbourhood of it, however small, is said to be a point of non-uniform converges of the sequence.

9.3.2 SUM FUNCTION OF A SERIES

Consider the series $\sum_{n=1}^{\infty} u_n(x) = u_1(x) + u_2(x) + ... + u_n(x) + ..., x \in X$ of real valued function defined

on a set X. This series gives rise to a sequence of function $\langle f_n(x) \rangle$ where $f_n(x) = u_1(x) + u_2(x) + ... + u_n(x)$

The series $\Sigma u_n(x)$ is said to be convergent on X if the corresponding sequence $\langle f_n(x) \rangle$ is convergent on X and the limit function $f(x)$ of the sequence is said to the sum function or the sum of the series.

9.4 UNIFORM CONVERGENCE OF A SERIES OF FUNCTIONS

The series $\sum_{n=1}^{\infty} u_n(x)$ is said to converge uniformly on X if the sequence $\langle f_n(x) \rangle$, where

$f_n(x) = u_1(x) + u_2(x) + ... + u_n(x)$ converges uniformly on X.

9.5 CAUCHY'S GENERAL PRINCIPLE OF UNIFORM CONVERGENCE

THEOREM I.   *Let $\langle f_n \rangle$ be a sequence of real valued function defined on X. Then $\langle f_n \rangle$ converges uniformly on X if and only if for every $\varepsilon > 0$, there exists a positive integer m such that*

$$n \geq m, \ p \geq m, \ x \in X \Rightarrow |f_n(x) - f_p(x)| < \varepsilon.$$

PROOF.   **The only if part.** Let us first suppose, the sequence $\langle f_n \rangle$ converges uniformly to the function f on X. Then, by definition, for given $\varepsilon > 0 \ \exists$, a positive integer m such that

$$|f_n(x) - f(x)| < \varepsilon / 2 \ \forall n \geq m, \ \forall x \in X \qquad ....(1)$$

Therefore, if $p, n \geq m$ we have for any $x \in X$.

$$|f_n(x) - f_p(x)| = |f_n(x) - f(x) + f(x) - f_p(x)|$$
$$\leq |f_n(x) - f(x)| + |f(x) - f_p(x)|$$
$$< \varepsilon / 2 + \varepsilon / 2 = \varepsilon.$$

Hence (1) holds for this m.

The if part. Let $\langle f_n \rangle$ be a sequence of function from X to **R** such that for given $\varepsilon > 0$, $\exists$ a positive integer m such that (1) holds.

To show $\exists$ a function f on X such that the sequence $\langle f_n \rangle$ converges uniformly to f on X.

Now, for each fixed $x \in X$, the sequence of real numbers $\langle f_n(x) \rangle$ is a Cauchy sequence and therefore $\lim_{n \to \infty} f_n(x)$ exists for every $x \in X$.

          ($\because$ Every Cauchy sequence of real numbers is convergent.)

Define $f : X \to \mathbf{R}$ by $f(x) = \lim_{n \to \infty} f_n(x), \forall x \in X$

We want to show that the sequence $<f_n>$ converges uniformly to f.

If $x \in \mathbf{R}$, $\varepsilon > 0$, then there is some $m \in \mathbf{N}$ such that

$$n, p \geq m \Rightarrow |f_n(x) - f_p(x)| < \varepsilon / 2 \text{ for all } x \in X.$$

For any fixed p, $p \geq m$ and fixed $x \in X$, consider that sequence $\langle |f_n(x) - f_p(x)| : n \in \mathbb{N} \rangle$
Since

$$\lim_{n \to \infty} f_n(x) = f(x) \text{ and } |f_n(x) - f_p(x)| < \varepsilon / 2 \text{ for } n \geq p,$$

we have $\lim_{n \to \infty} |f_n(x) - f_p(x)| = |f_n(x) - f_p(x)| < \varepsilon / 2$

Therefore, if $p \geq m$, then

$$|f(x) - f_p(x)| < \varepsilon, \ \forall \ x \in X.$$

Hence, the sequence $\langle f_n(x) \rangle$ converges uniformly to f on X.

THEOREM 2. *The series $\Sigma u_n(x)$ converges uniformly on X if and only if for every $\varepsilon > 0$, $\exists$ a positive integer m such that*

$$n \geq m \Rightarrow |u_{n+1}(x) + u_{n+2}(x) + ... + u_{n+p}(x)| < \varepsilon, p = 1, 2. \ \textit{for all } x \in X.$$

PROOF. Let $s_n(x)$ denotes the sequence of partial sum of the given series such that

$$s_n(x) = u_1(x) + u_2(x) + ... + u_n(x), x \in X.$$

Then, $s_{n+p}(x) - s_n(x) = u_{n+1}(x) + u_{n+2}(x) + ... + u_{n+p}(x)$

The series $\sum_{n=1}^{\infty} u_n(x)$ converges uniformly to $f(x)$ on X if and only if $\langle f_n \rangle$ converges uniformly on X. But $\langle s_n(x) \rangle$ converges to $s(x)$ on X if and only if for given $\varepsilon > 0$, $\exists$ a positive integer m such that $n \geq m \Rightarrow |s_{n+p}(x) - s_n(x)| < \varepsilon, p = 1, 2$ for all $x \in X$. Therefore,

$$n \geq m \Rightarrow |u_{n+1}(x) + u_{n+2}(x) + ... + u_{n+p}(x)| < \varepsilon, \ p = 1, 2,$$

for all $x \in X$.

9.6 DINI'S CRITERION FOR UNIFORM CONVERGENCE OF A SEQUENCE OF CONTINUOUS FUNCTIONS

THEOREM 1. *Let $\langle f_n \rangle$ be a sequence of continuous real valued function defined on the compact metric space (X, d) such that*

$$f_1(x) \geq f_2(x) \geq ... \geq f_n(x) \geq \qquad ...(1)$$

for every $x \in X$. If $\langle f_n \rangle$ pointwise converges on X to the continuous function f on X, then $\langle f_n \rangle$ converges uniformly to f on X.

PROOF. Let $g_n = f_n - f$ for each $n \in N$ Then, from (1), we get

$$g_1(x) \geq g_2(x) \geq g_n(x) \geq ... \geq 0 \qquad ...(2)$$

Also, since $\langle f_n \rangle$ converges to f on X, we have

$$\lim_{n \to \infty} g_n(x) = 0 \forall x \in X \qquad ... (3)$$

To show, $\langle g_n \rangle$ converges uniformly to 0 on X. Let $\varepsilon > 0$ be given.

If $x \in X$, then from (3), $\exists$ a positive integer $m(x)$ such that $0 \leq g_m(x) \leq \varepsilon / 2$

Since $g_m(x)$ is continuous at x, therefore, $\exists$ an open sphere $S(x, r)$ such that $y \in S(x, r) \Rightarrow g_{m(x)}(y) < \varepsilon$. Therefore, the collection $C = \{S(x, r) : x \in X, r > 0\}$ forms an open cover of X. Since X is compat, therefore, by definition $\exists$ a finite subcover of C i.e., $\exists$ a finite number of open spheres $S(x, r)$ say $S(x_1, r_1), S(x_2, r_2), ... S(x_k, r_k)$, which also cover X.

Now, let $m = \max\{m(x_1), m(x_2), ..., m(x_k)\}$,

If y is any point of X, then $y \in S(x_i, r)$ for some $i = 1, 2, ..., k$.

Therefore, $\qquad g_{m(x_i)}(y) < \varepsilon$

But since $m(x_i) \leq m$, therefore from (2), we have $g_m(y) = g_m(x_i)(y)$

$$\Rightarrow \qquad 0 \leq g_m(y) < \varepsilon, \ \forall y \in X$$

Thus, from (2), we have $0 \leq g_n(y) < \varepsilon, \ \forall \ n \geq m, y \in X$.

Hence, $\langle g_n \rangle$ converges uniformly to 0 on X. This implies that $\langle f_n \rangle$ converges uniformly on Y to the function f.

9.7 TESTS FOR UNIFORM CONVERGENCE

THEOREM 1. (M_n-test). *Let $\langle f_n \rangle$ be a sequence of function defined on a metric space X.*
Let $\lim\limits_{n \to \infty} f_n(x) = f(x)$ for all $x \in X$ and let $M_n = \sup\{|f_n(x) - f(x)| : x \in X\}$
Then $\langle f_n \rangle$ converges uniformly to f if and only if $M_n \to 0$ as $n \to \infty$.

PROOF. **Necessary condition.** Let us suppose, the sequence $\langle f_n \rangle$ of functions converges uniformly to f on X. Then by definition, for a given $\varepsilon > 0$, $\exists$ a positive integer m (independent of x) such that $n \geq m \Rightarrow |f_n(x) - f(x)| < \varepsilon \; \forall \, x \in X$.

Also, M_n is the supremum of $|f_n(x) - f(x)|$.
Therefore

$$|f_n(x) - f(x)| < \varepsilon \, \forall n \geq m, \forall x \in X$$

$$\Rightarrow \qquad M_n = \sup_{x \in X} |f_n(x) - f(x)| < \varepsilon \, \forall n \geq m$$

$$\Rightarrow \qquad M_n \to 0, \text{as } n \to \infty$$

Sufficient condition. Let us assume that $M_n \to 0$ as $n \to \infty$. Then for a given $\varepsilon > 0$, $\exists$ a positive integer m such that

$$|M_n - 0| < \varepsilon \, \forall n \geq m, \forall x \in X$$

$$\Rightarrow \qquad M_n < \varepsilon \, \forall n \geq m, \forall x \in X$$

$$\Rightarrow \qquad \sup_{x \in X} |f_n(x) - f(x)| = M_n < \varepsilon, \forall n \geq m$$

$$\Rightarrow \qquad |f_n(x) - f(x)| \leq M_n < \varepsilon \, \forall n \geq m, \forall x \in X|$$

$\Rightarrow \quad \langle f_n \rangle$ converges uniformly to f on X.

THEOREM 2. (T_n-test for uniform convergence). *Let $\langle f_n(x) \rangle$ be a sequence of functions defined on $X \subset R$ and let on X*

$$\lim_{n \to \infty} f_n(x) = f(x) \qquad \qquad \text{...(1)}$$

Let for $n \geq p \in N$, $T_n > 0$ exists such that

$$|f_n(x) - f(x)| \leq T_n, \forall \, x \in X \text{ and } n > p \qquad \qquad \text{...(2)}$$

Then $\langle f_n(x) \rangle$ converges uniformly to f(x) on X if

$$\lim_{n \to \infty} T_n = 0 \qquad \qquad \text{...(3)}$$

PROOF. Let $\varepsilon > 0$ be given. Then by (3), $\exists \, m = m \, (\varepsilon) \in N$ such that
$$|T_n - 0| < \varepsilon \Rightarrow T_n < \varepsilon, \, \forall \, n \geq m \qquad \qquad \text{...(4)}$$
Let $n_0 = \max \{m, p\}$, then by (2), we get

$$|f_n(x) - f_n| \leq T_n, \forall x \in X \text{ and } n \geq n_0 \qquad \qquad \text{[Using(4)]}$$
$$< \varepsilon$$

Hence, the sequence $\langle f_n(x) \rangle$ converges uniformly to $f(x)$ on X.

☞ **REMARKS**

➠ Let $\langle f_n(x) \rangle$ be a sequence of functions defined on $X \subset \boldsymbol{R}$ and let on X, $\lim\limits_{n \to \infty} f_n(x) = f(x)$

If $\lim\limits_{n \to \infty} T_n \neq 0$ where $T_n = |f_n(x_n) - f(x_n)| \, \forall n \geq n_0 \in \boldsymbol{N}, x_n = x(n) \in X$

Then $\langle f_n(x) \rangle$ does not converges uniformly to $f(x)$.

➠ The choice of T_n is not unique as that of M_n.

SOLVED EXAMPLES

EXAMPLE 1. *Show that the sequence $\langle f_n \rangle$ where $f_n(x) = nx(1-x)^n$ does not converge uniformly on [0, 1]*

SOLUTION. Here, we have

$$f(x) = \lim_{n \to \infty} f_n(x) = \lim_{n \to \infty} \frac{nx}{(1-x)^{-n}} \qquad \left| \text{Form} \frac{\infty}{\infty} \right.$$

$$= \lim_{n \to \infty} \frac{x}{-(1-x)^{-n} \log(1-x)} \qquad \text{[Using L-Hospital rule]}$$

$$= \lim_{n \to \infty} \left(-\frac{x(1-x)^n}{\log(1-x)} \right) = 0 \qquad [\because (1-x)^n \to 0 \, \forall x \in [0,1]]$$

$\Rightarrow \qquad f(x) = 0, \forall x \in [0,1]$

Now $\qquad M_n = \sup\{ | f_n(x) - f(x) | \} : x \in [0,1]$

$\qquad\qquad = \sup\{ nx(1-x)^n : x \in [0,1] \}$

$\qquad\qquad = \sup (1-x)^n nx, \, x \in [0, 1]$

Therefore,

$$M_n \geq n . \frac{1}{n} \left(1 - \frac{1}{n} \right)^n \qquad \left(\text{Taking } x = \frac{1}{n} \in [0,1] \right)$$

$$= \left(1 - \frac{1}{n} \right)^n \to \frac{1}{e} \text{ as } n \to \infty$$

Hence, by M_n-test, $\langle f_n \rangle$ does not converge uniformly on [0, 1]. Therefore, 0 is a point of non-uniform convergence, since $x = \frac{1}{n} \to 0$ as $n \to \infty$.

EXAMPLE 2. *Test for unifrom and non-uniform convergence of the sequence of functions $\langle f_n(x) \rangle$ where $f_n(x) = e^{-nx}, x \geq 0$*

SOLUTION. Here, we have

$$f(x) = \lim_{n \to \infty} f_n(x) = \begin{cases} 1, & x = 0 \\ 0, & x > 0. \end{cases}$$

Clearly, f is discontinuous at $x = 0$. Now, let $\delta > 0$ be any small number and let $X = \{x \in \mathbf{R} : 0 \leq x \leq \delta\}$.

For $n \geq n_0 \in \mathbf{N}$, let us take $x_n = \frac{1}{n}$ in such a way that $0 < \frac{1}{n} < \delta$, then for $n \geq n_0$, $x_n \in X$

and $x_n \neq 0$. Therefore $\forall n \geq n_0 \Rightarrow T_n = f_n(x_n) - f(x_n) = e^{-n/n} = e^{-1}$

$\Rightarrow \qquad\qquad \lim_{n \to \infty} T_n \neq 0$

Hence, by T_n test $\langle f_n(x) \rangle$ does not converge uniformly on X. We now discuss on the interval $0 < \delta \leq x$, where δ is fixed. Then for $x \geq \delta$

$$| f_n(x) - f(x) | = e^{-n\delta} \geq e^{-n\delta} = T_n \text{ (say)}.$$

Also, $\qquad\qquad \lim_{n \to \infty} T_n = \lim_{n \to \infty} e^{-n\delta} = 0$

Hence, by T_n-test, the given sequence converges uniformly for every $x \geq \delta > 0$.

THEOREM 3. (Weierstrass's M-test). *A series* $\sum\limits_{n=1}^{\infty} u_n(x)$ *of functions will converge uniformly on X if there exists a convergent series* $\sum\limits_{n=1}^{\infty} M_n$ *of positive constant such that* $|u_n(x)| \le M_n \forall n$ *and* $\forall x \in X$.

PROOF. Since ΣM_n is convergent, therefore, by definition, for a given $\varepsilon > 0$ we can find a positive integer m such that

$$n \ge m \Rightarrow M_{n+1} + M_{n+2} + ... + M_{n+p} < \varepsilon \qquad ... (1)$$

$$(\text{for } p = 1, 2, 3, ...)$$

Since $\qquad |u_n(x)| \le M_n \forall n$ and $\forall x \in X$ $\qquad\qquad ... (2)$

From (1) and (2), we conclude that

$$|u_{n+1}(x) + u_{n+2}(x) + ... + u_{n+p}(x)|$$

$$\le |u_{n+1}(x)| + |u_{n+2}(x)| + ... + |u_{n+p}(x)|$$

$$\le M_{n+1} + M_{n+2} + ... + M_{n+p}$$

$$< \varepsilon, \text{ for every } n \ge m \text{ and } \forall x \in X.$$

Hence, $\Sigma u_n(x)$ converges uniformly on X.

☛ **REMARKS**

⟹ The above series is also converges absolutely.

⟹ It is clear that same proof would hold if M_n were a function of x and if the series $\Sigma M_n(x)$ is uniformly convergent on X.

⟹ The series which satisfy M-test are called normally convergent.

🔺 SOLVED EXAMPLES

EXAMPLE 1. *Show that the series*

$$\frac{\cos x}{1^p} + \frac{\cos 2x}{2^p} + \frac{\cos 3x}{3^p} + ... + \frac{\cos nx}{n^p} +$$

converges uniformly on R if p > 1.

SOLUTION. Here, we have

$$\left|\frac{\cos nx}{n^p}\right| < \frac{1}{n^p} \forall x \subset R$$

Also, the series $\Sigma \dfrac{1}{n^p}$ is known to be convergent for $p > 1$.

Hence, by Weierstrass's M-test the given series converges uniformly on R for $p > 1$.

THEOREM 4. (Abel's Test). *The series* $\Sigma u_n(x)v_n(x)$ *will converges uniformly in* [a, b] *if*

(i) $\Sigma u_n(x)$ *is uniformly convergent in* [a, b]

(ii) *the sequence* $\langle v_n(x) \rangle$ *is monotonic for every* $x \in$ [a, b]

(iii) *the sequence* $\langle v_n(x) \rangle$ *is uniformly bounded in* [a, b] *by k i.e.,*

$$|v_n(x)| < k, \forall x \in [a, b] \quad and \quad \forall n \in N.$$

PROOF. Let $R_{n,p}(x)$ be the partial remainder of the series $\Sigma u_n(x)v_n(x)$ and $r_{n,p}(x)$ that of the series $\Sigma u_n(x)$. Then

$$R_{n,p}(x) = u_{n+1}(x)v_{n+1}(x) + u_{n+2}(x)v_{n+2}(x) + ... + u_{n+p}(x)v_{n+p}(x)$$

$$= r_{n,1}(x)v_{n+1}(x) + \{r_{n,\,2}(x) - r_{n,1}(x)\}v_{n+2}(x)$$
$$+ \{r_{n,\,3}(x) - r_{n,2}(x)\}v_{n+3}(x) + \ldots + \{r_{n,\,p}(x) - r_{n,p-1}(x)\}v_{n+p}(x)$$
$$= r_{n,\,1}(x)\{v_{n+1}(x) - v_{n+2}(x)\} + r_{n,2}(x)\{v_{n+2}(x) - v_{n+3}(x)\} +$$
$$+ \{r_{n,\,p-1}(x)\{v_{n+p-1}(x) - v_{n+p}(x)\} + r_{n,p}(x)v_{n+p}(x). \qquad \ldots(A)$$

Given that $\langle v_n(x)\rangle$ is monotonic, therefore,

$$\{v_{n+1}(x) - v_{n+2}(x)\}, \{v_{n+2}(x) - v_{n+3}(x)\}, \ldots, \{v_{n+p-1}(x) - v_{n+p}(x)\} \qquad \ldots(1)$$

all have the same sign for fixed value of x in $[a, b]$.

Also, given that $\langle v_n(x)\rangle$ is uniformly bounded by k, therefore

$$|v_n(x)| < k \text{ for all } x \in [a, b] \text{ and } \forall n \in \mathbf{N}. \qquad \ldots(2)$$

Also, since the given series $\sum u_n(x)$ is uniformly convergent in $[a, b]$, for a given $\varepsilon > 0$, $\exists$ a positive integer m, independent of x such that for $n \geq m$.

$$|r_{n,p}(x)| = |u_{n+1}(x) + u_{n+2}(x) + \ldots + u_{n+p}(x)| < \frac{\varepsilon}{3k} \qquad \ldots(3)$$

From (1) and (3), we have

$$|R_{n,p}(x)| < \frac{\varepsilon}{3k}|v_{n+1}(x) - v_{n+2}(x)| + \frac{\varepsilon}{3k}|v_{n+2}(x) - v_{n+3}(x)| + \ldots$$
$$+ \frac{\varepsilon}{3k}|v_{n+p-1}(x) - v_{n+p}(x)| + \frac{\varepsilon}{3k}|v_{n+p}(x)|$$
$$= \frac{\varepsilon}{3k}|v_{n+1}(x) - v_{n+p}(x)| + \frac{\varepsilon}{3k}|v_{n+p}(x)|. \qquad \ldots(4)$$

Using (A), we have
$$|v_{n+1}(x) - v_{n+2}(x)| + |v_{n+2}(x) - v_{n+3}(x)| + \ldots + |v_{n+p-1}(x) - v_{n+p}(x)|$$
$$= |v_{n+1}(x) - v_{n+2}(x) + v_{n+2}(x) - v_{n+3}(x) + \ldots$$
$$+ v_{n+p-1}(x) - v_{n+p}(x)|$$
$$= |v_{n+1}(x) - v_{n+p}(x)|.$$

Now $|v_{n+1}(x) - v_{n+p}(x)| \leq |v_{n+1}(x)| + |-v_{n+p}(x)|$
$$\leq k + k < 2k. \qquad \ldots(5)$$

Then (4) can be written as

$$|R_{n,\,p}(x)| < \frac{\varepsilon}{3k}.2k + \frac{\varepsilon}{3k}.k = \varepsilon \qquad \ldots(6)$$

i.e., $|u_{n+1}(x).v_{n+1}(x) + \ldots + u_{n+p}(x)v_{n+p}(x)| < \varepsilon \; \forall n \geq m \; \forall x \in [a,b]$.

Hence, from (6), the given series $\sum u_n(x)v_n(x)$ converges uniformly on $[a, b]$.

SOLVED EXAMPLES

EXAMPLE 1. **Test the series** $\sum \dfrac{(-1)^{n-1}}{n}.x^n$ **for uniform convergence in** $[0, 1]$.

SOLUTION. Let us suppose $v_n(x) = x^n$ and $u_n(x) = \dfrac{(-1)^{n-1}}{n}$

Clearly, the sequence $\langle v_n(x)\rangle$ is uniformly bounded and monotonically decreasing on $[0, 1]$.

Also, the series $\sum u_n(x) = \dfrac{\sum (-1)^{n-1}}{n}$ is convergent. Hence, by Abel's test the series

$$\sum u_n(x)v_n(x) = \frac{\sum (-1)^{n-1}}{n} \cdot x^n$$

is uniformly convergent on $[0, 1]$.

THEOREM 5. **(Dirichlet's Test). The series $\sum u_n(x)v_n(x)$ will be uniformly convergent on $[a, b]$ if**

    (i) The sequence $\langle v_n(x) \rangle$ is a positive monotonic decreasing sequence converging uniformly to zero for all $x \in [a, b]$.

    (ii) $f_n(x) = \sum\limits_{r=1}^{n} u_r(x)$ is uniformly bounded in $[a, b]$ i.e.,

$$|f_n(x)| = \left| \sum_{r=1}^{n} u_r(x) \right| < k$$

for every value of x in $[a, b]$ and for all positive integral values of n, where k is a fixed number, independent of x.

PROOF. Proceed as in previous theorem, we have

$R_{n,\,p}(x) = u_{n+1}(x)v_{n+1}(x) + u_{n+2}(x)v_{n+2}(x) + \dots + u_{n+p}(x)v_{n+p}(x)$

$= [f_{n+1}(x) - f_n(x)]v_{n+1}(x) + [f_{n+2}(x) - f_{n+1}(x)]v_{n+2}(x) + \dots$

$\qquad\qquad\qquad + [f_{n+p}(x) - f_{n+p-1}(x)]v_{n+p}(x)$

$= f_{n+1}(x)[v_{n+1}(x) - v_{n+2}(x)] + f_{n+2}(x)[v_{n+2}(x) - v_{n+3}(x)] + \dots$

$\qquad\qquad\qquad + f_{n+p-1}(x)[v_{n+p-1}(x) - v_{n+p}(x)]$

$\qquad\qquad\qquad + f_{n+p}(x)v_{n+p}(x) - f_n(x)v_{n+1}(x). \qquad\qquad \dots(1)$

Now, since $\langle v_n(x) \rangle$ is a positive monotonic decreasing sequence, therefore, $v_1(x)$, $v_2(x)$, $v_3(x), \dots$ are all positive and

$$v_1(x) > v_2(x) > v_3(x) > \dots > v_n(x) > \dots$$

Also $|f_n(x)| < k$ for all x in $[a, b]$ and for all $n \in \mathbf{N}$. $\therefore$ From (1), we have

$|R_{n,p}(x)| \le |f_{n+1}(x)| [v_{n+1}(x) - v_{n+2}(x)] + \dots$

$\qquad\qquad + |f_{n+p-1}(x)| [v_{n+p-1}(x) - v_{n+p}(x)]$

$\qquad\qquad + |f_{n+p}(x)| [v_{n+p}(x)] + |s_n(x)| [v_{n+1}(x)]$

$\qquad < k[v_{n+1}(x) - v_{n+p}(x) + v_{n+p}(x) + v_{n+1}(x)]$

$\qquad = 2k\, v_{n+1}(x). \qquad\qquad\qquad\qquad\qquad\qquad \dots (2)$

Also, since $\langle v_{n+1}(x) \rangle$ converges to zero, we have

$$|v_{n+1}(x)| < \frac{\varepsilon}{2k} \ \forall n \ge m \quad i.e., \quad v_n(x) < \frac{\varepsilon}{2k} \ \forall n \ge m \qquad \dots (3)$$

From (2) and (3), we conclude that

$$|R_{n,p}(x)| < 2k.\frac{\varepsilon}{2k} \text{ for } n \ge m$$

$\Rightarrow \qquad\qquad |R_{n,p}(x)| < \varepsilon \text{ for } n \ge m \ \forall x \in [a, b]$

Hence, the series $\sum u_n(x)v_n(x)$ is uniformly convergent in $[a, b]$.

SOLVED EXAMPLES

EXAMPLE 1. **_Show that the series_**

$$\sum_{n=1}^{\infty} (-1)^{n-1}.x^n$$

converges uniformly in $0 \le x \le k < 1$.

SOLUTION. Let $u_n = (-1)^{n-1}, v_n(x) = x^n$

Since $\quad s_n(x) = \sum_{r=1}^{n} u_r = \begin{cases} 0 & \text{if } n \text{ is even} \\ 1 & \text{if } n \text{ is odd} \end{cases}$

$\Rightarrow s_n(x)$ is bounded for all $n \in \mathbf{N}$.

Also $\langle v_n(x) \rangle$ is positive monotonic-decreasing sequence, converging to zero for all values of x in $0 \le x \le k < 1$.

Hence, by Dirichlet's test, the given series is uniformly convergent in $0 \le x \le k < 1$.

EXAMPLE 2. **_Show that the sequence $\langle f_n \rangle$, where $f_n(x) = \dfrac{nx}{1+n^2 x^2}$ does not converges_**

uniformly on R.

SOLUTION. Here, we have

$$f(x) = \lim_{n \to \infty} f_n(x) = \lim_{n \to \infty} \frac{nx}{1+n^2 x^2} = 0 \ \forall \ x \in \mathbf{R}.$$

Let if possible, the sequence converges uniformly on $\mathbf{R}$, then for a given $\varepsilon > 0$, $\exists$ a positive integer m such that

$$n \ge m, x \in \mathbf{R} \Rightarrow |f_n(x) - f(x)| = \frac{nx}{1+n^2 x^2} < \varepsilon \qquad \qquad \dots (1)$$

If we take $\varepsilon = \dfrac{1}{3}$ and $x = \dfrac{1}{n}(n = 1, 2, 3, \dots)$, then

$$|f_n(x) - f(x)| = \frac{n\dfrac{1}{n}}{1+n^2 \dfrac{1}{n^2}} = \frac{1}{2} < \frac{1}{3} = \varepsilon$$

Thus, there is no single m such that (1) holds simultaneously for all $x \in \mathbf{R}$. For if, such

an m exist, we would have (on taking $n = m$), $|f_m(x) - f(x)| < \dfrac{1}{3} \forall \ x \in \mathbf{R}$

but if we take $x = \dfrac{1}{m}$, we get a contradiction $\left(\because \text{ in this case } \dfrac{1}{2} < \dfrac{1}{3} \right)$ and therefore, the

sequence is not uniformly convergent on $\mathbf{R}$. Also since $\dfrac{1}{m} \to 0$ therefore, 0 is a point of

non-uniform convergence.

EXAMPLE 3. **_Discuss the uniform convergence of the series_**

$$\sum_{n=1}^{\infty} \left[\frac{nx}{1+n^2 x^2} - \frac{(n-1)x}{1+(n-1)^2 x^2} \right]$$

SOLUTION. Here, we have

$$u_1(x) = \frac{x}{1+x^2} - 0$$

$$u_2(x) = \frac{2x}{1+2^2 x^2} - \frac{x}{1+x^2}$$

...   .

$$u_n(x) = \frac{nx}{1+n^2x^2} - \frac{(n-1)x}{(1+(n-1)^2x^2)}$$

On adding, we get

$$f_n(x) = \frac{nx}{1+n^2x^2}$$

Now do same as example (2).

EXAMPLE 4.   **Show that the sequence (f_n) where $f_n(x) = \dfrac{x}{1+nx^2}$ converges uniformly on R.**

SOLUTION.   Here, we have

$$y = \lim_{n\to\infty} \frac{x}{1+nx^2} = 0 \ \forall \, x \in \textbf{R}$$

Let

$$y = f_n(x) - f(x) = \frac{x}{1+nx^2}$$

For maxima and minima of y, we must have $\dfrac{dy}{dx} = 0$

$$\Rightarrow \qquad \frac{(1+nx^2) - 2nx^2}{(1 \ nx^2)^2} \qquad \Rightarrow \qquad \frac{1-nx^2}{(1+nx^2)^2} = 0$$

$$\Rightarrow \qquad\qquad x = \pm\frac{1}{\sqrt{n}}$$

Clearly, $\dfrac{d^2y}{dx^2}$ is negative when $x = \dfrac{1}{\sqrt{n}}$

$$\therefore \qquad \text{Maximum value of} \quad y = \frac{1/\sqrt{n}}{1+n\left(\frac{1}{n}\right)} = \frac{1}{2\sqrt{n}}$$

Also, $\dfrac{1}{2\sqrt{n}} - |y| = \dfrac{1}{2\sqrt{n}} - \dfrac{|x|}{1+nx^2} = \dfrac{1+nx^2 - 2\sqrt{n}\,|x|}{2\sqrt{n}(1+nx^2)}$

$$= \frac{(1-|x|\sqrt{n})^2}{2\sqrt{n}(1+nx^2)} \geq 0$$

Now, $\qquad M_n = \sup_{x\in R}|f_n(x) - f(x)| = \sup_{x\in R}\left|\dfrac{x}{1+nx^2}\right| = \sup_{x\in R}|y|$

$$= \max. y = \frac{1}{2\sqrt{n}} \to 0 \text{ as } n\to\infty$$

Hence, by M_n-test the sequence is uniformly convergent on **R**.

EXAMPLE 5.   **Show that 0 is a point of non-uniform convergence of the sequence $\langle f_n(x)\rangle$ where $f_n(x) = nx\,e^{-nx^2}, x \in R$**

SOLUTION.   Here, we have

$$f(x) = \lim_{n\to\infty} f_n(x) = \lim_{n\to\infty} nx\,e^{-nx^2}$$

$$= \lim_{n\to\infty} \frac{nx}{e^{nx^2}} \qquad\qquad \left| \text{Form}\frac{\infty}{\infty}\right.$$

$$= \lim_{n \to \infty} \frac{x}{x^2 e^{nx^2}} \qquad \text{(By L-Hospital rule)}$$

$$= 0.$$

Let if possible, the sequence be uniformly convergent in a neighbourhood $]0, k[$ of 0, where $k \in \mathbf{N}$.

Then, for a given $\varepsilon > 0$, $\exists$ a positive integer m such that

$$n \geq m, x \in]0, k[\Rightarrow |f_n(x) - f(x)| = nxe^{-nx^2}$$
$$< \varepsilon \qquad \qquad ...(1)$$

In particular, the inequality (1) must be true for $x = \frac{1}{\sqrt{n}}$, where n is a positive integer

greater than m such that

$$0 < \frac{1}{\sqrt{n}} < k.$$

Then (1) gives,

$$\frac{\sqrt{n}}{e} > \varepsilon$$

Now, since $x \to 0$, when $n \to \infty$, we see that on taking x sufficiently near 0, we can take

n so large that $\dfrac{\sqrt{n}}{e} > \varepsilon$, which is a contradiction.

Hence, 0 is a point of non-uniform convergence of the sequence.

Aliter. Let $\qquad\qquad y = f_n(x) - f(x) = nxe^{-nx^2} = 0$

For maxima and minima of y, we must have

$$\frac{dy}{dx} = 0 \Rightarrow ne^{-nx^2} - 2n^2x^2e^{-nx^2} = 0$$

$$\Rightarrow x = \pm \frac{1}{\sqrt{2n}}$$

Also, $\qquad\qquad \dfrac{d^2y}{dx^2} = -ve, \text{ when } x = \dfrac{1}{\sqrt{2n}}$

Therefore, $\qquad$ maximum $y = n - \dfrac{1}{\sqrt{2n}} e^{-n} . \dfrac{1}{2n} = \sqrt{\dfrac{n}{2e}}$

$$\Rightarrow \qquad\qquad M_n = \sup_{x \in R} |f_n(x) - f(x)|$$

$$= \sup_{x \in R} n|x|e^{-nx^2}$$

$$= \sup |y| = \text{Max.} \, y$$

$$= \sqrt{\frac{n}{2e}} \to \infty \text{ as } n \to \infty$$

$\Rightarrow \quad M_n$ does not tends to zero as $n \to \infty$.

Hence by M_n -test, the given sequence is not uniformly convergent.

Also $x \to 0$ as $n \to \infty$, therefore, 0 is a point of non-uniform convergence.

EXAMPLE 6. *Show that if δ is any fixed positive number less than unity, the series*

$\displaystyle\sum_{n=1}^{\infty} \dfrac{x^n}{n+1}$ *is uniformly convergent in $[-\delta, \delta]$.*

SOLUTION.   Let $u_n(x) = x^n$ and $v_n = \dfrac{1}{n+1}$

Since for all $x \in [-\delta, \delta]$, we have $|x| \le \delta < 1$

Therefore,

$$|s_n(x)| = |u_1(x) + u_2(x) + \ldots + u_n(x)|$$

$$\le |x + x^2 + \ldots x^n|$$

$$\le \delta + \delta^2 + \ldots + \delta^n = \frac{\delta(1-\delta^n)}{1-\delta} < \frac{\delta}{1-\delta}$$

Also, $\langle v_n \rangle$ is a positive, monotonic decreasing sequence converging to zero. Hence, by Dirichlet's test, the given series $\sum \dfrac{x^n}{n+1}$ is uniformly convergent in $[-\delta, \delta]$.

EXAMPLE 7.   *Show that the sequence $\langle f_n \rangle$ where $f_n(x) = x^{n-1}(1-x)$ converges uniformly in the interval [0, 1].*

SOLUTION.   Here, we have

$$f(x) = \lim_{n\to\infty} f_n(x) = \lim_{n\to\infty} x^{n-1}(1-x) = 0 \,\forall x \in [0, 1]$$

Let

$$y : |f_n(x) - f(x)| = x^{n-1}(1-x)$$

For maxima or minima of y, we must have $\dfrac{dy}{dx} = 0$

$$\Rightarrow (n-1)x^{n-2}(1-x) - x^{n-1} = 0 \qquad \Rightarrow \qquad x^{n-2}[(n-1)(1-x) - x] = 0$$

$$\Rightarrow \qquad x = 0, \frac{n-1}{n}$$

Also, we can see that $\dfrac{d^2y}{dx^2}$ is negative, when $x = \dfrac{n-1}{n}$

Now

$$M_n = \sup_{x\in[0,1]} |f_n(x) - f(x)| = \sup_{x\in[0,1]} |x^{n-1}(1-x)|$$

$$= \sup_{x\in[0,1]} .|y| = \text{Max. } y$$

$$= \left(1 - \frac{1}{n}\right)^{n-1}\left(1 - \frac{n-1}{n}\right)$$

$$\to \frac{1}{e} \times 0 = 0 \text{ as } n \to \infty.$$

Hence, by M_n-test, the sequence is uniformly convergent on [0,1].

EXAMPLE 8.   *Show that the sequence $\langle f_n \rangle$ where $f_n(x) = \tan^{-1} nx, x \ge 0$ is uniformly convergent in any interval [a, b], a > 0, but is only pointwise convergent in [a, b].*

SOLUTION.   We have

$$f(x) = \lim_{n\to\infty} f_n(x) = \begin{cases} \pi/2 & , \quad x > 0 \\ 0 & , \quad x = 0 \end{cases}$$

Let $\varepsilon > 0$ be given, so that for $x > 0$, we have

$$|f_n(x) - f(x)| = \left|\tan^{-1} nx - \frac{\pi}{2}\right| < \varepsilon$$

$$\Rightarrow \qquad \cot^{-1} nx < \varepsilon$$

i.e., if $n > \dfrac{\cot \varepsilon}{x}$ which clearly decreases with x, the maximum value being $\dfrac{\cot \varepsilon}{a}$ in $[a,b]$.

Let m be any integer $\geq \dfrac{\cot \varepsilon}{a}$ Therefore, for a given $\varepsilon > 0$, $\exists\, m \in N$ such that $\forall x \in [a,b]$.

However, it is clear that $\dfrac{\cot \varepsilon}{x} \to \infty$ as $x \to 0$ so that no such integer m exists such that

$n \geq m \Rightarrow |f_n(x) - f(x)| < \varepsilon$

Hence, the sequence is not uniformly convergent on $[0, b]$.

Also, $\qquad\qquad\qquad n \geq m \Rightarrow |f_n(x) - f(x)| < \varepsilon$

Hence, the sequence converges uniformly on $[a, b]$.

EXAMPLE 9. **Show that the sequence $\langle f_n \rangle$ where $f_n(x) = \dfrac{n}{n+x}, x \geq 0$ is uniformly convergent in any finite interval.**

SOLUTION. Here, we have $f(x) = \lim\limits_{n\to\infty} f(x) = \lim\limits_{n\to\infty} \dfrac{n}{n+x}$ $\qquad$ $\left|\text{Form } \dfrac{\infty}{\infty}\right.$

$$= \lim\limits_{n\to\infty} \dfrac{1}{1+x/n} = 1 \,\forall\, x \geq 0$$

For an arbitrary choosen positive number ε, we have

$$|f_n(x) - f(x)| < \varepsilon$$

if $\qquad\qquad \left|\dfrac{n}{n+x} - 1\right| < \varepsilon$ or *i.e.*, if $\left|\dfrac{-x}{n+x}\right| < \varepsilon$

or *i.e.*, if $\qquad\qquad \dfrac{x}{n+x} < \varepsilon$ or *i.e.*, if $n > x\left(\dfrac{1}{\varepsilon} - 1\right)$

Obviously, n increase with x and tends to ∞ as $x \to \infty$.

Therefore, convergence is not uniform in $[0, \infty[$.

But if $]0, k[$ is any finite interval, where $k > 0$, however large then m is any positive

integer $\geq k\left(\dfrac{1}{\varepsilon} - 1\right)$ such that $n \geq m, x \in [0, k] \Rightarrow |f_n(x) - f(x)| < \varepsilon$

Hence, the sequence is uniformly convergent on $[0, k]$.

EXAMPLE 10. **Let $g_n(x) = \dfrac{1}{n} e^{-nx} (0 \leq x < \infty)$. Show that $\langle g_n \rangle$ converges uniformly to 0 on $[0, \infty[$.**

SOLUTION. Here, we have $\qquad g(x) = \lim\limits_{n\to\infty} g_n(x) = \lim\limits_{n\to\infty} \dfrac{e^{-nx}}{n} = 0$

Define $\qquad\qquad y = |g_n(x) - g(x)| = \left|\dfrac{e^{-nx}}{n} - 0\right| = \dfrac{e^{-nx}}{n}$

Since e^{-nx} is monotonic decreasing with x ($0 \leq x \leq \infty$), therefore,

$$\max. y = \dfrac{1}{n} \qquad\qquad\qquad (\text{where } x = 0)$$

Thus $\qquad\qquad M_n = \sup\limits_{x\in[0,\infty[} |g_n(x) - g(x)| = \sup y = \max. y = \dfrac{1}{n}$

$$\to 0 \text{ as } n \to \infty$$

Hence, by M_n-test, the given sequence $\langle g_n \rangle$ is uniformly convergent on $[0, \infty[$.

EXAMPLE 11. **Show that the series**

$$\dfrac{\cos x}{1^2} + \dfrac{\cos 2x}{2^2} + \dfrac{\cos 3x}{3^2} + \dots$$

converges uniformly on R. Give the interval of uniform convergence.

SOLUTION. Let $\sum_{n=1}^{\infty} u_n(x) = \sum_{n=1}^{\infty} \frac{\cos nx}{n^2}$

Then, we have

$$|u_n(x)| = \left|\frac{\cos nx}{n^2}\right| \le \frac{1}{n^2} \,\forall\, x \in \mathbf{R}$$

Taking $M_n = \frac{1}{n^2}$, the series $\sum M_n = \sum \frac{1}{n^2}$ is convergent. Hence, by Weierstrass's M-test, the given series converges uniformly on $\mathbf{R}$.

Also, the interval of uniform convergence is $a \le x \le b$, where a and b are any finite unequal real numbers.

EXAMPLE 12. **Test the series $\sum \frac{\sin nx}{n^p}$ for uniform convergence in any interval.**

SOLUTION. Let $\sum u_n(x) = \sum \frac{\sin nx}{n^p}$

Now, there are following two cases:

CASE I. Suppose $p > 1$, then we have for all $n \in \mathbf{N}$.

$$|u_n(x)| = \left|\frac{\sin nx}{n^p}\right| \le \frac{1}{n^p} = M_n(\text{say})$$

Now, since $\sum M_n = \sum \frac{1}{n^p}$ is convergent if $p > 1$, therefore, by Weierstrass

M-test, the series $\sum \frac{\sin nx}{n^p}$ is uniformly convergent if $p > 1$.

CASE II. Suppose $0 < p \le 1$.

Take $u_n(x) = \sin nx, v_n = \frac{1}{n^p}$

Then $f_n(x) = \sum_1^n u_n(x) = \sin x + \sin 2x + ... + \sin nx$

$$= \frac{\sin\left\{x + \frac{(n-1)}{2}.x\right\}.\sin\left(\frac{nx}{2}\right)}{\sin\frac{x}{2}}$$

Therefore, $|f_n(x)| \le \operatorname{cosec}\frac{x}{2}$. Also, $\operatorname{cosec}\frac{x}{2}$ becomes infinite if $x = 2m\pi$ for $m = 0, 1, 2,$ ... . Evidently, with the exclusion of these points, $|f_n(x)|$ is bounded uniformly for the remaining points. Also, $\langle v_n \rangle = \left\langle \frac{1}{n^p} \right\rangle$ is monotonic decreasing sequence converging to zero.

Hence, by Dirichlet's test, the given series is uniformly convergent in every interval from which the points $x = 2m\pi$ for $m = 0, 1, 2, 3, ...$ are excluded in case of $0 < p \le 1$.

EXAMPLE 13. **Show that the series $\sum \frac{a_n x^n}{1 + x^{2n}}$ converges uniformly for all real values of x if $\sum a_n$ is absolutely convergent.**

SOLUTION. Let $u_n(x) = \frac{a_n x^n}{1 + x^{2n}}$

For maxima or minima of $u_n(x)$, we must have

$$\frac{du_n}{dx} = 0 \Rightarrow \frac{a_n\left[nx^{n-1}(1+x^{2n}) - 2nx^{2n-1}.x^n\right]}{(1+x^{2n})^2}$$

$$\Rightarrow \qquad 1 + x^{2n} - 2x^{2n} = 0 \Rightarrow x^n = 1.$$

We can easily verify that $\dfrac{d^2u_n}{dx^2} < 0$, when $x^n = 1$ and

$$M_n = \max |u_n| = \max\left|\frac{a_n x^n}{1+x^{2n}}\right| = \frac{|a_n|}{1+1} = \frac{|a_n|}{2}$$

Now, $|u_n(x)| \le M_n$ and $\sum M_n = \dfrac{1}{2}\sum|a_n|$ is convergent as $\sum a_n$ is given to be absolutely convergent. Hence, by Weierstrass M-test, the series $\sum|u_n(x)|$ is uniformly convergent for all $x \in \mathbf{R}$.

EXAMPLE 14. **Show that the series** $\displaystyle\sum_{n=1}^{\infty} \frac{1}{1+n^2x}$ **converges uniformly in** $[1, \infty[$

SOLUTION. Let $\sum u_n(x) = \sum \dfrac{1}{1+n^2x}$. Then, $u_n(x) = \dfrac{1}{1+n^2x}$

$$\Rightarrow \qquad |u_n(x)| \le \frac{1}{1+n^2} \quad \forall x \in [1,\infty]$$

$$< \frac{1}{n^2} = M_n \text{ (say)}.$$

Since, the series $\displaystyle\sum_{n=1}^{\infty} M_n = \sum_{n=1}^{\infty}\frac{1}{n^2}$ is convergent. Hence, by Weierstrass M-test the given series is uniformly convergent on $[1, \infty[$.

EXAMPLE 15. **Test for uniform convergence of the series** $\displaystyle\sum_{n=1}^{\infty} xe^{-nx}$ **in the closed interval [0, 1].**

SOLUTION. Here, we have

$$f_n(x) = \sum_{n=1}^{n-1} xe^{-nx} = x + xe^{-x} + xe^{-2x} + \ldots + xe^{-(n-1)x}$$

$$= \frac{x(1 - 1/e^{nx})}{1 - e^{-x}} = \frac{xe^x}{e^x - 1}\left(1 - \frac{1}{e^{nx}}\right)$$

$$\Rightarrow \qquad f(x) = \lim_{n\to\infty} f_n(x) = \begin{cases} 0 & , \text{ when } x = 0 \\ \dfrac{xe^x}{e^x - 1} & , \text{ when } 0 < x \le 1 \end{cases}$$

For $0 < x < 1$, we have

$$M_n - n \sup_{x\in]0,1]} |f_n(x) - f(x)| = \sup_{x\in]0,1]} \frac{xe^x}{(e^x - 1)e^{nx}}$$

$$\ge \frac{\dfrac{1}{n}.e^{1/n}}{(e^{1/n} - 1)e} \qquad\qquad \left(\text{Taking } x = \frac{1}{n} \in]0,1]\right)$$

Now, $\qquad \displaystyle\lim_{n\to\infty} \frac{(1/n)e^{1/n}}{(e^{1/n} - 1).e} \qquad\qquad \left|\text{Form } \frac{0}{0}\right.$

$$= \lim_{n \to \infty} \frac{(1/n).e^{1/n}\left(-\dfrac{1}{n^2}\right) + \left(-\dfrac{1}{n^2}\right)e^{1/n}}{e.e^{1/n}\left(-\dfrac{1}{n^2}\right)}$$

(By Ľ Hospital Rule)

$$= \lim_{n \to \infty} \frac{\left(\dfrac{1}{n}+1\right)}{e} = \frac{(0+1)}{e} = \frac{1}{e}$$

Therefore, M_n does not tends to zero as $n \to \infty$. Hence, by M_n-test the sequence $\langle f_n(x) \rangle$ is non-uniformly convergent. Also, 0 is a point of non-uniform convergence.

EXAMPLE 16. *The sum to n terms of a series is*

$$f_n(x) = \frac{n^2 x}{1 + n^4 x^2}$$

Show that it converges non-uniformly in the **interval [0, 1].**

Solution. Here, we have

$$f(x) = \lim_{n \to \infty} f_n(x) = \lim_{n \to \infty} \frac{n^2 x}{1 + n^4 x^2} = 0 \,\forall x \in [0, 1],$$

Let if possible, the sequence $\langle f_n(x) \rangle$ converges uniformly on [0,1]. Then, by definition for a given $\varepsilon > 0$, $\exists\, m \in \mathbf{N}$ such that

$$n \geq m, x \in [0,1] \Rightarrow |f_n(x) - f(x)| = \frac{n^2 |x|}{1 + n^4 x^2} < \varepsilon \qquad \qquad ...(1)$$

If $x = \dfrac{1}{n^2} (n \in \mathbf{N})$, then $|f_n(x) - f(x)| = \dfrac{n^2 . \dfrac{1}{n^2}}{1 + n^4 . \dfrac{1}{n^4}} = \dfrac{1}{2}$

If we take $\varepsilon = \dfrac{1}{2}$, there is no single m such that (1) holds simultaneously for all $x \in [0,1]$. For if such m exists, we would have

$$|f_m(x) - f(x)| < \frac{1}{2}, x \in [0,1]$$

In particular, when $x = \dfrac{1}{m^2}$, we get a contradiction

$$\left(\because \text{ in this case we would have } \frac{1}{2} < \frac{1}{2}\right)$$

Hence, convergence is non-uniform on [0,1].

EXAMPLE 17. *Show that the series*

$$\frac{x}{1 + x^2} + \left(\frac{2^2 x}{1 + 2^3 x^2} - \frac{x}{1 + x^2}\right) + \left(\frac{3^2 x}{1 + 3^3 x^2} - \frac{2^2 x}{1 + 2^3 x^2}\right) + \, ...$$

does not converge uniformly on **[0,1].**

SOLUTION. Here, we have

$$u_1(x) = \frac{x}{1 + x^2}$$

$$u_2(x) = \frac{2^2 x}{1 + 2^3 x^2} - \frac{x}{1 + x^2}$$

$$... \quad ... \quad ... \quad ... \quad ... \quad ... \quad ... \quad ... \quad ...$$

$$u_n(x) = \frac{n^2 x}{1+n^3 x^2} - \frac{(n-1)^2 x}{1+(n-1)^3 x^2}$$

On adding, we get

$$f_n(x) = \frac{n^2 x}{1+n^3 x^2}$$

Therefore, $\quad f(x) = \lim_{n\to\infty} f_n(x) = \lim_{n\to\infty} \frac{n^2 x}{1+n^3 x^2}$ $\qquad \left|\text{Form}\,\frac{\infty}{\infty}\right.$

$$= \lim_{n\to\infty} \frac{x}{\dfrac{1}{n^2} + nx^2} = 0,\ \forall x \in [0,1]$$

Now, $\qquad M_n = \sup_{x\in[0,1]} |f_n(x) - f(x)| = \sup_{x\in[0,1]} \frac{n^2 x}{1+n^3 x^2}$

$$= \frac{n^2 \cdot \dfrac{1}{n^{3/2}}}{1 + n^3 \dfrac{1}{n^3}} = \frac{\sqrt{n}}{2} \qquad \left(\text{Taking } x = \frac{1}{n^{3/2}} \in]0,1[\right)$$

$$\to \infty \text{ as } n \to \infty.$$

Since M_n does not tend to zero as $n \to \infty$, the series is non-uniformly convergent on $[0,1]$ by M_n-test. Also, 0 is a point of non-uniform convergence.

EXAMPLE 18. *Show that 0 is a point of non-uniform convergence of the series*

$$\sum_{n=1}^{\infty} \frac{x}{[(n-1)x+1][nx+1]}$$

SOLUTION. Here, we have $\quad u_n(x) = \dfrac{x}{[(n-1)x+1][nx+1]} = \dfrac{1}{(n-1)x+1} - \dfrac{1}{nx+1}$

$$\Rightarrow \qquad u_1(x) = 1 - \frac{1}{x+1}$$

$$u_2(x) = \frac{1}{x+1} - \frac{1}{2x+1}$$

$$\cdots \quad \cdots \quad \cdots \quad \cdots \quad \cdots \quad \cdots \quad \cdots$$

$$u_n(x) = \frac{1}{(n-1)x+1} - \frac{1}{nx+1}$$

On adding, we get $f_n(x) = 1 - \dfrac{1}{nx+1}$

Therefore, $\qquad f(x) = \lim_{n\to\infty} f_n(x) = \begin{cases} 0 & , \quad \text{when } x = 0 \\ 1 & , \quad \text{when } x \neq 0 \end{cases}$

If $x \neq 0$, then we have

$$M_n = \sup_{x\in R_0} |f_n(x) - f(x)| = \sup_{x\in R_0} \frac{1}{|nx+1|}$$

$$\geq \frac{1}{\left|n.\dfrac{1}{n}+1\right|} = \frac{1}{2} \qquad \left(\text{Taking } x = \frac{1}{n}\right)$$

$\Rightarrow \quad M_n$ does not tends to zero $n \to \infty$.

Therefore, by M_n-test the given series is not uniformly convergent. Also zero is a point of non-uniform convergence.

EXAMPLE 19. *Test for uniform convergence of the series*

$$\sum x \left(\frac{n}{1+n^2x^2} - \frac{n+1}{1+(n+1)^2x^2} \right)$$

SOLUTION. Here, we have

$$u_n(x) = x \left\{ \frac{n}{1+n^2x^2} - \frac{n+1}{1+(n+1)^2.x^2} \right\}$$

$$\Rightarrow \qquad u_1(x) = \frac{x}{1+x^2} - \frac{2x}{1+2^2x^2}$$

$$u_2(x) = \frac{2x}{1+2^2x^2} - \frac{3x}{1+3^2x^2}$$

$$\cdots \quad \cdots \quad \cdots \quad \cdots \quad \cdots \quad \cdots \quad \cdots \quad \cdots \quad \cdots \quad \cdots$$

$$u_n(x) = \frac{nx}{1+n^2x^2} - \frac{(n+1)x}{1+(n+1)^2x^2}$$

On adding, we get $f_n(x) = \dfrac{x}{1+x^2} - \dfrac{(n+1).x}{1+(n+1)^2.x^2}$

Therefore, $\qquad f(x) = \lim\limits_{n\to\infty} f_n(x) = \begin{cases} 0 & , \quad \text{when } x = 0 \\ \dfrac{x}{1+x^2} & , \quad \text{when } 0 < x \le 1 \end{cases}$

If $0 < x \le 1$, then, we have

$$M_n = \sup_{x \in]0,1]} |f_n(x) - f(x)| = \sup_{x \in]0,1]} \frac{(n+1).x}{1+(n+1)^2x^2}$$

$$\ge \frac{(n+1).\dfrac{1}{n+1}}{1+(n+1)^2.\dfrac{1}{(n+1)^2}} \qquad \left(\text{Taking } x = \frac{1}{n+1} \in]0,1] \right)$$

$$= \frac{1}{2}.$$

Since M_n does not tends to zero as $n \to \infty$ and also $x \to 0$ as $n \to \infty$ therefore, the sequence $\langle f_n(x) \rangle$ and so the series $\sum u_n(x)$ does not converges uniformly in [0, 1]. Also, zero is a point of non-uniform convergence.

EXAMPLE 20. *If (x) denotes the pointwise or negative excess of x over the greatest integer and if x is midway between two integers, let (x) be zero. Test the uniform convergence of the series*

$$\sum \frac{(nx)}{n^2} = \frac{(x)}{1} + \frac{(2x)}{2^2} + \frac{(3x)}{3^2} + \dots + \frac{(nx)}{n^2} + \dots$$

SOLUTION. We can define

$$(x) = \begin{cases} x & , \quad \text{for } -\dfrac{1}{2} < x < \dfrac{1}{2} \\ 0 & , \quad \text{for } x = \dfrac{1}{2} \\ x-1 & , \quad \text{for } \dfrac{1}{2} < x < 1\dfrac{1}{2} \\ 0 & , \quad \text{for } x = 1\dfrac{1}{2} \\ x-2 & , \quad \text{for } 1\dfrac{1}{2} < x < 2\dfrac{1}{2} \text{ and so on.} \end{cases}$$

$$(2x) = \begin{vmatrix} 2x & , & \text{for} & -\frac{1}{4} < x < \frac{1}{4} \\ 0 & , & \text{for} & x = \frac{1}{4} \\ 2x-1 & , & \text{for} & \frac{1}{4} < x < \frac{3}{4} \\ 0 & , & \text{for} & x = \frac{3}{4} \\ 2x-2 & , & \text{for} & \frac{3}{4} < x < \frac{5}{4} \text{ and so on.} \end{vmatrix}$$

Similarly, we can write the expression for $(3x), (4x),...,$ etc.

Clearly $\qquad |(nx)| < \frac{1}{2}$

We, then have

$$|u_n(x)| = \left| \frac{(nx)}{n^2} \right| < \frac{1}{2n^2} < \frac{1}{n^2}.$$

Since, $\sum \frac{1}{n^2}$ is convergent. Hence by Weierstrass's M-test the series $\sum \frac{(nx)}{n^2}$ is uniformly convergent.

EXAMPLE 21. *Test the uniform convergence of the series* $\sum \dfrac{x}{(n+x^2)^2}.$

SOLUTION. We have $u_n(x) = n^{th}$ term of the series $= \dfrac{x}{(n+x^2)^2}$

For the maxima and minima of $u_n(x)$, we must have $\dfrac{du_n(x)}{dx} = 0$

$$\Rightarrow \qquad \frac{(n+x^2)^2 - 4x^2(n+x^2)}{(n+x^2)^4} = 0$$

$$\Rightarrow \qquad 3x^4 + 2nx^2 - n^2 = 0 \ i.e., x = \pm\sqrt{\frac{n}{3}}$$

We can easily show that at $x = \sqrt{\dfrac{n}{3}}, u_n(x)$ is maximum.

Hence, $\qquad \max . u_n(x) = \dfrac{\sqrt{\dfrac{n}{3}}}{\left(n+\dfrac{n}{3x}\right)^2} = \dfrac{3\sqrt{3}}{16n^{3/2}} = M_n \text{ (say)}$

$$\Rightarrow \qquad |u_n(x)| \leq M_n$$

Since, $\sum M_n$ is convergent, therefore by Weierstrass's M-test, the given series is uniformly convergent for all values of x.

EXERCISE 9.1

1. Test the series $\sum\limits_{n=1}^{\infty} \dfrac{(-1)^{n-1}}{n+x^2}$ for uniform convergence for all values of x.

2. Show that 0 is the point of non-uniform convergence of the sequence $\langle f_n(x) \rangle$ when $f_n(x) = e^{-nx}, x \geq 0$.

3. Show that 0 is a point of non-uniform convergence of the sequence $\langle f_n \rangle$ where $f_n(x) = 1 - (1-x^2)^n$

4. Test for uniform convergence the series

$$\sum_{n=0}^{\infty} a^n \cos nx = 1 + a\cos x + a^2 \cos 2x +$$

$$\ldots + a^n \cos nx \ldots$$

5. Show that the sequence $\langle f_n(x) \rangle$ on $X = [0, 1]$ is convergent on every point of the metric space X but is not uniformly convergent on X, when $f_n(x) = x_n$ and

$$\lim_{n \to \infty} x^n = 0, \text{ when } 0 < x \leq 1$$

and $\lim_{n \to \infty} x^n = 1$, when $x = 1$

6. Show that the sequence $\langle f_n \rangle$ where
$$f_n(x) = x^n (1 - x)$$
converges uniformly in $[0, 1]$.

7. Show that the series

$$\frac{1}{a} - \frac{2a}{a^2 - 1} \cos x + \frac{2a}{a^2 - 2^2} \cos 2x - \ldots \text{ is}$$

uniformly convergent in any finite interval.

8. Show that the series

$$1 + \frac{e^{-2x}}{2^2 - 1} - \frac{e^{-4x}}{4^2 - 1} + \frac{e^{-6x}}{6^2 - 1} - \ldots \quad \text{converges}$$

uniformly for all $x \geq 0$.

9. Show that the series

$$\sum_{n=0}^{\infty} \frac{(-1)^n x^{2n+1}}{(2n+1)!} = x - \frac{x^3}{3!} + \frac{x^5}{5!} - \ldots \quad \text{is}$$

uniformly convergent in every interval.

10. Show that the series

$$\frac{1}{1+x^2} - \frac{1}{2+x^2} + \frac{1}{3+x^2} - \ldots \quad \text{converges}$$

uniformly in the interval $x \geq 0$.

11. Show that the series $\sum \dfrac{x}{n(n+1)}$ is non-

uniformly convergent in $]0, \infty[$.

12. Show that the series

$$\frac{x^2}{1+x} + \left(\frac{2x^2}{1+2x} - \frac{x^2}{1+x} \right) + \ldots$$

$$+ \left(\frac{nx^2}{1+nx} - \frac{(n-1)x^2}{1+(n-1).x} \right) + \ldots$$

converges uniformly on $[0, 1]$.

13. Show that the sequence $\left\langle (\sin x)^{1/n} \right\rangle$

converges but not uniformly on $[0, \pi]$.

14. Show that the sequence $\left\langle \left(\dfrac{\sin x}{x} \right)^{1/n} \right\rangle$ converges but not uniformly on $[0, \pi]$.

15. Show that the series $\sum_{n=1}^{\infty} x^n(1 - x^n)$ is not uniformly convergent on $[0, 1]$

16. Show that the series $\sum_{n=1}^{\infty} \dfrac{1}{n^3 + x^3}$ converges uniformly in $[0, k]$ for $k > 0$.

17. Show that 0 is a point of non-uniform convergence of the series

$$\sum \left\{ \frac{2n^2 x^2}{e^{n^2 x^2}} - \frac{2(n-1)^2 . x^2}{e^{(n-1)^2} . x^2} \right\}$$

18. Show that the series

$$\frac{2x}{1+x^2} + \frac{4x^3}{1+x^4} + \frac{8x^7}{1+x^6} + \ldots \quad \text{is uniformly}$$

convergent in $-1 < x < 1$.

19. Show that $\sum \dfrac{1}{n^p + n^q x^2}$ is uniformly convergent for all values of x if $p + q > 2$.

20. Show that the given series $\sum \dfrac{x}{n(1+nx^2)}$ is uniformly convergent for all x.

21. Show that the series $\sum_{n=1}^{\infty} \dfrac{nx^2}{n^3 + x^3}$ is uniformly convergent in $[0, k]$ for any $k > 0$.

22. Show that the sequence $\left\langle \dfrac{nx}{1+n^2 x^2} \right\rangle$ converges to zero on closed interval $[0, 2]$, but does not uniformly convergent in $]0, 2[$.

23. Show that the series

$$\frac{1}{x+1} - \frac{1}{(x+1)(x+2)} - \frac{1}{(x+2)(x+3)} -$$

$$\ldots - \frac{1}{(x+n-1)(x+n)}$$

converges uniformly in $[0, 1]$.

24. Show that the following series converges uniformly in $[-1, 1]$

(i) $\sum \dfrac{x^n}{n^2}$ (ii) $\sum \dfrac{x^2}{n(n+1)}$

(iii) $\sum \dfrac{x^{2n}}{n + x^{2n}}$

25. Show that the following series converges

uniformly in $[-k, k]$ for all real k.

$\qquad$ (i) $\sum \dfrac{1}{n^4 + n^2 x^2}$ $\qquad$ (ii) $\sum \dfrac{1}{n^2 + n^4 x^2}$

ANSWERS

1. Uniformly convergent for all x. $\qquad$ **4.** Uniformly convergent for $0 < a < 1$.

9.8 UNIFORM CONVERGENCE AND CONTINUITY

THEOREM 1. *Let $f_n\{n = 1, 2, 3, ...,\}$ be the real valued function defined on a metric space (X, d) and let the sequence $\langle f_n \rangle$ converges uniformly to f on X. Let x_0 be the limit point of X, and suppose that $\lim\limits_{x \to x_0} f(x) = c_n (n = 1, 2, 3, ...)$*

Then, the sequence $\langle c_n \rangle$ of real constants converges, and $\lim\limits_{x \to x_0} f(x) = \lim\limits_{n \to \infty} c_n$

or $\qquad \lim\limits_{x \to x_0} \lim\limits_{n \to \infty} f_n(x) = \lim\limits_{n \to \infty} \lim\limits_{x \to x_0} f_n(x).$

PROOF. $\quad$ Let $\varepsilon > 0$ be given. Since $\langle f_n \rangle$ converges uniformly on X, therefore, there exists a positive integer m such that

$$n \geq m, p \geq m, x \in X \Rightarrow |f_n(x) - f(x)| < \varepsilon \qquad \text{... (1)}$$

Letting $x \to x_0$ in (1), we get $|c_n - c_p| < \varepsilon \ \forall \ n \geq m, p \geq m$.
Then by Cauchy's general principle of convergence, we have

The sequence $\langle c_n \rangle$ converges to c i.e., $\lim\limits_{n \to \infty} c_n = c$

Now, since $\langle c_n \rangle$ converges to c and $\langle f_n \rangle$ converges uniformly to f therefore, there exists a positive integer k such that

$$|c_k - c| < \varepsilon / 3 \qquad \text{...(2)}$$

and $\qquad |f_k(x) - f(x)| < \varepsilon / 3 \ \forall \ x \in X. \qquad \text{... (3)}$

Also, $\lim\limits_{x \to x_0} f_k(x) = c_k$ therefore $\exists \ \delta > 0$ such that

$$d(x, x_0) < \delta \Rightarrow |f(x) - c| < \varepsilon$$

$\Rightarrow \qquad \lim\limits_{x \to x_0} f(x) = c = \lim\limits_{n \to \infty} c_n$

$\Rightarrow \qquad \lim\limits_{x \to x_0} \lim\limits_{n \to \infty} f_n(x) = \lim\limits_{n \to \infty} \lim\limits_{x \to x_0} f_n(x)$

☛ **REMARK**

⇒ Let $\sum\limits_{n=1}^{\infty} u_n(x)$ be a series of real valued functions defined on a metric space (X, d) and let

$\lim\limits_{x \to x_0} u_n(x)$ exist $(n = 1, 2, 3, ...)$ where x_0 is a limit point of X. If the series $\sum u_n(x)$

converges uniformly on X, then $\lim\limits_{x \to x_0} \sum\limits_{n=1}^{\infty} u_n(x) = \sum\limits_{n=1}^{\infty} \left[\lim\limits_{x \to x_0} u_n(x) \right]$

THEOREM 2. $\quad$ **(Continuity of limit function).** *Let (f_n) be a sequence of real valued function on a set $X \subseteq R$ which converges uniformly to the function f on X. If each f_n $(n = 1, 2, 3 ...)$ is continuous on X, then f is also continuous on X.*

PROOF. $\quad$ Let $a \in X$ be arbitrary. We shall show that f is continuous at a.
Since each f_n is continuous on $X \quad \Rightarrow \quad f_n$ is continuous at $x = a$.

Also, since $\langle f_n \rangle$ converges uniformly to f on X, therefore, for a given $\varepsilon > 0 \; \exists \; m \in N$ such that $n \geq m \Rightarrow |f_n(x) - f(x)| < \varepsilon / 3 \; \forall \; x \in X$

In particular, we have

$$|f_m(x) - f(x)| < \varepsilon / 3 \qquad \qquad \text{... (1)}$$

and

$$|f_m(a) - f(a)| < \varepsilon / 3 \qquad \qquad \text{... (2)}$$

Again, since f_m is continuous at a, $\exists \; \delta > 0$ such that

$$d(x, a) < \delta \Rightarrow |f_m(x) - f_m(a)| < \varepsilon / 3 \qquad \qquad \text{... (3)}$$

Hence, if $d(x, a) < \delta$, we have

$$|f(x) - f(a)| = |f(x) - f_m(x) + f_m(x) - f_m(a) + f_m(a) - f(a)|$$

$$\leq |f(x) - f_m(x)| + |f_m(x) - f_m(a)| + |f_m(a) - f(a)|$$

$$< \varepsilon / 3 + \varepsilon / 3 + \varepsilon / 3 = \varepsilon.$$

Therefore, for a given $\varepsilon > 0 \; \exists \; \delta > 0$ such that

$$d(x, a) < \delta \Rightarrow |f(x) - f(a)| < \varepsilon$$

Hence, f is continuous at a.

☛ **REMARK**

➠ The above theorem can also be restated as

'*The limit function of a uniformly convergent sequence of continuous function is itself continuous.*'

THEOREM 3.　　**(Continuity of sum function).** *Let* $\overset{\infty}{\Sigma} \; u \; (x)$ *be a series of real valued continuous function defined on a set* $X \subseteq R$. *If the series converges uniformly to the function* $f(x)$ *on* X, *then* $f(x)$ *is continuous on* X.

PROOF.　　Let　　　　$f_n(x) = u_1(x) + u_2(x) + ... + u_n(x)$

By definition of the uniform convergence of the series $\overset{\infty}{\underset{n=1}{\Sigma}} u_n(x)$ depend upon the uniform convergence of the sequence $\langle f_n(x) \rangle$. Since $f_n(x)$ is the sum of finite number of continuous function, therefore, it is continuous. Now, since $\langle f_n(x) \rangle$ is a sequence of continuous function which converges uniformly to $f(x)$. Therefore, for given $\varepsilon > 0$, $\exists \; m \in N$ such that

$$n \geq m \Rightarrow |f_n(x) - f(x)| < \varepsilon / 3 \; \forall \; x \in X \qquad \qquad \text{... (1)}$$

Let $a \in X$ be arbitrary. Then by (1)

$$n \geq m \Rightarrow |f_n(a) - f(a)| < \varepsilon / 3 \qquad \qquad \text{... (2)}$$

Also, by continuity of f_n at $x = a$, we have for a given $\varepsilon > 0 \; \exists \; \delta > 0$ such that

$$d(x, a) < \delta \Rightarrow |f_n(x) - f_n(a)| < \varepsilon / 3 \qquad \qquad \text{...(3)}$$

Therefore,

$$|f(x) - f(a)| = |f(x) - f_n(x) + f_n(x) - f_n(a) + f_n(a) - f(a)|$$

$$\leq |f(x) - f_n(x)| + |f_n(x) - f_n(a)| + |f_n(a) - f(a)|$$

$$< \varepsilon / 3 + \varepsilon / 3 + \varepsilon / 3 = \varepsilon$$

Hence, 　　$d(x, a) < \delta \Rightarrow |f(x) - f(a)| < \varepsilon$

Therefore, $f(x)$ is continuous at $x = a$.

☞ **REMARKS**

➠ The above theorem can be restated as

'*The sum function of a uniformly convergent series of continuous function is itself continuous.*'

➠ Uniform convergence of $\langle f_n \rangle$ is a sufficient but not a necessary condition for the continuity of the sum function *i.e.*, if the limit function is continuous on X, it is not necessary that $\langle f_n \rangle$ converges uniformly on X.

➠ If each f_n is continuous on X and if limit function is not continuous on X, then $\langle f_n \rangle$ does not converge uniformly on X.

➠ The above theorem also shows that if the series of continuous function defined on a metric space X has discontinuous sum, it cannot be uniformly convergent on a subset Y of X which contains a point of discontinuity.

THEOREM 4.   *Let E be a compact set, and let*
     (i) $\langle f_n \rangle$ be a sequence of continuous function on E,
     (ii) $\langle f_n \rangle$ converges pointwise to a continuous function f on E
   (iii) $f_n(x) \geq f_{n+1}(x) \; \forall \; x \in E, \; n = 1, 2, 3, \dots$
     Then, f_n converges uniformly to f on E.

PROOF.   Define $g_n(x) = f_n(x) - f(x)$

Given that f_n and f are continuous, therefore, being the difference of two continuous function, $g_n(x)$ is also continuous.

Also, since $f_n \to f$, we have $g_n = f_n - f \to 0$

Morever, $g_n - g_{n+1} = (f_n - f) - (f_{n+1} - f) = f_n - f_{n+1}$

and therefore $f_n \geq f_{n+1} \Rightarrow g_n \geq g_{n+1}$

Now, to show g_n converges uniformly on E. Let $\varepsilon > 0$ be arbitrary.

Since $g_n \to E$. Therefore, for each $x \in E \; \exists n_x \in N$, such that $|g_{n_x}(x)| < \varepsilon / 2$ ...(1)

Since g_{n_x} is continuous, $\exists$ an open nbd G_x, of $x \in E$ such that for every $y \in G_x$, we have,

$$|g_{n_x}(y) - g_{n_x}(x)| < \varepsilon / 2 \Rightarrow |g_{n_x}(y)| - |g_{n_x}(x)| < \varepsilon / 2$$

$\Rightarrow$ $|g_{n_x}(y)| < |g_{n_x}(x)| + \varepsilon / 2 < \varepsilon / 2 + \varepsilon / 2$ [Using (1)]

Therefore $|g_{n_x}(y)| < \varepsilon$ ... (2)

Now, since $g_n \geq g_{n+1} \; \forall \; n \in N$, we have

$$n \geq n_x \Rightarrow g_{n_x}(x) \geq g_n(x)$$ ... (3)

From (2) and (3), we conclude that $|g_n(y)| < \varepsilon$ for every $y \in G_x$ and $n \geq n_x$.

Clearly, the collection $\{G_x : x \in E\}$ **is an open** cover of E. Since E is compact, there exists a finite set of points $x_1, x_2, \dots x_m$ such that $E \subset \overset{m}{\underset{i=1}{\cup}} G_{x_i}$

Let $n_0 = \max \{n_{x_1}, n_{x_2}, \dots n_{x_m}\}$

Therefore from (2) and (3), we conclude that

$$|g_n(y)| < \varepsilon \text{ for every } y \in E \text{ and } n \geq n_0.$$

∴ g_n converges uniformly on E. Hence f_n converges uniformly to f on E.

SOLVED EXAMPLES

EXAMPLE 1.   **Let $f_n(x) = n^2 x(1-x)^n$, $x \in R$ for each $n \in N$. Show that the limit function f is continuous, but $\langle f_n \rangle$ does not converge to f uniformly.**

SOLUTION.   We have

$$\lim_{n \to \infty} n^2 x(1-x)^n = 0, \text{ if } x \in [0, 1]$$

Therefore, $\langle f_n \rangle$ converges on $[0, 1]$.

For maxima and minima of $f_n(x)$, we must have $\dfrac{df_n(x)}{dx} = 0$ which implies

$$-n^3 x(1-x)^{n-1} + n^2(1-x)^n = 0$$

$$\Rightarrow \qquad n^2(1-x)^{n-1}[-nx + (1-x)] = 0 \Rightarrow x = 1, \frac{1}{n+1}$$

Obviously, $\dfrac{d^2 f_n(x)}{dx^2}$ is negative when $x = \dfrac{1}{n+1} \in [0,1]$.

Therefore, $\qquad \max. f_n(x) = f_n\left(\dfrac{1}{n+1}\right) = n\left(\dfrac{n}{n+1}\right)^{n+1}$

Now $\qquad\qquad M_n = \sup_{x \in [0,1]} |f_n(x) - f(x)| = \sup_{x \in [0,1]} n^2 x(1-x)^n$

$$= \max. \{n^2 x(1-x)^n\} = n\left(\frac{n}{n+1}\right)^{n-1}$$

and $\qquad \lim_{n \to \infty} M_n = \lim_{n \to \infty} n\left(1 - \dfrac{1}{n+1}\right)^{n+1} = \lim_{n \to \infty} \dfrac{n}{e} = \infty \neq 0$

Therefore, by Weierstrass's M-test, the sequence $\langle f_n \rangle$ is not uniformly convergent in $[0,1]$, Hence, $\langle f_n \rangle$ converges to a continuous limit function on $[0,1]$, but $\langle f_n \rangle$ does not converge uniformly on $[0, 1]$.

EXAMPLE 2.   **Show that the sum of the series**

$$\Sigma\left(\frac{nx}{1+n^2 x^2} - \frac{(n-1)x}{1+(n-1)^2 . x^2}\right)$$

is continuous for all values of x, although 0 is a point of non-uniform convergence of the series.

SOLUTION.   Here, we have

$$u_n(x) = \frac{nx}{1+n^2 x^2} - \frac{(n-1)x}{1+(n-1)^2 . x^2}$$

Then, $\qquad f_n(x) = u_1(x) + u_2(x) + \dots + u_n(x)$

$$= \left(\frac{x}{1+x^2} - 0\right) + \left(\frac{2x}{1+2x^2} - \frac{x}{1+x^2}\right) + \dots + \left(\frac{nx}{1+n^2 x^2} - \frac{(n-1).x}{1+(n-1)^2 . x^2}\right)$$

$$= \frac{nx}{1+n^2 x^2}$$

Therefore, $f(x) = \lim_{n \to \infty} f_n(x) = \lim_{n \to \infty} \dfrac{nx}{1+n^2 x^2}$ $\qquad\qquad\left|\text{Form } \dfrac{\infty}{\infty}\right.$

$$= \lim_{n \to \infty} \frac{x}{(1/n) + nx^2} = 0, \forall x$$

Hence, the sum function $f(x)$ is continuous for all values of x.

EXAMPLE 3. *Test for uniform convergence and continuity of the sum function of the series for which* $f_n(x) = nx(1-x)^n$, *for* $0 \le x \le 1$

SOLUTION. Here, we have
$$f_n(x) = nx(1-x)^n, 0 \le x \le 1$$
Therefore, when $0 < x < 1$, we have
$$f(x) = \lim_{n\to\infty} nx(1-x)^n = \lim_{n\to\infty} \frac{nx}{(1-x)^{-n}}$$
$$= \lim_{n\to\infty} \frac{x}{(1-x)^{-n} \log(1-x)}$$
$$= \lim_{n\to\infty} \frac{x.(1-x)^n}{\log(1-x)} = 0$$
Therefore, $\quad f(x) = \lim_{n\to\infty} f_n(x) = 0$, when $0 < x < 1$

Also, $\quad f_n(x) = 0 \, \forall x \in [0,1]$
Hence, the sum function $f(x)$ is continuous for all $x \in [0, 1]$, but the sequence $\langle f_n(x) \rangle$ is not uniformly convergent on $[0,1]$.

EXAMPLE 4. *If the series* $\sum_{n=0}^{\infty} a_n$ *is convergent and has the sum s, then the series*

$\sum_{n=0}^{\infty} a_n x^n$ *is uniformly convergent for* $0 \le x \le 1$ *and* $\lim_{x\to 1^-} \sum_{n=0}^{\infty} a_n x^n = s$

SOLUTION. Given that the series $\sum_{n=0}^{\infty} a_n$ converges to s.
Therefore, by definition, for given $\varepsilon > 0 \, \exists \, m = m(\varepsilon) \in N$ such that
$$\left| \sum_{k=n+1}^{p} a_k \right| < \varepsilon \, \forall p > n \ge m$$
Since $\langle x^n \rangle$ is a monotonic decreasing sequence, then we have
$$\left| \sum_{k=n+1}^{p} a_k x^k \right| < \varepsilon.x^{n+1} < \varepsilon \, \forall x \in [0,1]$$
Therefore, the series $\sum_{n=0}^{\infty} a_n x^n$ is uniformly convergent on $[0,1]$.

Now, for each $n \in N$, $a_n x^n$ is continuous in $[0, 1]$ and $\sum_{n=0}^{\infty} a_n x^n$ is uniformly convergent on $[0, 1]$. Hence, the sum function is also continuous in $[0, 1]$. Therefore
$$\lim_{x\to 1^-} \left(\sum_{n=0}^{\infty} a_n x^n \right) = \sum_{n=0}^{\infty} a_n \left(\lim_{x\to 1^-} x^n \right) = \sum_{n=0}^{\infty} a_n = s$$

EXAMPLE 5. *Examine the continuity of the sum function of the series* $\sum x^2(1-x^2)^{n-1}$ *in the interval* $-\sqrt{2} < -\delta \le x \le \delta < \sqrt{2}$.

SOLUTION. Here, we have $\quad f_n(x) = \sum_{1}^{n} u_n(x) = x^2 \sum_{1}^{n} (1-x^2)^{n-1}$
$$= x^2[1 + (1-x^2) + (1-x^2)^2 + ... + n \text{ terms}]$$
$$= \frac{x^2[1-(1-x^2)^n]}{1-(1-x^2)} = 1 - (1-x^2)^n, \text{if } x \ne 0$$

and
$$f(x) = \lim_{n \to \infty} f_n(x) = 0, \text{if } x = 0$$
$$= 1 \text{ if } 0 < |x| < \sqrt{2}$$
$\Rightarrow f(x)$ is discontinuouns at $x = 0$. Hence, 0 is a point of non-uniform convergence.

9.9 EQUICONTINUOUS FAMILY OF FUNCTIONS

Definition 1. *Let $\langle f_n \rangle$ be a sequence of function defined on a set E. Then $\langle f_n \rangle$ is said to be pointwise bounded on E, if the sequence $\langle f_n(x) \rangle$ is bounded for every $x \in E$ i.e., if there exists a finite-valued function ϕ defined on E such that*

$$|f_n(x)| < \phi(x), x \in E, n \in N$$

Definition 2. *The sequence of functions $\langle f_n \rangle$ defined on a set E is said to be uniformly bounded on E if there exists a number M such that $|f_n(x)| < M, x \in E, n \in N$*

☛ REMARKS

➡ If $\langle f_n \rangle$ is pointwise bounded on E and F a countable subset of E then it is always possible to find a subsequence $\langle f_{n_k}(x) \rangle$ converges for every $x \in E$.

➡ If $\langle f_n \rangle$ is uniformly bounded sequence of continuous function on E, then it is not necessary that there exists a subsequence which converges pointwise on E.

Definition 3. *Let $X \subseteq R$ be non empty. A family F of complex functions f defined on a set E in X is said to be equicontinuous on E if for given $\epsilon > 0 \exists \delta > 0$ such that*

$$x \in E, y \in E, d(x, y) < \delta, f \in F \Rightarrow |f(x) - f(y)| < \epsilon$$

<u>THEOREM I.</u> **Let K be a compact set $f_n \in F(K)$ for n= 1, 2, 3 ... and let $\langle f_n \rangle$ converges uniformly on K, Then $\langle f_n \rangle$ is equicontinuous on K.**

<u>PROOF.</u> Let $\epsilon > 0$ be given. Then by definition of uniformly convergent of $\langle f_n \rangle$, there exists a positive integer N such that
$$n > N \Rightarrow |f_n - f_N| < \epsilon/3 \qquad \dots (1)$$
We know that on the compact metric space, every continuous function is uniformly continuous, therefore $\exists \delta > 0$ such that
$$1 \le i \le N, d(x, y) < \delta \Rightarrow |f_i(x) - f_i(y)| < \epsilon/3 \qquad \dots (2)$$
If $n > N$ and $d(x, y) < \delta$, if follows that
$$|f_n(x) - f_n(y)| = |f_n(x) - f_N(x) + f_N(x) - f_N(y) + f_N(y) - f_n(y)|$$
$$\le |f_n(x) - f_N(x)| + |f_N(x) - f_N(y)| + |f_N(y) - f_n(y)|$$
$$< \epsilon/3 + \epsilon/3 + \epsilon/3$$
$$= \epsilon \qquad \text{[By (1) and (2)]} \qquad \dots(3)$$
From (2) and (3), we conclude that the sequence $\langle f_n \rangle$ is equicontinuous on K.

<u>THEOREM 2.</u> **(Arzele Ascoli Theorem). Let K be a compact metric space, $f_n \in F(K)$ for n = 1,2,3,... and let $\langle f_n \rangle$ be pointwise bounded and equicontinuous on K. Then**

(i) $\langle f_n \rangle$ is uniformly bounded on K.

(ii) $\langle f_n \rangle$ contains a uniformly convergent subsequence.

<u>PROOF.</u> (i) Let ϵ be given. Now, since $\langle f_n \rangle$ is equicontinuous on K, we can choose $\delta > 0$ such that
$$d(x, y) < \delta, x \in K, y \in K \Rightarrow |f_n(x) - f_n(y)| < \delta \, \forall n \qquad \dots (1)$$
Since, K is compact, therefore there are finitely many points $p_1, p_2, ..., p_r$ in K such that to every $x \in K$, there corresponds at least one p_i with $d(x, p_i) < \delta$.

Also, since $\langle f_n \rangle$ is pointwise bounded, $\exists M_i < \infty$ such that
$$| f_n(p_i) | < M_i \ \forall \ n = 1, 2, 3, \ldots$$
Set $M = \max\{M_1, M_2, \ldots M_r\}$. Then, we have
$$| f_n(x) | < M + \varepsilon \ \forall \ x \in K, n \in N$$
$\Rightarrow \langle f_n \rangle$ is uniformly bounded on K.

(ii) Since (X, d) is a compact metric space, so there always exist a countable dense subset. Now, for every positive integer n, there are finitely many neighbourhoods of radius $\dfrac{1}{n}$ whose union covers K. The collection of such neighbourhoods is a countable base for K. Choose one point of K from each member of this countable base. This set is countably dense in K.

Let E be such countable dense subset of K. Then evidently $\langle f_n \rangle$ has a subsequence $\langle f_{n_i} \rangle$ such that $\langle f_{n_i}(x) \rangle$ converges for every $x \in E$.

Put, $f_{n_i} = g_i$ Now, we shall show that $\langle g_i \rangle$ converges uniformly on K.

Let $\varepsilon > 0$ and choose $\delta > 0$ as in (1). Let $V(x, \delta)$ denote the set of all $y \in K$ with $d(x, y) < \delta$. Since E is a dense subset in K and K is compact, there exists finitely many points $x_1, x_2, \ldots x_m$ in E such that
$$K \subset V(x_1, \delta) \cup V(x_2, \delta) \cup \ldots \cup V(x_m, \delta) \quad \ldots(2)$$
Since $\langle g_i(x) \rangle$ converges for every $x_s (s = 1, 2, \ldots m)$, there exist finitely many points $x_1, x_2, \ldots x_m$ in E such that
$$i \geq N, j \geq N, 1 \geq s \geq m \Rightarrow | g_i(x_s) - g_j(x_s) | < \varepsilon / 3. \quad \ldots (3)$$
If $x \geq K$, then (2) shows that $x \in V(x_s, \delta)$ for some s. Therefore
$$| g_i(x) - g_i(x_s) | < \varepsilon / 3 \text{ for every } i.$$
If $i \geq N, j \geq N$, then from (3), we have
$$| g_i(x) - g_j(x) | = | g_i(x) - g_i(x_s) + g_i(x_s) - g_j(x_s) + g_j(x_s) - g_j(x) |$$
$$\leq | g_i(x) - g_i(x_s) | + | g_i(x_s) - g_j(x_s) | + | g_j(x_s) - g_j(x) |$$
$$< \varepsilon / 3 + \varepsilon / 3 + \varepsilon / 3 = \varepsilon$$
We have shown that for a given $\varepsilon > 0$, $\exists$ a positive integer N such that
$$i \geq N, j \geq N, x \in E \Rightarrow | f_{n_i}(x) - f_{n_j}(x) | < \varepsilon$$
$\Rightarrow \langle f_n \rangle$ converges uniformly on K.

Hence, $\langle f_n \rangle$ contains a uniformly convergent subsequence.

9.9.1 BERNSTEIN POLYNOMIALS

If n is a positive integer and k is an integer such that $0 \leq k \leq n$, then the polynomial B_n of each n is defined by

$$B_n(x) = \sum_{k=0}^{n} (^n c_k) x^k (1-x)^{n-k} f(^k c_n) ; x \in [0,1]$$

are called the Bernstein polynomials associated with the real function f

☛ REMARK

➠ Bernstein polynomials continuously converge to f on [0, 1].

9.10 WEIRSTRASS'S APPROXIMATION THEOREM

Let f be a real valued continuous function defined on a closed interval [a, b]. Then there exists a sequence of real polynomials $\langle P_n \rangle$ which converges uniformly to $\langle f_n \rangle$ on [a, b] i.e.,

$$\lim_{n \to \infty} p_n(x) = f_n(x) \text{ uniformly on } [a, b].$$

PROOF.   If $a = b$, then taking $P_n(x)$ to be a constant polynomial, result is obvious. Now assume that $a < b$.

Consider a linear transformation $h(x) = \dfrac{x-a}{b-a}$

Clearly, the transfromation is continuous mapping of [a, b] onto [0, 1]. Without loss of any generality, we may assume $a = 0, b = 1$.

Consider, $g(x) = f(x) - f(0) - x [f(1) - f(0)]$

Then $g(0) = g(1) = 0$

Also, g can be obtained as the limit function of a uniformly convergent sequence of polynomials, it is clear that same is true for f, since $f-g$ is a polynomial.

We define $f(x) = 0$ for x outside [0, 1], Then f would be uniformly continuous on **R**.

Let us define a polynomial (non-negative for $|x| \leq 1$)

$$Q_n(x) = c_n(1-x^2)^n \qquad (n=1, 2, 3, ...) \qquad ...(1)$$

where c_n independent of x, is so chosen that

$$\int_{-1}^{1} Q_n(x)\,dx = 1 \qquad ...(2)$$

Therefore, $1 = \int_{-1}^{1} c_n(1-x^2)^n\, dx = 2c_n \int_{0}^{1}(1-x^2)^n\, dx$

$$\geq 2c_n \int_{0}^{1/\sqrt{n}}(1-x^2)^n\, dx$$

$$\geq 2c_n \int_{0}^{1/\sqrt{n}}(1-nx^2)\, dx$$

($\because$ By Bernaulli's inequality if $y > -1$, then $(1+y)^n \geq 1 + ny \ \forall \ n \in N$)

$$= 2c_n \left[x - \frac{nx^3}{3} \right]_{0}^{1/\sqrt{n}}$$

$$= \frac{4c_n}{3\sqrt{n}} > \frac{c_n}{\sqrt{n}}$$

$\Rightarrow$ $c_n < \sqrt{n}$ ... (3)

Therefore, for any $\delta > 0$, equation (3) gives

$$Q_n(x) \leq \sqrt{n}(1-\delta^2)^n \text{ when } \delta \leq |x| \leq 1 \qquad ... (4)$$

so that $Q_n \to 0$ uniformly in $\delta \leq |x| \leq 1$

Let $P_n(x) = \int_{-1}^{1} f(x+t)Q_n(t)dt$

$$= \int_{-1}^{-x} f(x+t)Q_n(t)dt + \int_{-x}^{1-x} f(x+t)Q_n(t)dt + \int_{1-x}^{1} f(x+t)Q_n(t)dt$$

For $|x| \leq 1, -1+x \leq x+t \leq 0$, for $-1 \leq t \leq -x$ so that $x+t$ lies outside [0, 1] and therefore $f(x+t) = 0$ and hence the first integral on R.H.S vanishes. Similarly, the third integral is also vanishes.

$\therefore$ $P_n(x) = \int_{-x}^{1-x} f(x+t)Q_n(t)dt$

$$= \int_{0}^{1} f(t)Q_n(t-x)dt, \text{ which is a polynomial in } x.$$

Therefore $\langle P_n \rangle$ is a sequence of real polynomials. Now, to show that $\langle P_n(x) \rangle$ converges uniformly to f on $[0,1]$. Since the continuous function defined on a compact set $[0,1]$ is bounded and uniformly continuous, therefore, f is uniformly continuous on $[0,1]$

$\Rightarrow \exists\, M$ such that

$$M = \sup_{x \in [0,1]} |f(x)| \qquad \qquad ...(5)$$

and for any given $\varepsilon > 0$, we can choose $\delta > 0$ such that for any two points $x_1, x_2 \in [0,1]$

$$|f(x_1) - f(x_2)| < \varepsilon / 2, \qquad \text{whenever } |x_1 - x_2| < \delta \le 1. \qquad ... (6)$$

For, $0 \le x \le 1$, we have

$$|P_n(x) - f(x)| = \left| \int_{-1}^{1} f(x+t) Q_n(t)(dt) - f(x) \right|$$

$$= \left| \int_{-1}^{1} \{f(x+t) - f(x)\} Q_n(t)(dt) \right| \qquad \text{[Using (2)]}$$

$$\le \int_{-1}^{1} |f(x+t) - f(x)| Q_n(t)(dt) \qquad (\because Q_n(t) \ge 0)$$

$$= \int_{-1}^{-\delta} |f(x+t) - f(x)| Q_n(t)dt + \int_{-\delta}^{\delta} |f(x+t) - f(x)| Q_n(t)dt$$

$$\qquad \qquad + \int_{\delta}^{1} |f(x+t) - f(x)| Q_n(t)dt$$

$$\le 2M \int_{-1}^{-\delta} Q_n(t)dt + \frac{\varepsilon}{2} \int_{-\delta}^{\delta} Q_n(t)dt + 2M \int_{\delta}^{1} Q_n(t)dt \qquad \text{[By (5) and (6)]}$$

$$\le 2M \sqrt{n}(1-\delta^2)^n \left\{ \int_{-1}^{-\delta} dt + \int_{\delta}^{1} dt \right\} + \frac{\varepsilon}{2} \qquad \text{[By (2) and (4)]}$$

$$\le 4M \sqrt{n}(1-\delta^2)^n + \frac{\varepsilon}{2} + \varepsilon, \text{ for large } n.$$

Therefore, for any given $\varepsilon > 0 \,\exists\, N$, such that

$$|P_n(x) - f(x)| < \varepsilon \;\forall\; n \ge N$$

$\Rightarrow \qquad \lim_{n \to \infty} P_n(x) = f(x)$ uniformly on $[0, 1]$.

SOLVED EXAMPLES

EXAMPLE I.   **Show that the series, for which**

$$f_n(x) = \frac{nx}{1 + n^2 x^2}, 0 \le x \le 1$$

cannot be differentiated term by term at x = 0.

SOLUTION.   For $0 \le x \le 1$

$$f(x) = \lim_{n \to \infty} f_n(x) = \lim_{n \to \infty} \frac{nx}{1 + n^2 x^2} = 0$$

Also, $\qquad f_n'(0) = \lim_{h \to 0} \frac{f_n(0+h) - f_n(0)}{h}$

$$= \lim_{h \to 0} \frac{\frac{nh}{1 + n^2 h^2} - 0}{h}$$

$$= \lim_{h \to 0} \frac{n}{1 + n^2 h^2} = n \to \infty \text{ as } n \to \infty$$

Also, $\qquad f'(0) = 0$

Hence $\qquad f'(0) \neq \lim_{n \to \infty} f_n'(0)$

which shows that the given series cannot be differentiated term by term at $x = 0$.

EXAMPLE 2. *Show that the series*

$$\sum_{n=1}^{\infty} \frac{1}{n^3 + n^4 x^2}, x \in R$$

uniformly convergent for all values of x and is differentiable term by term.

SOLUTION. Let

$$u_n(x) = \frac{1}{n^3 + n^4 x^2}$$

Since, for all values of x,

$$u_n(x) = \frac{1}{n^3} \cdot \left(\frac{1}{1 + nx^2}\right) \le \frac{1}{n^3}$$

and $\sum \dfrac{1}{n^3}$ is convergent, then by Weierstrass M-test the series $\sum u_n$, converges uniformly.

Now

$$u_n'(x) = \frac{1}{n^3} \cdot \frac{-2nx}{(1 + nx^2)^2} = -\frac{2}{n^2} \cdot \frac{x}{(1 + nx^2)^2}$$

we know that $\dfrac{2x}{n^2 (1 + nx^2)^2}$ maximum if $2(1 + nx^2) - 2x.2(1 + nx^2).2nx = 0$

or if

$$1 + nx^2 - 4nx^2 = 0 \Rightarrow x = \frac{1}{\sqrt{3n}}$$

Therefore, maximum value of $\dfrac{2x}{n^2 (1 + nx^2)^2} = \dfrac{2.\dfrac{1}{\sqrt{3n}}}{n^2 \left(1 + \dfrac{1}{3}\right)^2} = \dfrac{3\sqrt{3}}{8n^{5/2}}$

$$\Rightarrow \qquad |u_n'(x)| \le \frac{3\sqrt{3}}{8} . \sum \frac{1}{n^{5/2}}$$

$$\le M_n, \text{ where } M_n = \frac{3\sqrt{3}}{8} . \frac{1}{n^{5/2}}$$

But $\displaystyle\sum_{n=1}^{\infty} M_n = \frac{3\sqrt{3}}{8} \sum \frac{1}{n^{5/2}}$ is absolutely convergent.

Hence, by Weierstrass's M-test, series $\sum u_n'(x)$ is uniformly convergent in all finite intervals. Therefore given series is term by term differentiable.

EXAMPLE 3. *If $f_n(x) = \dfrac{x^2}{x^2 + (1 - nx)^2}, (0 \le x \le 1), n = 1, 2, 3 \dots$ show that*

(i) *the sequence $\langle f_n \rangle$ is uniformly bounded on $[0,1]$.*

(ii) *no subsequence of $\langle f_n(x) \rangle$ can converge uniformly on $[0,1]$.*

(iii) *the sequence $\langle f_n(x) \rangle$ is not equicontinuous on $[0,1]$.*

SOLUTION. (i) Since, we have $x^2 + (1 - nx)^2 \ge x^2$

Therefore, $|f_n(x)| = \dfrac{x^2}{x^2 + (1 - nx)^2} \le 1 \ (0 \le x \le 1, n = 1, 2, 3 \dots)$

$$\Rightarrow \qquad \langle f_n \rangle \text{ is uniformly bounded on } [0,1].$$

(ii) Let $\langle f_{n_k} \rangle$ be any subsequence of $\langle f_n \rangle$, then

$$f_{n_k}(x) = \frac{x^2}{x^2 + (1 - n_k x^2)} = \frac{1}{1 + \left(\dfrac{1}{x} - n_k\right)^2} \quad (0 \le x \le 1, k = 1, 2, 3, ...)$$

Now, we have

$$\lim_{k \to \infty} f_{n_k}(x) = 0, 0 \le x \le 1$$

Also

$$f_{n_k}(x) = 0 \ \forall k$$

$\therefore$

$$\lim_{k \to \infty} f_{n_k}(x) = 0, \forall \ k$$

Further if

$$f(x) = \sum_{n=1}^{\infty} \frac{1}{n^3 + n^4 x^2}$$

Then

$$f'(x) = \sum_{n=1}^{\infty} u_n'(x) = -2x \sum_{n=1}^{\infty} \frac{1}{n^2 (1 + n x^2)^2}$$

Hence, the subsequence $\langle f_{n_k} \rangle$ converges pointwise to 0 on [0, 1] but

$$f_{n_k}\left(\frac{1}{n_k}\right) = 1, k = 1, 2, 3, ...$$

$\Rightarrow$ $\langle f_{n_k} \rangle$ cannot converge uniformly to 0 on [0, 1]. If we take $\varepsilon = \dfrac{1}{2}$, then for

$x = \dfrac{1}{n_k}$ we have $\left| f_{n_k}\left(\dfrac{1}{n_k}\right) - 0 \right| = |1 - 0| = 1 > \varepsilon \quad (k = 1, 2, 3, ...)$

Hence, no subsequence of $<f_n(x)>$ can converge uniformly on [0, 1].

(iii) To show that $\langle f_n \rangle$ is not equicontinuous on [0,1]. For this, we must find an $\varepsilon > 0$ corresponding to which there exists no $\delta > 0$ satisfying the definition of equicontinuity.

Let us take $\varepsilon = \dfrac{1}{4}, x = \dfrac{1}{n}$ and $y = \dfrac{1}{n+1}$. Then

$$|x - y| = \left| \frac{1}{n} - \frac{1}{n+1} \right| = \frac{1}{n(n+1)}$$

Since

$$f_n(x) = \frac{x^2}{x^2 + (1 - nx)^2} = \frac{1}{1 + \left(\dfrac{1}{x} - n\right)^2}$$

$\Rightarrow$ $\qquad f_n\left(\dfrac{1}{n}\right) = 1 \qquad$ and $\qquad f_n\left(\dfrac{1}{n+1}\right) = \dfrac{1}{1 + \{(n+1) - n\}^2} = \dfrac{1}{2}$

$\Rightarrow \left| f_n\left(\dfrac{1}{n}\right) - f_n\left(\dfrac{1}{n+1}\right) \right| = \left| 1 - \dfrac{1}{2} \right| = \dfrac{1}{2}$

$\Rightarrow$ For $\delta > 0$, we can always find a positive integer n such that $\dfrac{1}{n(n+1)} < \delta$.

Therefore, in this ease, we have $|x - y| < \delta$ but $|f_n(x) - f_n(y)| < \varepsilon$

Hence, the sequence $\langle f_n \rangle$ is not equicontinuous on [0, 1].

EXERCISE 9.2

1. Show by an example that for term by term differentiation, the condition of uniform convergence is sufficient but not necessary.

2. Show that the series $\Sigma \dfrac{1}{n^2 + n^4 . x^2}$ is uniformly convergent for all real values of x and that it can be differentiated term by term.

3. Show that the series

$$a^x = 1 + \frac{\log a}{1!}.x + \frac{(\log a)^2}{2!}.x^2 + ...$$
$$+ \frac{(\log a)^{n-1}}{(n-1)!} x^{n-1} + ...$$

can be integrated and differentiated term by term.

4. Let $f_n(x) = \dfrac{1}{2n^2} \log(1 + n^4 x^2)$.

Show that the series $\Sigma u'(x)$ does not converges uniformly but the given series can be differentiated term by term.

5. Show that the function represented by $\sum\limits_{n=1}^{\infty} \dfrac{\sin nx}{n^3}$ is differentiable for every x and its derivative is $\sum\limits_{n=1}^{\infty} \dfrac{\cos nx}{n^2}$

6. Let $\left\langle f_n(x) = \dfrac{x}{1 + nx^2} \right\rangle$, where $n = 1,2,3,...$ and x is real. Show that $\langle f_n \rangle$ converges uniformly to a function f and the equation $f'(x) = \lim\limits_{n \to \infty} f_n'(x)$ is correct if $x \ne 0$ and false if $x = 0$.

7. Show that $f(x) = \Sigma \dfrac{1}{n^p + n^q . x^2}, p > 1$ is uniformly convergent for all values of x and can be differentiated term by term if $q < 3p - a$.

8. Show that the series
$$\sin x = x - \frac{x^3}{3!} + \frac{x^5}{5!} - \frac{x^7}{7!} + ...$$
can be differented to obtained the expansion of cos x.

9. If f is continuous on [0, 1] and if $\int_0^1 x^n f(x)dx = 0$ for $n = 0, 1, 2, ...$ show that $f(x) = 0$ on [0, 1].

10. Show that, if f is continuous on **R**, then there exists a sequence $\langle P_n \rangle$ of polynomials converging uniformly to f each bounded subset of **R**.

REVIEW QUESTIONS AND ARCHIVE

1. Define the following :
 (i) Pointwise convergence
 (ii) Uniform convergence
 (iii) Dini's criteria
 (iv) Weirstrass approximation theorem

2. Show that the sequence $\langle f_n \rangle$ where $f_n(x) = \dfrac{x}{n + x}$ is uniformly convergent in $[0, k]$, $k < \infty$ but only pointwise convergent when the interval extends to ∞.

3. Show that the series
$$\frac{x}{1+x} + \frac{x}{(1+x)(1+2x)} + \frac{x}{(1+2x)(1+3x)} + ...$$
is uniformly convergent on [a, b] a > 0 but only pointwise in [0, b].

4. Show that the series $\Sigma \dfrac{x}{n^p + x^2 n^q}$ converges uniformly over any finite interval [a, b] for
 (i) $p > 1, q \ge 0$
 (ii) $0 < p \le 1, p + q > 2$

5. Show that the series $\Sigma(-1)^n \dfrac{x^2 + n}{n^2}$ converges uniformly in every bounded interval but does not converge absolutely for any value of x.

6. Show that $\Sigma \dfrac{\log n}{n^x}$ converges uniformly for all real $x > 1 + \alpha > 1$.

7. Show that $\Sigma \dfrac{1}{n^x}$ converges uniformly for all real $x > 1$.

8. Show that the series $e^x + e^{2x} + e^{3x} + ..., |x| \le \dfrac{1}{4}$ converges uniformly.

9. Show that the sequence $\left\langle (\sin x)^{1/n} \right\rangle$ converges but not uniformly on $[0, \pi]$.

10. If f is continuous on $[0, 1]$ and if $\int_0^1 x^n f(x)dx = 0$ for $n = 0,1,2...$ then show that $f(x) = 0$ on $[0, 1]$.

OBJECTIVE EVALUATION

▶ FILL IN THE BLANKS

1. The term of a sequence need not be _____.

2. Every convergent sequence in a _____ cauchy.

3. The sequence $\langle f_n \rangle$ where $f_n = nx(1-x)^n$ _____ converges uniformly.

4. The series which satisfy M_n-test is called _____ convergent.

▶ TRUE / FALSE

1. A point such that the sequence does not converges uniformly in any nbd of it, however small is said to be a point of non-uniform convergence. **(T/F)**

2. Uniform convergence of a sequence $\langle f_n(x) \rangle$ is a property associated with a domain of x comprising more than one point whereas simple convergence is meaningful at each point of that domain. **(T/F)**

3. The series which satisfy M_n-test is called normally convergent. **(T/F)**

▶ MULTIPLE CHOICE QUESTIONS (CHOOSE THE MOST APPROPRIATE ONE)

1. A sequence $\langle f_n \rangle$ of real valued functions is said to be uniformly bounded if :
   (a) $f_n(x) < M$
   (b) $|f_n(x)| \le M$
   (c) $f_n(x) \ge M$
   (d) none of the above

2. If a sequence $\langle f_n \rangle$ is convergent then limit function f of $< f_n (x) >$ is called
   (a) sum function
   (b) product function
   (c) convergent function
   (d) none of the above

3. The convergence of the sequence ensures it:
   (a) uniform convergence
   (b) not necessarily uniform convergence
   (c) pointwise convergence
   (c) both (a) and (b) are true

4. A point such that the sequence does not converge uniformly in any neighbourhood is called :
   (a) limit point
   (b) peak point
   (c) point of non-uniform convergence

   (d) none of the above

5. If $\lim_{n \to \infty} f_n(x) = f(x) \forall\ x \in X$ and $M_n = \sup\{|f_n(x) - f(x)| : x \in X\}$, then $\langle f_n \rangle$ converges uniformly to f if and only if
   (a) $M_n = 0 \ \forall\ n$
   (b) $M_n \to 0$ as $n \to 0$
   (c) $M_n \to 0$ as $n \to \infty$
   (d) none of the above

6. A series $\Sigma u_n(x)$ of functions will converge uniformly on x if there exists a convergent series ΣM_n of positive constant such that:
   (a) $u_n (x) = M_n$
   (b) $|u_n (x)| \ge M_n$
   (c) $|u_n(x)| \le M_n$
   (d) none of the above

7. If the given series $\Sigma u_n(x)$ converges uniformly, then :
   (a) f is integrable
   (b) $\int_a^b \left[\sum_{n=1}^{\infty} u_n(x) \right] dx = \sum_{n=1}^{\infty} \int_a^b u_n(x)dx$
   (c) both (a) and (b) are true

(d) none of the above

8. The condition of uniform convergence of the series $\Sigma u_n(x)$ for the validity of term by term integration is:

(a) only necessary condition

(b) only sufficient condition

(c) necessary and sufficient both

(d) none of the above

ANSWERS

▶ **FILL IN THE BLANKS**

1. distinct   **2.** not necessarily   **3.** does not   **4.** normally

▶ **TRUE/FALSE**

1. T   **2.** T   **3.** T

▶ **MULTIPLE CHOICE QUESTIONS**

1. (b)   **2.** (a)   **3.** (d)   **4.** (c)   **5.** (c)   **6.** (c)   **7.** (c)   **8.** (b)

CHAPTER SUMMARY

⮑ Every uniformly convergent sequence is point-wise convergent and the uniform limit function is same as the pointwise limit function.

⮑ Uniform convergence implies pointwise converegence but not vice-versa.

⮑ A sequence which is not pointwise convergent cannot be uniformly convergent *i.e.,* non-pointwise convergence implies non-uniform convergence.

⮑ (*Cauchy criterion for uniform convergence*). A sequence of functions $\langle f_n(x)\rangle$ converges uniformly on $\lfloor a, b \rfloor$ if and only if for given $\varepsilon > 0$ and for all $x \in [a, b]$ ∃ an integer m such that $|f_{n+p}(x) - f_n(x)| < \varepsilon \ \forall\, n \geq m, p > 1$

⮑ Let $\langle f_n \rangle$ be a sequence of functions such that

$$\lim_{n\to\infty} f(x) = f(x), x \in [a,b] \qquad \text{and} \qquad \text{let}$$

$$M_n = \sup_{x\in[a,b]} |f_n(x) - f(x)|. \text{ Then } f_n \to f$$

uniformly on $[a, b]$ if and only if $M_n \to 0$ as $n \to \infty$.

⮑ (*Weirstrass's M-test*): A series of function Σf_n will **converge uniformly and absolutely** on $[a, b]$ if there exists a convergent series ΣM_n of positive numbers such that for all

$x \in [a, b]$, $|f_n(x)| \leq M_n \ \forall\, n$.

⮑ *Abel's test:* The series $\Sigma u_n(x)\, v_n(x)$ will converges uniformly in $[a, b]$ if

(i) $\Sigma u_n(x)$ is uniformly convergent in $[a, b]$

(ii) The sequence $\langle v_n(x) \rangle$ is monotonic for every $x \in [a, b]$

(iii) The sequence $\langle v_n(x) \rangle$ is uniformly bounded in $[a, b]$ by k *i.e.,* $|v_n(x)| < k \ \forall\, x \in (a,b)$

⮑ A uniformly convergent series $\Sigma u_n(x)$ remains uniformly convergent on $[a, b]$ if its each term is multiplied by a function $a_n(x)$, $a \leq x \leq b$ provided that the sequence $<a_n(x)>$ is uniformly bounded on (a, b).

⮑ If $\sum\limits_{n=1}^{\infty} a_n x^n$ is a series which converges for all valued of x when $|x| < R$ then $\sum\limits_{n=1}^{\infty} a_n x^n$ is uniformly convergent in $[0, R]$ if and only if $\Sigma a_n R^n$ is convergent.

⮑ *Dirichlet's test:* The series $\Sigma u_n(x)\, v_n(x)$ will be uniformly convergent on $[a, b]$ if

(i) The sequence $\langle v_n(x) \rangle$ is a positive monotonic decreasing sequence converging uniformly to zero for all $x \in (a, b)$.

(ii) $f_n(x) = \sum\limits_{r=1}^{n} u_r(x)$ is uniformly bounded in (a, b).

➲ If $v_n(x)$ is a monotonic function of n for each fixed value of $x \in [a, b]$ and $b_n(x)$ tends to zero for $a \le x \le b$ and if $\sum u_n(x)$ either uniformly converges or oscillate finitely in $[a, b]$ then the series $\sum u_n(x)V_n(x)$ is uniformly convergent on $[a, b]$.

➲ If $\langle v_n \rangle$ is a monotonic sequence of real numbers that converges to zero, then each of the series $\sum v_n \sin n\theta$, $\sum v_n \cos n\theta$ is uniformly convergent with regard to θ in the interval $[\alpha, 2\pi - \alpha]$ where α is any fixed positive number less than π.

➲ If a sequence $\langle f_n(x) \rangle$ converges uniformly on $[a, b]$ and $x_0 \in [a, b]$ such that $\lim\limits_{x \to x_0} f_n(x) = a_n, n = 1, 2, 3, \ldots$

then (i) $\langle a_n \rangle$ converges

and (ii) $\lim\limits_{x \to x_0} f(x) = \lim\limits_{n \to \infty} a_n$

➲ The limit of the sum function of a series = the sum of the series of limit of functions.

➲ If $\langle f_n \rangle$ is a sequence of continuous function on an interval $[a, b]$ and if $f_n \to f$ uniformly on $[a, b]$ then f is continuous on $[a, b]$.

➲ If a series $\sum f_n$ converges uniformly to f in an interval $[a, b]$ and its term f_n are continuous at a point x_0 of the interval then the sum function f is also continuous at x_0.

➲ If the sum function of a series or limit function of a sequence of continuous terms is not continuous on an interval the convergence can not be uniform.

➲ *Dini's theorem* :

(i) If a sequence of continuous function $\langle f_n \rangle$ defined on $[a, b]$ is monotonic increeasing and converges pointwise to a continuous function f then the convergence is uniform on $[a, b]$.

(ii) If the sum function of a series $\sum f_n$ with non-negative continuous terms defined on an interval $[a, b]$ is continuous on $[a, b]$ then the series is uniformly convergent on $[a, b]$.

➲ *Weierstrass Approximation theorem*: If f is a real valued continuous function defined on a closed interval $[a, b]$ then there exists a sequence of real polynomials $\langle P_n \rangle$ which converges uniformly to $f(x)$ on $[a, b]$ i.e., $\lim\limits_{x \to 0} P_n(x) = f(x)$ converges uniformly on $[a, b]$.

➲ For any interval $[-a, a]$ there is a sequence of real polynomials P_n such that $P_n(0) = 0$ and that   uniformly on $[-a, a]$.

FOR ADVANCED LEARNERS

- The choice of m in the definition of pointwise convergence of a sequence of function is made after x and ε have been chosen so that in general m depends on x and ε.

- The pointwise limit of sequence of continuous function need not be continuous.

- $C[a, b]$ denote the set of all continuous real valued functions defined on $[a, b]$.

- The important fact about the metric space $C[a, b]$ is that convergence in this metric space coincides with uniform convergence.

- Let $<f_n>$ be the sequence in $C[a, b]$. Let $f \in C[a, b]$ then $<f_n>$ converges to f in $C[a, b]$ if and only if $<f_n>$ converges to f uniformly on $[a, b]$.

- Since, convergence in the metric space $C[a, b]$ is the same as uniform convergence, Weirstrass approximation theorem show that the set of all polynomials defined on $[a, b]$ is a dense subset of $[a, b]$.

- If $<f_n> \to f$ uniformly on X and each f_n is continuous on X then f is continuous on X.

- The uniform limit of a sequence of discontinuous function can be continuous.

- $C[a, b]$ is a complete metric space.

CHAPTER 10

Functions from R² to R

10.1 INTRODUCTION

Let f be a function from a set of ordered pair of real numbers to a set of real numbers; then f is said to be a real valued function of two real variables or, briefly, a real function of two variables. The value that f assumes at the arguments (x, y) is naturally written as $f(x, y)$. Let us suppose this value is called z. Then we write $z = f(x, y)$, where x and y are the independent variables and z is the dependent variable.

We shall write $z = z(x, y)$, which means that we are considering some function of two variables, where the independent variables are x and y and the dependent variable is z.

If to each pair of values of x and y there exists only one value of z, then the function is said to be single valued function. On the other hand, if there are two or more values of z correspond to some x and y or all of the values assigned to x and y, the function is called multiple valued.

Definition 1. *Let $f(x, y)$ be a function of two variables x and y, then we say* $\lim\limits_{\substack{x \to x_0 \\ y \to y_0}} f(x, y)$ *exists and is equal to l, if for every $\varepsilon > 0$, $\exists$ a $\delta > 0$ such that $|f(x, y) - l| < \varepsilon$ for all values of x and y in the neighbourhood of (x_0, y_0) defined by*
$$|x - x_0| < \delta, \ |y - y_0| < \delta.$$

10.2 LIMIT

Let $f(x, y)$ is a function of two variables x and y, we define several kind of limits. If (x_0, y_0) is the limiting point of a set of values on two dimensional space, then we have the following limits
$$\lim_{\substack{x \to x_0 \\ y \to y_0}} f(x, y), \ \lim_{x \to x_0} \lim_{y \to y_0} f(x, y), \ \lim_{y \to y_0} \lim_{x \to x_0} f(x, y).$$

Then limit of the first kind is known as simultaneous limit and the last two types are known as iterated limits.

☛ **REMARK**

⟹ It is not necessary that, the two repeated limits, even if they exist, may be equal *e.g.*, consider a function
$$f(x, y) = \frac{x^2 - y^2}{x^2 + y^2}; \quad (x, y) \neq (0, 0)$$

We have $\lim\limits_{x \to 0} f(x, y) = -1 \forall y$ $\Rightarrow$ $\lim\limits_{y \to 0}\left[\lim\limits_{x \to 0} f(x, y)\right] = -1$

Now $\lim\limits_{y \to 0} f(x, y) = 1 \ \forall x$ $\Rightarrow$ $\lim\limits_{x \to 0}\left[\lim\limits_{y \to 0} f(x, y)\right] = 1$

Thus, while the two repeated limit exist in this case, they are unequal. It may also be seen that $\lim\limits_{(x, y) \to (0, 0)} f(x, y)$ does not exist.

10.2.1 NON-EXISTENCE OF A LIMIT

To determine whether a simultaneous limit exists or not, it is a difficult matter but a simple consideration, which we describe, says us to decide about the non-existence of a limit.

If $\lim\limits_{(x,y)\to(a,b)} f(x,y) = l$ and if ϕ is any function of a single real variable such that

$$\lim_{x\to a} \phi(x) = b$$

Then

$$\lim_{x\to a} f[x, \phi(x)] = l$$

Thus, we can determine two functions ϕ_1 and ϕ_2 such that

$$\lim_{x\to a} f[x, \phi_1(x)] \neq \lim_{x\to a} f[x, \phi_2(x)]$$

Then, we can say that the simultaneous limit $\lim\limits_{(x,y)\to(a,b)} f(x,y)$ does not exist.

SOLVED EXAMPLES

EXAMPLE 1. **Show that** $\lim\limits_{(x,y)\to(0,0)} f(x,y)$, **where** $f(x,y) = \dfrac{x^2 - y^2}{x^2 + y^2}$ **does not exist.**

SOLUTION. Here $f(x,y) = \dfrac{x^2 - y^2}{x^2 + y^2}$

Now taking $y = mx$, then we have $\lim\limits_{x\to0} f(x,mx) = \dfrac{1 - m^2}{1 + m^2}$, which depend upon m. Therefore it is not unique.

Since, $\lim\limits_{x\to0} f(x,mx)$ is not unique. Hence $\lim\limits_{(x,y)\to(0,0)} f(x,y)$ does not exist.

EXAMPLE 2. **Show that the simultaneous limit,** $\lim\limits_{\substack{x\to0\\y\to0}} \dfrac{xy^3}{x^2 + y^6}$ **does not exist.**

SOLUTION. Let (x, y) tends to $(0, 0)$ through the line $y = x$, which is a line through the origin. Put $y = x$, in the given function, we get

$$\lim_{x\to0} \frac{x^4}{x^2 + x^6} = \lim_{x\to0} \frac{x^2}{1 + x^4} = 0$$

Again, let $(x, y) \to (0, 0)$ through the curve $x = y^3$.

Put $x = y^3$, in the given function, we obtain $\lim\limits_{y\to0} \dfrac{y^6}{y^6 + y^6} = \dfrac{1}{2}$

$\Rightarrow$ The limit obtained by two different methods are different. Hence, the simultaneous limit does not exist.

EXAMPLE 3. **Let** $f(x,y) = \dfrac{y-x}{y+x} \cdot \dfrac{1+x}{1+y}, (x,y) \neq (0,0)$ **Show that two repeated limits exist at origin but are unequal.**

SOLUTION. We have

$$\lim_{x\to0} \lim_{y\to0} f(x,y) = \lim_{x\to0} -\left(\frac{1+x}{1}\right) = -1$$

Again, $\lim\limits_{y\to0} \lim\limits_{x\to0} f(x,y) = \lim\limits_{y\to0}\left(\dfrac{1}{1+y}\right) = 1$

Therefore, the two repeated limits exist at the origin are not equal.

EXAMPLE 4. **Show that** $\lim\limits_{(x,y)\to(0,0)}\dfrac{3x-2y}{2x-3y}$ **does not exist.**

SOLUTION. When $(x,y)\to(0,0)$ along the straight line $y=x$, we have

$$\lim_{(x,y)\to(0,0)} f(x,y)=\lim_{x\to0}\frac{3x-2x}{2x-3x}=\lim_{x\to0}\frac{x}{-x}=\lim_{x\to0}-1=-1$$

Now, again when $(x,y)\to(0,0)$ along the straight line $y=0$, we have

$$\lim_{(x,y)\to(0,0)} f(x,y)=\lim_{x\to0}\frac{3x-0}{2x-0}=\lim_{x\to0}\frac{3x}{2x}=\lim_{x\to0}\frac{3}{2}=\frac{3}{2}$$

Since two methods of approach to the limiting point give different limiting values, hence the limit does not exist.

EXAMPLE 5. **Show that** $\lim\limits_{(x,y)\to(0,0)}\dfrac{2xy^2}{x^2+y^4}$ **does not exist.**

SOLUTION. Let (x,y) tends $(0,0)$ along the curve $x=my^2$, we put $x=my^2$ in the function and then allow y to tends to zero. Thus, we have

$$\lim_{(x,y)\to(0,0)}\frac{2xy^2}{x^2+y^4}=\lim_{y\to0}\frac{2my^4}{(m^2+1)y^4}=\lim_{y\to0}\frac{2m}{1+m^2}=\frac{2m}{1+m^2}$$

which is different for different values of m.

e.g., if $m=1$, then the limt $=\dfrac{2.1}{1+1^2}=1$ and if $m=2$, then this limit $=\dfrac{2.2}{1+2^2}=\dfrac{4}{5}$.

Therefore, the two methods of tends to the limiting point give different limiting values,

hence the limit $\lim\limits_{(x,y)\to(0,0)}\dfrac{2xy^2}{x^2+y^4}$ does not exist.

EXAMPLE 6. **Find that** $\lim\limits_{(x,y)\to(0,0)}\dfrac{x^3+y^3}{x-y}$ **does not exist.**

SOLUTION. Let (x,y) tends to $(0,0)$ through the curve $y=x-mx^3$, we have

$$\lim_{(x,y)\to(0,0)}\frac{x^3+y^3}{x-y}=\lim_{x\to0}\frac{x^3+(x-mx^3)^3}{x-(x-mx^3)}$$

$$=\lim_{x\to0}x^3\left[\frac{1+(1-mx^2)^3}{mx^3}\right]$$

$$=\lim_{x\to0}\frac{2-3mx^2+3m^2x^4-m^3x^6}{m}=\frac{2}{m}$$

which is different for different value of m.

Hence, the simultaneous limit $\lim\limits_{(x,y)\to(0,0)}\dfrac{x^3+y^3}{x-y}$ does not exist.

EXAMPLE 7. **Show that the limit, when $(x,y)=(0,0)$, exist in each case and equal to 0.**

(i) $\lim\dfrac{x^3y^3}{x^2+y^2}$ (ii) $\dfrac{x^4+y^4}{x^2+y^2}$

SOLUTION. (i) Here, we show that $\lim\limits_{(x,y)\to(0,0)}\dfrac{x^3y^3}{x^2+y^2}=0$

Take any given $\varepsilon>0$, for all $(x,y)=(0,0)$, we have

$$\left|\frac{x^3y^3}{x^2+y^2}-0\right|=\left|\frac{x^3y^3}{x^2+y^2}\right|$$

$$= |r^4 \cos^3 \theta \sin^3\theta| \qquad \text{[Putting } x=r \cos, \theta\, y = r \sin \theta]$$

$$= r^4 |\cos \theta|^3 |\sin \theta|^3$$

$$\leq r^4 \qquad \qquad [\because |\cos \theta| \leq 1 \text{ and } |\sin \theta| \leq 1]$$

$$= (x^2 + y^2)^2$$

$$< \varepsilon \text{ if } x^2 < \sqrt{\varepsilon}\,/2 \text{ and } y^2 < \sqrt{\varepsilon}\,/2$$

$$\text{i.e., if } |x| < \left(\frac{\sqrt{\varepsilon}}{2}\right)^{1/2} \text{ and } |y| < \left(\frac{\sqrt{\varepsilon}}{2}\right)^{1/2}$$

Now if we take $\delta = \dfrac{\sqrt{\varepsilon}}{2}$, then for any given $\varepsilon > 0$, there exists $\delta > 0$ such that

$$\left| \frac{x^3 y^3}{x^2 + y^2} - 0 \right| < \varepsilon \text{ whenever } |x| < \delta \text{ and } |y| < \delta$$

Hence, $\quad \lim\limits_{(x,y)\to(0,0)} \dfrac{x^3 y^3}{x^2 + y^2} = 0$

(ii) Take any given $\varepsilon > 0$ and for all $(x, y) \neq (0, 0)$, we have

$$\left| \frac{x^4 + y^4}{x^2 + y^2} - 0 \right| = \frac{x^4 + y^4}{x^2 + y^2}$$

$$= r^2 (\cos^4 \theta + \sin^4\theta) \qquad \text{[Putting } x=r \cos \theta, y = r \sin \theta]$$

$$\leq 2r^2 \qquad \qquad [\because \cos^4 \theta \leq 1 \text{ and } \sin^4 \theta \leq 1]$$

$$= 2 (x^2 + y^2)$$

$$< \varepsilon \text{ if } x^2 < \varepsilon/4 \text{ and } y^2 < \varepsilon/4 \text{ i.e., if } |x| < \sqrt{\varepsilon}\,/2 \text{ and } |y| < \sqrt{\varepsilon}\,/2$$

Now, if we take $\delta = \dfrac{\sqrt{\varepsilon}}{2}$, then for any given $\varepsilon > 0$, there exists $\delta > 0$, such that

$$\left| \frac{x^4 + y^4}{x^2 + y^2} - 0 \right| < \varepsilon \text{ , whenever } |x| < \delta \text{ and } |y| < \delta$$

Hence, $\quad \lim\limits_{(x,y)\to(0,0)} \dfrac{x^4 + y^4}{x^2 + y^2} = 0$

EXAMPLE 8.   **Prove that** $\quad \lim\limits_{(x,y)\to(0,0)} \dfrac{\sqrt{x^2 y^2 + 1} - 1}{x^2 + y^2} = 0$

SOLUTION.   Since x, y are small in absolute values, we have

$$\frac{\sqrt{x^2 y^2 + 1} - 1}{x^2 + y^2} = \frac{(1 + x^2 y^2)^{1/2} - 1}{x^2 + y^2} \approx \frac{\frac{1}{2} x^2 y^2}{x^2 + y^2}$$

Take any given $\varepsilon > 0$, we have

$$\left| \frac{\sqrt{x^2 y^2 + 1} - 1}{x^2 + y^2} - 0 \right| = \frac{\sqrt{x^2 y^2 + 1} - 1}{x^2 + y^2} = \frac{\frac{1}{2} x^2 y^2}{x^2 + y^2}$$

$$= \frac{1}{2} \frac{r^4 \cos^2 \theta \sin^2 \theta}{r^2} \qquad \qquad [\because x = r \cos \theta, y = r \sin \theta]$$

$$= \frac{1}{2} r^2 \cos^2 \theta \sin^2 \theta$$

$$\leq \frac{1}{2}r^2 \qquad\qquad [\because \cos^2\theta < 1 \text{ and } \sin^2\theta \leq 1]$$

$$\leq \frac{1}{2}(x^2 + y^2) \qquad\qquad [r^2 = x^2 + y^2]$$

$< \varepsilon$ if $x^2 < \varepsilon$ and $y^2 < \varepsilon$ i.e., $|x| < \sqrt{\varepsilon}$ and $|y| < \sqrt{\varepsilon}$.

Now if we take $\delta = \sqrt{\varepsilon}$, then for any given $\varepsilon > 0$, there exists $\delta > 0$ such that

$$\left| \frac{\sqrt{x^2 y^2 + 1} - 1}{x^2 + y^2} - 0 \right| < \varepsilon \text{ whenever } |x| < \delta \text{ and } |y| < \delta.$$

Hence, $\quad \displaystyle\lim_{(x,y)\to(0,0)} \frac{\sqrt{x^2 y^2 + 1} - 1}{x^2 + y^2} = 0$

☛ REMARKS

➡ If the simultaneous limit, $\displaystyle\lim_{\substack{x\to x_0 \\ y\to y_0}} f(x, y)$ exist, then the single limit, $\displaystyle\lim_{x\to x_0} f(x, y_0) \ \lim_{y\to y_0} f(x_0, y)$

also exist. But it is necessary that the single limits, $\displaystyle\lim_{x\to x_0} f(x, y), \ \lim_{\substack{x\to x_0 \\ y\to y_0}} f(x, y)$, exist for

$y \neq y_0, x \neq x_0$ respectively.

➡ If two iterated limits are equal such that $\displaystyle\lim_{x\to 0}\lim_{y\to 0} f(x, y) = \lim_{y\to 0}\lim_{x\to 0} f(x, y)$ then the limit of

the funciton need not exist.

10.2.2 SQUARE NEIGHBOURHOOD OF A POINT (a, b)

Let neighbourhood of a point (a, b) in the xy-plane determined by a positive number δ be the square bounded by the lines.

$$x = a - \delta, \quad x = a + \delta,$$
$$y = b - \delta, \quad y = b + \delta.$$

If a point (x, y) lies in the neighbourhood, we have

$$a - \delta < x < a + \delta \implies |x - a| < \delta$$
$$b - \delta < y < b + \delta \implies |y - b| < \delta$$

The centre of the square is at the point (a, b). This square is called the neighbourhood of the point (a, b). For every value of δ, we will get a neighbourhood.

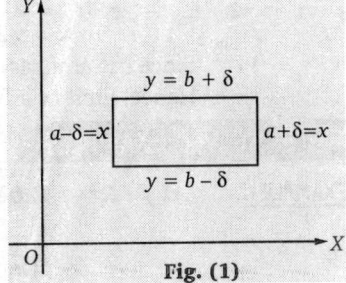

Fig. (1)

10.2.3 CIRCULAR NEIGHBOURHOOD OF A POINT (a, b)

A circular neighbourhood of a point (a, b) in R^2 is the set of all points (x, y) whose distance from the point (a, b) is less than some given $\delta > 0$ i.e., the set of all point (x, y) such that

$$\sqrt{(x - a)^2 + (y - b)^2} < \delta$$

i.e., $\qquad |(x, y) - (a, b)| < \delta$

Here, $|(x, y) - (a, b)|$ stands for distance betwen the points (x, y) and (a, b) i.e.,

Fig. (2)

$$\sqrt{(x - a)^2 + (y - b)^2}$$

10.2.4 ALGEBRA OF LIMITS

If $\lim\limits_{(x,y)\to(a,b)} f(x,y) = l_1$ and $\lim\limits_{(x,y)\to(a,b)} g(x,y) = l_2$, then

(i) $\lim\limits_{(x,y)\to(a,b)} [f(x,y) + g(x,y)] = l_1 + l_2$ (ii) $\lim\limits_{(x,y)\to(a,b)} [f(x,y) - g(x,y)] = l_1 - l_2$

(iii) $\lim\limits_{(x,y)\to(a,b)} [f(x,y).g(x,y)] = l_1.l_2$ (iv) $\lim\limits_{(x,y)\to(a,b)} \left[\dfrac{f(x,y)}{g(x,y)} \right] = \dfrac{l_1}{l_2}$ (provided $l_2 \neq 0$)

THEOREM 1. **Let $z = f(x, y)$ be a function, then $\lim\limits_{(x,y)\to(a,b)} f(x,y)$ if exists, is unique.**

PROOF. Let $z = f(x, y)$ be a function.
Let if possible

$$\lim_{(x,y)\to(a,b)} f(x,y) = l_1 \text{ and } \lim_{(x,y)\to(a,b)} f(x,y) = l_2$$

Now, to prove $l_1 = l_2$

Let us first suppose $\lim\limits_{(x,y)\to(a,b)} f(x,y) = l_1$, then by the definition of limit, we have "for

given $\varepsilon > 0 \; \exists \; \delta_1 > 0$ such that

$$|f(x,y) - l_1| < \varepsilon/2 \text{ whenever } |(x,y) - (a,b)| < \delta_1 \text{"} \qquad \text{... (1)}$$

Now suppose $\lim\limits_{(x,y)\to(a,b)} f(x,y) = l_2$

"for given $\varepsilon > 0 \; \exists \; \delta_2 > 0$ such that

$$|f(x,y) - l_2| < \varepsilon/2 \text{ whenever } |(x,y) - (a,b)| < \delta_2 \text{"} \qquad \text{... (2)}$$

Let $\delta = \min \{\delta_1, \delta_2\}$
Hence, we have

$|f(x,y) - l_1| < \varepsilon/2$ and $|f(x,y) - l_2| < \varepsilon/2$ whenever $|(x,y) - (a,b)| < \delta$

Now, consider

$$
\begin{aligned}
|l_1 - l_2| &= |l_1 - f(x,y) + f(x,y) - l_2| \\
&\leq |l_1 - f(x,y)| + |f(x,y) - l_2| \qquad \text{[By triangular inequality]} \\
&\leq |f(x,y) - l_1| + |f(x,y) - l_2| \\
&< \varepsilon/2 + \varepsilon/2 = \varepsilon
\end{aligned}
$$

Since ε is arbitrary and small, therefore $l_1 - l_2 = 0 \Rightarrow l_1 = l_2$.
Hence, limit of a function is unique.

SOLVED EXAMPLES

EXAMPLE 1. **Let $f : R^2 \to R$ be defined by**

$$f(x,y) = \begin{cases} \dfrac{xy}{x^2 + y^2} & , \; (x,y) \neq (0,0) \\ 0 & , \; (x,y) = (0,0) \end{cases}$$

Prove that $\lim\limits_{(x,y)\to(0,0)} f(x,y)$ does not exist.

SOLUTION. Since, if $\lim\limits_{(x,y)\to(0,0)} f(x,y)$ exists, then this limit is independent of the path along which

we approach the point (a, b). Let $(x, y) \to (0, 0)$ along the path $y = mx$ where $m \in R$. As
$x \to 0$, from $y = mx$, we have $y \to 0$.

Consider $\lim\limits_{(x,y)\to(a,b)} f(x,y) = \lim\limits_{(x,y)\to(0,0)} \dfrac{xy}{x^2 + y^2}$

$$= \lim_{x\to 0} \frac{x\,mx}{x^2 + m^2 x^2} = \lim_{x\to 0} \frac{mx^2}{x^2(1 + m^2)} \qquad \text{(Putting } y = mx)$$

$$= \lim_{x \to 0} \frac{m}{1+m^2} = \frac{m}{1+m^2}$$

which will be different for different values of m.

Therefore, $\lim_{(x,y) \to (0,0)} f(x,y)$ does not exist.

EXAMPLE 2. *Let* $A=\{(x, y) : 0 < x < 1,\ 0 < y < 1,\ x, y \in R\}$. *If* $f : A \to R$ *defined by* $f(x, y) = x + y$, *Show that*

$$\lim_{(x,y) \to \left(0, \frac{1}{2}\right)} f(x,y) = \frac{1}{2} \text{ where } x, y \in A.$$

SOLUTION. Let $\varepsilon > 0$ be given (an arbitrary positive small number)

Also, let $|x - 0| < \varepsilon/2$ and $\left| y - \frac{1}{2} \right| < \varepsilon/2$

Consider $\left| f(x,y) - \frac{1}{2} \right| = \left| x + y - \frac{1}{2} \right|$

$$\leq |x| + \left| y - \frac{1}{2} \right| < \varepsilon/2 + \varepsilon/2 = \varepsilon$$

Then, by the definition of limit, we have

$$\lim_{(x,y) \to \left(0, \frac{1}{2}\right)} f(x,y) = \frac{1}{2}$$

EXAMPLE 3. *If* $f(x,y) = y \sin \frac{1}{x} + x \sin \frac{1}{y}$ *where* $x \neq 0,\ y \neq 0$. *Then prove that* $f(x, y) \to 0$ *as* $(x, y) \to (0, 0)$

SOLUTION. Let ε be any given arbitrary small positive number since, $\varepsilon > 0$ let us take $\delta = \varepsilon$.

Also, let $|x - 0| < \varepsilon/2$, $|y - 0| < \varepsilon/2$

$$\therefore \quad |(x, y) - (0, 0)| = \sqrt{(x-0)^2 + (y-0)^2} \leq |x - 0| + |y - 0|$$

$$< \varepsilon/2 + \varepsilon/2 = \varepsilon$$

$$\Rightarrow \quad |f(x, y) - 0| = \left| y \sin \frac{1}{x} + x \sin \frac{1}{y} \right| \leq \left| y \sin \frac{1}{x} \right| + \left| x \sin \frac{1}{y} \right|$$

$$\leq |y| \left| \sin \frac{1}{x} \right| + |x| \left| \sin \frac{1}{y} \right|$$

$$\leq |y| + |x| \qquad \left[\because \left| \sin \frac{1}{x} \right| \leq 1 \text{ and } \left| \sin \frac{1}{y} \right| \leq 1 \right]$$

$$< \varepsilon/2 + \varepsilon/2 = \varepsilon$$

$$\Rightarrow \quad |f(x, y) - 0| < \varepsilon. \text{ Hence, } \lim_{(x,y) \to (0,0)} f(x,y) = 0$$

EXAMPLE 4. *Show that* $\lim_{(x,y) \to (0,1)} \tan^{-1} \frac{y}{x}$ *does not exist.*

SOLUTION. Let $y = mx$

$$\therefore \quad \lim_{(x,y) \to (0,1)} \tan^{-1} \frac{y}{x} = \lim_{x \to 0} \tan^{-1} \left(\frac{mx}{x} \right) = \tan^{-1} m$$

i.e., limit of the function depends on m; which means limit of the function at $(0, 1)$ is not unique. Hence, the limit of the function does not exist.

EXAMPLE 5. **If** $f(x,y) = \begin{cases} 2xy & , \textbf{if} \quad (x,y) \neq (1,2) \\ 0 & , \textbf{if} \quad (x,y) = (1,2) \end{cases}$ **then, show that** $\lim\limits_{(x,y) \to (1,2)} f(x,y) = 4$

SOLUTION. By definition of limit, we know that for $\varepsilon > 0$, we have to find $\delta > 0$ such that

$$|2xy - 4| < \varepsilon \quad \text{whenever} \quad \sqrt{(x-1)^2 + (y-2)^2} < \delta$$

$\Rightarrow \quad |2xy - 4| < \varepsilon \quad \text{whenever} \quad |x-1| < \delta \text{ and } |y-2| < \delta$

$\Rightarrow \quad 1 - \delta < x < 1 + \delta \text{ and } 2 - \delta < y < 2 + \delta$

Hence, $\quad (1 - \delta)(2 - \delta) < xy < (1+\delta)(2+\delta)$

$\Rightarrow \quad 2(1-\delta)(2-\delta) < 2xy < 2(1+\delta)(2+\delta)$

$\Rightarrow \quad (2 - 2\delta)(2 - \delta) < 2xy < (2 + 2\delta)(2 + \delta)$

$\Rightarrow \quad 2\delta^2 - 6\delta + 4 < 2xy < 2\delta^2 + 6\delta + 4$

$\Rightarrow \quad 2\delta^2 - 6\delta < 2xy - 4 < 2\delta^2 + 6\delta$

$\Rightarrow \quad |2xy - 4| < 8\delta \text{ if } \delta \leq 1$

Let $\quad \delta = \dfrac{\varepsilon}{8}$. Then $|2xy - 4| < \varepsilon$ whenever $|x-1| < \delta$ and $|y-2| < \delta$

Hence, we have for given $\varepsilon > 0$, there exists a $\delta > 0$ such that $|2xy - 4| < \varepsilon$

$\Rightarrow \quad \lim\limits_{(x,y) \to (1,2)} (2xy) = 4$

EXAMPLE 6. **Give an example to show that the order of iterated limits can be interchanged although the simultaneous limit does not exist.**

SOLUTION. Consider the function $f(x,y) = \dfrac{xy}{x^2 + y^2}$ for iterated and simulaneous limit at $(0, 0)$. Let us first suppose the variables approach the origin along the line $y = x$. Then putting $y = x$, we have

$$\lim\limits_{x \to 0} \frac{x^2}{x^2 + x^2} = \frac{1}{2}$$

Now consider

$$\lim\limits_{y \to 0} \lim\limits_{x \to 0} \frac{xy}{x^2 + y^2} = \lim\limits_{y \to 0} \frac{0}{0 + y^2} = 0$$

Since, the results obtained by two methods of approaches are different. Hence, the simultaneous limit does not exist.

Now, for iterated limits, we have

$$\lim\limits_{x \to 0} \left[\lim\limits_{y \to 0} \frac{x.y}{x^2 + y^2} \right] = \lim\limits_{x \to 0} \left[\frac{x.0}{x^2 + 0} \right] = 0$$

and $\quad \lim\limits_{y \to 0} \left[\lim\limits_{x \to 0} \frac{x.y}{x^2 + y^2} \right] = \lim\limits_{y \to 0} \left[\frac{0.y}{0 + y^2} \right] = 0$

Hence, the order of iterated limits can be interchanged.

10.3 CONTINUITY OF A FUNCTION OF TWO VARIABLES

(i) A function $f(x, y)$ is said to be continuous at the point (a, b) if $\lim\limits_{(x,y) \to (a,b)} f(x,y)$ exists and equal to $f(a, b)$.

(ii) A function $f(x, y)$ is said to be continuous at (a, b), if for every $\varepsilon > 0 \ \exists \ \delta > 0$ such that $|f(x, y) - f(a, b)| < \varepsilon$ whenever $|x - a| < \delta, |y - b| < \delta$.

SOLVED EXAMPLES

EXAMPLE 1. *Show that the function $f : R^2 \to R$ defined by*

$$f(x, y) = \begin{cases} \dfrac{xy(x^2 - y^2)}{x^2 + y^2} & , \ (x, y) \neq (0, 0) \\ 0 & , \ otherwise \end{cases}$$

is continuous at (0, 0)

SOLUTION. Let $\varepsilon > 0$ be given. Now, let us suppose $|x - 0| < \sqrt{\varepsilon}$ and $|y - 0| < \sqrt{\varepsilon}$
Consider,

$$|f(x, y) - f(0,0)| = \left| \frac{xy(x^2 - y^2)}{x^2 + y^2} - 0 \right| \leq |xy| \left| \frac{x^2 - y^2}{x^2 + y^2} \right|$$

$$\leq |xy| \qquad \left[\because |x^2 - y^2| \leq |x^2 + y^2| \Rightarrow \left| \frac{x^2 - y^2}{x^2 + y^2} \right| \leq 1 \right]$$

$$\Rightarrow \qquad |f(x, y) - f(0,0)| \leq |x| \, |y|$$
$$\Rightarrow \qquad |f(x, y) - f(0,0)| < \sqrt{\varepsilon} \cdot \sqrt{\varepsilon}$$
$$\Rightarrow \qquad |f(x, y) - f(0,0)| < \varepsilon.$$

Hence, we have $\lim\limits_{(x,y) \to (0,0)} f(x, y)$ exists and equal to $(0, 0)$.

EXAMPLE 2. *Show that the function, $f(x, y) = x^2 + 3y$ is continuous at (1, 2).*

SOLUTION. Here, we have

$$f(x, y) = x^2 + 3y$$
$$\Rightarrow \qquad f(1, 2) = 1 + 3 \times 2 = 7$$

We know, that if limit of the function $f(x, y) = x^2 + 3y$ at $(1, 2)$ is also 7, then function is continuous at $(1, 2)$.

Now to show $\lim\limits_{(x,y) \to (1,2)} f(x, y) = 7$

For given $\varepsilon > 0$, there must exist $\delta > 0$, such that

$$0 < \sqrt{(x - 1)^2 + (y - 2)^2} < \delta \Rightarrow |x^2 + 3y| < \varepsilon$$

Therefore

$$|x^2 + 3y - 7| = |(x^2 - 1) + 3(y - 2)|$$
$$= |(x - 1)(x + 1) + 3(y - 2)|$$
$$\leq |(x - 1)| \cdot |(x + 1)| + 3|(y - 2)| \qquad \dots (1)$$

But $$|y - 2| < \sqrt{(x - 1)^2 + (y - 2)^2} < \frac{\delta}{6} \qquad \dots (2)$$

and $$|x - 1| < \sqrt{(x - 1)^2 + (y - 2)^2} < \frac{\delta}{6} \qquad \dots (3)$$

$$\therefore \qquad 0 < x - 1 < 1 \Rightarrow 0 < x < 2 \Rightarrow 1 < x + 1 < 3$$

Hence, from (1)

$$|x^2 + 3y - 7| \leq 3 |x - 1| + 3 |y - 2|$$

Using (1) and (2), we get

$$|(x^2 + 3y) - 7| \leq 3|\sqrt{(x - 1)^2 + (y - 2)^2}| + 3|\sqrt{(x - 1)^2 + (y - 2)^2}|$$
$$= 6\sqrt{(x - 1)^2 + (y - 2)^2}$$

$$< \frac{\delta}{6} = \varepsilon$$

$$\Rightarrow \qquad |(x^2 + 3y) - 7| < \varepsilon$$

$$\Rightarrow \qquad \lim_{(x,y) \to (1,2)} (x^2 + 3y) = 7$$

Hence, the limit and value of the function at point (1, 2) are equal therefore the function is continuous at (1, 2).

EXAMPLE 3. *Show that the function*

$$f(x, y) = \frac{xy^3}{x^2 + y^6}, x \neq 0, y \neq 0 \ and \ f(0,0) = 0$$

is not continuous at (0, 0) in (x, y).

SOLUTION. Here $\qquad f(0, 0) = 0$ (given)

Let us suppose $(x, y) \to (0, 0)$ through the curve $x = y^3$.

Then $\qquad \lim_{(x,y) \to (0,0)} f(x, y) = \lim_{y \to 0} \frac{y^6}{y^6 + y^6} = \frac{1}{2}$

Aganin, let $(x, y) \to (0, 0)$ through the line $y = x$, then

$$\lim_{(x,y) \to (0,0)} f(x, y) = \lim_{x \to 0} \frac{x.x^3}{x^2 + x^6} = \lim_{x \to 0} \frac{x^2}{1 + x^4} = 0$$

Since, the limit obtained by two different approaches are different, hence $\lim_{(x,y) \to (0,0)} f(x, y)$ does not exist. Hence, the given function is not continuous.

EXAMPLE 4. *Let X = {(x, y) : 0 < x < 1, 0 < y < 1} and f : X → R be defined by f(x, y) = x + y. Prove that f is continuous at every point of the domain X.*

SOLUTION. Let $\ X = \{(x, y) : 0 < x < 1, 0 < y < 1\}$

Let (a, b) be any point of X.

To prove, that $f(x, y)$ is continuous at (a, b).

Let $\varepsilon > 0$ be given.

Also, let $\qquad |x - a| < \varepsilon/2$ and $|y - b| < \varepsilon/2$ $\qquad\qquad\qquad$ ... (1)

Therefore $|f(x, y) - (a, b)| = |x + y - (a + b)|$

$$= |(x - a) + (y - b)|$$

$$\leq |x - a| + |y - b| \quad \text{[By triangular inequality]}$$

$$< \varepsilon/2 + \varepsilon/2 \qquad\qquad\qquad \text{[From (1)]}$$

$$= \varepsilon$$

$$\Rightarrow \qquad\qquad |f(x, y) - (a, b)| < \varepsilon$$

$$\Rightarrow \qquad \lim_{(x,y) \to (a, b)} f(x, y) \ \text{exist and hence equal to } (a, b).$$

Therefore f is continuous at (a, b) and f is continuous at every point of X.

EXAMPLE 5. *Show that the function*

$$f(x, y) = \frac{xy}{\sqrt{(x^2 + y^2)}}, x \neq 0, y \neq 0 \ and \ f(0,0) = 0$$

is continuous at the origin in (x, y) together.

SOLUTION. Let $\varepsilon > 0$ be given

Let us suppose $x = r \cos \theta, y = r \sin \theta$.

Then $\qquad f(r \cos \theta, r \sin \theta) = \frac{r^2 \cos \theta \sin \theta}{r\sqrt{\sin^2 \theta + \cos^2 \theta}} = r \cos \theta \sin \theta$

$$= \frac{1}{2} r \sin 2\theta$$

Now, consider

$$| f(r \cos \theta, r \sin \theta) - f(0, 0)| = |f(r \cos \theta, r \sin \theta)|$$

$$= \left| \frac{1}{2} r \sin 2\theta \right|$$

$$= \frac{1}{2} r | \sin 2\theta |$$

$$\leq \frac{1}{2} r \qquad\qquad [\because \ |\sin 2\theta| \leq 1]$$

Now, if we choose $r = 2\varepsilon$.

Therefore we have $\varepsilon > 0$ such that

$$| f (r \cos \theta, r \sin \theta)| < \varepsilon \text{ for all values of } \theta \qquad\qquad ... (1)$$

Equation (1) is true for all points within a circle about the origin and radius $r=2\varepsilon$. Therefore $f(r \cos \theta, r \sin \theta)$ is uniformly continuous in r for all values of θ. Hence $f(x, y)$ is continuous in (x, y) at the origin.

EXAMPLE 6. *Check the continuity at (1, 2) of the function*

$$f(x,y) = \begin{cases} x^2 + 4y & , \text{ when } (x,y) \neq (1,2) \\ 0 & , \text{ when } (x,y) = (1,2) \end{cases}$$

SOLUTION. We have

$$\lim_{(x,y) \to (1,2)} x^2 + 4y = (1)^2 + 4 \cdot 2 = 9$$

Therefore, the limit exists and is equal to 9.

Since it is given that $\quad f(1, 2) = 0 \ \text{ and } \ \lim_{(x,y)} f(x,y) = 9$

Thus $\qquad\qquad \lim_{(x,y) \to (1,2)} f(x,y) \neq f(1,2)$

Hence, the function is not continuous at (1, 2).

EXAMPLE 7. *Show that the function f defined as follows has a removable discontinuity at (2, 3).*

$$f(x,y) = \begin{cases} 3xy & , \ (x,y) \neq (2,3) \\ 6 & , \ (x,y) = (2,3) \end{cases}$$

Again redefine the function to make it continuous at (2, 3)

SOLUTION. We have

$$\lim_{(x,y) \to (2,3)} f(x,y) = \lim_{(x,y) \to (2,3)} 3xy = 3 \times 2 \times 3 = 18 \qquad\qquad ... (1)$$

Since, given that $\qquad f(2, 3) = 6 \qquad\qquad ... (2)$

Therefore, from equation (1) and (2), we get

$$\lim_{(x,y) \to (2,3)} f(x,y) \neq f(2, 3).$$

Hence, the function $f(x, y)$ is discontinuous at (2, 3).

Since $\lim_{(x,y) \to (2,3)} f(x,y)$ exists but is not equal to $f(2, 3)$ therefore the discontinuity is removable.

We can remove the discontinuity by redefining the function as follows

$$f(x,y) = \begin{cases} 3xy & , \ (x,y) \neq (2,3) \\ 18 & , \ (x,y) = (2,3) \end{cases}$$

EXAMPLE 8. **Show that the following function are discontinuous at the origin**

$$f(x,y) = \begin{cases} \dfrac{x^4 - y^4}{x^4 + y^4} & , \ (x,y) \neq (0,0) \\ 0 & , \ (x,y) = (0,0) \end{cases}$$

Solution. Since, we know that the function $f(x, y)$ is continuous at a point (a, b) of its domain if $\lim\limits_{(x,y)\to(a,b)} f(x,y)$ exists and is equal to $f(a, b)$ otherwise is discontinuous at (a, b).

Now, let $(x, y) \to (0, 0)$ along the straight line $y = mx$. Then

$$\lim_{(x,y)\to(0,0)} f(x,y) = \lim_{(x,mx)\to(0,0)} \frac{x^4 - y^4}{x^4 + y^4}$$

Putting $y = mx$

$$= \lim_{x\to0} \frac{x^4 - m^4 x^4}{x^4 + m^4 x^4} = \lim_{x\to0} \frac{1 - m^4}{1 + m^4} = \frac{1 - m^4}{1 + m^4}$$

which is different for different values of m. Therefore, $\lim\limits_{(x,y)\to(0,0)} f(x,y)$ does not exist.

Hence, $f(x, y)$ is discontinuous at the origin.

EXAMPLE 9. **Show that the following function is discontinuous at (0, 0)**

$$f(x,y) = \begin{cases} \dfrac{x^3 + y^3}{x - y} & , \ x \neq y \\ 0 & , \ x = y \end{cases}$$

SOLUTION. Since we know that a function $f(x, y)$ is continuous at a point (a, b) of its domain if $\lim\limits_{(x,y)\to(a,b)} f(x,y)$ exists and is equal to $f(a, b)$, otherwise $f(x, y)$ is discontinuous at (a, b).

Now, we shall show that $\lim\limits_{(x,y)\to(0,0)} f(x,y)$ does not exist.

Let (x, y) tends to $(0, 0)$ through the curve $y = x - mx^3$. Then

$$\lim_{(x,y)\to(0,0)} f(x,y) = \lim_{(x,y)\to(0,0)} \frac{x^3 + y^3}{x - y}$$

$$= \lim_{x\to0} \frac{x^3 + (x - mx^3)^3}{x - (x - mx^3)} \qquad [\because \ y = x - mx^3]$$

$$= \lim_{x\to0} \frac{x^3[1 + (1 - mx^3)^3]}{mx^3} = \lim_{x\to0} \frac{2 - 3mx^2 + 3m^2 x^4 - m^3 x^6}{m}$$

$$= \frac{2}{m}, \quad \text{(which is different for different values of } m.)$$

Therefore, $\lim\limits_{(x,y)\to(0,0)} f(x,y)$ does not exist.

Hence, the function $f(x, y)$ is discontinuous at the origin.

EXAMPLE 10. **Show that the function is continuous at the origin**

$$f(x,y) = \begin{cases} \dfrac{x^3 y^3}{x^2 + y^2} & , \ (x,y) \neq (0,0) \\ 0 & , \ (x,y) = (0,0) \end{cases}$$

SOLUTION. We shall show that

$$\lim_{(x,y)\to(0,0)} f(x,y) = 0$$

Let $\varepsilon > 0$. For all $(x, y) \neq (0, 0)$, we have

$$| f(x,y) - 0 | = \left| \frac{x^3 y^3}{x^2 + y^2} \right|$$

$$= \frac{x^3 y^3}{x^2 + y^2} \qquad \qquad \text{[Putting } x = r\cos\theta, y = r\sin\theta]$$

$$= r^4 \cos^3\theta \sin^3\theta \qquad \qquad [\because \ \cos^3\theta \leq 1, \sin^3\theta \leq 1]$$

$$\leq r^4 = (x^2 + y^2)^2$$

$$< \varepsilon \text{ if } x^2 < \sqrt{\varepsilon/2} \text{ and } y^2 < \sqrt{\varepsilon/2}.$$

Now, if we take $\delta = (\varepsilon/2)^{1/4}$, then, we see that for any given $\varepsilon > 0$, there exists $\delta > 0$, such that

$$|f(x,y) - 0| < \varepsilon, \text{ whenever } |x| < \delta, |y| < \delta.$$

Therefore, $\lim\limits_{(x,\ y) \to (0,\ 0)} f(x,y) = 0$

Since given that the $f(0, 0) = 0$, therefore $\lim\limits_{(x,\ y) \to (0,\ 0)} f(x,y) = f(0,0)$

Hence, $f(x, y)$ is continuous at the origin.

EXAMPLE 11. *Let $f : R^2 \to R$ be a function, defined by*

$$f(x,y) = \begin{cases} \dfrac{xy}{x^2 + y^2} & \text{when} \ \ (x,y) \neq (0,0) \\ 0 & \text{when} \ \ (x,y) = (0,0) \end{cases}$$

Show that f is not continuous at (0, 0) but is continuous in each variable seperately.

SOLUTION. For point (x, y) on the x-axis, we have $y = 0$ and $f(x, y) = f(x, 0) = 0$, so the function has the constant value, 0, every where on the x-axis, which gives that $f(x, y)$ is continuous at $x = 0$. Similarly $f(x, y)$ has the constant value, 0, at all points on the y-axis, so if we put $x = 0$, the function $f(x, y)$ is continuous at $y = 0$. Now, we shall show that $f(x, y)$ is not continuous at origin.

Let $\qquad\qquad\qquad y = x \quad$ Then $\quad f(x,y) = f(x,x) = \dfrac{x^2}{2x^2} = \dfrac{1}{2}$

Also $\qquad\qquad f(0,0) = 0$ (given)

Since there are points on the line arbitrarily close to the origin and since $f(0, 0) \neq \dfrac{1}{2}$, the function of two variable $f(x, y)$ is not continuous at the origin.

EXAMPLE 12. *Show that the function $f(x, y) = xy$ is continuous at the point (2, 3).*

SOLUTION. Here, we have $\quad f(x, y) = xy \quad \Rightarrow \quad f(2, 3) = 2 \times 3 = 6$

Now, we shall find the limit of $f(x, y) = xy$.

Let $\varepsilon > 0$ be given.

For a given $\varepsilon > 0$, we want to find $\delta > 0$ such that

$$0 < \sqrt{(x-2)^2 + (y-3)^2} < \delta \Rightarrow |(xy) - 6| < \varepsilon$$

$$|x - 2| \leq \sqrt{(x-2)^2 + (y-3)^2} < \delta \qquad \qquad \dots (1)$$

and $\qquad |y - 3| \leq \sqrt{(x-2)^2 + (y-3)^2} < \delta \qquad \qquad \dots (2)$

Now, consider

$$|xy - 6| = |xy - 2y + 2y - 6| = |y(x-2) + 2(y-3)|$$

$$\leq |y| \cdot |x - 2| + 2|y - 3| \qquad \qquad \dots (3)$$

Now, from (2) and (3), we have

$$|y - 3| < \delta < 1 \Rightarrow 2 < |y| < 4$$

$$\therefore \qquad |xy - 6| < 4.|x - 2| + 2|y - 3| < 4\delta + 2\delta = 6\delta$$

If $\delta = \dfrac{\varepsilon}{6}$, then

$$|xy - 6| < 6\frac{\varepsilon}{6} = \varepsilon$$

$$\Rightarrow \qquad |xy - 6| < \varepsilon$$

$$\Rightarrow \qquad \lim_{(x,y)\to(2,3)} xy = 6$$

Since, $\qquad f(2, 3) = \lim_{(x,y)\to(2,3)} f(x,y)$

Hence, the function $f(x, y) = xy$ is continuous at (2, 3).

EXAMPLE 13. *Let $f : R^2 \to R$ be a function defined by*

$$f(x,y) = \begin{cases} \dfrac{xy(x^2 - y^2)}{x^2 + y^2} & , \ (x,y) \neq (0,0) \\ 0 & , \ (x,y) = (0,0) \end{cases}$$

show that the function f(x, y) is continuous at (0, 0).

SOLUTION. We know that the function $f(x, y)$ will be continuous at (0, 0) if limit of the function= value of the function at (0, 0). It is given that, the value of the function $f(x, y)$ at (0, 0) is equal to zero *i.e.*, $f(0, 0) = 0$

Now, we shall find the limit of $f(x, y)$.

Let $\varepsilon > 0, \exists \delta > 0$, such that

$$|f(x, y) - f(0, 0)| < \varepsilon \text{ whenever } |(x, y) - (0, 0)| < \delta$$

or $\qquad \left| \dfrac{xy(x^2 - y^2)}{x^2 + y^2} - 0 \right| < \varepsilon$

whenever $\quad \sqrt{(x - 0)^2 + (y - 0)^2} = \sqrt{x^2 + y^2} < \delta$ ... (1)

Now $\qquad \sqrt{x^2 + y^2} < \delta \Rightarrow |x| < \delta < \sqrt{\varepsilon}$ and $|y| < \delta < \sqrt{\varepsilon}$

The function (a, b), but naturally the point itself, must belong to the domain.

$$\therefore \quad \text{from (1)} \quad \left| \frac{xy(x^2 - y^2)}{x^2 + y^2} \right| \leq \left| \frac{xy(x^2 + y^2)}{x^2 + y^2} \right| = |x|.|y| < \sqrt{\varepsilon}.\sqrt{\varepsilon} = \varepsilon$$

which gives that

$$\left| \frac{xy(x^2 - y^2)}{x^2 + y^2} - 0 \right| < \varepsilon \text{ whenever } |(x, y) - (0, 0)| < \delta.$$

$$\Rightarrow \quad \lim_{(x,y)\to(0,0)} \left(\frac{xy(x^2 - y^2)}{(x^2 + y^2)} \right) = 0.$$

Hence, $f(x, y)$ is continuous at (0, 0).

☛ **REMARKS**

➡ For functions of two variables, continuous at some point (a, b), it is true that addition, subtraction, multiplication and division, produce new functions which are also continuous at this point [if only the domains of the resulting functions satisfy the condition, mentioned above in (1)].

➡ Every polynomial and every rational function of two variables will be continuous in all of its domain.

➡ For a function of two variables defined and continuous in a closed region it can be shown that the range of the function is a closed interval. Such a function thus has an absolute maximum value and an absolute minimum value.

➡ In order that function of two variables, $f(x, y)$ shall be continuous in both the variables together, it must have the same limiting value by all possible approaches to the critical point. Therefore, a necessary and sufficient condition involves the condition that the function is not only continuous in each direction, but the continuity is uniform for all direction.
If we put $x = a + r\cos\theta, y = b + r\sin\theta$, then by the definition of continuity, we have
$$|f(a + r\cos\theta, b + r\sin\theta) - f(a, b)| < \varepsilon$$
which must hold for all values of r, which is independent of θ i.e., the transformed function must be uniformly continuous in r for all values of $|\theta| \le 2\pi$.

➡ If $\lim\limits_{r\to 0} f(a + r\cos\theta, b + r\sin\theta) = f(a, b)$ for every value of θ, then, it is not necessary that the function is continuous at (a, b).

For example, Let $f(x, y) = \dfrac{xy^3}{x^2 + y^6}, x \ne 0, y \ne 0$ and $f(0, 0) = 0$ does not exist as x and y approaches zero. Thus function is therefore discontinuous at the origin. However, putting $x = r\cos\theta, y = r\sin\theta$, we have

$$\lim_{r\to 0}\frac{r^4\cos\theta\sin^3\theta}{r^2\cos^2\theta + r^6\sin^6\theta} = \lim_{r\to 0} r^2 \cdot \frac{\cos\theta\sin^3\theta}{\cos^2\theta + r^4\sin^6\theta}$$
$$= 0 = f(0, 0) \text{ for each constant value of } \theta.$$

EXERCISE 10.1

1. Let $f : R^2 \to R$ be defined by $f(x, y) = x^2 + y^2$, show that
$$\lim_{(x,y)\to(0,0)} f(x,y) = 0$$

2. Show that $\lim\limits_{(x,y)\to(0,0)} \dfrac{2x^3 - y^3}{x^2 + y^2} = 0$.

3. Show that $\lim\limits_{(x,y)\to(0,0)} \dfrac{2x - y}{x^2 + y^2}$ does not exist.

4. Prove that $\lim\limits_{(x,y)\to(0,0)} f(x,y)$ does not exist,

where $f(x, y) = \dfrac{xy^2}{x^2 + y^4}, (x, y) \ne (0, 0)$

5. Let
$$f(x, y) = \frac{x^2 y^2}{x^2 y^4 + (x - y)^2} (x^2 y^2 + (x - y)^2 \ne 0)$$

Show that
$$\lim_{x\to 0}\left[\lim_{y\to 0} f(x,y)\right] = \lim_{y\to 0}\left[\lim_{x\to 0} f(x,y)\right] = 0$$

6. Show that $f(x, y) = \dfrac{2y}{x}$ exists at $(0, 0)$.

7. Show that the iterated limit of the function
$$f(x, y) = y\sin\frac{1}{x}$$
exist at $(0, 0)$, but the limit of the function does not exist.

8. Show that $\lim\limits_{(x,y)\to(0,0)}\left[\log\dfrac{a(1 - e^x)}{(a - y^x)}\right]$ does not exist.

9. Show that iterated limit of the function
$$f(x, y) = y\sin\frac{1}{x}$$
exists at $(0, 0)$ but the limit of the function does not exist.

10. Show that $\lim\limits_{(x,y)\to(0,0)}\left[\dfrac{x-y}{\sqrt{x^2-y^2}}\right]$ does not exist.

11. Let

$$f(x,y)=\begin{cases}\dfrac{xy}{\sqrt{x^2-y^2}} & \text{,when } (x,y)\neq(0,0)\\ 0 & \text{, when } (x,y)=(0,0)\end{cases}$$

Show that $f(x,y)$ is continuous at $(0,0)$

12. Let $f(x,y)=\begin{cases}x\sin\dfrac{1}{y} & \text{, if } y\neq 0\\ 0 & \text{, if } y=0\end{cases}$

is continuous at $(0,0)$.

13. Show that the function xy is continuous at each point in the x-y plane.

14. Find the point where the function

$$f(x,y)=\dfrac{x^2}{x^2-2y} \text{ is discontinuous.}$$

15. Show that $f(x,y)=\sqrt{|xy|}$ is continuous at $(0,0)$.

16. Let $f:R^2\to \mathbf{R}$ be defined as

$$f(x,y)=\begin{cases}1 & \text{, if } x \text{ is irrational}\\ 0 & \text{, if } x \text{ is rational}\end{cases}$$

Show that for any point (a,b), $\lim\limits_{(x,y)\to(a,b)} f(x,y)$ does not exist.

17. Show that the function $f(x,y)$ defined by

$$f(x,y)=\begin{cases}\dfrac{x^2-y^2}{x^2+y^2} & \text{if } x\neq 0, y\neq 0\\ 0 & \text{if } x=y=0\end{cases}$$

is discontinuous at that point $(0,0)$.

18. Show that the function defined by

$$f(x,y)=\dfrac{xy^2}{x^2+y^4}, f(0,0)=0$$

is discontinuous in (x,y) at the origin. Also, show that this function is continuous along the radius vector $\theta=\pi/2$.

19. Let $f:R^2\to\mathbf{R}$ be a continuous function. Define a function $g:R^2\to \mathrm{R}$ as

$$g(x,y)=\begin{cases}f(x,y) & \text{if } (x,y)\neq(0,0)\\ f(x,y)+1 & \text{if } (x,y)=(0,0)\end{cases}$$

Show that g is not continuous at $(0,0)$.

20. If $f:R^2\to\mathbf{R}$ be a function defined by

$$f(x,y)=\begin{cases}\dfrac{xy^2+x^2y}{x^3+y^3} & \text{if } (x,y)\neq(0,0)\\ 0 & \text{if } (x,y)=(0,0)\end{cases}$$

is not continuous at $(0,0)$.

HINTS TO SELECTED PROBLEMS

2. $\lim\limits_{\substack{x\to0\\y\to0}}\dfrac{2x^3-y^3}{(x^2+y^2)}=\lim\limits_{x\to0}\dfrac{2x^3-m^3x^3}{x^2+m^2x^2}$

$$=\lim\limits_{x\to0}x\left(\dfrac{2-m^3}{1+m^2}\right)=0$$

9. $\lim\limits_{x\to0}\lim\limits_{y\to0}f(x,y)=\lim\limits_{x\to0}\lim\limits_{y\to0}y\sin\dfrac{1}{x}$

$$=\lim\limits_{x\to0}.0=0$$

Also $\lim\limits_{\substack{x\to0\\y\to0}}f(x,y)=\lim\limits_{\substack{x\to0\\y\to0}}y\sin\dfrac{1}{x}=\lim\limits_{x\to0}mx\sin\dfrac{1}{x}$

13. $\lim\limits_{\substack{x\to a\\y\to b}}f(x,y)=\lim\limits_{\substack{x\to a\\y\to b}}xy=\lim\limits_{x\to a}bx=ab=f(a,b)$

17. Taking the simultaneous limit along the line $y=mx$.

19. $\lim\limits_{\substack{x\to0\\y\to0}}g(x,y)=\lim\limits_{\substack{x\to0\\y\to0}}f(x,y)$

Therefore, $\lim\limits_{\substack{x\to0\\y\to0}}g(x,y)$ exists because $f(x,y)$ is

continuous and since $\lim\limits_{\substack{x\to0\\y\to0}}f(x,y)=f(0,0)$.

But $g(0,0)=f(0,0)+1$

$\therefore\ \lim\limits_{\substack{x\to0\\y\to0}}g(x,y)=\lim\limits_{\substack{x\to0\\y\to0}}f(x,y)=f(0,0)\neq g(0,0)$.

10.4 PARTIAL DIFFERENTIATION OF FUNCTION OF TWO VARIABLES

Let $z=f(x,y)$ be a continuous function of two independent variables x and y. If we treat y as constant and only x changes, then the partial differential coefficient of z with respect to x, is denoted by

$$\dfrac{\partial z}{\partial x} \text{ or } \dfrac{\partial f}{\partial x} \text{ or } f_x \text{ or } D_x f.$$

Thus, $\dfrac{\partial f}{\partial x} = \lim\limits_{\delta x \to 0} \dfrac{f(x + \delta x, y) - f(x, y)}{\delta x}$ provided this limit exists and is unique.

Similarly, if x is regarded as constant and only y-varies, then the differential coefficient of z with respect to y is denoted by

$$\frac{\partial f}{\partial y} \text{ or } \frac{\partial z}{\partial y} \text{ or } f_y \text{ or } D_y f.$$

Thus, $\dfrac{\partial f}{\partial y} = \lim\limits_{\delta y \to 0} \dfrac{f(x, y + \delta y) - f(x, y)}{\delta y}$ provided this limit exists and is unique.

10.4.1 SECOND ORDER PARTIAL DIFFERENTIAL COEFFICIENTS

Partial differential coefficients $\dfrac{\partial f}{\partial x}$ and $\dfrac{\partial f}{\partial y}$ can be differentiated further, partially with

respect to x and y.

Thus, partial differential coefficients of $\dfrac{\partial f}{\partial x}$ with respect to x and y are $\dfrac{\partial}{\partial x}\left(\dfrac{\partial f}{\partial x}\right)$ and $\dfrac{\partial}{\partial y}\left(\dfrac{\partial f}{\partial x}\right)$

respectively.

or $\dfrac{\partial^2 f}{\partial x^2}$ and $\dfrac{\partial^2 f}{\partial y \partial x}$ respectively

By means of the limit, the derivative of the second order are defined by

$$\frac{\partial^2 f}{\partial x^2} = \lim_{\delta x \to 0} \frac{f(x_0 + 2\delta x, y_0) - 2f(x_0 + \delta x, y_0) + f(x_0, y_0)}{(\delta x)^2}$$

$$\frac{\partial^2 f}{\partial y^2} = \lim_{\delta y \to 0} \frac{f(x_0, y_0 + 2\delta y) - 2f(x_0, y_0 + \delta y) + f(x_0, y_0)}{(\delta y)^2}$$

$$\frac{\partial^2 f}{\partial y \partial x} = \lim_{\delta y \to 0} \lim_{\delta x \to 0} \frac{f(x_0 + \delta x, y_0 + \delta y) - f(x_0, y_0 + \delta y) - f(x_0 + \delta x, y_0) + f(x_0, y_0)}{\delta y \delta x}$$

and $\dfrac{\partial^2 f}{\partial x \partial y} = \lim\limits_{\delta x \to 0} \lim\limits_{\delta y \to 0} \dfrac{f(x_0 + \delta x, y_0 + \delta y) - f(x_0 + \delta x, y_0) - f(x_0, y_0 + \delta y) + f(x_0, y_0)}{\delta x \, \delta y}$

☞ **REMARK**

⇒ It is clear that $\dfrac{\partial^2 f}{\partial x \partial y} = \dfrac{\partial^2 f}{\partial y \partial x}$

SOLVED EXAMPLES

EXAMPLE 1.   *Let f(x, y) be a function, defined by*

$$f(x, y) = x \sin\frac{1}{x} + y \sin\frac{1}{y}, x \neq 0, y \neq 0$$

$$f(0, y) = y \sin\frac{1}{y}, y \neq 0$$

$$f(x, 0) = x \sin\frac{1}{x}, x \neq 0$$

$$f(0, 0) = 0$$

Examine the existence of f_x and f_{yx} at x = 0, y = 0.

SOLUTION. Here, we have

$$f_x(0,0) = \lim_{\delta x \to 0} \frac{f(\delta x, 0) - f(0,0)}{\delta x}$$

$$= \lim_{\delta x \to 0} \frac{\delta x \sin \dfrac{1}{\delta x} - 0}{\delta x}$$

$$= \lim_{\delta x \to 0} \sin \frac{1}{\delta x}$$

Since, limit does not exist

$\Rightarrow \quad f_x(0, 0)$ does not exists.

Now, $\quad f_{yx} = \lim_{\delta y \to 0} \lim_{\delta x \to 0} \dfrac{f(\delta x, \delta y) - f(0, \delta y) - f(\delta x, 0) + f(0,0)}{\delta x \delta y}$

$$= \lim_{\delta y \to 0} \lim_{\delta x \to 0} \frac{\delta x \sin \dfrac{1}{\delta x} + \delta y \sin \dfrac{1}{\delta y} - \delta y \sin \dfrac{1}{\delta y} - \delta x \sin \dfrac{1}{\delta x}}{\delta x \delta y}$$

$$= \lim_{\delta y \to 0} \lim_{\delta x \to 0} \frac{0}{\delta x \delta y} = 0$$

In spite of the fact that limit is zero, the derivative $f_{yx}(0, 0)$ cannot be said to exist, since $f_y(0, 0)$ does not exist.

☛ **REMARK**

⇒ The existence of higher derivatives implies the existence of the corresponding derivatives of lower order. Thus, in order that f_{xy} should exist at a point, it is necessary that the partial derivatives f_x should exist in the neighbourhood of that point. However, it is possible for the limit, defining f_{xy} to exist without the partial derivative f_x existing. In such cases higher derivatives cannot be said to exist.

EXAMPLE 2. Let $f(x,y) = \dfrac{x^2 y}{x^4 + y^2}$, for $x \neq 0$, $y \neq 0$ and $f(0, 0) = 0$

Show that the partial derivatives f_x, f_y exist everywhere in the region $-1 \le x \le 1$, $-1 \le y \le 1$, although $f(x, y)$ is discontinuous in (x, y) at the origin.

SOLUTION. Here, we have

$$f(x,y) = \frac{x^2 y}{x^4 + y^2} \qquad\qquad (x \neq 0, y \neq 0)$$

$\Rightarrow \qquad\qquad f_x = 2xy \cdot \dfrac{y^2 - x^4}{(x^4 + y^2)^2} \qquad\qquad (x \neq 0, y \neq 0)$

and $\qquad\qquad f_y = x^2 \cdot \dfrac{x^4 - y^2}{(x^4 + y^2)^2} \qquad\qquad (x \neq 0, y \neq 0)$

Also, for $x = y = 0$, we obtain

$$f_x = \lim_{\delta x \to 0} \frac{\delta x \cdot 0}{(\delta x)^4 + 0} = 0 \ \text{ and } f_y = 0$$

Similarly, the following results can be prove

$$f_x(x, y) = 0 \text{ for } x = 0, y \neq 0$$
$$f_x(x, y) = 0 \text{ for } x \neq 0, y = 0$$
$$f_y(x, y) = 0 \text{ for } x = 0, y \neq 0$$

$$f_y(x,y) = \frac{1}{x^2} \text{ for } x \neq 0, y = 0$$

Hence, the partial derivative f_x, f_y exist at all points of the given region.
Now, we shall check the continuity of the given function.
The limiting value of $f(x)$, along the line $y = 0$, is given by $\lim\limits_{\delta x \to 0} .0 = 0$.

and, limiting value of $f(x)$, along the line $y = x^2$ is given by $\lim\limits_{x \to 0} \dfrac{x^4}{x^4 + x^4} = \dfrac{1}{2}$.

Now, since limit obtained by two different approaches are different. Hence $f(x)$ is discontinuous in (x, y) at the origin.

EXAMPLE 3. **If $u = f(y / x)$, then prove that $x\dfrac{\partial u}{\partial x} + y\dfrac{\partial u}{\partial y} = 0$**

SOLUTION. Here, we have $u = f\left(\dfrac{y}{x}\right)$

$$\therefore \qquad \dfrac{\partial u}{\partial x} = f'\left(\dfrac{y}{x}\right)\left(-\dfrac{y}{x^2}\right)$$

$$\Rightarrow \qquad x\dfrac{\partial u}{\partial x} = -\dfrac{y}{x} f'\left(\dfrac{y}{x}\right) \qquad\qquad \text{... (1)}$$

Also $\qquad \dfrac{\partial u}{\partial y} = f'\left(\dfrac{y}{x}\right).\dfrac{1}{x}$

$$\therefore \qquad y\dfrac{\partial u}{\partial y} = \dfrac{y}{x} f'\left(\dfrac{y}{x}\right) \qquad\qquad \text{... (2)}$$

Now, from (1) and (2), we get

$$x\dfrac{\partial u}{\partial x} + y\dfrac{\partial u}{\partial y} = 0$$

EXAMPLE 4. **If $z = x^2 \tan^{-1}\left(\dfrac{y}{x}\right) - y^2 \tan^{-1}\left(\dfrac{x}{y}\right)$, then prove that $\dfrac{\partial^2 z}{\partial y \partial x} = \dfrac{x^2 - y^2}{x^2 + y^2}$**

SOLUTION. Here, we have

$$z = x^2 \tan^{-1}\left(\dfrac{y}{x}\right) - y^2 \tan^{-1}\left(\dfrac{x}{y}\right)$$

$$\therefore \quad \dfrac{\partial z}{\partial x} = 2x \tan^{-1}\dfrac{y}{x} + x^2 . \dfrac{1}{1 + \dfrac{y^2}{x^2}}\left(-\dfrac{y}{x^2}\right) - y^2 \dfrac{1}{1 + \dfrac{x^2}{y^2}} . \dfrac{1}{y}$$

$$= 2x \tan^{-1}\dfrac{y}{x} - \dfrac{x^2 y}{x^2 + y^2} - \dfrac{y^3}{x^2 + y^2}$$

Now, $\dfrac{\partial^2 z}{\partial y \partial x} = \dfrac{\partial}{\partial y}\left(\dfrac{\partial z}{\partial x}\right)$

$$= \dfrac{\partial}{\partial y}\left[2x \tan^{-1}\dfrac{y}{x} - \dfrac{x^2 y}{x^2 + y^2} - \dfrac{y^3}{x^2 + y^2}\right]$$

$$= 2x.\dfrac{1}{1 + \dfrac{y^2}{x^2}}.\dfrac{1}{x} - x^2\left[\dfrac{(x^2 + y^2).1 - y.2y}{(x^2 + y^2)^2}\right] - \dfrac{(x^2 + y^2)3y^2 - y^3.2y}{(x^2 + y^2)^2}$$

$$= \dfrac{2x^2}{x^2 + y^2} - x^2\left[\dfrac{x^2 - y^2}{(x^2 + y^2)^2}\right] - \dfrac{y^2(3x^2 + 3y^2 - 2y^2)}{(x^2 + y^2)^2}$$

$$= \frac{2x^2}{x^2+y^2} - x^2\left[\frac{x^2-y^2}{(x^2+y^2)^2}\right] - y^2\left[\frac{3x^2+y^2}{(x^2+y^2)^2}\right]$$

$$= \frac{2x^2(x^2+y^2) - x^2(x^2-y^2) - y^2(3x^2+y^2)}{(x^2+y^2)^2}$$

$$= \frac{2x^4 + 2x^2y^2 - x^4 + x^2y^2 - 3x^2y^2 - y^4}{(x^2+y^2)^2}$$

$$= \frac{x^4-y^4}{(x^2+y^2)^2} = \frac{(x^2-y^2)(x^2+y^2)}{(x^2+y^2)^2}$$

$$= \frac{x^2-y^2}{(x^2+y^2)}$$

EXAMPLE 5. *If $x^x y^y z^z = c$, then show that $\dfrac{\partial^2 z}{\partial x \partial y} = -[x\log(ex)]^{-1}$ when $x = y = z$.*

SOLUTION. Here, we have

$$x^x y^y z^z = c$$

Taking logarithm, we get

$$x\log x + y\log y + z\log z = \log c.$$

Now, differentiating partially w.r.t. x, we get

$$\left[x.\frac{1}{x} + 1.\log x\right] + \left[z.\frac{1}{z} + 1.\log z\right]\frac{\partial z}{\partial x} = 0 \qquad [\because z \text{ is a function of } x \text{ and } y]$$

$$\Rightarrow \quad (1+\log x) + (1+\log z)\frac{\partial z}{\partial x} = 0$$

$$\Rightarrow \qquad\qquad \frac{\partial z}{\partial x} = -\frac{1+\log x}{1+\log z}$$

Similarly, we get

$$\frac{\partial z}{\partial y} = -\frac{1+\log y}{1+\log z}$$

Now

$$\frac{\partial^2 z}{\partial x\,\partial y} = \frac{\partial}{\partial x}\left(\frac{\partial z}{\partial y}\right) = \frac{\partial}{\partial x}\left[-\frac{1+\log y}{1+\log z}\right] = -(1+\log y)\frac{\partial}{\partial x}\left(\frac{1}{1+\log z}\right)$$

$$= -(1+\log y)\left[\frac{-1}{(1+\log z)^2}.\frac{1}{z}.\frac{\partial z}{\partial x}\right] = \frac{(1+\log y)}{(1+\log z)^2}.\frac{1}{z}.\frac{\partial z}{\partial x}$$

$$= \frac{(1+\log y)}{(1+\log z)^2}\left[-\frac{1}{z}\left(\frac{1+\log x}{1+\log z}\right)\right] = -\frac{(1+\log y)(1+\log x)}{z(1+\log z)^3}$$

$$= -\frac{(1+\log x)(1+\log x)}{z(1+\log x)^3}, \text{at } x = y = z$$

$$= -\frac{1}{x(1+\log x)} = -\frac{1}{x[\log e + \log x]}$$

$$= -\frac{1}{x\log[ex]} = -[x\log(ex)]^{-1}$$

EXAMPLE 6. *If $u = x^y$, then show that $\dfrac{\partial^2 u}{\partial y\,\partial x} = \dfrac{\partial^2 u}{\partial x\,\partial y}$.*

SOLUTION. Here, we have $u = x^y$

$$\Rightarrow \qquad \frac{\partial u}{\partial y} = x^y . \log x$$

$$\frac{\partial}{\partial x}\left(\frac{\partial u}{\partial y}\right) = \frac{\partial^2 u}{\partial x \partial y} = yx^{y-1}\log x + x^y x$$

$$= yx^{y-1}\log x + x^{y-1} \qquad \qquad \dots (1)$$

Now $\qquad \frac{\partial u}{\partial x} = yx^{y-1}$

$$\therefore \qquad \frac{\partial^2 u}{\partial y \partial x} = \frac{\partial}{\partial y}\left(\frac{\partial u}{\partial x}\right) = \frac{\partial}{\partial y}(yx^{y-1})$$

$$= x^{y-1} + yx^{y-1}\log x \qquad \qquad \dots (2)$$

From (1) and (2), we conclude that

$$\frac{\partial^2 u}{\partial x \partial y} = \frac{\partial^2 u}{\partial y \partial x}.$$

EXAMPLE 7. *Let f(x, y) be a function defined by*

$$f(x,y) = \begin{cases} xy\dfrac{x^2-y^2}{x^2+y^2} & \text{for } (x,y) \neq (0,0) \\ 0 & \text{for } (x,y) = (0,0) \end{cases}$$

Prove that

(i) f, f_x, f_y are continuous in (x, y),

(ii) f_{xy} and f_{yx} exist at every point (x, y) and are continuous except at (0,0)

(iii) $f_{xy}(0, 0) = 1$ and $f_{yx}(0, 0) = -1$.

SOLUTION. Firstly, we examine the continuity of f at (0, 0)

Putting $x = r \cos \theta, y = r \sin \theta$, we get

$$f(r\cos\theta, r\sin\theta) = r^2 \cos\theta \sin\theta.\left(\frac{r^2\cos^2\theta - r^2\sin^2\theta}{r^2\cos^2\theta + r^2\sin^2\theta}\right)$$

$$= r^2 \sin\theta\cos\theta\,(\cos^2\theta - \sin^2\theta)$$

$$= \frac{1}{2}r^2 \sin 2\theta\cos 2\theta = \frac{1}{4}r^2 \sin 4\theta$$

$$\le \varepsilon \text{ where } r < 2\sqrt{(\varepsilon \operatorname{cosec} 4\theta)}$$

Since, $\sqrt{(\operatorname{cosec} 4\theta)}$ is never less than one, it follows that if we put $r = 2\sqrt{\varepsilon}$ then ε may be chosen such that $f(r \cos\theta, r \sin \theta)$ is always less than ε for all values of θ and r. Thus, the transformed function is uniformly continuous in r for all values of θ. Hence, the function f is continuous in (x, y) together, at the origin.

Now, we show that f_x and f_y are continuous at (0, 0).

For $(x, y) \neq (0, 0)$, we have

$$f_x = y\left(\frac{x^2-y^2}{x^2+y^2} + \frac{4x^2y^2}{(x^2+y^2)^2}\right), f_y = x\left(\frac{x^2-y^2}{x^2+y^2} - \frac{4x^2y^2}{(x^2+y^2)^2}\right)$$

For $(x, y) = (0, 0)$ we have

$$f_x(0,0) = \lim_{\delta x \to 0} \frac{\delta x.0}{\delta x}\left[\frac{(\delta x)^2 - 0}{(\delta x)^2 + 0}\right] = 0$$

$$f_y(0,0) = \lim_{\delta y \to 0} \frac{0.\delta y}{\delta y}\left[\frac{0 - (\delta y)^2}{0 + (\delta y)^2}\right] = 0$$

Now, we shall show that f_x and f_y are continuous in (x, y) at the origin. We only prove the continuity of f_x at the origin. Now, putting $x = r \cos\theta$, $y = r \sin\theta$, we get

$$f_x(r\cos\theta, r\sin\theta) = r\sin\theta\left[\frac{r^2\cos^2\theta - r^2\sin^2\theta}{r^2\cos^2\theta + r^2\sin^2\theta} + \frac{4r^4\cos^2\theta\sin^2\theta}{(r^2\cos^2\theta + r^2\sin^2\theta)^2}\right]$$

$$= r\sin\theta(\cos 2\theta + \sin^2 2\theta)$$
$$\leq 2r, \text{ for all values of } \theta.$$

Hence, if we take $r = \dfrac{\varepsilon}{2}$ then ε may be chosen such that
$$f_x(r\cos\theta, r\sin\theta) < \varepsilon \text{ for } r \text{ and for all values of } \theta.$$
Hence, f_x is continuous at the origin.
Similarly, we can shown that f_y is continuous at the origin.
Now,

$$f_{xy}(0,0) = \lim_{\delta x \to 0}\lim_{\delta y \to 0}\left[\frac{f(0+\delta x, 0+\delta y) - f(0+\delta x, 0) - f(0, 0+\delta y) + f(0,0)}{\delta x.\delta y}\right]$$

$$= \lim_{\delta x \to 0}\lim_{\delta y \to 0}\left[\frac{(\delta x)^2 - (\delta y)^2}{(\delta x)^2 + (\delta y)^2}\right]$$

$$= \lim_{\delta x \to 0}\left[\frac{(\delta x)^2 - 0}{(\delta x)^2 + 0}\right] = 1$$

and $f_{yx}(0,0) = \lim_{\delta y \to 0}\lim_{\delta x \to 0}\left[\frac{(\delta x)^2 - (\delta y)^2}{(\delta x)^2 + (\delta y)^2}\right]$

$$= \lim_{\delta y \to 0}\left[\frac{0 - (\delta y)^2}{0 + (\delta y)^2}\right] = -1$$

Hence, $f_{xy}(0, 0) \neq f_{yx}(0, 0)$, in the case. It may be shown that the order of differentiation can be interchanged at every other point of the finite region.

EXAMPLE 8.   Let $f(x,y) = \begin{cases} \dfrac{1}{4}(x^2 + y^2)\log(x^2 + y^2) & , \text{ when } (x,y) \neq (0,0) \\ 0 & , \text{ when } (x,y) = (0,0) \end{cases}$

Show that $f_{xy} = f_{yx}$ at all points (x, y). Also, show that neither of the derivatives is continuous in (x, y) at the origin.

SOLUTION.   For $x \neq 0$ $y \neq 0$, we have

$$f_x = \frac{1}{2}x\{1 + \log(x^2 + y^2)\} \text{ and } f_y = \frac{1}{2}y\{1 + \log(x^2 + y^2)\}$$

and $\qquad f_{xy} = f_{yx} = \dfrac{xy}{x^2 + y^2}$

For $x = 0$, $y = 0$, we have
$$f_x = f_y = f_{xy} = f_{yx} = 0$$
Hence, $\qquad f_{xy} = f_{yx}$ at every point.
Now, we show that $f_{xy} = f_{yx}$ is not continuous at $(0, 0)$.

Since $\quad \lim_{(x,y)\to(0,0)} \dfrac{xy}{x^2 + y^2}$ does not exist.

(Because, if we put $y = mx$, then limit of the function, depends upon m, i.e., limit is not unique). Hence, the limit does not exist. It follows that $f_{xy} = f_{yx}$ is not continuous at the origin.

EXAMPLE 9. Let $f(x,y) = \begin{cases} x^2 y^2 \cdot \cos\dfrac{1}{x}, & \text{for all values of } y \text{ so long as } x \neq 0 \\ 0, & \text{for } x = 0 \end{cases}$

Show that

(i) $f_{xy} = f_{yx}$ at all points (x, y),

(ii) neither f_{xy} nor f_{yx} is continuous in x at x = 0, if y ≠ 0.

and (iii) both f_{xy} and f_{yx} are continuous in (x, y), together at the origin.

SOLUTION. (i) For $x \neq 0, y \neq 0$, we have

$$f_{xy} = f_{yx} = 4xy\cos\frac{1}{x} + 2y\sin\frac{1}{x}$$

For $(x=0, y\neq 0)$, $(x=0, y=0)$ and $(x\neq 0, y=0)$, it can be easily shown that

$$f_{xy} = f_{yx} = 0$$

Hence, $f_{xy} = f_{yx}$ at every point.

(ii) If $y \neq 0$, then

$$\lim_{x\to 0}\left(4xy\cos\frac{1}{x} + 2y\sin\frac{1}{x}\right) \text{ does not exist.}$$

Hence, neither f_{xy} nor f_{yx} is continuous in x at x = 0 if y ≠ 0.

(iii) Both f_{xy} and f_{yx} are continuous in (x, y) together at the origin. Since, the simultaneous limit

$$\lim_{(x,y)\to(0,0)}\left\{4xy\cos\frac{1}{x} + 2y\sin\frac{1}{x}\right\} = 0 = f_{xy}(0,0) = f_{yx}(0,0)$$

EXAMPLE 10. Find $\dfrac{\partial f}{\partial x}, \dfrac{\partial f}{\partial y}$ at (1, 2) if $f(x, y) = 2x^2 - xy + 2y^2$

SOLUTION. We have

$$\left(\frac{\partial f}{\partial x}\right)_{(1,2)} = \lim_{h\to 0}\frac{f(1+h,2) - f(1,2)}{h}$$

$$= \lim_{h\to 0}\frac{\{2(1+h)^2 - (1+h).2 + 2(2)^2\} - \{2(1)^2 - 1.2 + 2(2)^2\}}{h}$$

$$= \lim_{h\to 0}\frac{2h^2 + 2h}{h} = \lim_{h\to 0}(2h+2) = 2$$

Similarly $\left(\dfrac{\partial f}{\partial y}\right)_{(1,2)} = \lim_{k\to 0}\dfrac{f(1,2+k) - f(1,2)}{k}$

$$= \lim_{k\to 0}\frac{\{2-(2+k)+2(2+k)^2\} - \{2-2+8\}}{k}$$

$$= \lim_{k\to 0}\frac{2k^2 + 7k}{k} = \lim_{k\to 0}2k+7 = 7.$$

EXERCISE 10.2

1. If $z = \left[\dfrac{x^2+y^2}{x+y}\right]$, show that

$$\left(\frac{\partial z}{\partial x} - \frac{\partial z}{\partial y}\right)^2 = 4\left(1 - \frac{\partial z}{\partial x} - \frac{\partial z}{\partial y}\right).$$

2. If $u = f\left(\dfrac{y}{x}\right)$, show that $x\dfrac{\partial u}{\partial x} + y\dfrac{\partial u}{\partial y} = 0$.

3. If $u = \sin^{-1}\dfrac{x}{y} + \tan^{-1}\dfrac{y}{x}$ show that

$$x\frac{\partial u}{\partial x} + y\frac{\partial u}{\partial y} = 0.$$

4. If $u = \sin^{-1} \dfrac{\sqrt{x} - \sqrt{y}}{\sqrt{x} + \sqrt{y}}$, show that $x\dfrac{\partial u}{\partial x} + y\dfrac{\partial u}{\partial y} = 0$

5. If $u = \tan^{-1} \dfrac{xy}{\sqrt{1 + x^2 + y^2}}$, show that

$$\dfrac{\partial^2 u}{\partial x \partial y} = \dfrac{1}{(1 + x^2 + y^2)^{3/2}}.$$

6. If $u = \sin(\sqrt{x} + \sqrt{y})$, show that

$$x\dfrac{\partial u}{\partial x} + y\dfrac{\partial u}{\partial y} = \dfrac{1}{2}(\sqrt{x} + \sqrt{y})\cos(\sqrt{x} + \sqrt{y}).$$

7. If $u = f(r)$, where $r^2 = x^2 + y^2$, show that

$$\dfrac{\partial^2 u}{\partial x^2} + \dfrac{\partial^2 u}{\partial y^2} = f''(r) + \dfrac{f'(r)}{r}.$$

8. If $f(x,y) = \begin{cases} \dfrac{xy}{x^2 + y^2} & , \quad x \neq 0, y \neq 0 \\ 0 & , \quad x = 0, y = 0 \end{cases}$

Then , show that f_x, f_y exist at $(0, 0)$ and examine the continuity of f_x, f_y with respect to x, y and (x, y) together.

9. In any given region **R**, where $f(x, y)$ is continuous in x and y together and $f_x = 0$, $f_y = 0$ for all values of x and y in **R**. Show that $f(x, y)$ is constant in **R**.

10. Show that $f(x, y) = |x| + |y|$ is continuous at $(0, 0)$.

11. If $f(x,y) = (x^2 + y^2)\tan^{-1} \dfrac{y}{x}$ when $x \neq 0$

and $\qquad f(0, y) = \dfrac{\pi y^2}{2}$

Show that $f_{xy}(0,0) \neq f_{yx}(0,0)$.

12. If $f(x,y) = \begin{cases} xy & \text{if } |y| \leq |x| \\ -xy & \text{if } |y| > |x| \end{cases}$

Show that $f_{xy}(0, 0) \neq f_{yx}(0, 0)$

HINTS TO SELECTED PROBLEMS

1. $\dfrac{\partial z}{\partial x} = \dfrac{x^2 - y^2 + 2xy}{(x + y)^2}, \dfrac{\partial z}{\partial y} = \dfrac{y^2 - x^2 + 2xy}{(x + y)^2}$

2. $\dfrac{\partial u}{\partial x} = \left[f'\left(\dfrac{y}{x}\right) \right]\left(-\dfrac{y}{x^2}\right)$

$\Rightarrow x\dfrac{\partial u}{\partial x} = \left(-\dfrac{y}{x}\right)f'\left(\dfrac{y}{x}\right)$ ... (1)

Similarly, $y\dfrac{\partial u}{\partial y} = \dfrac{y}{x}f'\left(\dfrac{y}{x}\right)$ ... (2)

Adding (1) and (2) to get the required results.

10. The function $f(x, y)$ will be continuous at $(0, 0)$ if the simultaneous limit of $f(x, y)$ as $x \to 0, y \to 0$ exists.

$\therefore \lim\limits_{\substack{x \to 0 \\ y \to 0}} f(x,y) = \lim\limits_{\substack{x \to 0 \\ y \to 0}} (|x| + |y|)$

$= \lim\limits_{\substack{x \to 0 \\ y \to 0}} (|x| + |mx|)$

$= \lim\limits_{x \to 0} |x|.(1 + m) = 0 = f(0,0).$

ANSWERS

8. $f_x(0, 0) = f_y(0, 0) = 0, f_x$ is continuous in x, discontinuous in y at $(0, 0)$ and discontinuous in (x, y) at $(0, 0)$; f_y is continuous in y, discontinuous in x at $(0,0)$ and discontinuous in (x, y) at $(0,0)$.

10.5 DIFFERENTIABILITY OF FUNCTION OF TWO VARIABLES

A function $f(x, y)$ defined on an open interval $]a, b[$ is said to be totally differentiable or simply differentiable in the open interval $]a, b[$, if there exist two constant α and β depending on f and $]a, b[$ such that

$$\lim\limits_{\substack{\delta x \to 0 \\ \delta y \to 0}} \dfrac{f(a + \delta x, b + \delta y) - f(a,b) - \alpha \delta x - \beta \delta y}{\sqrt{(\delta x)^2 + (\delta y)^2}} = 0$$

or $\qquad \lim\limits_{\substack{h \to 0 \\ k \to 0}} \dfrac{f(a + h, b + k) - f(a,b) - \alpha h - \beta k}{\sqrt{h^2 + k^2}} = 0$

or $\qquad \lim\limits_{\substack{h \to 0 \\ k \to 0}} \dfrac{f(a + h, b + k) - f(a,b) - hf_x(a,b) - kf_y(a,b)}{\sqrt{h^2 + k^2}} = 0$

THEOREM 1. **If the function f(x, y) is totally differentiable at the point (a, b), then it is continuous at (a, b).**

PROOF. Since $f(x, y)$ is differentiable at (a, b) we have

$$\lim_{\substack{h \to 0 \\ k \to 0}} \frac{f(a+h, b+k) - f(a,b) - hf_x(a,b) - kf_y(a,b)}{\sqrt{h^2 + k^2}} = 0$$

$$\Rightarrow \lim_{\substack{h \to 0 \\ k \to 0}} [f(a+h, b+k) - f(a,b) - hf_x(a,b) - kf_y(a,b)] = 0$$

$$\Rightarrow \lim_{\substack{h \to 0 \\ k \to 0}} f(a+h, b+k) = f(a,b).$$

Hence, $f(x, y)$ is continuous at $[a, b]$.

THEOREM 2. **If a function f(x, y) is totally differentiable, then the partial derivatives f_x and f_y both exist and finite.**

PROOF. Let $f(x, y)$ be a function, which is totally differentiable , then there exist constants α and β such that

$$\lim_{\substack{h \to 0 \\ k \to 0}} \frac{f(a+h, b+k) - f(a,b) - \alpha h - \beta k}{\sqrt{h^2 + k^2}} \quad \text{exists.}$$

$$\Rightarrow \lim_{k \to 0} \frac{f(a, b+k) - f(a,b) - \beta k}{k} \quad \text{exists.}$$

$$\Rightarrow f_y(a, b) = \beta$$

Similarly, we get $f_x(a, b) = \alpha$

☛ **REMARKS**

➡ Continuity in two variables is a necessary condition for total differentiability. It is not a sufficient condition.

➡ If $z = f(x, y)$, then $z + \delta z = f(x + \delta x, y + \delta y)$ so that $\delta z = f(x + \delta x, y + \delta y) - f(x, y)$. If z is differentiable, then the value of dz is given by

$$\delta z = \frac{\partial z}{\partial x} \delta x + \frac{\partial z}{\partial y} \delta y + \varepsilon \delta(x, y) \quad \text{where } \varepsilon \to 0 \text{ as } \delta x \text{ and } \delta y \text{ approaches zero}$$

simultaneously.

➡ The increment $\{f(x, y) - f(a, b)\}$ itself may depend on x and y, the differentiability expresss that as an approximation we can replace the increment by the very simple linear function $A(x - a) + B(y - b)$ i.e., the differentiable of dz. The error, thus, introduce as $\varepsilon \delta(x, y)$, which is vanishing of a higher order than the linear expression, or in general small in comparision with this.

➡ Geometrically, the approximation , discussed above means that, the given surface $z = f(x, y)$ is replaced by the plane. $z' - z_0 = A(x - a) + B(y - b)$
which is called the tangent plane to the surface at the point $P[x_0, y_0, f(x_0, y_0)]$. A normal vector to the surface (i.e., to its tangent plane) at the point P, is therefore

$$(A, B, -1) = (f_x'(a, b), f_y'(u, b), -1)$$

10.5.1 CONDITION FOR DIFFERENTIABILITY IN POLAR COORDINATES

Let $f(x, y)$ be a differentiable function at a point (a, b), then

$$\lim_{(h,k) \to (0,0)} \frac{f(a+h, b+k) - f(a,b) - hf_x(a,b) - kf_y(a,b)}{\sqrt{h^2 + k^2}} = 0$$

Put $h = r\cos\theta, y = r\sin\theta$ and take limit $r \to 0$, then we have

$$\lim_{r \to 0}\left[\frac{f(a + r\cos\theta, b + r\sin\theta) - f(a,b)}{r} - \cos\theta f_x(a,b) - \sin\theta f_y(a,b)\right] = 0$$

$$\Rightarrow \lim_{r \to 0}\left[\frac{f(a + r\cos\theta, b + r\sin\theta) - f(a,b)}{r}\right] = \cos\theta f_x(a,b) + \sin\theta f_y(a,b)$$

THEOREM I. *Let $f(x, y)$ be a function such that $f_x(x, y)$, $f_y(x, y)$ exist at the point (x_0, y_0) and let one of those derivative say $f_y(x, y)$ exist for all values of x, y in the neighbourhood of (x_0, y_0) and be continuous at that point in two variables together. Then, $f(x, y)$ is totally differentiable at (x_0, y_0).*

PROOF. Let $\varepsilon > 0$ be given. Since, the partial derivative $f_x(x, y)$ exist at (x_0, y_0), there exist a $\delta_1 > 0$ such that for $|\delta x| < \delta_1$, we have

$$\left|\frac{f(x_0 + \delta x, y_0) - f(x_0, y_0)}{\delta x} - f(x_0 y_0)\right| < \frac{\varepsilon}{2}$$

Let $\dfrac{f(x_0 + \delta x, y_0) - f(x_0, y_0)}{\delta x} - f(x_0 y_0) = k$, where $|k| < \dfrac{\varepsilon}{2}$ ... (1)

Since, $f_y(x, y)$ exists for all values of x and y in the neighbourhood of (x_0, y_0) and is continuous in (x, y) together at this point, then by mean value theorem, we have

$$f(x_0 + \delta x, y_0 + \delta y) - f(x_0 + \delta x, y_0) = \delta y \, f_y(x_0 + \delta x, y_0 + \theta\delta y) \quad \text{... (2)}$$

Therefore,

$$\left|\frac{1}{\delta(x,y)}\left[f(x_0 + \delta x, y_0 + \delta y) - f(x_0, y_0) - \delta x f_x(x_0, y_0) - \delta y \cdot f_y(x_0, y_0)\right]\right|$$

$$= \left|\frac{1}{\delta(x,y)}[f(x_0 + \delta x, y_0 + \delta y) - f(x_0 + \delta x, y_0) + f(x_0 + \delta x, y_0)\right.$$

$$\left. - f(x_0, y_0) - \delta x f_x(x_0, y_0) - \delta y f_y(x_0, y_0)]\right|$$

$$= \left|\frac{1}{\delta(x,y)}[\delta y \, f_y(x_0 + \delta x, y_0 + \theta\delta y) + \delta x \, f_x(x_0, y_0) + k\,\delta x - \delta x \, f_x(x_0, y_0)\right.$$

$$\left. - \delta_y f_y(x_0, y_0)]\right|$$

$$= \left|\frac{\delta y}{\delta(x,y)}\{f_y(x_0 + \delta x, y_0 + \theta\delta y) - f_y(x_0, y_0)\} + \frac{\delta x}{\delta(x,y)}.k\right|$$

$$\leq \left|\frac{\delta y}{\delta(x,y)}\right| |f_y(x_0 + \delta x, y_0 + \theta\delta y) - f_y(x_0, y_0)| + \left|\frac{\delta x}{\delta(x,y)}.k\right| \quad \text{... (3)}$$

Now, since $f_y(x, y)$ is continuous in (x, y) at (x_0, y_0), there exist a $\delta_2 > 0$ such that

$$|f_y(x_0 + \delta x, y_0 + \theta\delta y) - f_y(x_0, y_0)| < \varepsilon / 2 \quad \text{...(4)}$$

For $|\delta x| < \delta_2$ and $|\delta y| < \delta_2$.

Since, $\delta x \neq 0$, $\delta y \neq 0$, we have

$$\left|\frac{\delta x}{\delta(x,y)}\right| < 1, \left|\frac{\delta x}{\delta(x,y)}\right| < 1 \quad \text{... (5)}$$

Let $\delta_3 = \min\{\delta_1, \delta_2\}$, then for $|\delta x| < \delta_3$ and $|\delta y| < \delta_3$ the inequality (3), with the help of (4) and (5), becomes

$$\left|\frac{1}{\delta(x,y)}[f(x_0 + \delta x, y_0 + \delta y) - f(x_0, y_0) - \delta x f_x(x_0, y_0) - \delta y f_y(x_0, y_0)]\right|$$

$$< \frac{\varepsilon}{2} + \frac{\varepsilon}{2} = \varepsilon$$

$$\therefore \quad \lim_{\substack{\delta x \to 0 \\ \delta y \to 0}} \frac{f(x_0 + \delta x, y_0 + \delta y) - f(x_0, y_0) - \delta x\, f_x(x_0, y_0) - \delta y\, f_y(x_0, y_0)}{\delta(x, y)} = 0$$

It follows that, $f(x, y)$ is totally differentiable at (x_0, y_0).

SOLVED EXAMPLES

EXAMPLE 1.   **Let $f(x, y) = \begin{cases} \dfrac{xy}{\sqrt{x^2 + y^2}} &, \quad (x, y) \neq (0, 0) \\ 0 &, \quad (x, y) = (0, 0) \end{cases}$**

Show that $f(x, y)$ is continuous but not differentiable at $(0, 0)$.

SOLUTION.   (i) *Test for continuity.* Let us suppose (x, y) approaches $(0, 0)$ through the line $y = mx$, then

$$\lim_{\substack{x \to 0 \\ y \to 0}} \frac{xy}{\sqrt{x^2 + y^2}} = \lim_{x \to 0} \frac{x\,mx}{\sqrt{x^2 + m^2 x^2}} = 0 \qquad \dots (1)$$

Again, let (x, y) approaches through the path $x = y^3$, then

$$\lim_{\substack{x \to 0 \\ y \to 0}} \frac{xy}{\sqrt{x^2 + y^2}} = \lim_{y \to 0} \frac{y^4}{\sqrt{y^6 + y^2}} = 0 \qquad \dots (2)$$

Now, from (1) and (2), it is obvious that

$$\lim_{\substack{x \to 0 \\ y \to 0}} \frac{xy}{\sqrt{x^2 + y^2}} \text{ exists and equal to } 0.$$

Since $f(0, 0) = 0$ (given)

Hence, the function $f(x, y)$ is continuous at $(0, 0)$.

(ii) *Test for differentiability.* Here, we have

$$f_x(0, 0) = \lim_{h \to 0} \frac{f(0 + h, 0) - f(0, 0)}{h}$$

$$= \lim_{h \to 0} \frac{f(h, 0) - f(0, 0)}{h} = 0$$

and $\quad f_y(0, 0) = \lim_{k \to 0} \dfrac{f(0, 0 + k) - f(0, 0)}{k} = \lim_{k \to 0} \dfrac{f(0, k) - f(0, 0)}{k} = 0$

Now $\quad \displaystyle\lim_{(h,k) \to (0,0)} \frac{f(0 + h, y + k) - f(0, 0) - h f_x(0, 0) - k f_y(0, 0)}{\sqrt{h^2 + k^2}}$

$$= \lim_{(h,k) \to (0,0)} \frac{f(h, k)}{\sqrt{h^2 + k^2}} \qquad [\because\ f_x(0, 0) = f_y(0, 0) = 0]$$

$$= \lim_{(h,k) \to (0,0)} \frac{hk}{h^2 + k^2}$$

$$= \lim_{(h,k) \to (0,0)} \frac{mh^2}{h^2(1 + m^2)} \qquad [\text{Let } (h, k) \to (0, 0) \text{ along } k = mh]$$

$$= \frac{m}{1 + m^2}$$

which implies that the required limit is depend upon m *i.e.*, it depends upon the path along with $(h, k) \to (0, 0)$, so this limit does not exist.

Hence, $f(x, y)$ is not differentiable at $(0, 0)$.

EXAMPLE 2. *Examine the continuity and differentiablility of the function*

$$f(x, y) = \begin{cases} \dfrac{xy^2}{x^2 + y^2} & \textbf{when} \quad (x, y) \neq (0, 0) \\ 0 & \textbf{when} \quad (x, y) = (0, 0) \end{cases}$$

SOLUTION. (i) *Test for continuity.* Firstly, we test the continuity of $f(x, y)$ at $(0, 0)$.

Let $\varepsilon > 0$ be given.

Let $x = r\cos\theta$ and $y = r\sin\theta$, then

$$f(r\cos\theta, y\sin\theta) = \frac{r\cos\theta . r^2\sin^2\theta}{r^2(\cos^2\theta + \sin^2\theta)} = r\cos\theta\sin^2\theta$$

Now, consider

$$|f(r\cos\theta, r\sin\theta) - f(0,0)| = |r\cos\theta\sin^2\theta - 0|$$

$$= |r\cos\theta|\sin^2\theta$$

$$\leq r \text{ for all values of } \theta.$$

If we set $r_0 = \varepsilon$, then for all values of θ and for $r < r_0$, we have

$$|f(r\cos\theta, r\sin\theta) - f(0, 0)| < \varepsilon$$

$\therefore$ the transformed function is uniformly continuous in r for all values of θ.

Hence, $f(x, y)$ is continuous in (x, y) together at the origin.

(ii) *Test for differentiability.* Now, we discuss the differentiability of function $f(x, y)$ at the origin

$$f_x(0,0) = \lim_{h \to 0} \frac{f(h, 0) - f(0, 0)}{h} = \lim_{h \to 0} \frac{1}{\delta x}\left[\frac{\delta x . 0}{(h)^2 + 0} - 0\right] = 0$$

Similarly $\quad f_y(0, 0) = 0$

Now, consider

$$\lim_{\substack{h \to 0 \\ k \to 0}} \frac{f(h, k) - f(0, 0) - h f_x(0, 0) - k f_y(0, 0)}{\delta(x, y)}$$

$$= \lim_{\substack{h \to 0 \\ k \to 0}} \frac{1}{\sqrt{h^2 + k^2}}\left[\frac{hk^?}{h^2 + k^2} - 0 - h . 0 - k . 0\right]$$

$$= \lim_{\substack{h \to 0 \\ k \to 0}} \frac{hk^2}{(h^2 + k^2)^{3/2}}$$

Now, if we put $k = mh$, then

$$\lim_{\substack{h \to 0 \\ k \to 0}} \frac{h . k^2}{(h^2 + k^2)^{3/2}} = \lim_{h \to 0} \frac{h . m^2 h^2}{h^3 (1 + m^2)^{3/2}} = \frac{m^2}{(1 + m^2)^{3/2}}$$

Since, this limit is not unique (it depends upon m), thus the limit does not exist.

Hence, the given function $f(x, y)$ is not differentiable at $(0,0)$.

EXAMPLE 3. *Show that the function $f(x, y) = \sin x + \cos y$ is differentiable every where.*

SOLUTION. Let (a, b) be any arbitrary point of $\boldsymbol{R}^2$, then

$$f_x(a, b) = \lim_{h \to 0} \frac{f(a + h, b) - f(a, b)}{h}$$

$$= \lim_{h \to 0} \frac{\sin(a+h) + \cos b - \sin a - \cos b}{h}$$

$$= \lim_{h \to 0} \frac{\sin(a+h) - \sin a}{h}$$

$$= \lim_{h \to 0} \frac{2\cos(a+h/2)\sin h/2}{h}$$

$$= \lim_{h \to 0} \cos\left(a + \frac{h}{2}\right)\frac{\sin h/2}{h/2} = \cos a$$

Similarly $\quad f_y(a,b) = -\sin b$

Now, consider

$$\lim_{(h,k) \to (0,0)} [f(a+h, b+k) - f(a,b) - hf_x(a,b) - kf_y(a,b)]$$

$$= \lim_{(h,k) \to (0,0)} [\sin(a+h) + \cos(b+k)]$$

$$-\sin a - \cos b - h\cos a + k\sin b]$$

$$= \lim_{(h,k) \to (0,0)} \left[h\left\{ \frac{\sin(a+h) - \sin a}{h} - \cos a \right\} \right.$$

$$\left. +k\left\{ \frac{\cos(b+k) - \cos b}{k} + \sin b \right\} \right] = 0$$

Hence, the given function is differentiable at every point.

EXAMPLE 4. *Let $f(x, y) = xy + x + y^2$. Show that f is differentiable at the origin.*

SOLUTION. Here, we have

$$f(x, y) = xy + x + y^2.$$

It can be easily verified that

$$f_x(0, 0) = 1 \text{ and } f_y(0, 0) = 0$$

$$\therefore \quad \lim_{(h,k) \to (0,0)} [f(h,k) - f(0,0) - hf_x(0,0) - kf_y(0,0)]$$

$$= \lim_{(h,k) \to (0,0)} (hk + k^2) = 0$$

Hence, $f(x, y)$ is differentiable at $(0, 0)$.

EXAMPLE 5. *Let $f(x,y) = \begin{cases} \dfrac{x^3 - y^3}{x^2 + y^2} & when \quad (x,y) \neq (0,0) \\ 0 & when \quad (x,y) = (0,0) \end{cases}$*

Show that the function f is continuous but not differentiable at the origin.

SOLUTION. (i) *Test for continuity.* If we put $x = r\cos\theta$ and $y = r\sin\theta$, then

$$|f(r\cos\theta, r\sin\theta) - f(0,0)| = \left| \frac{r^3(\cos^3\theta - \sin^3\theta)}{r^2(\cos^2\theta + \sin^2\theta)} - 0 \right|$$

$$= r|(\cos^3\theta - \sin^3\theta)|$$

$$\leq r(|\cos^3\theta| + |\sin^3\theta|)$$

$$\leq 2r, \text{ for all values of } \theta$$

Now, set $r_0 = \dfrac{\varepsilon}{2}$, then for all values of θ and $r < r_0$, we have

$$|f(r\cos\theta, r\sin\theta) - f(0,0)| < \varepsilon$$

$\therefore$ The transformed function is uniformly continuous in r for all values of θ. Hence $f(x, y)$ is continuous in (x, y) together at the origin.

(ii) *Test for differentiability.* Here, we have

$$f_x(0,0) = \lim_{h \to 0} \frac{f(0+h,0) - f(0,0)}{h} = \lim_{h \to 0} \frac{f(h,0) - f(0,0)}{h}$$

$$= \lim_{h \to 0} \frac{1}{h} \left[\frac{(h)^3 - 0}{(h)^2 + 0} - 0 \right] = 1$$

and $$f_y(0,0) = \lim_{k \to 0} \frac{f(0,0+k) - f(0,0)}{k} = \lim_{k \to 0} \frac{f(0,k) - f(0,0)}{k}$$

$$= \lim_{k \to 0} \frac{1}{k} \left[\frac{0 - (k)^3}{0 + (k)^2} - 0 \right] = -1$$

Now, consider

$$\lim_{(h,k) \to (0,0)} \frac{f(h,k) - f(0,0) - h f_x(0,0) - k f_y(0,0)}{\sqrt{h^2 + k^2}}$$

$$= \lim_{(h,k) \to (0,0)} \frac{1}{\sqrt{h^2 + k^2}} \left[\frac{h^3 - k^3}{h^3 + k^3} - h + k \right]$$

$$= \lim_{(h,k) \to (0,0)} \frac{(h - k)[(h)^2 + hk + (k)^2 - h^2 - k^2]}{[(h)^2 + (k)^2]^{3/2}}$$

$$= \lim_{(h,k) \to (0,0)} \left[\frac{(h-k)(h+k)}{(h^2 + k^2)^{3/2}} \right]$$

If we put $k = mh$, then the above limit

$$= \lim_{(h \to 0)} \frac{(h - mh)(hmh)}{(h)^3 (1 + m^2)^{3/2}} = \frac{(1 - m)m}{(1 + m^2)^{3/2}}$$

Hence, this limit does not exist, since it depends upon m, It follows that the given function is not differentiable at (0, 0).

EXAMPLE 6.   *Discuss the differentiability of f defined by*

$$f(x,y) = \begin{cases} x^2 \sin\dfrac{1}{x} + y^2 \sin\dfrac{1}{y} & when \quad (x,y) \neq (0,0) \\ 0 & when \quad (x,y) = (0,0) \end{cases}$$

SOLUTION.   We first calculate the partial derivatives at (0, 0), we have

$$f_x(0,0) - \lim_{h \to 0} \frac{f(0+h,0) - f(0,0)}{h}$$

$$= \lim_{h \to 0} \frac{\left[(h)^2 \sin\dfrac{1}{h} - 0 \right]}{h} = \lim_{h \to 0} h \sin\frac{1}{h} = 0$$

Similarly $f_y(0, 0) = 0$

Since, we know that, the function will be differentiable or not at (0, 0) according as

$$\frac{1}{r} \{ f(r \cos\theta, r \sin\theta) - f(0,0) \}$$

converges uniformly to the limit $\cos\theta f_x(0,0) + \sin\theta f_y(0,0) = 0$

Now, $\dfrac{1}{r} \{ f(r \cos\theta, r \sin\theta) - f(0, 0) \}$

$$= \frac{1}{r} \left\{ r^2 \cos^2\theta \sin\left(\frac{1}{r \cos\theta} \right) + r^2 \sin^2\theta \sin\left(\frac{1}{r \sin\theta} \right) - 0 \right\}$$

$$= \frac{1}{r}\left\{\cos^2\theta \sin\left(\frac{1}{r\cos\theta}\right) + \sin^2\theta \sin\left(\frac{1}{r\sin\theta}\right)\right\}$$

which converges uniformly to 0 in the closed interval $[0, \pi]$ as r tends to 0.

Hence, $f(x, y)$ is differentiable at $(0, 0)$.

EXAMPLE 7. **Show that the function**

$$f(x,y) = \begin{cases} x^2\sin(1/x) + y^2\sin(1/y) & , & x, y \neq 0 \\ x^2\sin(1/x) & , & x \neq 0, y = 0 \\ y^2\sin(1/y) & , & x = 0, y \neq 0 \\ 0 & , & x = y = 0 \end{cases}$$

is differentiable at the origin.

SOLUTION. Since we know that function $f(x, y)$ is said to be differentiable at any point (a, b) of its domain if $f(a+h, b+k) - f(a, b)$ can be expressed in the form of

$$f(a+h, b+k) - f(a,b) = h.f_x(a,b) + k.f_y(a,b) + \sqrt{h^2 + k^2}.g(h,k)$$

where $g(h, k) \to 0$ as $(h, k) \to (0, 0)$

Hence, $\quad f_x(0,0) = \lim_{h \to 0} \dfrac{f(0+h, 0) - f(0,0)}{h}$

$$= \lim_{h \to 0} \frac{f(h,0) - f(0,0)}{h} = \lim_{h \to 0} \frac{h^2\sin(1/h) - 0}{h}$$

$$= \lim_{h \to 0} h\sin(1/h) = 0$$

Again $\quad f_y(0,0) = \lim_{k \to 0} \dfrac{f(0, 0+k) - f(0,0)}{k} = \lim_{k \to 0} \dfrac{f(0,k) - f(0,0)}{k}$

$$= \lim_{k \to 0} \frac{k^2\sin 1/k}{k} = 0$$

Now, again

$$f(0+h, 0+k) - f(0,0) = f(h,k) - f(0,0)$$

$$= h^2\sin 1/h + k^2\sin 1/k - 0$$

$$= 0.h + 0.k + \sqrt{h^2+k^2}\left[\frac{h^2}{\sqrt{h^2+k^2}}\sin\frac{1}{h} + \frac{k^2}{\sqrt{h^2+k^2}}\sin\frac{1}{k}\right]$$

$$= f_x(0,0).h + f_y(0,0).k + \sqrt{h^2+k^2}.g(h,k)$$

where $\quad g(h,k) = \dfrac{h^2}{\sqrt{h^2+k^2}}\sin\dfrac{1}{h} + \dfrac{k^2}{\sqrt{h^2+k^2}}\sin\dfrac{1}{k}$

We have

$$\lim_{(h,k)\to(0,0)} g(h,k) = \lim_{(h,k)\to(0,0)}\left[\frac{h}{\sqrt{h^2+k^2}}.h\sin\frac{1}{h} + \frac{k}{\sqrt{h^2+k^2}}.k\sin\frac{1}{k}\right]$$

$$= 0 \qquad \left[\begin{array}{l} \because \lim_{h\to 0} h\sin\dfrac{1}{h} = 0, \lim_{k\to 0} k\sin\dfrac{1}{k} = 0 \\[2mm] \text{and both } \dfrac{h}{\sqrt{h^2+k^2}} \text{ and } \dfrac{k}{\sqrt{h^2+k^2}} \text{ are bounded} \end{array}\right]$$

Hence, $f(x, y)$ is differentiable at $(0, 0)$.ss

EXERCISE 10.3

1. Show that the function $f(x, y)$ defined by

$$f(x,y) = \begin{cases} \dfrac{x^3 - y^3}{x^3 + y^3} & , \ (x,y) \neq (0,0) \\ 0 & , \ (x,y) = (0,0) \end{cases}$$

is continuous but not differentiable at $(0, 0)$.

2. Show that the function $f(x, y) = \{|xy|\}^{1/2}$ is not totally differentiable at $(0, 0)$ but that $\dfrac{\partial f}{\partial x}$ and $\dfrac{\partial f}{\partial y}$ both exist at the origin and have the value 0.

3. Show that the function

$$f(x,y) = \begin{cases} \dfrac{x^2 y^2}{x^2 + y^2} & \text{when } (x,y) \neq (0,0) \\ 0 & \text{when } (x,y) = (0,0) \end{cases}$$

is not totally differentiable at the origin.

4. Show that the function $f(x, y)$ defined by

$$f(x,y) = \begin{cases} \dfrac{2xy}{x^2 + y^2} & , \ (x,y) \neq (0,0) \\ 1 & , \ (x,y) = (0,0) \end{cases}$$

is not differentiable at $(0, 0)$.

5. Show that the function $f(x, y)$ defined by

$$f(x,y) = \begin{cases} \dfrac{x^3 + y^3}{x - y} & , \ \text{if } x \neq y \\ 1 & , \ \text{if } x = y \end{cases}$$

is not differentiable at $(0, 0)$.

6. If $u = x^2 + y^2, \ v = x^3 + y^3$, prove that, if x is considered as a function of u and v, then

$$\frac{\partial x}{\partial u} = -\frac{y}{2x(x-y)}, \frac{\partial x}{\partial v} = \frac{1}{3x(x\ y)}.$$

7. Show that the function, $f(x, y)$ defined by

$$f(x,y) = \begin{cases} \dfrac{x^2 y^2}{x^4 + y^4} & \text{for } x \neq y \neq 0 \\ 0 & \text{for } x = y = 0 \end{cases}$$

is not totally differentiable at origin.

8. Verify the following function for differentiability

(a) $f(x, y) = e^x + y$ at $(1, 3)$

(b) $f(x, y) = \cos x + \sin y$ at $(0, 0)$

(c) $f(x, y) = \cos(xy)$ at $\left(\dfrac{\pi}{4}, \dfrac{\pi}{4}\right)$

(d) $f(x, y) = |x^2 - y^2|$ at origin.

(e) $f(x, y) = y^2 \sin \dfrac{x}{y}$ at origin.

9. Examine the following function for total differentiability at the origin.

$$f(x,y) = \begin{cases} xy \cdot \dfrac{x^2 - y^2}{\sqrt{x^2 + y^2}}, & \text{when } (x,y) \neq (0,0) \\ 0 & , \ \text{when } (x,y) = (0,0) \end{cases}$$

10. If $f(x, y, z) = 0$, where f is a differentiable function of x, y, z prove that

$$\left(\frac{\partial z}{\partial y}\right)_x \left(\frac{\partial x}{\partial z}\right)_y \left(\frac{\partial y}{\partial x}\right)_z = -1$$

where $\left(\dfrac{\partial z}{\partial y}\right)_x$ denotes the derivative of z with respect to y when x is constant.

HINTS TO SELECTED PROBLEMS

1. Along $y = mx$

$$\lim_{\substack{x \to 0 \\ y \to 0}} f(x,y) = \frac{1 - m^3}{1 + m^3}, \text{ not unique.}$$

Also $f_x(0,0) = \infty, f_y(0,0) = -\infty$

3. $f_x(0,0) = 0, f_y(0,0) = 0$

Also among $k = mh$, simultaneous double limit

exists.

6. Since $\dfrac{\partial x}{\partial u} = -\dfrac{y}{2x(x-y)}, \dfrac{\partial x}{\partial v} = \dfrac{1}{3x(x-y)}$

Now using the result

$$\frac{\partial x}{\partial x} = 1 = \frac{\partial x}{\partial u} \cdot \frac{\partial u}{\partial x} + \frac{\partial x}{\partial v} \cdot \frac{\partial v}{\partial x}$$

and $\dfrac{\partial x}{\partial y} = \dfrac{\partial x}{\partial u} \cdot \dfrac{\partial u}{\partial y} + \dfrac{\partial x}{\partial v} \cdot \dfrac{\partial v}{\partial y}$

10.6 DIRECTIONAL DERIVATIVES OF A FUNCTION OF TWO VARIABLES

Let $f(x, y)$ be a function of two variables x and y and a line S is inclined at an angle θ, with x-axis, then the directional derivative of $f(x, y)$ along this line at a point $P(x, y)$ is defined as

$$\frac{\partial f}{\partial S} = f_x \cos \theta + f_y \sin \theta$$

☛ **REMARK**

⇒ The existence of all the directional derivatives at a point may not imply the continuity of the function at the point e.g. the function $f(x,y) = \dfrac{xy}{x^2 + y^2}$ for $x \neq y \neq 0$ and $f(0, 0) = 0$ then $f(x, y)$ admits of every directional derivatives at the origin $(0, 0)$, but $f(x, y)$ is not continuous at $(0,0)$.

10.7 COMPOSITE FUNCTIONS

If z is a function of two variables x and y and these variables themselves are given to be the function of the variable t, then z is said to be the composite function of t.

i.e., the relation $\begin{cases} z = f(x,y) \\ x = \phi(t) \\ y = \Psi(t) \end{cases}$, define z as composite function of t.

Here, $\dfrac{dz}{dt}$ is called the total differential coefficients of z with respect to t.

THEOREM I. *If z is a composite function of t, defined by the relations*
$$z = f(x, y), \ x = \phi(t), \ y = \psi(t)$$
where z, possesses first order partial derivatives with respect to x and y, and also x and y possesses continuous derivatives w.r.t., 't', then
$$\frac{dz}{dt} = \frac{\partial z}{\partial x} \cdot \frac{dx}{dt} + \frac{\partial z}{\partial y} \cdot \frac{dy}{dt}$$

PROOF. Let $z = f(x, y)$ ... (1)

Let us suppose δt, δx and δy be the corresponding changes in t, x and y respectively, then
$$z + \delta z = f(x + \delta x, y + \delta y)$$ ... (2)

From (1) and (2), we get
$$\delta z = f(x + \delta x, y + \delta y) - f(x, y)$$
$$= [f(x + \delta x, y + \delta y) - f(x, y + \delta y) + f(x, y + \delta y) - f(x, y)]$$
$$\Rightarrow \frac{\delta z}{\delta t} = \frac{f(x + \delta x, y + \delta y) - f(x, y + \delta y)}{\delta x} \cdot \frac{\delta x}{\delta t} + \frac{f(x, y + \delta y) - f(x, y)}{\delta y} \cdot \frac{\delta y}{\delta t}$$ ... (3)

Now, proceeding to limit $\delta t \to 0$ and consequently δx and δy also tends to zero, we have
$$\lim_{\delta t \to 0} \frac{\delta z}{\delta t} = \frac{dz}{dt}, \quad \lim_{\delta t \to 0} \frac{\delta x}{\delta t} = \frac{dx}{dt}, \quad \lim_{\delta t \to 0} \frac{\delta y}{\delta t} = \frac{dy}{dt}$$

and $$\lim_{\delta x \to 0} \frac{f(x + \delta x, y + \delta y) - f(x, y + \delta y)}{\delta x} = \frac{\delta z}{\delta x}$$

[Because, while x change to $x + \delta x$, $y + \delta y$ remains unchanged.]

Similarly $$\lim_{\delta y \to 0} \frac{f(x, y + \delta y) - f(x, y)}{\delta y} = \frac{\partial z}{\partial y}$$

Hence, (3) gives $$\frac{dz}{dt} = \frac{\partial z}{\partial x} \cdot \frac{dx}{dt} + \frac{\partial z}{\partial y} \cdot \frac{dy}{dt}$$

☛ **REMARK**

⇒ In general, if $z = f(x_1, x_2, \dots, x_n)$ where $x_1, x_2, \dots, x_n$ all are functions of t, we have
$$\frac{dz}{dt} = \frac{\partial z}{\partial x_1} \cdot \frac{dx_1}{dt} + \frac{\partial z}{\partial x_2} \cdot \frac{dx_2}{dt} + \dots + \frac{\partial z}{\partial x_n} \cdot \frac{dx_n}{dt}$$

10.8 MEAN VALUE THEOREM FOR TWO VARIABLES

Let $f(x, y)$ be a function, such that

 (i) *$f(x, y)$ is continuous in the closed domain D*

(ii) *first order differential coefficient of $f(x, y)$ exists in the open domain D'.*

Then $\qquad f(a+h, b+k) = f(a, b)+h f_x (a+\theta h, b+\theta k)+k f_y(a+\theta h, b+\theta k)$

where, $\quad (a, b), (a+h), (b+k) \in D$ *and* $0 < \theta < 1$.

Proof. Let us define a new variable t as $x = a+ht, y = b+kt, 0 < t \le 1$

where, h and k are constants, then we get the function of single variable t as

$$F(t) = f(x, y) = f(a+ht, b+kt) \qquad \qquad ...(1)$$

By Lagrange's mean value theorem of single variable, we have

$$\frac{F(1)-F(0)}{1-0} = F'(\theta), 0 < \theta < 1$$

$\Rightarrow \qquad\qquad F(1) = F(0) + F'(\theta), 0 < \theta < 1 \qquad\qquad ... (2)$

But, from (1)

$\therefore \qquad\qquad F(1) = f(a+h, b+k) \quad$ and $\quad F(0) = f(a,b)$

$\qquad f(a+h, b+k) - f(a,b) = F'(\theta), 0 < \theta < 1 \qquad\qquad ... (3)$

If $x = a + ht \qquad$ and $\qquad y = b+kt$, then $F(t) = f(x,y)$

$\Rightarrow \qquad\qquad F'(t) = \dfrac{d}{dt}F(t) = \dfrac{d}{dt}f(x,y)$

$$= \frac{\partial f}{\partial x}\cdot\frac{dx}{dt} + \frac{\partial f}{\partial y}\cdot\frac{dy}{dt} \qquad \text{[By definition of Total derivatives]}$$

$$= h f_x(x, y) + k f_y(x, y) \qquad\qquad \left[\because \frac{dx}{dt} = h, \frac{dy}{dt} = k\right]$$

or $\qquad\qquad F'(t) = h f_x(a + ht, b+kt) + k f_y(a + ht, b+kt) \qquad\qquad ... (4)$

Now substitute θ in place of t, in (4), we get

$$F'(\theta) = h f_x(a + \theta h, b + \theta k) + k f_y(a + h\theta, b + k\theta) \qquad\qquad ... (5)$$

Now, from (3) and (5), we have

$$f(a+h, b+k) - f(a, b) = h f_x(a+\theta h, b+\theta k) + k f_y(a+\theta h, b+\theta k)$$

$\Rightarrow \qquad f(a+h, b+k) = f(a, b) + h f_x(a+\theta h, b+\theta k) + k f_y(a+\theta h, b+\theta k)$

10.9 TAYLOR'S THEOREM FOR FUNCTION OF TWO VARIABLES

If $f(x, y)$ possesses continuous partial derivatives upto nth order inclusive for all points (x, y) in the region $a \le x \le a-k, b \le y \le b+k$

Then $\qquad f(x+h, y+k) = f(x,y) + \left(h\dfrac{\partial}{\partial x} + k\dfrac{\partial}{\partial y}\right)f(x,y)$

$$+\frac{1}{2!}\left(h\frac{\partial}{\partial x} + k\frac{\partial}{\partial y}\right)^2 f(x,y) + ... + \frac{1}{(n-1)!}\left(h\frac{\partial}{\partial x} + k\frac{\partial}{\partial y}\right)^1 f(x,y)$$

$$+\frac{1}{n!}\left(h\frac{\partial}{\partial x} + k\frac{\partial}{\partial y}\right)^n f(x+\theta h, y+\theta k) \text{where } 0 < \theta < 1.$$

PROOF. Let us suppose that x varies while y remains constant.

Thus, regarding $f(x+h, y-k)$ as a function of one variable only, say that of x. Then using

Taylor's theorem for one variable we have

$$f(x+h,y+k) = f(x,y-k-h)\frac{\partial f(x,y+k)}{\partial x} + \frac{h^2}{2!}\frac{\partial^2 f(x,y+k)}{\partial x^2} + \dots \qquad \dots (1)$$

Further, expanding each term of R.H.S. of (1) by Taylor's regarding y as variable and x as constant, we have

$$f(x+h,y+k) = f(x,y) + k\frac{\partial f(x,y)}{\partial y} + \frac{k^2}{2!}\frac{\partial^2 f(x,y)}{\partial y^2} +$$

$$+h\frac{\partial}{\partial x}\left\{f(x,y)+k\frac{\partial f(x,y)}{\partial y}+\dots\right\} + \frac{h^2}{2!}\frac{\partial^2}{\partial x^2}\left\{f(x,y)+k\frac{\partial f(x,y)}{\partial y}+\dots\right\}+\dots$$

Hence, $f(x+h,y+k) = f(x,y) + \left[h\frac{\partial f}{\partial x}+k\frac{\partial f}{\partial y}\right] + \frac{1}{2!}\left[h^2\frac{\partial^2 f}{\partial x^2}+2hk\frac{\partial^2 f}{\partial x\,\partial y}+k^2\frac{\partial^2 f}{\partial y^2}\right] + \dots$

or $\quad f(x+h,y+k) = f(x,y) + \left(h\frac{\partial}{\partial x}+k\frac{\partial}{\partial y}\right)f(x,y)$

$$+\frac{1}{2!}\left(h\frac{\partial}{\partial x}+k\frac{\partial}{\partial y}\right)f(x,y)+\dots+\frac{1}{n!}\left(h\frac{\partial}{\partial x}+k\frac{\partial}{\partial y}\right)^n f(x,y)+\dots$$

THEOREM I.   *Suppose E is an open set in R^n, f maps E into R^m, f is differentiable at $x_0 \in E$. g maps an open set containing $f(E)$ into R^k, and g is differentiable at $f(x_0)$. Then mapping F of E into R^k defined by $F(x) = g(f(x))$ is differentiable at t_0 and that*

$$F'(x_0) = g'(x_0))f'(x_0) \qquad \dots (1)$$

PROOF.   Suppose that $\;y_0 = f(x_0), A = f'(x_0), B = g(y_0)$, and define

$$u(h) = f(x_0+h) - f(x_0) - A.h,$$
$$v(k) = g(h_0+k) - g(y_0) - B.k,$$

for all $h \in R^n$ and $k \in R^m$ for which $f(x_0+h)$ and $g(y_0+k)$ are defined. By definition

$$\lim_{h\to 0}\frac{|f(x_0+h)-f(x_0)-Ah|}{|h|} = 0$$

and $\qquad \lim_{h\to 0}\frac{|g(y_0+k)-g(y_0)-Bk|}{|k|} = 0$

It follows that

$$\lim_{h\to 0}\frac{|u(h)|}{|h|} = 0 \quad \text{and} \quad \lim_{k\to 0}\frac{|v(k)|}{|k|} = 0. \text{ Then}$$

$$|u(h)| = \varepsilon(h)|h|,\, |v(k)| = \eta(k)|k|, \qquad \dots (2)$$

where $\varepsilon(h) \to 0$, as $h \to 0$ and $\eta(k) \to 0$ as $k \to 0$.
For given h, put $k = f(x_0+h) - f(x_0)$, then

$$|k| = |Ah+u(h)| \le |Ah|+|u(h)|$$
$$\le [\|A\|+\varepsilon(h)]|h|, \text{ by (2)}$$

and $F(x_0+h) - F(x_0) - BAh = g(f(x_0+h)) - g(f(x_0)) - BAh$

$$= g([f(x_0)+f(x_0+h)-f(x_0)]-g(f(x_0))-BAh$$
$$= g(y_0+k)-g(y_0)-BAh = v(k)+Bk-BAh$$
$$= B(k-Ah)+v(k) = Bu(h)+v(k)$$

Hence (2) and (3) imply, for $h \ne 0$ that

$$\frac{|F(x_1+h)-F(x_0)-BAh|}{|h|} = \frac{|Bu(h)+v(k)|}{|h|}\text{s}$$

$$\leq \frac{|Bu(h)| + |v(k)|}{|h|} \leq \frac{\|B\| \|u(h)| + \eta(k)| (k)}{|h|}$$

$$\leq \|B\| \varepsilon(h) + [\|A\| + \varepsilon(h)| \eta(k)]$$

Let $h \to 0$ then $\varepsilon(h) \to 0$. Also $k \to 0$, by (3), so that $\eta(k) \to 0$. Thus

$$\lim_{h \to 0} \frac{|f(x_0 + h) - F(x_0) - BAh|}{|h|} = 0$$

It follows that $F'(x_0) = BA$, i.e.,

$$F'(x_0) = g'(y_0) + f'(x_0) = g'(f(x_0))f'(x_0)$$

SOLVED EXAMPLES

EXAMPLE 1.   **If $f(x) = x^2 + y^2 + 2x + 3y$, then find the directional derivatives of $f(x)$ at (2, 1) in the direction of the vector $2i + j$.**

SOLUTION.   Directional derivatives $= \dfrac{\partial f}{\partial x} \cos \alpha + \dfrac{\partial f}{\partial y} \sin \beta$.

It is obvious that $\cos \alpha = \dfrac{2}{\sqrt{5}}$ and $\sin \beta = \dfrac{1}{\sqrt{5}}$

and $\qquad\qquad \dfrac{\partial f}{\partial x} = 2x + 2, \qquad \dfrac{\partial f}{\partial y} = 2y + 3$

$$\left(\frac{\partial f}{\partial x} \right)_{(2,1)} = 6, \left(\frac{\partial f}{\partial y} \right)_{(2,1)} = 5$$

Hence, the directional derivatives of $f(x)$ at (2, 1) in the given direction

$$= 6 \times \frac{2}{\sqrt{5}} + 5 \times \frac{1}{\sqrt{5}} = \frac{17}{\sqrt{5}}$$

EXAMPLE 2.   **If $f = xy^2 + 2xy$, then find the directional derivatives of f at (1, 2) in the direction $\theta = \dfrac{\pi}{2}$.**

SOLUTION.   Let $\qquad\qquad\qquad f = xy^2 + 2xy$

$$\Rightarrow \qquad\qquad \frac{\partial f}{\partial x} = y^2 + 2y, \quad \frac{\partial f}{\partial y} = 2xy + 2x$$

$$\therefore \qquad\qquad \left(\frac{\partial f}{\partial x} \right)_{(1,2)} = 8 \text{ and } \left(\frac{\partial f}{\partial y} \right)_{(1,2)} = 6$$

Hence, the directional derivative at (1, 2) in the direction $\theta = \dfrac{\pi}{2}$

$$= \frac{\partial f}{\partial x} \cos \frac{\pi}{2} + \frac{\partial f}{\partial y} \sin \frac{\pi}{2}$$

$$= 8 \cos \frac{\pi}{2} + 6 \sin \frac{\pi}{2} = 6.$$

EXAMPLE 3.   **Check the continuity of the function $x^3 + y^3 - 3xy + y = 0$, near the point (0,0).**

SOLUTION.   Here, we have $\quad f(x, y) = x^3 + y^3 - 3xy + y = 0$, at (0, 0)

Now check the continuity of f_x and f_y at (0, 0)

(i) *Continuity of f_x at (0, 0)*

The simultaneously limit $\lim\limits_{\substack{x \to 0 \\ y \to 0}} (3x^2 - 3y) = 0$, exists.

$\therefore$ f_x is continuous.

The simultaneous limit $\lim\limits_{\substack{x \to 0 \\ y \to 0}} (3y^2 - 3x + 1) = 1$, exists.

$\therefore$ f_y is continuous.

EXAMPLE 4.   ***Check the continuity of the derivative of the following function***:

     (a) $xy \sin x + \cos y = 0$ **at** $\left(0, \dfrac{\pi}{2}\right)$ **(b)** $y^3 \cos x + y^2 \sin^2 x = 7$ **at** $\left(\dfrac{\pi}{3}, 2\right)$.

SOLUTION.   (a) $f(x, y) = xy \sin x + \cos y$, $\left(0, \dfrac{\pi}{2}\right)$

$\therefore$ $f\left(0, \dfrac{\pi}{2}\right) = 0 + \cos\dfrac{\pi}{2} = 0$

Now $f_x = y[x \cos x + \sin x]$, $f_y = x \sin x - \sin y$.

Further we check the continuity of f_x and f_y at $\left(0, \dfrac{\pi}{2}\right)$.

Continuity at $\left(0, \dfrac{\pi}{2}\right)$ *of* f_x.

The simultaneous limit of $f_x(x, y)$.

i.e., $\lim\limits_{\substack{x \to 0 \\ y \to (\pi/2)}} (xy \cos x + \sin x) = 0$, exists

and the simultaneous limit of $f_y(x, y)$, *i.e.*,

$\therefore$ $\lim\limits_{\substack{x \to 0 \\ y \to (\pi/2)}} (x \sin x - \sin y) = \lim\limits_{y \to (\pi/2)} (-\sin y) = -1$, exists.

Hence both f_x and f_y are continuous

and $f_y\left(0, \dfrac{\pi}{2}\right) = \lim\limits_{k \to 0} \dfrac{f\left(0, \dfrac{\pi}{2} + k\right) - f\left(0, \dfrac{\pi}{2}\right)}{k}$

$= \lim\limits_{k \to 0} \dfrac{\cos\left(\dfrac{\pi}{2} + k\right)}{k} = \lim\limits_{k \to 0} \dfrac{-\sin k}{k} = -1$

$\therefore$ $f_y\left(0, \dfrac{\pi}{2}\right) \neq 0$

(b) $f(x, y) = y^3 \cos x + y^2 \sin^2 x - 7 = 0$

$\therefore$ $f\left(\dfrac{\pi}{3}, 2\right) = 8\cos\dfrac{\pi}{3} + 4\sin^2\dfrac{\pi}{3} - 7 = 4 + 3 - 7 = 0$

Now $f_x(x, y) = -y^3 \sin x + 2y^2 \sin x \cos x$

and $f_y(x, y) = 3y^2 \cos x + 2y \sin^2 x$

Next, we have to examine the **continuity** of f_x and f_y.

(i)   *Continuity of* $f_x (x, y)$ *at* $\left(\dfrac{\pi}{3}, 2\right)$.

The simultaneous limit is

$\lim\limits_{\substack{x \to (\pi/3) \\ y \to 2}} [-y^3 \sin x + 2y^2 \sin x \cos x] = \lim\limits_{y \to 2}\left[-y^3 \sin\dfrac{\pi}{3} + 2y^2 \sin\dfrac{\pi}{3}\cos\dfrac{\pi}{3}\right]$

$= -8 \cdot \dfrac{\sqrt{3}}{2} + 8 \cdot \dfrac{\sqrt{3}}{2} \cdot \dfrac{1}{2}$

$$= -4\sqrt{3} + 4\sqrt{3} = 0 \text{, exists}$$

∴ The simultaneous limit is given by

$$\lim_{\substack{x \to (\pi/3) \\ y \to 2}} [3y^2 \cos x + 2y \sin^2 x] = \lim_{y \to 2} \left[3y^2 \cos\frac{\pi}{3} + 2y \sin^2\frac{\pi}{3} \right]$$

$$= 12.\frac{1}{2} + 4.\frac{3}{4} = 6 + 3 = 9 \text{ exists}$$

∴ f_x and f_y both are continuous.

EXAMPLE 5. *Examine the following equations for the existence of a unique implicit function near the point indicated:*

(a) $y^2 + 2x^2y + x^5 = 0$ *at* (1, –1)   (b) $x^2 + xy + y^2 - 1 = 0$ *at* (1, 0)

(c) $y^2 - yx^2 - 2x^5 = 0$ *at* (0, 0)

SOLUTION.   (a) Let $f(x, y) = y^2 + 2x^2y + x^5$

Then $f_x = 4xy + 5x^4$

$f_y = 2y + 2x^2$

Also, we have $f(1, -1) = 1 - 2 + 1 = 0$

and $f_y(1, -1) = -2 + 2 = 0$

which implies that both the partial derivatives f_x and f_y are continuous functions in a neighbourhood of (1, –1).

(b) Let $f(x, y) = x^2 + xy + y^2 - 1$

Then, we have

$f_x = 2x + y$, $f_y = x + 2y$

Also, $f(1, 0) = 1 + 0 + 0 - 1 = 0$

$f_y(1, 0) = 1 \neq 0$

Clearly, both the derivatives f_x and f_y are continuous in the nbd of (1,0).

(c) Let $f(x, y) = y^2 - yx^2 - 2x^5$

Then, we have $f_x = -2xy - 10x^4$, $f_y = 2y - x^2$

Also, $f(0, 0) = 0 \Rightarrow f_y(0, 0) = 0$

We have both the partial derivatives f_x and f_y are continuous functions in a neighbouhood of (0, 0).

EXERCISE 10.4

1. If $(x, y) = (x^2 + y^2) \log (x^2 + y^2)$ for $(x, y) \neq (0, 0)$ and $f(0, 0) = 0$; show that f_{xy} and f_{yx} are not continuous at (0, 0) but $f_{xy}(0,0) = f_{yx}(0,0)$.

2. Find the first derivative of the equation. $2xy - \log xy = 2$ at (1, 1).

10.10 SCHWARZ'S THEOREM

Let f be a real valued function defined on a domain $D \subset \mathbf{R}^2$ and (a, b) be a point of the domain D, such that

(i) f_{xy} is continuous at (a, b).

(ii) f_x exists in a certain neighbourhood of (a, b)

then, $f_{yx}(a, b)$ exists and equal to $f_{xy}(a, b)$.

PROOF. Let $f(x, y)$ be a real valued function in a domain D. Condition (i) and (ii) implies that there exist a certain neighbourhood of (a, b) at every point (x, y) of which $f_x(x, y), f_y(x, y)$ and $f_{xy}(x, y)$ exists.

Let us suppose $(a+h, b+k)$ be any point of this neighbourhood, then we may define two functions ϕ and g such that

$$\phi(h, k) = f(a+h, b+k) - f(a+h, b) - f(a, b+k) + f(a, b)$$

and
$$g(y) = f(a+h, y) - f(a, y)$$

so that
$$\phi(h, k) = g(b+k) - g(b) \qquad \ldots (1)$$

Since, f_y, exists in a neighbourhood of (a, b), therefore the function g of one variable is differentiable in $[b, b + k]$, then by Lagrange's mean value theorem, we have

$$\frac{g(b+k) - g(b)}{b+k-b} = g'(b+\theta k), \; 0 < \theta < 1 \qquad \ldots (2)$$

Using (1) and (2), we have

$$\phi(h, k) = kg'(b+\theta k)$$
$$= k[f_y(a+h, b+\theta k) - f_y(a, b+\theta k)] \qquad \ldots (3)$$

Now, since, f_{xy} exist in the neighbourhood of (a, b), then again apply the Lagrange's mean value theorem to the right of (3), we get

$$\phi(h,k) = hk f_{xy}(a+\theta'h, b+\theta k) \qquad (0 < \theta' < 1)$$

$$\Rightarrow \frac{1}{k}\left[\frac{f(a+h,b+k)-f(a,b+k)}{h} - \frac{f(a+h,b)-f(a,b)}{h}\right] = f_{xy}(a+\theta'h, b+\theta k)$$

Since, f_{xy} exist in a neighbourhood of (a, b), this gives when $h \to 0$

$$\frac{f_x(a,b+k)-f_x(a,b)}{k} = \lim_{h\to 0} f_{xy}(a+\theta'h, b+\theta k)$$

Letting $k \to 0$, we get

$$\lim_{k\to 0}\frac{f_x(a,b+k)-f_x(a,b)}{k} = \lim_{k\to 0}\left[\lim_{h\to 0} f_{xy}(a+\theta'h, b+\theta k)\right]$$

$$\Rightarrow \quad f_{yx}(a,b) = \lim_{k\to 0}\lim_{h\to 0} f_{xy}(a+\theta'h, b+\theta k)$$

$$= f_{xy}(a, b)$$

Hence, we get $\quad f_{yx}(a, b) = f_{xy}(a, b)$

☞ **REMARK**

⟹ If f_{xy} and f_{yx} are both continuous at (a, b), then $f_{xy}(a, b) = f_{yx}(a, b)$

10.11 YOUNG'S THEOREM

Let f be a real valued function defined on a domain $D \subset R^2$ and (a, b) be any point of the domain D, such that f_x and f_y are both differentiable at (a, b), then $f_{xy}(a, b) = f_{yx}(a, b)$.

PROOF. Since, f_x and f_y are both differentiable at (a, b) therefore, the derivatives of f_x and f_y exist in a certain neighbourhood of (a, b) and given by $f_{xx}, f_{yx}, f_{xy}, f_{yy}$.

Let $(a+h, b+h)$ be a point of this neighbourhood.

Then, we may define two functions ϕ and g such that

$$\phi(h, h) = f(a + h, b + h) - f(a + h, b) - f(a, b + h) + f(a, b)$$
$$g(y) = f(a+h, h) - f(a, y) \qquad \ldots (1)$$
$$\Rightarrow \quad \phi(h, h) = g(b+h) - g(b)$$

Since, f_y exist in a neighbourhood of (a, b), apply the mean value theorem to the expression on the right of (1), we get

$$\frac{g(b+h) - g(b)}{b+h-b} = g'(b+\theta h); \qquad 0 < \theta < 1$$

$\Rightarrow \qquad\qquad g(b+h) - g(b) = hg'(b+\theta h) ; \qquad 0 < \theta < 1$

$\Rightarrow \qquad\qquad \phi(h,h) = hg'(b+\theta h) ; \qquad 0 < \theta < 1$

$$= h[f_y(a+h, b+\theta h) - f_y(a, b+\theta h)] \qquad \dots (2)$$

Since, f_y is differentiable at (a, b), we have by definition,

$$f_y(a+h, b+\theta h) - f_y(a,b) = hf_{xy}(a,b) + \theta hf_{yy}(a,b) + |h|\sqrt{(1+\theta^2)}\phi_1(h,h) \qquad \dots (3)$$

where $\phi_1 \to 0$ as $h \to 0$

and $\qquad f_y(a, b+\theta h) - f_y(a,b) = \theta hf_{yy}(a,b) + \theta|h|\phi_2(h,h) \qquad \dots (4)$

where $\phi_2 \to 0$ as $h \to 0$

Now, from (2), (3) and (4), we get

$$\frac{\phi(h,h)}{h^2} = f_{xy}(a,b) + \sqrt{(1+\theta^2)}\phi_1(h,h) - \theta\phi_2(h,h) \qquad \dots (5)$$

In a similar way considering $F(x) = f(x, b+h) - f(x, b)$, we can easily show that

$$\frac{\phi(h,h)}{h^2} = f_{yx}(a,b) + \sqrt{(1+\theta'^2)}.\psi_1(h,h) - \theta'\psi_2(h,h) \qquad \dots (6)$$

where $\psi_1, \psi_2 \to 0$ as $h \to 0$.

Now equating the right hand sides of (5) and (6) and letting $h \to 0$, we get

$$f_{xy}(a, b) = f_{yx}(a, b).$$

THEOREM 1. *Suppose f maps a convex open set $E \subset R^n$ into R^m, f is differentiable in E, and there is a real number M such that : $\|f'(x)\| \le M$, for every x. Then, for $a, b \in E$,*

$$|f(b) - f(a)| \le M|b - a|$$

What would happen if $f'(x) = 0$ for all x?

PROOF. Let $a \in E, b \in E$ and define

$$\gamma(t) = (1 - t)a + tb \qquad \dots (1)$$

for all $t \in R'$ such that $\gamma(t) \in E$.

If $0 \le t \le 1$, then by convexity of $E, \gamma(t) \in E$

Putting $\qquad g(t) = f(\gamma(t))$. Then by chain rule, we have

$$g'(t) = f'(\gamma(t')(\gamma')(t)) = f'\gamma(t)(b-a), \text{by}(1)$$

So that

$$|g'(t)| = |f'\gamma(t)(b-a)| \le \|f'\gamma(t)\| |(b-a)|$$

$$\le M|(b-a)| \text{ for all } t \in [0, 1].$$

Now, by the mean value theorem for vector valued function, we get

$$|g(1) - g(0)| \le (1-0)|g'(t)|$$

or $\qquad |g(1) - g(0)| \le M|b-a|$ by (2).

But $g(1) = f(\gamma)(1) = f(b)$ and $g(0) = f\gamma(0) = f(a)$. Thus $|f(b) - f(a)| \le M(b-a)$

Further $f' = 0$ for all $x \in E$, then f is constant.

By (2) if $f'(x) = 0$ for all $x \in E$, we have $g'(t) = 0$

So that $\qquad\qquad g'(t) = 0 \qquad \Rightarrow |g(1) - g(0)| \le 0$

$\Rightarrow \qquad\qquad g(1) = g(0)$

$\Rightarrow \qquad\qquad f(b) = f(a) \qquad \Rightarrow f$ is constant.

SOLVED EXAMPLES

EXAMPLE 1.   **Explain the inequality**

$$f_{xy}(0,0) \ne f_{yx}(00)$$

for the function $f(x,y) = \begin{cases} \dfrac{xy(x^2-y^2)}{x^2+y^2} & , \ (x,y) \ne (0,0) \\ 0 & , \ (x,y) = (0,0) \end{cases}$

in view of the Schwariz's and Young's theorems.

SOLUTION.   Here we have

$$f(x,y) = \frac{xy(x^2-y^2)}{x^2+y^2}; (x,y) \ne (0,0)$$

$$f(0,0) = 0$$

According to Schwarz's iheorem. If $f_x(0,0)$ exists and f_{xy} is continuous at $(0, 0)$. Then $f_{yx}(0,0)$ exists and equal to $f_{xy}(0, 0)$.

First we check that $f_x(0, 0)$ exists or not

$$f_x(0,0) = \lim_{\delta x \to 0} \frac{f(\delta x,0) - f(0,0)}{\delta x} = \lim_{\delta x \to 0} \frac{0-0}{\delta x} = 0$$

$\therefore \quad f_x(0, 0)$ exists

Now, $f_x = y \left[\dfrac{x^2-y^2}{x^2+y^2} + \dfrac{4x^2y^2}{(x^2+y^2)^2} \right]$

$\therefore \quad f_{xy} = \dfrac{x^2-y^2}{x^2+y^2} + \dfrac{4x^2y^2}{(x^2+y^2)^2} + y \left[\dfrac{(x^2+y^2)(-2y) - (x^2-y^2)2y}{(x^2+y^2)^2} \right.$

$$\left. + \frac{(x^2+y^2)^2 8x^2y - 4x^2y^2 . 2(x^2+y^2) . 2y}{(x^2+y^2)^4} \right]$$

$$= \frac{x^2-y^2}{x^2+y^2} + \frac{4x^2y^2}{(x^2+y^2)^2} + y \left[\frac{-4x^2y}{(x^2+y^2)^2} + \frac{(x^2+y^2)8x^2y - 16x^2y^3}{(x^2+y^2)^3} \right]$$

$$= \frac{x^2-y^2}{x^2+y^2} + \frac{4x^2y^2}{(x^2+y^2)^2} + y \left[\frac{-4x^2y}{(x^2+y^2)^2} + \frac{8x^4y + 8x^2y^3 - 16x^2y^3}{(x^2+y^2)^3} \right]$$

$$= \frac{x^2-y^2}{x^2+y^2} + \frac{y(8x^4y - 8x^2y^3)}{(x^2+y^2)^3}$$

$\Rightarrow \quad f_{xy} = \dfrac{x^2-y^2}{x^2+y^2} + \dfrac{8x^2y^2(x^2-y^2)}{(x^2+y^2)^3}$

Now find the simultaneous limit of f_{xy} i.e.,

$$\lim_{\substack{x \to 0 \\ y \to 0}} \left[\frac{x^2-y^2}{x^2+y^2} + \frac{8x^2y^2(x^2-y^2)}{(x^2+y^2)^3} \right]$$

Taking $y = mx$, we get

$$\lim_{\substack{x \to 0 \\ y \to 0}} \left[\frac{x^2-y^2}{x^2+y^2} + \frac{8x^2y^2(x^2-y^2)}{(x^2+y^2)^3} \right] = \lim_{x \to 0} \left[\frac{1-m^2}{1+m^2} + \frac{8m^2(1-m^2)}{(1+m^2)^3} \right]$$

$$= \left[\frac{1-m^2}{1+m^2} + \frac{8m^2(1-m^2)}{(1+m^2)^3} \right]$$

Thus the limit does not exist since it depends upon m.

Hence, f_{xy} is not continuous at $(0, 0)$, and hence

$$f_{xy}(0, 0) \neq f_{yx}(0, 0)$$

According to Young's theorem, if f_x and f_y are both differentiable at $(0, 0)$ then f_{xy} $(0, 0) = f_{yx}(0, 0)$. Therefore first we check the differentiability of f_x and f_y at $(0, 0)$.

Since

$$f_x = y\left[\frac{x^2 - y^2}{x^2 + y^2} + \frac{4x^2y^2}{(x^2 + y^2)^2}\right] = g(x, y) \qquad \text{(say)}$$

and

$$f_y = x\left[\frac{x^2 - y^2}{x^2 + y^2} - \frac{4x^2y^2}{(x^2 + y^2)^2}\right] = h(x, y) \qquad \text{(say)}$$

Now find

$$g_x(0,0) = \lim_{\delta x \to 0} \frac{g(\delta x, 0) - g(0,0)}{\delta x} = \lim_{\delta x \to 0} \frac{0 - 0}{\delta x} = 0$$

$$g_y(0,0) = \lim_{\delta y \to 0} \frac{g(0, \delta y) - g(0,0)}{\delta y} = \lim - \frac{\delta y}{\delta y} = -1$$

$$\therefore \quad \lim_{\substack{\delta x \to 0 \\ \delta y \to 0}} \frac{g(\delta x, \delta y) - g(0,0) - \delta x\, g_x(0,0) - \delta y\, g_y(0,0)}{\sqrt{(\delta x)^2 + (\delta y)^2}}$$

$$= \lim_{\substack{\delta x \to 0 \\ \delta y \to 0}} \frac{\delta y\left[\dfrac{(\delta x)^2 - (\delta y)^2}{(\delta x)^2 + (\delta y)^2} + \dfrac{4(\delta x)^2(\delta y)^2}{[(\delta x)^2 + (\delta y)^2]^2}\right] + \delta y}{\sqrt{(\delta x)^2 + (\delta y)^2}}$$

If we put $\delta y = m\delta x$, then above limit

$$= \lim_{\delta x \to 0} \frac{m\,\delta x\left[\dfrac{1 - m^2}{1 + m^2} + \dfrac{4m^2}{(1 + m^2)^2}\right] + m\,\delta x}{\delta x\sqrt{1 + m^2}}$$

$$= \frac{m\left[\dfrac{1 - m^2}{1 + m^2} + \dfrac{4m^2}{(1 + m^2)^2}\right] + m}{\sqrt{1 + m^2}}$$

This limit depends upon m. Hence the limit does not exist.

Therefore, $g(x, y) = f_x(x, y)$ is not differentiable at $(0, 0)$.

Hence, $f_{xy}(0, 0) \neq f_{yx}(0, 0)$

EXAMPLE 2. Let $f(x, y) = \dfrac{x^2 y^2}{x^2 + y^2}, (x, y) \neq (0, 0)$ and $f(0, 0) = 0$

Verify that f_{xy} and f_{yx} exist in a neighbourhood of $(0, 0)$ but are not continuous at $(0, 0)$ and yet are equal at $(0, 0)$.

SOLUTION. We have $f_{xy}(0,0) = \lim_{\delta x \to 0} \lim_{\delta y \to 0} \dfrac{f(\delta x, \delta y) - f(\delta x, 0) - f(0, \delta y) + f(0,0)}{\delta x \delta y}$

$$= \lim_{\delta x \to 0} \lim_{\delta y \to 0} \frac{\dfrac{(\delta x)^2 (\delta y)^2}{(\delta x)^2 + (\delta y)^2} - 0 - 0 + 0}{\delta x \delta y} = \lim_{\delta x \to 0} \lim_{\delta y \to 0} \frac{\delta x \delta y}{(\delta x)^2 + (\delta y)^2}$$

$$= \lim_{\delta x \to 0}\left[\frac{\delta x . 0}{(\delta x)^2 + 0}\right] = 0$$

$\therefore \quad f_{xy}$ exists and $f_{xy}(0, 0) = 0$

and
$$f_{yx}(0,0) = \lim_{\delta y \to 0} \lim_{\delta x \to 0} \frac{f(\delta x, \delta y) - f(0, \delta y) - f(\delta x, 0) + f(0,0)}{\delta x \delta y}$$

$$= \lim_{\delta y \to 0} \lim_{\delta x \to 0} \frac{\delta x \delta y}{(\delta x)^2 + (\delta y)^2}$$

$$= \lim_{\delta y \to 0} \left[\frac{0.\delta y}{0 + (\delta y)^2} \right] = 0$$

$\therefore$ f_{yx} exist and $f_{yx}(0,0) = 0$

$\therefore$ $\qquad f_{xy}(0,0) = f_{yx}(0,0) = f_{xy}(0,0)$

Now find $\qquad f_x = \dfrac{(x^2 + y^2).2xy^2 - x^2 y^2 (2x)}{(x^2 + y^2)^2}$

$$= \frac{2x^3 y^2 + 2xy^4 - 2x^3 y^2}{(x^2 + y^2)^2} = \frac{2xy^4}{(x^2 + y^2)^2}$$

$\therefore$ $\qquad f_{yx} = \dfrac{(x^2 + y^2)^2.(8xy^3) - 2xy^4.2(x^2 + y^2).2y}{(x^2 + y^2)^4}$

$$= \frac{8x^3 y^3 + 8xy^5 - 8xy^5}{(x^2 + y^2)^3} = \frac{8x^3 y^3}{(x^2 + y^2)^3}$$

Similarly $\qquad f_{xy} = \dfrac{8x^3 y^3}{(x^2 + y^2)^3}$

Now we check the continuity at $(0, 0)$.

The simulaneous limit $= \lim\limits_{\substack{x \to 0 \\ y \to 0}} \dfrac{8x^3 y^3}{(x^2 + y^2)^3}$

Put $y = mx$.

$\therefore$ $\qquad \lim\limits_{x \to 0} \cdot \dfrac{8m^3 x^6}{x^6 (1 + m^2)^3} = \dfrac{8m^3}{(1 + m^2)^3}$

Thus the limit depends upon m hence $f_{xy} = f_{yx}$ is not continuous at $(0, 0)$.

10.12 INVERTIBLE FUNCTION

Let f be a real valued function with domain D and range E as subset of $\mathbf{R}^n$.
Here, we can write
$$y = f(x),\ x \in D \quad \text{and}\ y \in E \qquad \qquad \text{... (1)}$$
Let y_n can be written as $y_n = f_n(x_1, x_2, \dots x_n)$
Then, the function f is a transformation, which transforms the set D to the set E. Now, to each point of D, there corresponds a point of E. Now there are two cases :
(i) One-one transformation
(ii) Many-one transformation.
Then, a one-one function f with domain D and range E is called invertible.
If $y = f(x) \Leftrightarrow x = g(y)$, then the function g is called the inverse of the function f.

10.13 LINEAR TRANSFORMATION

Let V_1 and V_2 be two vector spaces. A mapping $f : V_1 \to V_2$ is said to be linear transformation.
if $\qquad\qquad f_1(x_1 + x_2) = f(x_1) + f(x_2)$
and $\qquad\qquad f(cx) = cf(x)\ \forall\ x, x_1, x_2 \in V_1,\ c \in R$

☛ REMARKS

⟹ Linear transformation V_1 into V_2 will be called linear operator on V_1.
⟹ Linear operator A on V_1 is said to be invertible if A is
  (i) one to one
  (ii) f maps V_1 onto V_2
⟹ If A is invertiable, then an operator A^{-1} can be defined by setting
$$A^{-1}(Ax) = x \ \forall x \in V_1$$

THEOREM 1.   *If A is a linear operator on a finite dimensional vector space X. Then A is one to one iff range of A is full of X.*

PROOF.   Let $B = \{x_1, ... x_n\}$ be a basis of X. Let $R(A)$ denote the range of A. We shall show that the set $Q = (Ax_1, ... Ax_n)$ spans $R(A)$. Let $y \in R(A)$. Then $y = Ax$. For some $x \in X$. Since B spans X, therefore there exists scalars $c_1, ... c_n$ such that
$$x = c_1 x_1 + c_2 x_2 + ... + c_n x_n$$
Therefore, $y = A(c_1 x_1 + ... + c_n x_n) = c_n Ax_n$, by linearity of A. This shows that Q spans $R(A)$, then $R(A) = X$ if and only if Q is linearly independent. We have to prove that this happens if and only if A is one-to-one. Let us suppose Q is linearly independent and let $x \in X$ be arbitrary. Since B is a basis. Therefore,
$$x = \sum_{i=1}^{n} c_i x_i \text{ for some scalars } c_i, i=1, 2, ... n$$
Then $Ax = 0 \quad \Rightarrow \quad A\left(\sum_{i=1}^{n} c_i x_i\right) = 0$
$\Rightarrow \quad \sum c_i Ax_i = 0$, by linearity of A
$\Rightarrow \quad c_1 = c_2 = ... = c_n = 0$, since Q is linearly independent
$\Rightarrow \quad x = 0$
Therefore,
$\quad Ax = 0 \quad \Rightarrow \quad x = 0$
So, $\quad Ax = Ay \quad \Rightarrow \quad Ax - Ay = 0$ ...(1)
$\Rightarrow \quad A(x - y) = 0$, since A is linear.
$\Rightarrow \quad x - y = 0$ by (1)
$\Rightarrow \quad x = y$
$\Rightarrow \quad A$ is one-to-one.
Conversely, suppose A is one-to-one.
Then
$$\sum_{i=1}^{n} c_i Ax_i = 0 \Rightarrow \quad A\left(\sum_{i=1}^{n} c_i x_i\right) = 0 \text{ since } A \text{ is linear.}$$
$$\Rightarrow \quad \sum_{i=1}^{n} c_i x_i = 0, \text{ since } A \text{ is one-to-one}$$
$$\Rightarrow \quad c_1 = c_2 = ... = c_n = 0, \text{ since } B \text{ is independent.}$$
Thus we conclude that Q is linearly independent.

THEOREM 2.   *Let W be the set of all invertible operators on R^n. Then, if $A \in W$ and $B \in L(R^n)$ such that $\|B - A\|.\|A^{-1}\| < 1$, then $B \in w$.*

PROOF.   Put $\|A^{-1}\| = \dfrac{1}{\alpha}$ and $\|B - A\| = \beta$

Then $\|B - A\|\|A^{-1}\| < 1$ implies that $\beta < \alpha$. For every $x \in R^n$. we have
$$|x| = |A^{-1}Ax| \leq \|A^{-1}\| |Ax| = \frac{1}{\alpha}|Ax| \text{ so that}$$

$$(\alpha - \beta)\,|\,x\,| = \alpha\,|\,x\,| - \beta\,|\,x\,| \leq |\,Ax\,| - \|B - A\|\,|\,x\,|$$

$$\leq |\,Ax\,| - |\,(B - A)x\,|, \text{since } |\,Ax\,| \leq \|A\|\,|\,x\,|$$

$$= |\,Ax\,| - |\,Bx - Ax\,| = |\,Ax\,| \leq \|Ax - Bx\|$$

$$\leq |\,Ax\,| - (|\,Ax\,| - |\,Bx\,|), \text{since } |\,y_1\,| - |\,y_2\,| \leq |\,y_1 - y_2\,|$$

$$\text{for all } y_1, y_2 \in R^n$$

$$= |\,Bx\,|\,(x \in R\ )$$

Thus, $|\,Bx\,| \geq (\alpha - \beta)\,|\,x\,|$ for all $x \in R^n$ ...(2)

Note that $(\alpha - \beta)\,|\,x - y\,|$ cannot be negative since $\alpha - \beta > 0$.

Now $Bx = By \Rightarrow Bx - By = 0$

$$\Rightarrow \quad B(x - y) = 0 \Rightarrow |B(x - y)| = 0$$

$$\Rightarrow \quad (\alpha - \beta)\,|\,x - y\,| = 0 \text{ by (1)}$$

$$\Rightarrow \quad |\,x - y\,| = 0 \text{ since } \alpha \neq \beta$$

$$\Rightarrow \quad x - y = 0 \Rightarrow x = y$$

This show that B is one-to-one. B is also onto. Hence B is an invertible operator, and so $B \in W$.

10.13.1 INVERSE FUNCTION THEOREM

Let f be a vector valued function with its domain D and range E as subset of R^n. Let $a \in D$, $f(a) = b \in E$. Let f admit of continuous first order partial derivatives in a neighbourhood of a and let the Jacobian $J_f(a), \neq 0$ then the function f is locally invertible at a. Also the local inverse g of f admits of continuous first order partial derivatives in a neighbourhood of b.

PROOF.   Let f be a vector valued function with its domain D and range E as subset of R^n, thus, we show that, under the given conditions on f, there exist neighbourhoods P and Q of a and b respectively such that

(i) $f(P) = \{f(x) : x \in P\} = Q$

(ii)   No two different points of P correspond to the same point of V.

(iii) $f(a) = b$.

If g is the inverse of f with domain Q and range P so that

$$y = f(x) \Leftrightarrow x = g(y); \quad x \in P \quad \text{and} \quad y \in Q$$

the function g admits of continuous first order partial derivative in Q.

Now consider the determinant

$$\begin{vmatrix} D_1 f_1(x_1) & D_2 f_1(x_1) & \cdots & D_n f_1(x_1) \\ D_1 f_2(x_2) & D_2 f_2(x_2) & \cdots & D_n f_2(x_2) \\ \cdots & \cdots & \cdots & \cdots \\ \cdots & \cdots & \cdots & \cdots \\ D_1 f_n(x_n) & D_2 f_n(x_n) & \cdots & D_n f_n(x_n) \end{vmatrix}$$

where $x_1, x_2, ..., x_n$ are points in a neighbourhood of a. We see upon this determinant as a function of n^2 variables viz. The n^2 co-ordinates of the n-points $x_1, x_2, ..., x_n$, such that its value when $x_1, x_2, ..., x_n$ all take the same value a is non-zero. Also the determinant being a polynomial in its elements is a continuous function. Thus $\exists$ a neighbourhood of the point a such that if P_1, $P_2, P_3 .... P_n$ are any n arbitrary points of this neighbourhood, we have $|D_j f_i(P_i)| \neq 0$

Again there exist $s > 0$ s.t. the closed sphere $S\,(a, s)$ is contained in the neighbourhood.

(i)   Let $x \neq y$ be two points of the sphere $S(a, s)$. We shall show that

$$x \neq y \Rightarrow f(x) \neq f(y)$$

Let if possible, $\qquad f(x) = f(y)$

$\Leftrightarrow \qquad\qquad\qquad f_i(x) = f_i(y), \, 1 \le i \le n$

By the mean value theorem, there exist $\theta_i, \, 0 < \theta_i < 1$ s.t.

$$0 = f_i(x) - f_i(y) = \sum_{j=1}^{n} (y_i - x_j)D_j f_i[x + \theta_i(y - x)]$$

we write $x + \theta_i(y - x) = P_i$ therefore, $\displaystyle\sum_{j=1}^{n} (y_j - x_j)D_j f_i(P_i) = 0$ thus we obtain a system of n

linear equations

$$(y_1 - x_1)D_1 f_1(P_1) + (y_2 - x_2)D_2 f_1(P_1) + ... + (y_n - x_n)D_n f_1(P_1) = 0$$

$$\cdots \qquad \cdots \qquad \cdots \qquad \cdots \qquad \cdots \qquad \cdots \qquad \cdots \qquad \cdots \qquad \cdots$$

$$\cdots \qquad \cdots \qquad \cdots \qquad \cdots \qquad \cdots \qquad \cdots \qquad \cdots \qquad \cdots \qquad \cdots$$

$$\cdots \qquad \cdots \qquad \cdots \qquad \cdots \qquad \cdots \qquad \cdots \qquad \cdots \qquad \cdots \qquad \cdots$$

$$(y_1 - x_1)D_1 f_n(P_n) + (y_2 - x_2)D_2 f(P_n) + ... + (y_n - x_n)D_n f_n(P_n) = 0$$

whose determinant

$$|D_j f_i(P_i)| \ne 0, 1 \le i \le n, 1 \le j \le n$$

$\Rightarrow \qquad y_1 - x_1 = 0, ..., y_n - x_n = 0 \Rightarrow x = y$

which is a contradiction.

Thus, we prove that $x \ne y \Rightarrow f(x) \ne f(y)$.

(ii) Now. we shall show that $f(\bar{S})$ contains, a neighbourhood of $b = f(a)$.

Here $f(S) = \{f(x) : x \in S\}$

Let T denote the boundary of the sphere $\bar{S}$ i.e., the set $\{x : \|x - a\| \le s\}$.

Now consider $\|f(x) - b\| : x \in T$.

Here, we have a real valued function with values $\|f(x) - f(a)\| = \|f(x) - b\|$ with domain T, as the function f is one-one with domain T and $a \notin T$, we see that

$\|f(x) - b\|$ is positive $\forall \, x \in T$.

Since, T is a compact set and the real valued function is continuous $\forall \, x \in T$. We see that the infimum of the function is necessarily positive and be denoted by $2k$ so that we have

$$2k \le \|f(x) - b\| \, x \, \forall \, x \in T.$$

We now define a set Q s.t.

$$Q = \{y : \|y - P\| < k\} \text{ and show that } Q \text{ is contained in } f(\bar{S}).$$

Let $y \in Q$. We have to show that there exist $x \in \bar{S}$ s.t. $y = f(x)$. we have

$$\alpha k \le \|f(x) - b\| \le \|f(x) - y\| + \|y - b\|$$

$$\le \|f(x) - y\| + k$$

so that $\qquad \|f(x) - y\| > k \, \forall \, y \in Q, \forall x \in T$ $\qquad\qquad$ ... (1)

Also $\qquad \|f(a) - y\| = \|y - b\| < k$ $\qquad\qquad\qquad\qquad\qquad$ ... (2)

Taking y fixed, we consider the real valued function with value $\|f(x) - y\|^2, x \in \bar{S}$

From (1) and (2) we see that the infimum of .this function is assumed at an interior point of $\bar{S}$ and as such this infimum is a minimum value.

We have $\qquad \displaystyle\sum_{i=1}^{n} [f_i(x) - f_i]^2 = \|f(x) - y\|$

Thus, we see that

$$\sum_{i=1}^{n} [f_i(x) - y_i]D_i f_i(x) = 0, \, 1 \le j \le n$$

Rewritting these equations, we obtain

$$[f_1(x) - y_1]D_1f_1(x) + ... +, [f_n(x) - y_n]D_1f_n(x) = 0$$
$$... \quad ... \quad ... \quad ... \quad ... \quad ...$$
$$[f_1(x) - y_1]D_nf_1(x) + ... +, [f_n(x) - y_n]D_nf_n(x) = 0$$

the determinant of this set of linear equations being non-zero.

We see that

$$f_1(x) - y_1 = 0, f_2(x) - y_2 = 0, ... f_n(x) - y_n = 0$$

which imply $f(x) = y$.

Thus, Q is contained in $f\{S\}$.

(iii) Let P be the set of points x such that $f(x) \in P$.

The set P is open, therefore f is continuous and Q is an open set.

Thus, we see that there exist an open set P and Q containing a and $f(a) = b$ respectively such that f is a one-one function from P onto Q. i.e., $x \in P \Rightarrow f(x) \in Q$

and to $y \in Q$ there corresponds one and only one $x \in P$ s.t. $f(x) = y$.

Thus the function f with domain P and range Q is invertible. Let g denote the inverse of the function f so that Q is its domain and P its range.

Thus, we have $\quad y = f(x) \Leftrightarrow x = g(y), x \in P, y \in Q$.

(iv) Now, we have only show that function g itself have continuous first order partial derivatives.

Let $y = f(x) \Leftrightarrow x = g(y)$.

For sufficiently small values of $\lambda, y + \lambda u_k \in Q$.

Let $x' = g(y + \lambda u_k) \Leftrightarrow y + \lambda u_k = x'$

$\therefore \qquad f(x') - f(x) = \lambda u_k$

$\Rightarrow \quad f_i(x')f_i(x) = \delta_{ik}, 1 \le i \le n$.

Now by mean value theorem, we get

$$f_i(x') - f_i(x) = \sum_{j=1}^{n}(x'_j - x_j)D_jf_i[x + \theta_i(x' - x)]$$

so that we obtain

$$\sum_{j=1}^{n}(x'_j - x_j)D_jf_i[x + \theta_i(x' - x)] = \lambda\delta_{ik}$$

$\Leftrightarrow \qquad \sum_{j=1}^{n}\dfrac{[g_j(y + \lambda u_k) - g_i(y)]}{\lambda}D_i[x + \theta_i(x' - x) = \delta_{ik}]$

The determinant of this system of linear equations is non-zero. We see that

$$\dfrac{g_j(y + \lambda u_k) - g_j(y)]}{\lambda} \text{ is determined; } 1 \le j \le n$$

Consider these equations, solved and letting $x' \to x$ which is equivalent to λ tending to 0, we get

$$\sum_{j=1}^{n}D_kg_j(y)D_jf_i(x) = \delta_{ik}$$

Thus $D_kg_j(y)$ exist $\forall 1 \le k \le n, 1 \le j \le n$ and $\forall y \in Q$ so that g possess of partial derivatives in Q.

Because the partial derivatives $D_k g_j(y)$ can be expressed as linear combinations of $D_jf_i(x)$, the partial derivatives $D_kg_j(y)$ are continuous in Q.

10.14 IMPLICIT FUNCTION

Let f be a real valued function of two variables so that its domain D is a subset of R^2.
Let us define E as follows : $E = \{(x, y) : (x, y) \in D \text{ and } f(x, y) = 0\}$.

Now, there may or may not exist a real valued function g of single variable with domain $A \subset R$ such that $E = \{(x, g(x)) : x \in A\}$

If g exists, then $y = g(x)$ is a solution of $f(x, y) = 0$.

(i) *First differential coefficient of an implicit function. If $u = f(x, y)$ = constant and x and y, both are the functions of t and r, then*

$$\frac{dy}{dx} = -\frac{p}{q}$$

where

$$p = \frac{\partial u}{\partial x} \text{ and } q = \frac{\partial u}{\partial y}$$

$$\frac{\partial u}{\partial x} = \frac{\partial u}{\partial t} \cdot \frac{\partial t}{\partial x} + \frac{\partial u}{\partial r} \cdot \frac{\partial r}{\partial x}$$

and

$$\frac{\partial u}{\partial y} = \frac{\partial u}{\partial t} \cdot \frac{\partial t}{\partial y} + \frac{\partial u}{\partial r} \cdot \frac{\partial r}{\partial y}$$

(ii) *Second differential coefficient of an implicit function.* Let $u = f(x, y) = c$, be the implicit function. Then,

$$\frac{d^2y}{dx^2} = -\left[\frac{q^2 r - 2pqs + p^2 t}{q^3}\right]$$

where $p = \dfrac{\partial f}{\partial x}$, $q = \dfrac{\partial f}{\partial y}$, $r = \dfrac{\partial^2 f}{\partial x^2}$, $s = \dfrac{\partial^2 f}{\partial x \partial y}$ and $t = \dfrac{\partial^2 f}{\partial y^2}$

SOLVED EXAMPLE

EXAMPLE I. *If $x^y + y^x = c$, find $\dfrac{dy}{dx}$.*

SOLUTION. Let $f(x, y) = x^y + y^x - c$

$\therefore \quad \dfrac{\partial f}{\partial x} = yx^{y-1} + y^x \log y$

and $\quad \dfrac{\partial f}{\partial y} = x^y \log x + xy^{x-1} = x^y \log x + xy^{x-1}$

Hence, $\dfrac{dy}{dx} = -\dfrac{\partial f / \partial x}{\partial f / \partial y} = -\left[\dfrac{yx^{y-1} + y^x \log y}{x^y \log x + xy^{x-1}}\right]$

10.15 IMPLICIT FUNCTION THEOREM

Let $f(x, y)$ be a function of two variables x and y and let (a, b) be a point of its domain such that
(i) $f(a, b) = 0$

(ii) *f possesses continuous partial derivatives f_x and f_y in a certain neighbourhood of (a, b) and* (iii) $f_y(a, b) \neq 0$

then, there exist a rectangle $[a - h, a + h; b - k, b + k]$ about (a, b) such that $\forall x \in [a-h, a+h]$, the equation $f(x, y) = 0$ has one and only one solution $y = g(x)$, lying in the interval $[b - k, b + k]$, which have the following properties :

(i) $b = g(a)$

(ii) $f(x, g(x)) = 0 \ \forall \ x \in [a-h, a+h]$

(iii) *g is differentiable. and both g and g' are continuous in* $[a - h, a + h]$.

PROOF. Let us suppose $f_y(a, b) > 0$ (If $f_y(a, b) \neq 0$, then replace $f(x, y)$ by $-f(x, y)$). Consider the figure (3):

(i) *Existence and Uniqueness of the Solution.* Let f_x, f_y be continuous in a neighbourhood R_1 of (a, b), where

$$R_1 = [a - h_1, a + h_1; b - k_1, b + k_1]$$

Now, since f_x, f_y are continuous in R_1, therefore, f is also continuous in R_1.

Also, since f_y is continuous at (a, b) and $f_y(a,b) > 0$, there exists a rectangle

$R_2 = [a - h_2, a + h_2; b - k, b+k]$, $h_2 < h_1$, $k < k_1$ such that for every point (x, y) of the rectangle R_2, and $f_y(x, y) > 0$.

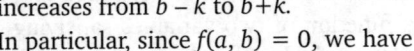

Now, since $f_y(x, y) > 0 \ \forall \ (x, y) \in R_2$, therefore for all $x \in [a - h_2, a+h_2]$ the function f of y strictly increasing as y increases from $b - k$ to $b+k$.

In particular, since $f(a, b) = 0$, we have

$f(a, b - k) < 0, f(a, b + k) > 0$ \hfill $[\because (b - k) < 0$ and $(b + k) > 0]$

Since, f is continuous, and $f(a, b - k) < 0, f(a, b + k) > 0$, therefore, there exists an interval $[a - h, a + h]$, $(h < h_2)$ such that for every x of this interval, we have

$f(x, b - k) < 0, f(x, b+k) > 0$, in some neighbourhood of (a, b).

Now, for every fixed value of x in $[a - h, a+h]$, the continuous function of y strictly increases from a negative to positive value as y increases from $b - k$ to $b + k$ and therefore, there exists one and only one value of y for which $f(x, y) = 0$.

Hence, for each value of x in $[a - h, a + h]$, there is a unique value $y = g(x)$ for which $f(x, y) = 0$. Hence we can say the equation $f(x, y) = 0$ has one and only one solution $y = g(x)$ lying in the interval $[b - k, b + k]$ such that

(a) $b = g(a)$

and (b) $f(x, g(x)) = 0 \ \forall x \in [a - h, a + h]$

(ii) *Test for Continuity.* Now, we shall prove g is continuous in $[a - h, a + h]$. Let x_0 be any point such that $x_0 \in [a - h, a + h]$.

Let $$y_0 = g(x_0)$$

Let $\varepsilon > 0$ be given. Now, consider a rectangle R', lying within the interval $R = [a - h, a + h; b - k, b + k]$ such that

$$R' = [x_0 - \delta_1, x_0 + \delta_1; y_0 - \varepsilon, y_0 + \varepsilon]$$

Since, $y = g(x)$ is the solution of $f(x, y) = 0$ in R which encloses R'. therefore $y = g(x)$ is also the solution of $f(x, y) = 0$ in R'.

Therefore, there exists an interval $]x_0 - \delta, x_0 + \delta[, (\delta \leq \delta_1)$ \hfill such that for every value of x

in the interval $]x_0 - \delta, x_0 + \delta[, g(x)$ lies between $y_0 - \varepsilon$ and $y_0 + \varepsilon$ *i.e.*,

$$|y - y_0| = |g(x) - g(x_0)| < \varepsilon \text{ whenever } |x - x_0| < \delta$$

Hence, g is continuous at x_0 and therefore in $[a - h, a + h]$.

(iii) *Test for Differentiability.* Let x be any point of $[a - h, a + h]$ and let $x+p$ be another point of $[a - h, a + h]$.

Let $$y = g(x), y + q = g(x+p)$$

$\Rightarrow$ $$f(x, y) = 0, f(x+p, y+q) = 0$$

$\Rightarrow$ $f(x+p, y+q) - f(x, y) = 0.$

$\Rightarrow$ $f(x+p, y+q) - f(x+p, y) + f(x+p, y) - f(x, y) = 0$

$\Rightarrow$ $q f_y (x + p, y + \theta_1 q) + p f_x (x + \theta_2 p, y) = 0$

(By mean value theorem, $0 < \theta_1 < 1$ and $0 < \theta_2 < 1$)

Since, $f_y \neq 0$ in R and $(x + p, y + \theta_1 q)$ is a point of R, we have

$$\frac{q}{p} = -\frac{f_x(x+\theta_2 p, y)}{f_y(x+p, y+\theta_1 q)}$$

Since, g is continuous and $q \to 0$ as $p \to 0$, therefore f_x and f_y being continuous, we get

$$g'(x) = -\frac{f_x(x,y)}{f_y(x,y)} \text{ when } (p, q) \to (0, 0)$$

Thus g is differentiable and $g'(x) = -\dfrac{f_x(x,y)}{f_y(x,y)}$

Also, since f_x and f_y are both continuous, therefore $g'(x)$ is continuous.

☛ REMARKS

⟹ Here, the function $y = g(x)$ is said to be the unique implicit function, determined by $f(x,y)=0$.

⟹ The implicit function theorem states that if a function of two variables satisfying certain assumptions of continuity and differentiability in a neighbourhood of a point (a, b) and If $(a,b) = 0$, then there exist a neighbourhood $[a - h, a + h; b - k, b + k]$ of (a, b), such that $x \in [a - h, a + h]$, there exist a unique y belonging to $[b - k, b + k]$ such that $f(x, y) = 0$. Hence, we can say that $f(x, y) = 0$ defines a functions $y = g(x)$ in $[a - h, a + h]$; $y \in [b - k, b+k]$, such that $g(x)$ is differentiable.

SOLVED EXAMPLES

EXAMPLE1. **Show that the equation $x^3 + y^3 - 3xy + y = 0$, determine unique solution near the point (0, 0). Also find the first derivative of the solution.**

SOLUTION. Here, we have

$$f(x,y) = x^3 + y^3 - 3xy + y = 0 \text{ at } (0, 0)$$

(i) $f(0, 0) = 0$

(ii) $f_x = 3x^2 - 3y, f_y = 3y^2 - 3x + 1$

Now check the continuity of f_x and f_y at $(0,0)$.

(i) *Continuity of f_x at $(0, 0)$.*

The simultaneously limit is $\lim_{\substack{x \to 0 \\ y \to 0}} (3x^2 - 3y) = 0$, exists.

$\therefore$ f_x is continuous.

The simultaneous limit is $\lim_{\substack{x \to 0 \\ y \to 0}} (3y^2 - 3x + 1) = 1$, exists.

$\therefore$ f_y is continuous.

and $$f_y(0,0) = \lim_{k \to 0} \frac{f(0,k) - f(0,0)}{k}$$

$$= \lim_{k \to 0} \frac{k^3 + k}{k}$$

$$= \lim_{k \to 0} k^2 + 1 = 1 \neq 0$$

$\therefore$ $f_y(0, 0) \neq 0$

Thus the equation $f(x, y) = 0$ determine a unique solution
$$y = \phi(x)$$
(By implicit function theorem)

The unique solution of $f(x, y) = 0$ is given by
$$x^3 + y^3 - 3xy + y = 0$$
$$y^3 - y(3x - 1) + x^3 = 0$$

Let $\quad y = \phi(x)$, therefore $[\phi(x)]^3 - \phi(x)[3x - 1] + x^3 = 0$

Since $\qquad\qquad\qquad \phi'(0) = 0$ ...(1)

Differentiating (1) , we get
$$3[\phi(x)]^2 . \phi'(x) - 3\phi(x) - (3x - 1)\phi'(x) + 3x^2 = 0$$
$$0 - 3 \times 0 - (-1)\phi'(0) + 0 = 0$$
$$\Rightarrow \qquad\qquad\qquad \phi'(0) = 0$$

EXAMPLE 2. *Show that the following equation determine unique solution near the point indicated.*

Also, find the derivative of the solution

(a) xy sin x + cos y = 0 at $\left(0, \dfrac{\pi}{2}\right)$ (b) $y^3 \cos x + y^2 \sin^2 x = 7$ at $\left(\dfrac{\pi}{3}, 2\right)$

SOLUTION. (a) $f(x, y) = xy \sin x + \cos y = 0$ Point $= \left(0, \dfrac{\pi}{2}\right)$

$\therefore \qquad\qquad f\left(0, \dfrac{\pi}{2}\right) = 0 + \cos\dfrac{\pi}{2} = 0$

Now $\qquad\qquad f_x = y[x \cos x + \sin x], f_y = x \sin x - \sin y$

Further we check the continuity of f_x and f_y at $\left(0, \dfrac{\pi}{2}\right)$.

Continuity of f_x at $\left(0, \dfrac{\pi}{2}\right)$.

The simultaneous limit of $f_x(x, y)$. *i.e.,* $\displaystyle\lim_{\substack{x \to 0 \\ y \to (\pi/2)}} (xy \cos x + \sin x) = 0$ exists.

and the simultaneous limit of $f_y (x, y)$ *i.e,*

$\displaystyle\lim_{\substack{x \to 0 \\ y \to (\pi/2)}} (x \sin x - \sin y) = \lim_{y \to (\pi/2)} (-\sin y) = -1$, **exists**

Hence both f_x and f_y are continuous.

and $\qquad f_y\left(0, \dfrac{\pi}{2}\right) = \lim_{k \to 0} \dfrac{f\left(0, \dfrac{\pi}{2} + k\right) - f\left(0, \dfrac{\pi}{2}\right)}{k}$

$$= \lim_{k \to 0} \dfrac{\cos\left(\dfrac{\pi}{2} + k\right)}{k} = \lim_{k \to 0} \dfrac{-\sin k}{k} = -1$$

$\therefore \qquad\qquad f_y\left(0, \dfrac{\pi}{2}\right) \neq 0$

Then by implicit function theorem, the equation $f(x, y) = 0$ determines a unique solulion. Now we have to find its first derivative
$$f(x, y) = 0$$
$$\Rightarrow \qquad xy \sin x + \cos y = 0$$

$$\therefore \qquad \phi'(x) = -\frac{f_x(x,y)}{f_y(x,y)} = -\frac{xy\cos x + y\sin x}{x\sin x - \sin y}$$

$$\phi'(x) = -\frac{0}{-1} = 0$$

$$\left[\begin{array}{l} \because \ f_x\left(0,\dfrac{\pi}{2}\right) = 0 \\[2mm] \text{and } f_y\left(0,\dfrac{\pi}{2}\right) = -1 \end{array}\right]$$

(b) $\qquad f(x,y) = y^3\cos x + y^2\sin^2 x - 7$; point $= \left(\dfrac{\pi}{3}, 2\right)$

$$\therefore \qquad f\left(\frac{\pi}{3},2\right) = 8\cos\frac{\pi}{3} + 4\sin^2\frac{\pi}{3} - 7 = 4 + 3 - 7 = 0$$

Now $\quad f_x(x,y) = -y^3\sin x + 2y^2\sin x\cos x$

and $\quad f_y(x,y) = 3y^2\cos x + 2y\sin^2 x$

Next, we have to examine the continuity of f_x and f_y

Continuity of $f_x(x,y)$ at $\left(0,\dfrac{\pi}{2}\right)$

The simultaneous limit is

$$\lim_{\substack{x\to(\pi/3)\\y\to 2}}[-y^3\sin x + 2y^2\sin x\cos x] = \lim_{y\to 2}\left[-y^3\sin\frac{\pi}{3} + 2y^2\sin\frac{\pi}{3}\cos\frac{\pi}{3}\right]$$

$$= -8.\frac{\sqrt3}{2} + 8.\frac{\sqrt3}{2}.\frac{1}{2}$$

$$= -4\sqrt3 + 2.\sqrt3 = -2\sqrt3 \neq 0, \text{ exists}$$

$\therefore$ The simultaneous limit is given by

$$\lim_{\substack{x\to(\pi/3)\\y\to 2}}[3y^2\cos x + 2y\sin^2 x] = \lim_{y\to 2}\left[3y^2\cos\frac{\pi}{3} + 2y\sin^2\frac{\pi}{3}\right]$$

$$= 12.\frac{1}{2} + 4.\frac{3}{4} = 6 + 3 = 9, \text{ exists}$$

$\therefore$ f_x and f_y both are continuous

$$f_y\left(\frac{\pi}{3},2\right) = 12\times\frac{1}{2} + 4\times\frac{3}{4} = 6 + 3 = 9 \neq 0.$$

Then by implicit function theorem, $f(x,y) = 0$ determines a unique solution. And its derivative is given by

$$\phi'(x) = -\frac{f_x(x,y)}{f_y(x,y)}$$

$$\phi'\left(\frac{\pi}{3}\right) = -\left[\frac{-8\sin\dfrac{\pi}{3} + 8\sin\dfrac{\pi}{3}.\cos\dfrac{\pi}{3}}{9}\right] = -\left[\frac{-8\dfrac{\sqrt3}{2} + 2\sqrt3}{9}\right] = \frac{2\sqrt3}{9}$$

Let $\qquad y = \phi(x)$

$$\therefore \qquad \phi\left(\frac{\pi}{3}\right) = 2$$

$$[\phi(x)]^3\cos x + [\phi(x)]^2\sin^2 x - 7 = 0$$

$$3[\phi(x)]^2.\phi'(x).\cos x - [\phi(x)]^3\sin x + 2\phi'(x)\phi(x)\sin^2 x + 2[\phi(x)]^2\sin x\cos x = 0$$

$$3\left[\phi\left(\frac{\pi}{3}\right)\right]^2 \phi'\left(\frac{\pi}{3}\right).\cos\frac{\pi}{3} - \left[\phi\left(\frac{\pi}{3}\right)\right]^3 .\sin\frac{\pi}{3}$$

$$+2\phi'\left(\frac{\pi}{3}\right).\phi\left(\frac{\pi}{3}\right).\sin^2\frac{\pi}{3} + 2\left[\phi\left(\frac{\pi}{3}\right)\right]^2 \sin\frac{\pi}{3}\cos\frac{\pi}{3} = 0$$

$$12.\frac{1}{2}\phi'\left(\frac{\pi}{3}\right) - 8.\frac{\sqrt{3}}{2} + 4.\frac{3}{4}\phi'\left(\frac{\pi}{3}\right) + 8.\frac{\sqrt{3}}{2}.\frac{1}{2} = 0$$

$$9\phi'\left(\frac{\pi}{3}\right) - 4\sqrt{3} + 2\sqrt{3} = 0$$

$$\phi'\left(\frac{\pi}{3}\right) = \frac{2\sqrt{3}}{9}$$

EXERCISE 10.5

1. If $f(x, y) = (x^2 + y^2) \log (x^2 + y^2)$ for $(x, y) \neq (0, 0)$ and $f(0, 0) = 0$; show that f_{xy} and f_{yx} are not continuous at $(0, 0)$ but $f_{xy}(0, 0) = f_{yx}(0, 0)$.

2. Show that the following equation determine unique solutions near the point indicated. Find also the first derivative of the solution $2xy - \log xy = 2$. at $(1, 1)$.

3. If f is a continuous function of each variable x and y separately in a certain neighbourhood of (a, b), and $f(a, b) = 0, f_y$ is continuous at (a, b) and $f_y(a, b) \neq 0$, then the equation $f(x, y) = 0$ determines a unique continuous implicit solution $y = \phi(x)$ near (a, b).

4. Show that the least positive root of $xy = \tan y$ is a continuous function of x throughout the interval $[1, \infty[$ and increases from 0 to $\frac{\pi}{2}$ as y increases from 1 towards ∞.

10.16 MAXIMA AND MINIMA OF A FUNCTION OF SEVERAL INDEPENDENT VARIABLES

Let $f(x, y, z, ...)$ be a function of several independent variables $x, y, z....$ If f is continuous and finite for all values of $x, y, z, ...$ in the neighbourhood of $x = a, y = b, z = c, ...$ respectively, then the value of $(a, b, c, ...)$ is said to be a maximum or minimum if $f(a+h, b+k, c+l, ...)$ is less than or greater than $f(a, b, c, ...)$ for all values of $h, k, l, ...$ (where $h, k, l, ...$) are sufficiently small, may be positive or negative provided they are not all zero.

In other words we can say, the value of $f(a, b, c, ...)$ is said to be a maximum or minimum if $f(a+h, b + k, c + l, ...) - f(a, b, c, ...)$ maintain an invariant sign (may be positive or negative) for all values of $h, k, l, ...$ positive or negative provided they are taken sufficiently small and finite.

10.16.1 STATIONARY AND EXTREME POINTS

A point $(a_1, a_2, ..., a_n)$ is called a stationary point, if all the first order partial derivative of the function $f(x_1, x_2, ..., x_n)$ vanish at the point. A stationary point, if it is maximum or minimum is known as extreme point and the value of the function at an extreme point is known as an extreme value.

☛ REMARK

➡ A stationary point may be a maximum or minimum or neither of these two.

10.17 NECESSARY CONDITION FOR THE EXISTENCE OF MAXIMA OR MINIMA

Let $f(x, y, z, ...)$ be a function of several independent variables $x, y, z,...$ It is clear from the definition of maxima and minima that maximum or minimum of $f(x, y, z, ..)$ will occur for those values of $x, y, z, ...$, for which the expression $f(x+h, y+k, z+l, ...) - f(x, y, z, ...)$ maintain an invariant sign for all sufficiently small and finite values of $h, k, l, ...$ positive or negative.

Now, expanding $f(x+h, y+k, z+l, ...)$ by Taylor's theorem, we have

$$f(x+h, y+k, z+l...) = f(x,y,z) + \left(h\frac{\partial f}{\partial x} + k\frac{\partial f}{\partial y} + l\frac{\partial f}{\partial z} + ... \right) + \text{terms of second and higher order.}$$

$$\Rightarrow f(x+h, y+k, z+l...) - f(x,y,z,...) = \left(h\frac{\partial f}{\partial x} + k\frac{\partial f}{\partial y} + l\frac{\partial f}{\partial z} + ... \right)$$

$$+ \text{terms of second and higher orders.} \qquad ...(1)$$

Now, since $h, k, l, ...$ are sufficiently small, the first degree expression

$$\left(h\frac{\partial f}{\partial x} + k\frac{\partial f}{\partial y} + l\frac{\partial f}{\partial z} + ... \right)$$

of the equation (1) can be made to govern the sign of right hand side and hence, of the left hand side as well as. Thus, by changing the sign of the left hand side of the equation (1) will also change.

Since, left hand side is to preserve an invariable sign for maxima or minima, therefore, as a necessary condition for maximum and minimum values, we must have

$$h\frac{\partial f}{\partial x} + k\frac{\partial f}{\partial y} + l\frac{\partial f}{\partial z} + ... = 0 \qquad ...(2)$$

Now, since $h, k, l, ...$ are arbitrary and independent of each other, we must have

$$\frac{\partial f}{\partial x} = 0, \frac{\partial f}{\partial y} = 0, \frac{\partial f}{\partial z} = 0, \text{ etc.} \qquad ...(3)$$

If the number of independent variables be n, we shall get n simultaneous equations in these n variables, which will give the values $a, b, c, ...$ of the n variables $x, y, z, ...$ respectively for which $f(x, y, z, ...)$ will have a maximum or a minimum values.

☛ **REMARKS**

⇒ The necessary condition for a function $f(x, y, z, ...)$ of the independent variables $x, y, z, ...$ to be maximum or minimum is given by

$$\frac{\partial f}{\partial x} = 0, \frac{\partial f}{\partial y} = 0, \frac{\partial f}{\partial z} = 0,$$

⇒ The conditions given above is only a necessary condition for the maxima and minima of the function $f(x, y, z, ...)$. These conditions are not sufficient.

10.17.1 MAXIMA AND MINIMA FOR A FUNCTION OF TWO INDEPENDENT VARIABLES

(1) *To find the condition which governs the sign of a quadratic expression.*

Consider, a binary expression

$$I = ax^2 + 2hxy + by^2$$

of two variables x and y. Then I can be written as

$$I = ax^2 + 2hxy + by^2$$

$$= \frac{1}{a}[(ax + hy)^2 + (ab - h^2)y^2].$$

If $(ab - h^2)$ is positive, the sign of I will be the same as that of a.

But if $(ab - h^2)$ is negative, then, the expression within the brackets may be positive or negative and therefore we cannot say anything about the sign of expression I.

(2) *Stationary and extreme points* (*For the function of two independent variables*):

Let $f(x, y)$ be a function of two independent variables x and y. A point (a, b) is called a stationary point, if both the first order partial derivatives $\left(\dfrac{\partial f}{\partial a} \text{ and } \dfrac{\partial f}{\partial b} \right)$ of the function $f(x, y)$ at (a, b) vanish.

A stationary point which is either a maximum or minimum is called an extreme point.

☞ **REMARKS**

➠ A stationary point is not necessarily an extreme point, hence a stationary point may be a maximum or a minimum or neither of these two.

➠ The value of the function at extreme point is called extreme value.

➠ A point at which function is neither maximum nor minimum, is known as saddle point.

10.18 NECESSARY CONDITION FOR MAXIMA OR MINIMA

Let $f(x, y)$ be a function of two independent variables x and y. Then, we have the maximum or minimum of $f(x, y)$ at $x = a$ and $x = b$ if the expression $f(a+h, b+k) - f(a, b)$ is of invariable sign for all sufficiently small independent variables h and k provided both of them are not equal to zero.

We observe that,

(i) If the sign of $f(a+h, b+k) - f(a, b)$ is negative, then we have a maximum of $f(x, y)$ at $x = a, y = b$.

(ii) If the sign of $f(a+h, b+k) - f(a, b)$ is positive, we have a minimum of $f(x, y)$ at $x = a, y = b$.
Expand $f(a+h, b+k)$ by Taylor's theorem, we have

$$f(a+h, b+k) = f(a,b) + \left(h\frac{\partial f}{\partial x} + k\frac{\partial f}{\partial y} \right)_{\substack{x=a \\ y=b}}$$

$$+ \frac{1}{2!}\left(h^2\frac{\partial^2 f}{\partial x^2} + 2hk\frac{\partial^2 f}{\partial x\,\partial y} + k^2\frac{\partial^2 f}{\partial y^2} \right)_{\substack{x=a \\ y=b}} + \dots$$

$$\Rightarrow \quad f(a+h, b+k) - f(a,b) = h\left(\frac{\partial f}{\partial x} \right)_{\substack{x=a \\ y=b}} + k\left(\frac{\partial f}{\partial y} \right)_{\substack{x=a \\ y=b}}$$

+ term of the second and higher orders in h and k.

Now, since h and k are sufficiently small, the expression $h\left(\dfrac{\partial f}{\partial x} \right)_{\substack{x=a \\ y=b}} + k\left(\dfrac{\partial f}{\partial y} \right)_{\substack{x=a \\ y=b}}$ of the equation (1) can be made to govern the sign of right hand side and hence of the left hand side as well. Thus by changing the sign of h and k, the sign of the left hand side of the equation (1) will also change.

Since L.H.S. is to preserve an invariable sign for maximum or minimum, therefore as a necessary condition for maximum and minimum values, we must have

$$h\left(\frac{\partial f}{\partial x} \right)_{\substack{x=a \\ y=b}} + k\left(\frac{\partial f}{\partial y} \right)_{\substack{x=a \\ y=b}} = 0. \qquad \dots(2)$$

If $k = 0$, we find that if $\left(\dfrac{\partial f}{\partial x} \right)_{\substack{x=a \\ y=b}} \neq 0$, the R.H.S. of (2) changes sign when h changes sign.

Therefore $f(x, y)$ cannot have a maximum or minimum at $x = a$, $y = b$ if $\left(\dfrac{\partial f}{\partial x}\right)_{\substack{x=a \\ y=b}} \neq 0$.

Similarly, taking $h = 0$, we see that $f(x, y)$ cannot have a maximum or a minimum at $x = a$,

$y = b$ if $\left(\dfrac{\partial f}{\partial y}\right)_{\substack{x=a \\ y=b}} \neq 0$. Thus, a set of necessary conditions that $f(x, y)$ should have a maximum or

minimum at $x=a$, $y=b$ is that

$$\left(\frac{\partial f}{\partial x}\right)_{\substack{x=a \\ y=b}} = 0 \ and \ \left(\frac{\partial f}{\partial y}\right)_{\substack{x=a \\ y=b}} = 0.$$

10.19 SUFFICIENT CONDITION FOR MAXIMA OR MINIMA: THE LAGRANGE'S CONDITION

Let $f(x, y)$ be a function of two variables x and y.

Let $\qquad r = \dfrac{\partial^2 f}{\partial x^2}, s = \dfrac{\partial^2 f}{\partial x \partial y}, t = \dfrac{\partial^2 f}{\partial y^2}$ at $x = a$ and $y = b$.

As a set of necessary conditions for a maximum or minimum at (a, b) we have

$$\frac{\partial f}{\partial x} = 0 \ and \ \frac{\partial f}{\partial y} = 0 \ at \ (a, b)$$

then $\qquad f(a + h, b+ k) - f(a, b) = \dfrac{1}{2!}[rh^2 + 2shk + tk^2] + R$ ...(1)

Where remainder, R consists of terms of third and higher order of small quantities h and k.

Now, by taking h and k sufficiently small, the second degree terms in R.H.S. of (1) may be made to govern the sign of R.H.S. and therefore of the L.H.S. also *i.e.*, for sufficiently small values of h and k, the sign of $\dfrac{1}{2}(rh^2 + 2shk + tk^2) + R$ is same as that of $rh^2 + 2shk + tk^2$.

If the sign is negative, then the function is maximum at (a, b) and if the sign is positive, then the function is minimum at (a, b).

Case (i) *If* $(rt - s^2) > 0$.

Then, neither r nor t can be zero. Hence, we can write

$$rh^2 + 2shk + tk^2 = \frac{1}{2}[r^2h^2 + 2rshk + rtk^2] = \frac{1}{2}[(rh + sk)^2 + (rt - s^2)k^2]$$

since $rt - s^2 > 0$, therefore $(rh + sk)^2 + (rt - s^2)k^2 > 0$ for all values of h and k except when $rh + sk = 0$, $k = 0$ *i.e.*, at $h = 0$, $k = 0$, which is not possible.

Hence, in this case the expression $rh^2 + 2shk + tk^2$ will have the same sign for all values of h and k, and the sign is determined by the sign of r.

Thus, the function $f(x, y)$ will have a maximum or minimum at $x = a$ and $y = b$. If $rt - s^2 > 0$. The function $f(x, y)$ is maximum or minimum according as r is negative or positive.

Case (ii) *If* $(rt - s^2) < 0$.

If $rt - s^2$ is negative, we are not sure about the sign of second degree term of R.H.S. of (1) and hence there is neither a maximum nor a minimum value.

Case (iii) *If* $rt - s^2 = 0$.

If $rt = s^2$, then quadratic expression $rh^2 + 2shk + tk^2$ becomes $\dfrac{1}{r}(hr + ks)^2$.

So that, the quadratic expression will be of the same sign as that of r or t unless

$$\frac{h}{k} = -\frac{s}{r} = \alpha \ (say) \ i.e., \ rh + sk = 0.$$

If this condition is safisfied, then the second degree expression in R.H.S. of (1) vanishes and hence, the sign of the R.H.S. of (1) depends upon third degree expression in h and k, which change sign with the change of sign of h and k and hence, the sign of L.H.S. of (1) will also change and hence, there will be neither maximum nor minimum.

Thus, the necessary condition for the existence of maxima and minima now is that the cubic terms must vanish collectively in R.H.S. of (1) when $\dfrac{h}{k} = -\dfrac{s}{r} = \alpha$; and then the biquadratic terms of R.H.S. of (1) must collecctively be of the same sign as r and t, when

$$\frac{h}{k} = -\frac{s}{r} = \alpha$$

i.e.,
$$hr + ks = 0$$

Hence, the case is doubtful.

Thus, if $rt - s^2 = 0$, the case is doubtful and further, investigation is needed to determine the maxima and minima of $f(x, y)$ at (a, b).

◢ WORKING PROCEDURE

To discuss the maxima and minima at $x = a, y = b$, we must find

$$r = \left(\frac{\partial^2 u}{\partial x^2}\right)_{\substack{x=0\\y=0}}, \ s = \left(\frac{\partial^2 u}{\partial x \partial y}\right)_{\substack{x=a\\y=b}}, \ t = \left(\frac{\partial^2 u}{\partial y^2}\right)_{\substack{x=a\\y=b}}$$

Then, calculate $rt - s^2$.

Now following cases arise :

(i) If $rt - s^2 > 0$, then

    (A) If r is negative then, $f(x, y)$ is maximum at $x = a, y = b$.

    (B) If r is positive then, $f(x, y)$ is minimum at $x = a, y = b$.

(ii) If $rt - s^2 < 0$, $f(x, y)$ is neither maximum nor minimum at $x = a, y = b$.

(iii) If $rt - s^2 = 0$, the case is doubtful, and further investigation will be required.

☞ REMARK

⇒ While solving problems, we frequently used the identity, given by Lagrange.

$$\{(a^2 + b^2 + c^2)(p^2 + q^2 + r^2) - (ap + bq + cr)^2\}$$
$$= \{(br - cq)^2 + (cp + ar)^2 + (aq - bp)^2\}.$$

◢ SOLVED EXAMPLES

EXAMPLE 1.   *Find all maximum or minimum values of the function :*

$$f(x, y) = y^2 + x^2 y + x^4.$$

SOLUTION.   Since, we have

$$f(x, y) = y^2 + x^2 y + x^4.$$

$$\therefore \qquad \frac{\partial f}{\partial x} = 2xy + 4x^3 \quad \text{and} \quad \frac{\partial f}{\partial y} = 2y + x^2.$$

For a maximum or minimum of $f(x, y)$, we must have

$$\frac{\partial f}{\partial x} = 0 \quad \text{and} \quad \frac{\partial f}{\partial y} = 0$$

$$\therefore \qquad \frac{\partial f}{\partial x} = 0 \implies 2xy + 4x^3 = 0$$

$$\implies 2x(y + 2x^2) = 0 \qquad \qquad \dots(1)$$

$$\frac{\partial f}{\partial y} = 0 \Rightarrow 2y + x^2 = 0$$

Solving (1) and (2), we get $x = 0, y = 0$.

Thus $(0, 0)$ is the only point of maximum or minimum.

Now
$$r = \left(\frac{\partial^2 f}{\partial x^2}\right)_{(0,0)} = [2y + 12x^2]_{(0,0)} = 0$$

$$s = \left(\frac{\partial^2 f}{\partial x \partial y}\right)_{(0,0)} = [2x]_{(0,0)} = 0$$

and
$$t = \left(\frac{\partial^2 f}{\partial y^2}\right)_{(0,0)} = [2]_{(0,0)} = 2$$

$\therefore$
$$rt - s^2 = 0\,(2) - 0^2 = 0.$$

Thus, the case is doubtful and further investigation will be required.

EXAMPLE 2. **Find the maximum or minimum values of the function $x^3y^2(1 - x - y)$.**

SOLUTION. Let
$$u = x^3 y^2 (1 - x - y)$$

$\Rightarrow$
$$\frac{\partial u}{\partial x} = 3x^2y^2(1 - x - y) - x^3y^2$$

and
$$\frac{\partial u}{\partial y} = 2x^3y(1 - x - y) - x^3y^2.$$

For a maximum or minimum of u, we must have $\dfrac{\partial u}{\partial x} = 0$ and $\dfrac{\partial u}{\partial y} = 0$

$\Rightarrow$
$$3x^2y^2(1 - x - y) - x^3y^2 = 0 \qquad \ldots(1)$$

and
$$2x^3y(1 - x - y) - x^3y^2 = 0. \qquad \ldots(2)$$

Now, subtracting (2) from (1), we have $x^2y(1 - x - y)(3y - 2x) = 0$

which gives
$$y = \frac{2}{3}x.$$

Putting the value of y in (1), we get $x = \dfrac{1}{2}$

So $\left(\dfrac{1}{2}, \dfrac{1}{3}\right)$ be the point of maxima or minima.

Now
$$r = \frac{\partial^2 u}{\partial x^2} = 6xy^2 - 12x^2y^2 - 6xy^3 = -\frac{1}{9}, \text{ at } \left(\frac{1}{2}, \frac{1}{3}\right)$$

$$t = \frac{\partial^2 u}{\partial y^2} = 2x^3 - 2x^4 - 6x^3y = -\frac{1}{8}, \text{ at } \left(\frac{1}{2}, \frac{1}{3}\right)$$

$$s = \frac{\partial^2 u}{\partial x \partial y} = 6x^2y - 8x^3y - 9x^2y^2 = -\frac{1}{12} \text{ at } \left(\frac{1}{2}, \frac{1}{3}\right).$$

Now,
$$rt - s^2 = \text{positive.}$$

Also, r is negative, hence the function u has a maximum at $x = \dfrac{1}{2}, y = \dfrac{1}{3}$.

The maximum value is $= \left(\dfrac{1}{2}\right)^3 \left(\dfrac{1}{3}\right)^2 \left(1 - \dfrac{1}{2} - \dfrac{1}{3}\right) = \dfrac{1}{432}.$

EXAMPLE 3. **Discuss the maximum or minimum values of u, where**
$$u = 2a^2xy - 3ax^2y - ay^3 + x^3y + xy^3.$$

SOLUTION.   We have $u = 2a^2xy - 3ax^2y - ay^3 + x^3y + xy^3$
which gives

$$\frac{\partial u}{\partial x} = 2a^2y - 6axy + 3x^2y + y^3$$

and $$\frac{\partial u}{\partial y} = 2a^2x - 3ax^2 - 3ay^2 + x^3 + 3xy^2$$

For a maximum and minima of u, we have

$$\frac{\partial u}{\partial x} = 0, \frac{\partial u}{\partial y} = 0$$

which gives,

$$y(2a^2 - 6ax + 3x^2 + y^2) = 0 \qquad \text{...(1)}$$

and $2a^2x - 3ax^2 - 3ay^2 + x^3 + 3xy^2 = 0$ ...(2)

Equation (1) and (2) gives the following values of x and y :

$$x = 0, y = 0; \; x = a, y = 0; \; x = 2a, \; y = 0; \; x = \frac{3}{2}a, \; y = \pm\frac{1}{2}a;$$

$$x = a, y = a, \; x = \frac{1}{2}, \; y = \frac{1}{2}a; x = a, y = -a; \; x = \frac{1}{2}a, \; y = -\frac{1}{2}a.$$

Then, we get the following pairs of values of x and y which make the function u stationary.

$$(0,0), (a,0), (2a,0), \left(\frac{3}{2}a, \frac{1}{2}a\right), \left(\frac{3}{2}a, -\frac{1}{2}a\right) (a,a), \left(\frac{1}{2}a, \frac{1}{2}a\right), (a,-a), \left(\frac{1}{2}a, -\frac{1}{2}a\right).$$

Also $$r = \frac{\partial^2 u}{\partial x^2} = -6ay + 6xy,$$

$$s = \frac{\partial^2 u}{\partial x \partial y} = 2a^2 - 6ax + 3x^2 + 3y^2,$$

and $$t = \frac{\partial^2 u}{\partial y^2} = -6ay + 6xy.$$

For (0, 0).

$$r = 0, s = 2a^2, t = 0$$

$\Rightarrow rt - s^2$, is negative.

Therefore, we have neither maximum nor a minimum of u at (0, 0).

Similarly, we can easily shown that u has neither a maximum nor a minimum at $(a, 0)$, $(2a, 0)$, (a, a), $(a, -a)$.

For $\left(\frac{3a}{2}, \frac{a}{2}\right)$.

$$r = \frac{3}{2}a^2, s = \frac{1}{2}a^2, t = \frac{3}{2}a^2, \quad \Rightarrow \quad rt - s^2 \text{ is positive.}$$

Here, since r is positive, therefore u has minimum at $\left(\frac{3a}{2}, \frac{a}{2}\right)$.

Similarly, we can check the maxima and minima at all other points.

EXAMPLE 4.   ***Find the maximum and minimum values of $xy(a - x - y)$.***

SOLUTION.   Let $u = xy(a - x - y)$

Then $\frac{\partial u}{\partial x} = ay - 2xy - y^2$ and $\frac{\partial u}{\partial y} = ax - x^2 - 2xy.$

For a maximum or minimum of u, we have

$$\frac{\partial u}{\partial x} = 0 \text{ and } \frac{\partial u}{\partial y} = 0.$$

Thus, we have

$$ay - 2xy - y^2 = 0 \Rightarrow y(a - 2x - y) = 0 \qquad \text{...(1)}$$
$$ax - x^2 - 2xy = 0 \Rightarrow x(a - x - 2y) = 0 \qquad \text{...(2)}$$

Solving (1) and (2), we get the following pairs of values x and y which makes the function stationary

$$(0,0), (0,a), (a,0), \left(\frac{1}{3}a, \frac{1}{3}a\right).$$

Here

$$r = \frac{\partial^2 u}{\partial x^2} = -2y, \quad s = \frac{\partial^2 u}{\partial x \, \partial y} = a - 2x - 2y,$$

and

$$t = \frac{\partial^2 u}{\partial y^2} = -2x.$$

For (0, 0). $\quad r = 0, s = a, t = 0 \quad \Rightarrow rt - s^2$ is negative.

$\therefore$ We have neither a maximum nor a minimum of u at $(0, 0)$.

For (0, a). $\quad r = -2a, s = -a, t = 0 \quad \Rightarrow rt - s^2$ is negative.

$\therefore$ We have neither a maximum nor a minimum of u at $(a, 0)$.

Similarly, we have neither a maximum nor a minimum of u at $(\alpha, 0)$.

For $\left(\dfrac{1}{3}a, \dfrac{1}{3}a\right)$.

$$r = -\frac{2}{3}a, s = -\frac{1}{3}a, t = -\frac{2}{3}a \qquad \Rightarrow rt - s^2 \text{ is positive.}$$

Since $\quad rt - s^2 > 0$.

$\therefore \quad u$ has an extreme value at $\left(\dfrac{1}{3}a, \dfrac{1}{3}a\right)$

$\Rightarrow \quad u$ has a maximum if r is negative, *i.e.*, if a is positive and u has a minimum if r is positive, *i.e.*, if a is negative.

EXAMPLE 5. ***Find a point within a triangle such that the sum of the squares of its distances from the vertices is a minimum.***

SOLUTION. Let us suppose $[(x_r, y_r) : r = 1, 2, 3]$ be the vertices of the triangle and (x, y) be any point inside the triangle.

Now, let us define a function

$$u = \sum_{r=1}^{3} [(x - x_r)^2 + (y - y_r)^2].$$

Then, we have

$$\frac{\partial u}{\partial x} = \Sigma 2(x - x_r) = 2[(x - x_1) + (x - x_2) + (x - x_3)]$$

and

$$\frac{\partial u}{\partial y} = \Sigma 2(y - y_r) = 2[(y - y_1) + (y - y_2) + (y - y_3)].$$

For a maximum or minimum of u, we must have

$$\frac{\partial u}{\partial x} = 0 \Rightarrow (x - x_1) + (x - x_2) + (x - x_3) = 0 \Rightarrow x = \frac{x_1 + x_2 + x_3}{3}$$

and

$$\frac{\partial u}{\partial y} = 0 \Rightarrow (y - y_1) + (y - y_2) + (y - y_3) = 0 \Rightarrow y = \frac{y_1 + y_2 + y_3}{3}.$$

Thus, we have $\left(\dfrac{x_1 + x_2 + x_3}{3}, \dfrac{y_1 + y_2 + y_3}{3}\right)$ is the only point at which u have a maximum or minimum.

Now $$r = \frac{\partial^2 u}{\partial x^2} = 6, \, s = \frac{\partial^2 u}{\partial x \, \partial y} = 0, \, t = \frac{\partial^2 u}{\partial y^2} = 6.$$

Now, at $\left[\dfrac{x_1 + x_2 + x_3}{3}, \dfrac{y_1 + y_2 + y_3}{3} \right]$

$$r = 6, \, s = 0, \, t = 6$$

$\Rightarrow$ $$rt - s^2 = 36 > 0.$$

Also, since $$r > 0.$$

Therefore u have a minimum value at $\left[\dfrac{x_1 + x_2 + x_3}{3}, \dfrac{y_1 + y_2 + y_3}{3} \right]$.

Hence, the point $\left(\dfrac{x_1 + x_2 + x_3}{3}, \dfrac{y_1 + y_2 + y_3}{3} \right)$ is the required point at which u is minimum.

☛ **REMARK**

⇒ The point $\left(\dfrac{x_1 + x_2 + x_3}{3}, \dfrac{y_1 + y_2 + y_3}{3} \right)$ is the centroid of the given triangle.

EXAMPLE 6. *Show that the minimum value of* $u = xy + \left(\dfrac{a^3}{x} \right) + \left(\dfrac{a^3}{y} \right)$ *is* $3a^2$.

SOLUTION. We have

$$u = xy + \left(\frac{a^3}{x} \right) + \left(\frac{a^3}{y} \right)$$

$\Rightarrow$ $$\frac{\partial u}{\partial x} = y - \frac{a^3}{x^2} \text{ and } \frac{\partial u}{\partial y} = x - \frac{a^3}{y^2}.$$

For a maximum or minimum of u, we have

$$\frac{\partial u}{\partial x} = 0 \text{ and } \frac{\partial u}{\partial y} = 0$$

Now, $$\frac{\partial u}{\partial x} = 0 \Rightarrow y - \frac{a^3}{x^2} = 0 \qquad \qquad \dots (1)$$

and $$\frac{\partial u}{\partial y} = 0 \Rightarrow x - \frac{a^3}{y^2} = 0. \qquad \qquad \dots (2)$$

Solving (1) and (2), we get

$$x - a, \, y = a$$

Now $r = \dfrac{\partial^2 u}{\partial x^2} = \dfrac{2a^3}{x^3}, \, s = \dfrac{\partial^2 u}{\partial x \, \partial y} = 1 \text{ and } t = \dfrac{\partial^2 u}{\partial y^2} = \dfrac{2a^3}{y^3}.$

At $x = y = a$, we have

$$r = 2, \, s = 1, \, t = 2$$

$\Rightarrow$ $$rt - s^2 = 3 > 0.$$

Thus, at (a, a), $rt - s^2 > 0$ and $r > 0$. Therefore u is minimum at $x = a$, $y = a$.

The minimum value of $u = a.a + \left(\dfrac{a^3}{a} \right) + \left(\dfrac{a^3}{a} \right) = 3a^2.$

EXAMPLE 7. *Determine the points where a function* $x^3 + y^3 - 3axy$ *has maximum or minimum.*

SOLUTION. Here, we have

$$u = x^3 + y^3 - 3axy$$

$$\Rightarrow \quad \frac{\partial u}{\partial x} = 3x^2 - 3ay \text{ and } \frac{\partial u}{\partial y} = 3y^2 - 3ax.$$

For a maximum or minimum of u, we must have

$$\frac{\partial u}{\partial x} = 0 \text{ and } \frac{\partial u}{\partial y} = 0$$

which gives, $\quad x^2 - ay = 0$ ... (1)

and $\quad y^2 - ax = 0$ ... (2)

Solving (1) and (2), we get

$$x = 0, y = 0; x = a, y = a.$$

Thus $(0, 0)$ and (a, a) are the stationary points of u.

Now $\quad r = \dfrac{\partial^2 u}{\partial x^2} = 6x, s = \dfrac{\partial^2 u}{\partial x \, \partial y} = -3a, t = \dfrac{\partial^2 u}{\partial y^2} = 6y.$

For $x = 0, y = 0$

$$r = 0, s = -3a \text{ and } t = 0$$

$\therefore \quad rt - s^2 = -9a^2 < 0$, for all values of a.

$\Rightarrow \quad u$ is neither maximum nor minimum at $x = 0, y = 0$.

For $x = a, y = a$

$$r = 6a, s = -3a \text{ and } t = 6a$$

$\Rightarrow \quad rt - s^2 = 27a^2 > 0$, for all values of a.

Also $r = 6a$, which is positive if $a > 0$.

Thus (i) u is maximum at $x = a, y = a$ if $a < 0$

and (ii) u is minimum at $x = a, y = a$ if $a > 0$.

EXAMPLE 8. **Discuss the maxima and minima of the function $u = \sin x \sin y \sin(x + y)$.**

SOLUTION. Here, we have

$$u = \sin x \sin y \sin (x + y)$$

$$\Rightarrow \quad \frac{\partial u}{\partial x} = \sin y [\sin x \cos(x + y) + \cos x \sin (x + y)]$$

and $\quad \dfrac{\partial u}{\partial y} = \sin x [\sin y \cos(x + y) + \cos y \sin (x + y)].$

For a maxima and minima of u, we must have

$$\frac{\partial u}{\partial x} = 0 \text{ and } \frac{\partial u}{\partial y} = 0.$$

$\Rightarrow \quad \sin y [\sin x \cos (x + y) + \cos x \sin (x+y)] = 0$

and $\quad \sin x [\sin y \cos (x + y) + \cos y \sin (x+y)] = 0.$

Equation (1) and (2) gives

$$\tan (x + y) = - \tan x \quad ...(1) \Rightarrow \quad \tan x = \tan y$$

and $\quad \tan (x + y) = - \tan y \quad ...(2) \Rightarrow \quad x = y$

From (1) and (2), we have

$$\tan 2x = - \tan x = \tan (\pi - x)$$

$\Rightarrow \quad 2x = \pi - x \quad \Rightarrow \quad 3x = \pi \quad \Rightarrow \quad x = \dfrac{\pi}{3} = y.$

Moreover, $\quad \dfrac{\partial u}{\partial x} = 0$, gives $\sin y = 0 \Rightarrow y = 0$

and $\quad \dfrac{\partial u}{\partial y} = 0$, gives $\sin x = 0 \Rightarrow x = 0.$

Thus, we get the following pair of values, which makes the function u stationary

$$(0,0), \left(\frac{\pi}{3}, \frac{\pi}{3}\right).$$

Now

$$r = \frac{\partial^2 u}{\partial x^2} = 2\sin y \cos(2x + y),$$

$$s = \frac{\partial^2 u}{\partial x \, \partial y} = \sin 2(x + y),$$

and

$$t = \frac{\partial^2 u}{\partial y^2} = 2\sin x \cos(2y + x).$$

For (0, 0). $r = 0, s = 0, t = 0$

$\Rightarrow$ $rt - s^2 = 0.$

$\therefore$ this case is doubtful and need further investigation.

For $\left(\dfrac{\pi}{3}, \dfrac{\pi}{3}\right).$

$$r = 2\sin\frac{1}{3}\pi . \cos\pi = -\sqrt{3},$$

$$s = \sin\left(\frac{4\pi}{3}\right) = -\sin\frac{\pi}{3} = -\frac{\sqrt{3}}{2},$$

and

$$t = 2\sin\frac{1}{3}\pi \cos\pi = -\sqrt{3}.$$

$\therefore$ $rt - s^2 = \frac{9}{4} = \text{positive}.$

Also $r = -\sqrt{3}.$

Hence, u has a maximum value at $\left(\dfrac{\pi}{3}, \dfrac{\pi}{3}\right).$

EXAMPLE 9.   *Discuss the maxima and minima of the function $u = x^2 y^2 - 5x^2 - 8xy - 5y^2$.*

SOLUTION.   Here, we have

$$u = x^2 y^2 - 5x^2 - 8xy - 5y^2$$

$\Rightarrow$

$$\frac{\partial u}{\partial x} = 2xy^2 - 10x - 8y$$

and

$$\frac{\partial u}{\partial y} = 2x^2 y - 8x - 10y.$$

For a maximum or minimum of u, we must have $\dfrac{\partial u}{\partial x} = 0$ and $\dfrac{\partial u}{\partial y} = 0$.

which implies $2xy^2 - 10x - 8y = 0,$ ... (1)

and $2x^2 y - 8x - 10y = 0.$ ... (2)

From equation (2) we have $y = \dfrac{4x}{x^2 - 5}$

Put this value of y in equation (1), we get

$$x . \frac{16x^2}{(x^2 - 5)^2} - 5x - \frac{16}{x^2 - 5} = 0$$

$\Rightarrow$ $x[-5x^4 + 50x^2 - 45] = 0$

$\Rightarrow$ $x[x^4 - 10x^2 + 9] = 0$

$\Rightarrow$ $x = 0, \pm1, \pm3.$

Also from (2), for $x = 0$, $y = 0$,
for $x = 1$, $y = -1$,
for $x = -1$, $y = 1$,
for $x = 3$, $y = 3$,
and   for $x = -3$, $y = -3$.

Hence, the function u is stationary at the points $(0, 0)$, $(1, -1)$, $(-1, 1)$, $(3, 3)$ and $(-3, -3)$.

Now
$$r = \frac{\partial^2 u}{\partial x^2} = 2y^2 - 10, \quad s = \frac{\partial^2 u}{\partial y\, \partial x} = 4xy - 8,$$

and
$$t = \frac{\partial^2 u}{\partial y^2} = 2x^2 - 10.$$

For $(0, 0)$. $r = -10, s = -8, t = -10$

$\Rightarrow$ $rt - s^2 = 36 = +$ ve.

Since $r = -10 < 0$. Hence, u is maximum at $(0, 0)$.

For $(1, -1)$. $r = -8, s = -12, t = -8$

$\Rightarrow$ $rt - s^2 = -80 < 0.$

Hence, the stationary value of u at $(1, -1)$ is neither maximum nor minimum.

Similarly at $(-1, 1)$, $(3, 3)$ and $(-3, 3)$ the function u is neither maximum nor minimum.

EXAMPLE 10. *Find the maximum value of $x^2 + y^2 + z^2$ when $ax + by + cz = p$.*

SOLUTION. Here, $u = x^2 + y^2 + z^2$ ...(1)

Also $ax + by + cz = p$

$\Rightarrow$ $z = \dfrac{p - ax - by}{c}.$

Put this value of z in equation (1), we get

$$u = x^2 + y^2 + \frac{(p - ax - by)^2}{c^2}$$

$\Rightarrow$ $\dfrac{\partial u}{\partial x} = 2x - \dfrac{2a}{c^2}(p - ax - by)$

$\Rightarrow$ $\dfrac{\partial u}{\partial y} = 2y - \dfrac{2b}{c^2}(p - ax - by).$

For a maxima and minima of u, we must have $\dfrac{\partial u}{\partial x} = 0$ and $\dfrac{\partial u}{\partial y} = 0$

$\Rightarrow$ $x = \dfrac{ap}{a^2 + b^2 + c^2}$ and $y = \dfrac{bp}{a^2 + b^2 + c^2}.$

Now, $r = \dfrac{\partial^2 u}{\partial x^2} = 2 + \dfrac{2a^2}{c^2}, \ s = \dfrac{\partial^2 u}{\partial x\, \partial y} = \dfrac{2ab}{c^2}$

and $t = \dfrac{\partial^2 u}{\partial y^2} = 2 + \dfrac{2b^2}{c^2}$

$\Rightarrow$ $rt - s^2 = 4\left(1 + \dfrac{a^2}{c^2}\right)\left(1 + \dfrac{b^2}{c^2}\right) - \dfrac{4a^2b^2}{c^4}$

$$= 4\left(1 + \frac{a^2}{c^2} + \frac{b^2}{c^2}\right) = \text{positive.}$$

Since r is positive and $rt - s^2 > 0$, therefore u is minimum for the above values of x and y.

The minimum value is $\dfrac{p^2}{a^2 + b^2 + c^2}.$

EXAMPLE 11. *Find the stationary point of $x^4 + y^4 - 2x^2 + 4xy - 2y^2$ and determine their nature.*

SOLUTION. Here, we have $\quad u = x^4 + y^4 - 2x^2 + 4xy - 2y^2$

$$\Rightarrow \qquad \frac{\partial u}{\partial x} = 4x^3 - 4x + 4y$$

and $\qquad \dfrac{\partial u}{\partial y} = 4y^3 + 4x - 4y.$

For, a maxima and minima of u, we must have

$$\frac{\partial u}{\partial x} = 0 \Rightarrow 4x^3 - 4x + 4y = 0 \qquad\qquad ...(1)$$

and $\qquad \dfrac{\partial u}{\partial y} = 0 \Rightarrow 4y^3 + 4x - 4y = 0. \qquad\qquad ...(2)$

Solving (1) and (2), we get

$$4x^3 + 4y^3 = 0 \Rightarrow x^3 + y^3 = 0$$

$$\Rightarrow \qquad (x + y)(x^2 - xy + y^2) = 0$$

$$\Rightarrow \qquad \text{either} \ \ x + y = 0 \ \text{or} \ x^2 - xy + y^2 = 0 \qquad\qquad ...(3)$$
$$x + y = 0 \ \Rightarrow y = -x.$$

Put $y = -x$ in equation (1), we get $4x^3 - 8x = 0$

$$\Rightarrow \qquad x(x^2 - 2) = 0$$

$$\Rightarrow \qquad x = 0, \sqrt{2}, -\sqrt{2}.$$

Then $\qquad y = 0, -\sqrt{2}, \sqrt{2}$

Also from (3) $\qquad x = 0, y = 0 \qquad\qquad$ (only real solution)

Hence, the stationary points u are given by $(0,0), (\sqrt{2}, -\sqrt{2}), (-\sqrt{2}, \sqrt{2}).$

Now, we have $\quad r = \dfrac{\partial^2 u}{\partial x^2} = 12x^2 - 4, \ s = \dfrac{\partial^2 u}{\partial x \partial y} = 4,$

and $\qquad t = \dfrac{\partial^2 u}{\partial y^2} = 12y^2 - 4.$

For (0, 0). $\qquad r = -4, s = 4, t = -4$

$$\Rightarrow \qquad rt - s^2 = 0.$$

$\Rightarrow \quad$ At the point $(0, 0)$ the case is doubtful, and there is a need of further investigation.

For $(\sqrt{2}, -\sqrt{2}).$

$$r = 20, s = 4, t = 20$$

$$\Rightarrow \qquad rt - s^2 = 400 - 16 = 384 > 0.$$

Also $\qquad r > 0.$

$\Rightarrow \quad u$ has a minimum value at $(\sqrt{2}, -\sqrt{2}).$

Similarly u has a minimum value at $(-\sqrt{2}, \sqrt{2}).$

EXAMPLE 12. *Prove that the maxima or minima of the function*

$$u = \left[\frac{ax^2 + by^2 + 2hxy + 2gx + 2fy + c}{a'x^2 + b'y^2 + 2h'xy + 2g'x + 2f'y + c'} \right]$$

are given by the roots of the equation

$$\begin{vmatrix} a - a'u & h - h'u & g - g'u \\ h - h'u & b - b'u & f - f'u \\ g - g'u & f - f'u & c - c'u \end{vmatrix} = 0.$$

SOLUTION.   Here, we have

$$u = \left[\frac{ax^2 + by^2 + 2hxy + 2gx + 2fy + c}{a'x^2 + b'y^2 + 2h'xy + 2g'x + 2f'y + c'} \right]$$

$$\Rightarrow \quad u[a'x^2 + b'y^2 + 2h'xy + 2g'x + 2f'y + c'] = [ax^2 + by^2 + 2hxy + 2gx + 2fy + c]. \qquad ...(1)$$

Differentiating (1) partially w.r.t. x and y, we have

$$\frac{\partial u}{\partial x}[a'x^2 + b'y^2 + 2h'xy + 2g'x + 2f'y + c'] + u[2a'x + 2h'y + 2g'] = 2ax + 2hy + 2g$$

$$...(2)$$

and $\quad \dfrac{\partial u}{\partial y}[a'x^2 + b'y^2 + 2h'xy + 2g'x + 2f'y + c'] + u[2b'y + 2h'x + 2f'] = 2by + 2hx + 2f.$

$$...(3)$$

For the maxima and minima of u, we must have

$$\frac{\partial u}{\partial x} = 0 \Rightarrow u[a'x + h'y + g'] = ax + hy + g \qquad ...(4)$$

and $\quad \dfrac{\partial u}{\partial y} = 0 \Rightarrow u[h'x + b'y + f'] = hx + by + f.$ $\qquad ...(5)$

Now, multiplying (4) by x, (5) by y and adding, we have

$$u[a'x^2 + b'y^2 + 2h'xy + g'x + f'y] = ax^2 + by^2 + 2hxy + gx + fy \qquad ...(6)$$

Subtracting (6) from (1), we get

$$u(g'x + f'y + c') = gx + fy + c \qquad ...(7)$$

Now, from (4), (5) and (7), we have

$$(a - a'u)x + (h - h'u)y + (g - g'u) = 0 \qquad ...(8)$$

$$(h - h'u)x + (b - b'u)y + (f - f'u) = 0 \qquad ...(9)$$

$$(g - g'u)x + (f - f'u)y + (c - c'u) = 0. \qquad ...(10)$$

By eliminating x and y from (8), (9) and (10), we get

$$\begin{vmatrix} a - a'u & h - h'u & g - g'u \\ h - h'u & b - b'u & f - f'u \\ g - g'u & f - f'u & c - c'u \end{vmatrix} = 0.$$

which is a cubic equation in u. The roots of this equation gives the required maxima and minima.

☞ **REMARK**

⟹ If there is a function of two variables x and y connected by a relation $g(x, y) = 0$. Then we find the maxima and minima of the function in the following manner.

Let $\qquad\qquad\qquad\qquad u = f(x, y)$ $\qquad\qquad\qquad\qquad\qquad\qquad$ ...(1)

and $\qquad\qquad\qquad\qquad g(x, y) = 0.$ $\qquad\qquad\qquad\qquad\qquad\qquad$ ...(2)

Generally, it is possible to eliminate one of the variables x and y from (1) and (2), then u is expressed in terms of a single variable and we can proceed in the usual way. But if it is not convenient to take the value of one variable in terms of the other from (2), then we should proceed as follows :

From (2), we get $\qquad\qquad\qquad \dfrac{dg}{dx} = -\dfrac{\partial g / \partial x}{\partial g / \partial y}$ $\qquad\qquad\qquad\qquad$ ...(3)

Now, differentiating (1) with respect to x, we get

$$\frac{du}{dx} = \frac{\partial f}{\partial x} + \frac{\partial f}{\partial y}\frac{dg}{dx} \qquad \text{... (4)}$$

Now, from (3) and (4), we get

$$\frac{du}{dx} = 0 \qquad \text{...(5)}$$

Solve (5) with the help of (2), and get the required values of x and y for which u will have maximum or minimum values.

EXAMPLE 13. **Test the function $u = x^2y - y^2x - x + y$ for maximum and minimum.**

SOLUTION. For maximum of u, we must have

$$\frac{\partial u}{\partial x} = 2xy - y^2 - 1 = 0$$

and

$$\frac{\partial u}{\partial y} = x^2 - 2xy + 1 = 0.$$

Solving these equations, we get $x = 1, y = 1, x = -1, y = -1$.

Now

$$r = \frac{\partial^2 u}{\partial x^2} = 2y, \quad s = \frac{\partial^2 u}{\partial x\, \partial y} = 2x - 2y,$$

$$t = \frac{\partial^2 u}{\partial y^2} = -2x.$$

For $x = 1, y = 1$, we have $r = 2, s = 0, t = -2$ so that $rt - s^2 = -4$, which is negative. Hence, u has neither a maximum nor a minimum at $(1, 1)$. Thus $(1, 1)$ is a saddle point.

For $x = -1, y = -1$. We have $r = -2, s = 0, t = 2$. so that $rt - s^2 = -4$, which is negative.

Hence, u has neither a maximum nor a minimum at $(-1, -1)$.

Thus $(-1, -1)$ is a saddle point.

EXAMPLE 14. **Show that distance l of any point (x, y, z) on the plane $2x + 3y - z = 12$ from the origin is given by**

$$l = \sqrt{[x^2 + y^2 + (2x + 3y - 12)^2]}.$$

Hence, find the point on the plane that is nearest to the origin.

SOLUTION. If l is the distance from $(0, 0, 0)$ of any point (x, y, z) then $l = \sqrt{(x^2 + y^2 + z^2)}$. If the point (x, y, z) lies on the plane $2x + 3y - z = 12$, then

$$l = \sqrt{[x^2 + y^2 + (2x + 3y - 12)^2]}$$

$$[\because z = 2x + 3y - 12, \text{ from the equation of the plane}]$$

$$\therefore \quad l^2 = x^2 + y^2 + (2x + 3y - 12)^2$$

$$= 5x^2 + 10y^2 + 12xy - 48x + 72y + 144 = u\,(\text{say}).$$

Now l is maximum or minimum according as l^2 i.e., u is maximum or minimum.
For a maximum or minimum of u, we get

$$\frac{\partial u}{\partial x} = 10x + 12y - 48 = 0$$

and

$$\frac{\partial u}{\partial y} = 20y + 12x - 72 = 0$$

Solving these equations, we get $x = \dfrac{12}{7}$ and $y = \dfrac{18}{7}$.

Also $\qquad r = \dfrac{\partial^2 u}{\partial x^2} = 10, s = \dfrac{\partial^2 u}{\partial x\,\partial y} = 12$ and $t = \dfrac{\partial^2 u}{\partial y^2} = 20.$

Therefore $\quad rt - s^2 = 10 \times 20 - (12)^2 = +$ ve, since $rt - s^2 > 0$
and $r > 0$, then u is minimum and hence l is minimum.

When $x = \dfrac{12}{7}$ and $y = \dfrac{18}{7}$. Putting these values of x and y in the equation of the plane, we get

$$z = 2 \cdot \left(\dfrac{12}{7}\right) + 3 \cdot \left(\dfrac{18}{7}\right) - 12 = -\dfrac{6}{7}.$$

Hence, the required point is $\left(\dfrac{12}{7}, \dfrac{18}{7}, -\dfrac{6}{7}\right)$.

EXAMPLE 15. *Find the points on $z^2 = xy + 1$ nearest to the origin.*

SOLUTION. Let l be the distance from the origin $(0, 0, 0)$ of any point (x, y, z) on the surface
$$z^2 = xy + 1 \qquad\qquad ...(1)$$

Then $\qquad l = \sqrt{x^2 + y^2 + z^2} \ = \sqrt{(x^2 + y^2 + xy + 1)} \qquad$ [Using equation (1)]

Since l is always greater than zero, therefore l is maximum or minimum according as l^2, i.e., u is maximum or minimum, where $u = l^2$.

For a maximum or minimum of u, we must have

$$\dfrac{\partial u}{\partial x} = 2x + y = 0 \qquad\qquad ...(2)$$

and $\qquad \dfrac{\partial u}{\partial y} = 2y + x = 0. \qquad\qquad ...(3)$

Solving the equation (2) and (3), we get
$$x = 0, y = 0$$

Also $\qquad r = \dfrac{\partial^2 u}{\partial x^2} = 2, s = \dfrac{\partial^2 u}{\partial x\,\partial y} = 1, t = \dfrac{\partial^2 u}{\partial y^2} = 2.$

$\therefore \qquad\qquad rt - s^2 = 2.\,2 - 1 = 3\ > 0.$

Since at $x = 0, y = 0$, then $rt - s^2 > 0$ and $r > 0$.

Therefore u is minimum at $x = 0, y = 0$. Hence l is minimum, when $x = 0, y = 0$.

Putting $x = 0, y = 0$ in the equation (1), we get $z^2 = 1$ i.e., $z = \pm 1$.

Hence, the required points are $(0, 0, 1)$ and $(0, 0, -1)$.

EXERCISE 10.6

1. Find the points (x, y) where the function $f(x, y) = xy(1-x-y)$ is maximum or minimum. Also find the maximum value of $f(x, y)$.

2. Discuss the maxima and minima of the function $f(x, y) = x^2 + y^2 + \dfrac{2}{x} + \dfrac{2}{y}$.

3. Find the values of x and y for which the expression
$$(a_1 x + b_1 y + c_1)^2 + (a_2 x + b_2 y + c_2)^2 + ...$$
$$+ (a_n x + b_n y + c_n)^2$$
is minimum.

4. Discuss the maxima and minima of the function $f(x, y) = x^4 + 2x^2 y - x^2 + 3y^2$.

5. Examine for maximum and minimum values of the function $f(x, y) = x^2 - 3xy + y^2 + 2x$.

6. Examine the function $f(x, y) = x^2 y - y^2 x - x + y$ for maxima and minima.

7. Discuss the maxima and minima of the function
$$f(x, y) = 2\sin\dfrac{1}{2}(x + y)\cos\dfrac{1}{2}(x - y) + \cos(x + y).$$

8. Find points on $z^2 = xy + 1$ nearest to the origin.

9. Show that the distance l of any point (x, y, z) on the plane $2x + 3y - z = 12$ from the orign

   is given by $l = \sqrt{(x^2 + y^2) + (2x + 3y - 12)^2}$.

10. Find the maximum and minimum values of $u = 6xy + (47 - x - y)(4x + 3y)$.

HINTS TO SELECTED PROBLEMS

1. $\dfrac{\partial f}{\partial x} = 0 \Rightarrow y - 2xy - y^2 = 0$

   $\dfrac{\partial f}{\partial y} = 0 \Rightarrow x - 2xy - x^2 = 0.$

   On solving above two equations, we get $(0, 0)$, $(1, 0)$, $(0, 1)$ and $(1/3, 1/3)$ are the extreme points.

   At $(0, 0)$, $rt - s^2$ is negative $\Rightarrow f(x, y)$ is neither maximum nor minimum.

   At $(1, 0)$, $rt - s^2$ is negative $\Rightarrow f(x, y)$ is neither maximum nor minimum.

   At $(0, 1)$, $rt - s^2$ is negative $\Rightarrow f(x, y)$ is neither maximum nor minimum.

   At $(1/3, 1/3)$, $rt - s^2$ is positive and $r\left(= -\dfrac{2}{3}\right)$ is negative.

   Hence, at $\left(\dfrac{1}{3}, \dfrac{1}{3}\right)$, $f(x, y)$ is maximum.

7. $\dfrac{\partial f}{\partial x} = \cos x - \sin(x + y), \dfrac{\partial f}{\partial y} = \cos y - \sin(x + y)$

   $\dfrac{\partial f}{\partial x} = 0$, $\dfrac{\partial f}{\partial y} = 0$, we get $\cos x = \sin(x + y)$, and $\cos y = \sin(x + y)$.

   The extreme points are given by

   $\left(-\dfrac{\pi}{2}, \dfrac{\pi}{2}\right), \left(\dfrac{3\pi}{2}, \dfrac{\pi}{2}\right)$ and $\left(\dfrac{\pi}{2}, \dfrac{\pi}{2}\right)$.

ANSWERS

1. $f(x, y)$ is maximum at the point $\left(\dfrac{1}{3}, \dfrac{1}{3}\right)$; maximum value $= \dfrac{1}{27}$.

2. $f(x, y)$ is minimum at $(1, 1)$.

3. $f(x, y)$ is minimum for the value of x and y which are obtained by $\Sigma(a_1^2)x + (a_1 b_1)y + a_1 c_1 = 0$ and $\Sigma(a_1 b_1)x + (b_1)y + b_1 c_1 = 0.$

4. $f(x, y)$ is minimum for $\left(\dfrac{\sqrt{3}}{2}, \dfrac{-1}{4}\right)$ and $\left(-\dfrac{\sqrt{3}}{2}, -\dfrac{1}{4}\right)$.

5. Stationary point is $x = \dfrac{4}{5}$, $y = \dfrac{6}{5}$. The function $f(x, y)$ is neither maximum nor minimum at $\left(\dfrac{4}{5}, \dfrac{6}{5}\right)$.

6. At $(1, 1)$ and $(-1, -1)$ function is neither maximum nor minimum.

7. $x = y = 2n\pi \pm \pi/2$; neither maximum nor minimum

   $x = y = n\pi + (-1)^n \pi/6$; f is maximum.

8. $(0, 0, 1)$ and $(0, 0, -1)$.   **10.** Maximum value is 3384.

10.20 MAXIMA AND MINIMA OF THE FUNCTION OF THREE INDEPENDENT VARIABLES

(1) *To find the condition, which governs the sign of the quadratic equation of three independent variables.*

Let I be the expression of three independent variables x, y and z given by

$$I = ax^2 + by^2 + cz^2 + 2fyz + 2gzx + 2hxy$$

I can be written as

$$I = \dfrac{1}{a}\left[a^2 x^2 + aby^2 + acz^2 + 2afyz + 2agzx + 2ahxy\right] (a \neq 0)$$

$$= \dfrac{1}{a}\left[a^2 x^2 + 2ax(gz + hy) + aby^2 + acz^2 + 2afyz\right]$$

$$= \frac{1}{a}\left[(ax+hy+gz)^2 + aby^2 + acz^2 + 2afyz - (gz+hy)^2\right]$$

$$= \frac{1}{a}\left[(ax+hy+gz)^2 + (ab-h^2)y^2 + 2yz(af-gh) + (ac-g^2)z^2\right]$$

Here, we observe that I be of the same sign as provided the expression within the square brackets is positive which will of course be so if $ab-h^2$ and $\{(ah-h^2)(ac-g^2)-(af-gh)^2\}$ are positive *i.e.*, if

$$ab-h^2 \quad \text{and} \quad a[abc+2fgh-af^2-bg^2-ch^2] \text{ are both positive.}$$

Hence, I will be positive if

$$a, \begin{vmatrix} a & h \\ h & b \end{vmatrix}, \begin{vmatrix} a & h & g \\ h & b & f \\ g & f & c \end{vmatrix}$$

be all positive and will be negative if these three expression are alternately negative and positive.

10.21 MAXIMA AND MINIMA FOR A FUNCTION OF THREE INDEPENDENT VARIABLES: THE LAGRANGE'S CONDITION

Let $f(x,y,z)$ be a given function of three independent variables x, y and z.
Let A, B, C, F, G, H stand for

$$\frac{\partial^2 f}{\partial x^2}, \frac{\partial^2 f}{\partial y^2}, \frac{\partial^2 f}{\partial z^2}, \frac{\partial^2 f}{\partial y \partial z}, \frac{\partial^2 f}{\partial z \partial x}, \frac{\partial^2 f}{\partial x \partial y} \text{ respectively.}$$

Let a set of the values of x, y, z obtained by solving the equations

$$\frac{\partial f}{\partial x} = \frac{\partial f}{\partial y} = \frac{\partial f}{\partial z} = 0 \text{ be } a, b, c.$$

By Taylor's theorem, we have

$$f(a+h, b+k, c+l) - f(a,b,c)$$

$$= \frac{1}{2!}\left[Ah^2 + Bk^2 + Cl + 2Fkl + 2Glh + 2Hhk\right] + R \qquad \dots(1)$$

where, remainder term R consist of third and higher order of same quantity (*i.e.*, h, k, l).

Now, by taking h, k, l sufficiently small the second term of R.H.S. of (1) can be made to govern the sign of R.H.S. and therefore of L.H.S. also.

If for all such values of h, k and l, these terms be of permanent sign, then we shall have a maximum or minimum of $f(x,y,z)$ according as that sign is negative or positive.

Hence, the function will be minimum if the expression

$$A, \begin{vmatrix} A & H \\ H & B \end{vmatrix}, \begin{vmatrix} A & H & G \\ H & B & F \\ G & F & C \end{vmatrix} \text{ be all positive.}$$

The function will have a maximum value, if the above three quantities are alternately negative and positive. If these conditions are not satisfied, we have neither a maximum nor a minimum.

WORKING PROCEDURE

Let us $f(x, y, z)$ be a function of three independent variables x, y and z. Find the values of triads (a,b,c) of the value x, y and z by putting $\frac{\partial f}{\partial x} = 0, \frac{\partial f}{\partial y} = 0, \frac{\partial f}{\partial z}$. The values of triads (a,b,c) will give the stationary values of $f(x, y, z)$.

Now, to discuss maximum and minimum values, at (a, b, c) we find the following six partial derivatives of second order

$$A = \frac{\partial^2 f}{\partial x^2}, B = \frac{\partial^2 f}{\partial y^2}, C = \frac{\partial^2 f}{\partial z^2}, F = \frac{\partial^2 f}{\partial y \partial z}, G = \frac{\partial^2 f}{\partial z \partial x}, \text{and } H = \frac{\partial^2 f}{\partial x \partial y}$$

Now, we have the following cases :

Case (i) The function $f(x,y,z)$ will be minimum at (a,b,c) if the expressions

$$A, \begin{vmatrix} A & H \\ H & B \end{vmatrix}, \begin{vmatrix} A & H & G \\ H & B & F \\ G & F & C \end{vmatrix} \text{ be all positive at } (a, b, c).$$

Case (ii) The function $f(x, y, z)$ will be maximum at (a, b, c) if the expressions

$$A, \begin{vmatrix} A & H \\ H & B \end{vmatrix}, \begin{vmatrix} A & H & G \\ H & B & F \\ G & F & C \end{vmatrix}$$

be alternately negative and positive.

Case (iii) If the expression, using in case (i) and (ii) neither be all positive nor having alternately negative and positive sign at (a,b,c). Then $f(x, y, z)$ is neither maximum nor minimum at (a,b,c).

☞ **REMARK**

⇒ To find the maximum and minimum of the function at stationary point, it is sufficient to find the value of a second order partial derivative of function with respect to any of the independent variables. Then, the value of the function is maximum or minimum according as the value of this second order partial derivative at the stationary point under consideration is negative or positive.

SOLVED EXAMPLES

EXAMPLE 1.   *Find the maximum value of u, where $u = \dfrac{xyz}{(a + x)(x + y)(y + z)(z + b)}$.*

SOLUTION.   We have

$$u = \frac{xyz}{(a + x)(x + y)(y + z)(z + b)}$$

Taking, log of both the sides, we have

$$\log u = \log x + \log y + \log z - \log(a+x) - \log(x+y) - \log(y+z) - \log(z+b).$$

Differentiating w.r.t. x, we have

$$\frac{1}{u}\frac{\partial u}{\partial x} = \frac{1}{x} - \frac{1}{a+x} - \frac{1}{x+y} = \frac{ay - x^2}{x(a+x)(x+y)}$$

$$\Rightarrow \qquad \frac{\partial u}{\partial x} = \frac{(ay - x^2)u}{x(a+x)(x+y)}$$

Similarly $\quad \dfrac{\partial u}{\partial y} = \dfrac{(xz - y^2)u}{y(x+y)(y+z)}$ and $\dfrac{\partial u}{\partial z} = \dfrac{(by - z^2)u}{z(y+z)(z+b)}$

For, a maxima and minima of u, we must have

$$\frac{\partial u}{\partial x} = 0 \quad \Rightarrow \quad ay - x^2 = 0 ; \quad \frac{\partial u}{\partial y} = 0 \quad \Rightarrow \quad xz - y^2 = 0$$

and $\qquad \dfrac{\partial u}{\partial z} = 0 \quad \Rightarrow \quad by - z^2 = 0$

Here, we observe that $x^2 = ay, y^2 = xz, z^2 = by$ which implies that a, x, y, z and b are in G.P. Let r be the common ratio of this G.P.

Then $\quad ar^4 = b \quad$ or $\quad r = \left(\dfrac{b}{a}\right)^{1/4}$

Also $\quad x = ar, y = ar^2, z = ar^3$.

Hence, we have

$$u = \frac{ar.ar^2.ar^3}{a(1+r)ar(1+r)ar^2(1+r)ar^3(1+r)}$$

$$= \frac{1}{a(1+r)^4} = \frac{1}{a\left[1 + \left(\dfrac{b}{a}\right)^{1/4}\right]^4} = \frac{1}{\left(a^{1/4} + b^{1/4}\right)^4}$$

which gives a stationary value of u. Now, to decide whether this value of u is a maximum or a minimum, we proceed to find the second order partial derivative of u.

Here $\quad \dfrac{\partial^2 u}{\partial x^2} = \dfrac{-2ux}{x(a+x)(x+y)} + \left(ay - x^2\right)\dfrac{\partial}{\partial x}\left[\dfrac{u}{x(a+x)(x+y)}\right]$

When $x = ar, y = ar^2, z = ar^3$, we have

$$A = \frac{\partial^2 u}{\partial x^2} = -\frac{2u}{a^2 r(1+r)^2} < 0$$

Hence, the above stationary value of u is maximum.

EXAMPLE 2. *Find the maxima and minima value of the function*

$$u = sin\ x\ sin\ y\ sin\ z$$

where x, y and z are the vertex angles of a triangle.

SOLUTION. Here, we have

$$u = \sin x \sin y \sin z\ ;\ \text{where } x + y + z = \pi \qquad \qquad \dots(1)$$

$\therefore \qquad u = \sin x \sin y \sin[\pi - (x+y)] = \sin x \sin y \sin(x+y)$

$\therefore \qquad \dfrac{\partial u}{\partial x} = \cos x \sin y \sin(x+y) + \sin x \sin y \cos(x+y)$

$$= \sin y \sin(2x+y). \qquad \qquad \dots(2)$$

Similarly $\quad \dfrac{\partial u}{\partial y} = \sin x \sin(2y+x). \qquad \qquad \dots(3)$

For a maxima and minima, we must have

$$\frac{\partial u}{\partial x} = 0, \frac{\partial u}{\partial y} = 0$$

So, $\quad \dfrac{\partial u}{\partial x} = 0 \qquad \Rightarrow \quad \sin y \sin(2x+y) = 0$

$\qquad \qquad \qquad \Rightarrow \quad \sin y = 0 \quad \text{or} \quad \sin(2x+y) = 0$

$\qquad \qquad \qquad \Rightarrow \quad y = 0 \quad \text{or} \quad \sin(x+x+y) = 0$

$\qquad \qquad \qquad \Rightarrow \quad y = 0 \text{ or } \sin x \cos(x+y) + \cos x \sin(x+y) = 0$

$\qquad \qquad \qquad \Rightarrow \quad \tan(x+y) = -\tan x$

$\qquad \qquad \qquad \Rightarrow \quad \tan(x+y) = \tan(-x) = \tan(\pi - x) \qquad \dots(4)$

$\qquad \qquad \qquad \Rightarrow \quad x + y = \pi - x$

$\qquad \qquad \qquad \Rightarrow \quad 2x + y = \pi \qquad \qquad \dots(5)$

Similarly, from (3)

$$x=0 \quad \text{or} \quad \tan(x+y)=-\tan y. \qquad \text{...(6)}$$

Now, by (4) and (6), we have

$$\tan x=\tan y \quad \Rightarrow \quad x = y.$$

Hence, by (5), we have

$$3y=\pi \quad \Rightarrow \quad y=\frac{\pi}{3} \text{ and } x=\frac{\pi}{3}$$

Therefore, the stationary points are $\left(\dfrac{\pi}{3},\dfrac{\pi}{3}\right)$ and $(0, 0)$.

For (0,0): $u=0$.

For $\left(\dfrac{\pi}{3},\dfrac{\pi}{3}\right)$

$$r=\frac{\partial^2 u}{\partial x^2} = 2\sin y \cos(2x+y) = 2\sin\frac{\pi}{3}\cos\left(\frac{2\pi}{3}+\frac{\pi}{3}\right)=-\sqrt{3}<0$$

$$s=\frac{\partial^2 u}{\partial x \partial y}=\sin(2x+2y)=\sin\left(\frac{2\pi}{3}+\frac{2\pi}{3}\right)=\sin\left(\frac{4\pi}{3}\right)=-\frac{\sqrt{3}}{2}<0$$

and

$$t=\frac{\partial^2 u}{\partial y^2} = 2\sin x \cos(x+2y) = 2\sin\frac{\pi}{3}\cos\pi = -\sqrt{3}<0$$

Now $\quad rt-s^2=\left(-\sqrt{3}\right)\left(-\sqrt{3}\right)-\left(\dfrac{\sqrt{3}}{2}\right)^2=\dfrac{9}{4}>0$

Thus $\quad rt-s^2>0$ and $r<0$.

Hence, the function u will be maximum at $\left(\dfrac{\pi}{3},\dfrac{\pi}{3}\right)$.

EXAMPLE 3. *Show that the points such that the sum of the squares of its distances from n given points shall be minimum, is the centre of the mean position of the given points.*

SOLUTION. Let n given points be $(a_1, b_1, c_1), (a_2, b_2, c_2),...,(a_n, b_n, c_n)$ and let (x, y, z) be the coordinates of the required point.

If u denotes the sum of the squares of the distances of (x, y, z) from the n given points, then

$$u=\Sigma[(x-a_1)^2+(y-b_1)^2+(z-c_1)^2]$$
$$=\Sigma(x-a_1)^2+\Sigma(y-b_1)^2+\Sigma(z-c_1)^2$$

$$\Rightarrow \qquad \left. \begin{aligned} \frac{\partial u}{\partial x} &= 2\Sigma(x-a_1) = 2nx - 2\Sigma a_1 \\[4pt] \frac{\partial u}{\partial y} &= 2\Sigma(y-b_1) = 2ny - 2\Sigma b_1 \\[4pt] \frac{\partial u}{\partial z} &= 2\Sigma(z-c_1) = 2nz - 2\Sigma c_1 \end{aligned} \right\} \qquad \text{...(1)}$$

and

For the maxima and minima of u, we must have

$$\frac{\partial u}{\partial x}=0, \frac{\partial u}{\partial y}=0 \text{ and } \frac{\partial u}{\partial z}=0 \qquad \text{...(2)}$$

Now from (1) and (2), we have

$$x=\frac{\Sigma a_1}{n}, y=\frac{\Sigma b_1}{n}, z=\frac{\Sigma c_1}{n}$$

Now
$$A = \frac{\partial^2 u}{\partial x^2} = 2n, B = \frac{\partial^2 f}{\partial y^2} = 2n, C = \frac{\partial^2 f}{\partial z^2} = 2n,$$

$$F = \frac{\partial^2 f}{\partial y \partial z} = 0, G = \frac{\partial^2 f}{\partial z \partial x} = 0, H = \frac{\partial^2 f}{\partial x \partial y} = 0.$$

Here, we have
$$A = 2n, \begin{vmatrix} A & H \\ H & B \end{vmatrix} = \begin{vmatrix} 2n & 0 \\ 0 & 2n \end{vmatrix} = 4n^2$$

and
$$\begin{vmatrix} A & H & G \\ H & B & F \\ G & F & C \end{vmatrix} = \begin{vmatrix} 2n & 0 & 0 \\ 0 & 2n & 0 \\ 0 & 0 & 2n \end{vmatrix} = 8n^3$$

Since, these expressions are all positive, therefore u is minimum when
$$x = \frac{\Sigma a_1}{n}, y = \frac{\Sigma b_1}{n}, z = \frac{\Sigma c_1}{n} .$$

Hence, the function u is minimum when the point (x, y, z) is the centre of the mean position of n given points.

EXAMPLE 4.   *Show that the function $u = (x+y+z)^3 - 3(x+y+z) - 24xyz + a^3$ has minimum at $(1,1,1)$ and maximum at $(-1,-1,-1)$.*

SOLUTION.   Here we have

$$u = (x+y+z)^3 - 3(x+y+z) - 24xyz + a^3$$

$$\Rightarrow \qquad \frac{\partial u}{\partial x} = 3(x+y+z)^2 - 3 - 24yz \qquad\qquad \ldots(1)$$

$$\frac{\partial u}{\partial y} = 3(x+y+z)^2 - 3 - 24xz \qquad\qquad \ldots(2)$$

and
$$\frac{\partial u}{\partial z} = 3(x+y+z)^2 - 3 - 24xy \qquad\qquad \ldots(3)$$

For the maxima and minima of u, we must have
$$\frac{\partial u}{\partial x} = 0, \frac{\partial u}{\partial y} = 0 \text{ and } \frac{\partial u}{\partial z} = 0$$

The equations (1), (2) and (3) are satisfied when $x = y = z$.
Putting $y = x$ and $z = x$ in (1), we get
$$27 . x^2 - 3 - 24x^2 = 0$$
$$\Rightarrow \qquad\qquad x = \pm 1$$
$$\Rightarrow \quad x = y = z = 1 \text{ and } x = y = z = -1 \text{ are the solutions of (1), (2) and (3).}$$

Hence, the stationary points are $(1,1,1)$ and $(-1,-1,-1)$.

Now,
$$A = \frac{\partial^2 u}{\partial x^2} = 6(x+y+z), \quad B = \frac{\partial^2 u}{\partial y^2} = 6(x+y+z), \quad C = \frac{\partial^2 u}{\partial z^2} = 6(x+y+z),$$

$$F = \frac{\partial^2 u}{\partial y \partial z} = 6(x+y+z) - 24x, \quad G = \frac{\partial^2 u}{\partial z \partial x} = 6(x+y+z) - 24y,$$

$$H = \frac{\partial^2 u}{\partial x \partial y} = 6(x+y+z) - 24z.$$

For $(1,1,1)$. $A = 18, B = 18, C = 18, F = -6, G = -6, H = -6.$

$\therefore$ At the point $(1,1,1)$, we have $A = 18. > 0$
$$\begin{vmatrix} A & H \\ H & B \end{vmatrix} = \begin{vmatrix} 18 & -6 \\ -6 & 18 \end{vmatrix} = 288 > 0$$

and $\quad \begin{vmatrix} A & H & G \\ H & B & F \\ G & F & C \end{vmatrix} = \begin{vmatrix} 18 & -6 & -6 \\ -6 & 18 & -6 \\ -6 & -6 & 18 \end{vmatrix} = 3426 > 0$

Since, all these three expressions are positive, therefore u is minimum at the point (1,1,1).

For (–1,–1,–1).

$$A = -18, B = -18, C = -18, F = 6, G = 6, H = 6.$$

∴   At the point (–1,–1,–1), we have $A = -18 < 0$

$$\begin{vmatrix} A & H \\ H & B \end{vmatrix} = \begin{vmatrix} -18 & 6 \\ 6 & -18 \end{vmatrix} = 288 > 0$$

and $\quad \begin{vmatrix} A & H & G \\ H & B & F \\ G & F & C \end{vmatrix} = \begin{vmatrix} -18 & 6 & 6 \\ 6 & -18 & 6 \\ 6 & 6 & -18 \end{vmatrix} = -3426 < 0$

Here, the above three expressions are alternately negative and positive. Hence, u is maximum at the point (–1,–1,–1).

EXERCISE 10.7

1. Prove that the function $u = x^2 + y^2 + x - 2z - xy$ is minimum at $\left(-\dfrac{2}{3}, -\dfrac{1}{3}, 1\right)$.

2. Find the maximum and minimum values of $u = y^2 + 2z^2 - 5x^4 + 4x^5$.

3. Find the maximum or minimum values of the function u, where
$$u = axy^2z^3 - x^2y^2z^3 - xy^3z^3 - xy^2z^4$$

4. Find the maximum value of
$$(ax + by + cz)\, e^{-\left(\alpha^2 x^2 + \beta^2 y^2 + \gamma^2 z^2\right)}.$$

5. A rectangle box is placed on x-y plane. The one end of the box is at the origin. If the vertex opposite to the origin be on the plane $6x + 4y + 3z = 24$, then find the maximum value of this box.

6. In a plane triangle xyz, find the maximum value of $\sin x \sin y \sin z$.

ANSWERS

2. Minimum at (1,0,0), neither maximum nor minimum at (0,0,0).

3. Maximum at $\left(\dfrac{a}{7}, \dfrac{2a}{7}, \dfrac{3a}{7}\right)$, max. value $= \dfrac{108a^7}{7^7}$

4. Maximum at $\left(\dfrac{a}{2\alpha^2 k}, \dfrac{b}{2\beta^2 k}, \dfrac{c}{2\gamma^2 k}\right)$ where $k = \sqrt{\left\{\dfrac{1}{2}\left(\dfrac{a^2}{\alpha^2} + \dfrac{b^2}{\beta^2} + \dfrac{c^2}{\gamma^2}\right)\right\}}$

Maximum value $= \sqrt{\left\{\dfrac{1}{2e}\left(\dfrac{a^2}{\alpha^2} + \dfrac{b^2}{\beta^2} + \dfrac{c^2}{\gamma^2}\right)\right\}}$

5. Maximum at $\left(\dfrac{4}{3}, 2\right)$. maximum value $= \dfrac{64}{9}$ cube units. Neither maximum nor minimum at (0,0).

6. Maximum at $\left(\dfrac{\pi}{3}, \dfrac{\pi}{3}, \dfrac{\pi}{3}\right)$, value $= \dfrac{3\sqrt{3}}{8}$

10.22 LAGRANGE'S METHOD OF UNDETERMINED MULTIPLIERS

Let $u = f(x_1, x_2, ..., x_n)$ be a function of n variables $x_1, x_2, ..., x_n$.
Let us suppose these variables $x_1, x_2, ..., x_n$ are connected by k equations
$$g_1(x_1, x_2, ..., x_n) = 0$$

$$g_2(x_1, x_2,..., x_n)=0$$

$$\cdots \quad \cdots \quad \cdots \quad \cdots \quad \cdots$$

$$g_k(x_1, x_2,..., x_n)=0$$

so, that there are $n - k$ independent variables out of these n variables. For the maxima and minima of u, we find

$$du = \frac{\partial u}{\partial x_1}dx_1 + \frac{\partial u}{\partial x_2}dx_2 +...+ \frac{\partial u}{\partial x_n}dx_n = 0 \qquad \qquad ...(1)$$

Also

$$dg_1 = \frac{\partial g_1}{\partial x_1}dx_1 + \frac{\partial g_1}{\partial x_2}dx_2 +...+ \frac{\partial g_1}{\partial x_n}dx_n = 0 \qquad \qquad ...(2)$$

$$dg_2 = \frac{\partial g_2}{\partial x_1}dx_1 + \frac{\partial g_2}{\partial x_2}dx_2 +...+ \frac{\partial g_2}{\partial x_n}dx_n = 0 \qquad \qquad ...(3)$$

$$\vdots \quad \vdots \quad \vdots \quad \vdots \quad \vdots \quad \vdots \quad \vdots \quad \vdots$$

$$dg_k = \frac{\partial g_k}{\partial x_1}dx_1 + \frac{\partial g_k}{\partial x_2}dx_2 +...+ \frac{\partial g_k}{\partial x_n}dx_n = 0 \qquad \qquad ...(k+1)$$

Multiplying equation (1), (2), (3) ... (k+1) by 1, l_1, l_2, ..., k respectively and adding, we get the result, which can be written as

$$P_1dx_1+P_2dx_2+P_3dx_3+...+P_ndx_n=0 \qquad \qquad ...(4)$$

where

$$P_k= \frac{\partial u}{\partial x_k}+l_1\frac{\partial g_1}{\partial x_k}+l_2\frac{\partial g_2}{\partial x_k}+...+l_k\frac{\partial g_k}{\partial x_k}$$

Now we have at our choice k multiple *viz* $l_1, l_2,...,l_k$ and can be chosen such that

$$P_1=0, P_2=0, ..., P_k=0$$

Then, the equation (4) reduces to

$$P_{k+1}dx_{k+1}+P_{k+2}dx_{k+2}+P_{k+3}dx_{k+3}+...+P_ndx_n=0 \qquad \qquad ...(5)$$

Now, let us suppose that out of n variables, the $(n-k)$ variables $x_{k+1}, x_{k+2}, ..., x_n$ are independent. Then, since $n-k$ quantities $dx_{k+1}, dx_{k+2}, ..., dx_n$ are independent so their coefficients must be separately zero. Hence, we have

$$P_{k+1}=0, P_{k+2}=0, ..., P_n=0$$

Thus, we have $k+n$ equations

$$P_1=0, P_2=0, ..., P_n=0$$

and

$$g_1=0, g_2=0, ..., g_k=0.$$

Hence, we get $(n+k)$ equations which determine the k multipliers $l_1, l_2,...,l_k$ and get the possible value of u.

☞ **REMARKS**

⟱ The Lagrange's method of undetermined multipliers is very convenient to apply. It gives the maximum and minimum values of the function without actually determining the values of the multipliers $l_1, l_2,...,l_k$.

⟱ It does not determine the nature of stationary point, which is the only drawback of this method.

10.22.1 APPLICATIONS OF THE METHOD OF UNDETERMINED MULTIPLIERS

The Lagrange's method of undetermined multipliers can be applied to determine the extreme values of the given functions, it does not detemine the nature of stationary point. Now, it is more convenient to find out the extreme values of a function F with the help of new function, given by

$$V=g+l_1f_1+l_2f_2+...+l_mf_m$$

and use the following method. Here, we give the method for four variables x,y,u,v connected by the following two relations.

Let $F = g(x, y, u, v)$ be subjected to the conditions

$$f_1(x,y,u,v) = 0 \qquad \qquad ...(1)$$

and $\qquad f_2(x,y,u,v) = 0. \qquad \qquad ...(2)$

For the maxima and minima of F, we have

$$dF = \frac{\partial g}{\partial x}dx + \frac{\partial g}{\partial y}dy + \frac{\partial g}{\partial u}du + \frac{\partial g}{\partial v}dv = 0 \qquad ...(3)$$

Now, from (1) and (2), we have

$$df_1 = \frac{\partial f_1}{\partial x}dx + \frac{\partial f_1}{\partial y}dy + \frac{\partial f_1}{\partial u}du + \frac{\partial f_1}{\partial v}dv = 0 \qquad ...(4)$$

and $\qquad df_2 = \frac{\partial f_2}{\partial x}dx + \frac{\partial f_2}{\partial y}dy + \frac{\partial f_2}{\partial u}du + \frac{\partial f_2}{\partial v}dv = 0 \qquad ...(5)$

Multiplying (4) by l_1, (5) by l_2 and adding their sum to (3), we get

$$\left(\frac{\partial g}{\partial x} + l_1\frac{\partial f_1}{\partial x} + l_2\frac{\partial f_2}{\partial x}\right)dx + \left(\frac{\partial g}{\partial y} + l_1\frac{\partial f_1}{\partial y} + l_2\frac{\partial f_2}{\partial y}\right)dy + \left(\frac{\partial g}{\partial u} + l_1\frac{\partial f_1}{\partial u} + l_2\frac{\partial f_2}{\partial u}\right)du + \left(\frac{\partial g}{\partial v} + l_1\frac{\partial f_1}{\partial v} + l_2\frac{\partial f_2}{\partial v}\right)dv = 0$$

$$...(6)$$

Here, we have l_1 and l_2 are arbitrary, therefore we can choose them to satisfy the two linear equations

$$\frac{\partial g}{\partial x} + l_1\frac{\partial f_1}{\partial x} + l_2\frac{\partial f_2}{\partial x} = 0 \qquad ...(7)$$

and $\qquad \frac{\partial g}{\partial y} + l_1\frac{\partial f_1}{\partial y} + l_2\frac{\partial f_2}{\partial y} = 0 \qquad ...(8)$

Using (7) and (8), equation (6) reduces to

$$\left(\frac{\partial g}{\partial u} + l_1\frac{\partial f_1}{\partial u} + l_2\frac{\partial f_2}{\partial u}\right)du + \left(\frac{\partial g}{\partial v} + l_1\frac{\partial f_1}{\partial v} + l_2\frac{\partial f_2}{\partial v}\right)dv = 0$$

Since, the given function contains four variables (namely x, y, u and v) and we are given two equations of conditions, therefore, only two of the variables are independent and it is immaterial which two of the four variables are regarded as independent. Let them be u and v then du and dv are also independent, therefore, their coefficients must be zero separately. Thus

$$\frac{\partial g}{\partial u} + l_1\frac{\partial f_1}{\partial u} + l_2\frac{\partial f_2}{\partial u} = 0 \qquad ...(9)$$

$$\frac{\partial g}{\partial v} + l_1\frac{\partial f_1}{\partial v} + l_2\frac{\partial f_2}{\partial v} = 0 \qquad ...(10)$$

Now, we have six equations namely (1), (2), (7), (8), (9) and (10) to determine the two multipliers l_1, l_2 and values of the four variables x, y, u and v for which maximum and minimum values of F are possible.

Now, defined a new function $V(x, y, u, v)$ such that

$$V(x,y,u,v) = g(x, y, u, v) + l_1 f_1(x, y, u, v) + l_2 f_2(x, y, u, v).$$

Assuming that x, y, u, v are now all independent variables. Hence, for the maxima and minima of V, we must have

$$\frac{\partial V}{\partial x} = \frac{\partial g}{\partial x} + l_1\frac{\partial f_1}{\partial x} + l_2\frac{\partial f_2}{\partial x} = 0 \qquad ...(11)$$

$$\frac{\partial V}{\partial y} = \frac{\partial g}{\partial y} + l_1\frac{\partial f_1}{\partial y} + l_2\frac{\partial f_2}{\partial y} = 0 \qquad ...(12)$$

$$\frac{\partial V}{\partial u} = \frac{\partial g}{\partial u} + l_1\frac{\partial f_1}{\partial u} + l_2\frac{\partial f_2}{\partial u} = 0 \qquad ...(13)$$

and
$$\frac{\partial V}{\partial v} = \frac{\partial g}{\partial v} + l_1 \frac{\partial f_1}{\partial v} + l_2 \frac{\partial f_2}{\partial v} = 0 \qquad \dots(14)$$

Equations (11), (12), (13) and (14) are exactly the same as the equations (7). (8), (9) and (10). Hence, the maxima and minima of $V(x, y, u, v)$ are same as those of $F(x, y, u, v)$ assuming that $V(x, y, u, v)$ the variables x, y, u, v are now all independent.

Now, we proceed to find whether the values of F obtained with the help of above equations are maximum or minimum. For this, adopt the procedure, which is discussed ahead.

From (3), we get

$$d^2F = \left(\frac{\partial}{\partial x}dx + \frac{\partial}{\partial y}dy + \frac{\partial}{\partial u}du + \frac{\partial}{\partial v}dv\right)^2 g + \left(\frac{\partial g}{\partial x}d^2x + \frac{\partial g}{\partial y}d^2y + \frac{\partial g}{\partial u}d^2u + \frac{\partial g}{\partial y}d^2v\right)\dots \qquad \dots(15)$$

Also $d^2f_1 = \left(\frac{\partial}{\partial x}dx + \frac{\partial}{\partial y}dy + \frac{\partial}{\partial u}du + \frac{\partial}{\partial v}dv\right)^2 f_1 + \frac{\partial f_1}{\partial x}d^2x + \frac{\partial f_1}{\partial y}d^2y + \frac{\partial f_1}{\partial u}d^2u + \frac{\partial f_1}{\partial v}d^2v = 0 \qquad \dots(16)$

and $d^2f_2 = \left(\frac{\partial}{\partial x}dx + \frac{\partial}{\partial y}dy + \frac{\partial}{\partial u}du + \frac{\partial}{\partial v}dv\right)^2 f_2 + \frac{\partial f_2}{\partial x}d^2x + \frac{\partial f_2}{\partial y}d^2y + \frac{\partial f_2}{\partial u}d^2u + \frac{\partial f_2}{\partial v}d^2v = 0 \qquad \dots(17)$

Multiplying (16) by l_1 and (17) by l_2 and adding their sum to (15) and using the result (11), (12),(13) and (14), we have

$$d^2F = \left(\frac{\partial}{\partial x}dx + \frac{\partial}{\partial y}dy + \frac{\partial}{\partial u}du + \frac{\partial}{\partial v}dv\right)^2 (g + l_1f_1 + l_2f_2)$$

$$= \left(\frac{\partial}{\partial x}dx + \frac{\partial}{\partial y}dy + \frac{\partial}{\partial u}du + \frac{\partial}{\partial v}dv\right)^2 V$$

$$= d^2V.$$

Hence d^2F is equal to d^2V, where d^2V is obtained by assuming all the variables x, y, u and v as independent. Therefore, it is clear that d^2F and d^2V have the same sign. Hence, F will be minimum or maximum according as V is minimum or maximum.

☛ **REMARK**

⇒ This method has the advantage over the Lagrange's methods that it enables us to decide whether the values are maximum or minimum.

SOLVED EXAMPLES

EXAMPLE 1. *Find the maxima and minima of $x^2 + y^2 + z^2$ subject to the conditions :*
$$ax^2 + by^2 + cz^2 = 1$$
and $$lx+my+nz = 0$$

SOLUTION. Here, we have
$$u = x^2 + y^2 + z^2 \qquad \dots(1)$$
where, the relations between the variables x, y and z are given by
$$ax^2+by^2+cz^2=1 \qquad \dots(2)$$
and $$lx+my+nz=0 \qquad \dots(3)$$
For the maxima and minima of u, we must have
$$du=0$$
$\Rightarrow \qquad 2xdx+2ydy+2zdz=0$
$\Rightarrow \qquad xdx+ydy+zdz=0 \qquad \dots(4)$
From (2) and (3), we get
$$ax\,dx+by\,dy+cz\,dz=0 \qquad \dots(5)$$
$$l\,dx+m\,dy+n\,dz=0 \qquad \dots(6)$$

Now, multiplying (4) by 1, (5) by l_1 and (6) by l_2 and adding, we get

$$(x\,dx + y\,dy + z\,dz) + l_1(ax\,dx + by\,dy + cz\,dz) + l_2(l\,dx + m\,dy + n\,dz) = 0$$

$$\Rightarrow \quad (x + al_1x + ll_2)dx + (y + bl_1y + ml_2)dy + (z + cl_1z + nl_2)dz = 0$$

Now equating the coefficient of dx, dy, dz to zero, we get

$$x + l_1ax + l_2l = 0 \qquad \qquad \text{...(7)}$$
$$y + bl_1y + ml_2 = 0 \qquad \qquad \text{...(8)}$$

and $\qquad \qquad z + cl_1z + nl_2 = 0 \qquad \qquad \text{...(9)}$

Multiplying the equations (7), (8) and (9) by x, y and z respectively, and adding we get

$$x^2 + y^2 + z^2 + l_1(ax^2 + by^2 + cz^2) + l_2(lx + my + nz) = 0$$

or $\qquad \qquad u + l_1.1 + l_2.0 = 0 \qquad \qquad$ [By using (1), (2) and (3)]

$$\Rightarrow \qquad \qquad l_1 = -u$$

Substituting for l_1 in the equations (7), (8) and (9), we get

$$x = \frac{l_2l}{au - 1}, y = \frac{l_2m}{bu - 1}, z = \frac{l_2n}{cu - 1} \qquad \qquad \text{...(10)}$$

Now from (10) and (3), we get

$$\frac{l_2l^2}{au - 1} + \frac{l_2m^2}{bu - 1} + \frac{l_2n^2}{cu - 1} = 0$$

or $\qquad \qquad \dfrac{l^2}{au - 1} + \dfrac{m^2}{bu - 1} + \dfrac{n^2}{cu - 1} = 0 \qquad \qquad \text{...(11)}$

which gives the maximum and minimum of $u = x^2 + y^2 + z^2$.

☛ **REMARKS**

⟼ Equation (11) is a quadratic in u. So it gives two stationary values of u.

⟼ Geometrically, the surface $ax^2 + by^2 + cz^2 = 1$ represents an ellipsoid whose centre is origin, and $lx + my + nz = 0$ represents a plane passing through the origin. The points (x, y, z) satisfying both the conditions (2) and (3) lies on the conic in which (2) and (3) intersect. $x^2 + y^2 + z^2$ gives the square of the distance (x, y, z) from the origin, which is also the centre of the conic of intersection. The maximum value of this distance is the major axis of this conic, and the minimum value of this distance is the minor axis of this conic. Hence, equation (11) gives the squares of the lengths of the semi-axis of the conic of intersection.

EXAMPLE 2. *Find the maxima and minima of $x^2 + y^2 + z^2$, where*
$$ax^2 + by^2 + cz^2 + 2fyz + 2gzx + 2hxy = 1.$$

SOLUTION. Let

$$u = x^2 + y^2 + z^2 \qquad \qquad \text{...(1)}$$

where the relation between the variables x, y and z is

$$ax^2 + by^2 + cz^2 + 2fyz + 2gzx + 2hxy = 1. \qquad \qquad \text{...(2)}$$

For a maximum or minima of u, we must have

$$du = 0$$

$$\Rightarrow \qquad \qquad x\,dx + y\,dy + z\,dz = 0. \qquad \qquad \text{...(3)}$$

From (2), we have

$$2ax\,dx + 2by\,dy + 2cz\,dz + 2fy\,dz + 2fz\,dy + 2gz\,dx + 2gx\,dz + 2hx\,dy + 2hy\,dx = 0$$

$$\Rightarrow \quad (ax + hy + gz)dx + (hx + by + fz)dy + (gx + fy + cz)dz = 0. \qquad \text{...(4)}$$

Now, multiplying (3) by 1 and (4) by l_1, adding, and then equating the coefficient of dx, dy, dz to zero, we have

$$x + l_1(ax + hy + gz) = 0. \qquad \qquad \text{...(5)}$$
$$y + l_1(hx + by + fz) = 0. \qquad \qquad \text{...(6)}$$
$$z + l_1(gx + fy + cz) = 0. \qquad \qquad \text{...(7)}$$

Multiplying (5) by x, (6) by y, (7) by z and adding, we get

$$x^2+y^2+z^2+l_1(ax^2+by^2+cz^2+2fyz+2gzx+2hxy)=0$$

$\Rightarrow \qquad\qquad\qquad u+l_1.1=0 \qquad\qquad\qquad$ [From (1) and (2)]

$\therefore \qquad\qquad\qquad l_1=-u.$

Hence, from (5), we have

$$x-u(ax+hy+gz)=0$$

$\Rightarrow \qquad\qquad \left(a-\dfrac{1}{u}\right)x+hy+gz=0 \qquad\qquad$ ...(8)

Similarly from (6) and (7), we get

$$hx+\left(b-\dfrac{1}{u}\right)y+fz=0 \qquad\qquad\qquad$$ ...(9)

and $\qquad\qquad gx+fy+\left(c-\dfrac{1}{u}\right)z=0 \qquad\qquad\qquad$ ...(10)

Eliminating x,y,z from (8), (9) and (10), we get

$$\begin{vmatrix} \left(a-\dfrac{1}{u}\right) & h & g \\ h & \left(b-\dfrac{1}{u}\right) & f \\ g & f & \left(c-\dfrac{1}{u}\right) \end{vmatrix}=0 \qquad\qquad$$ ...(11)

Hence, the maximum or minimum values of u are the roots of the equation (11).

EXAMPLE 3.   ***Find the maximum and minima of $u=x^2+y^2$ subject to the condition*** $ax^2+2hxy+by^2=1.$

SOLUTION.   Here, we have $\qquad\qquad u=x^2+y^2 \qquad\qquad\qquad$ ...(1)

where the relation between the variables x and y is

$$ax^2+2hxy+by^2=1. \qquad\qquad\qquad$$ ...(2)

For the maxima and minima of u, we must have

$$du=0$$

$\Rightarrow \qquad\qquad\qquad 2x\,dx+2y\,dy=0$

$\Rightarrow \qquad\qquad\qquad x\,dx+y\,dy=0. \qquad\qquad\qquad$ ...(3)

Now, from (2), we get

$$2ax\,dx+2hx\,dy+2hy\,dx+2by\,dy=0$$

$\Rightarrow \qquad\qquad (ax+hy)dx+(hx+by)dy=0 \qquad\qquad$ ...(4)

Now, multiplying (3) by 1, (4) by l_1, adding and then equating the coefficients of dx, dy to zero, we have

$$x+l_1(ax+hy)=0 \qquad\qquad\qquad$$ ...(5)

and $\qquad\qquad y+l_1(hx+by)=0 \qquad\qquad\qquad$ ...(6)

Multiplying (5) by x, (6) by y and adding, we get

$$x^2+y^2+l_1(ax^2+2hxy+by^2)=0$$

$\Rightarrow \qquad\qquad\qquad u+l_1.1=0 \qquad\qquad\qquad$ [Using (1) and (2)]

$\Rightarrow \qquad\qquad\qquad u=-l_1$

Therefore, from (5), we have

$$x-u(ax+hy)=0$$

$\Rightarrow \qquad\qquad \left(a-\dfrac{1}{u}\right)x+hy=0 \qquad\qquad\qquad$ ...(7)

Similarly from (6), we have

$$hx + \left(b - \frac{1}{u}\right)y = 0 \qquad \text{...(8)}$$

Eliminating x and y from (7) and (8), we get

$$\begin{vmatrix} a - \dfrac{1}{u} & h \\[2mm] h & b - \dfrac{1}{u} \end{vmatrix} = 0 \qquad \text{...(9)}$$

Hence, the maximum or minimum values of u are the roots of the equation (9).

EXAMPLE 4. **Find the maximum value of $u = x^m y^n z^p$ subject to the condition $x+y+z=a$.**

SOLUTION. Here, we have

$$u = x^m y^n z^p \qquad \text{...(1)}$$

and x, y, z connected by the relation given by

$$x+y+z = a \qquad \text{...(2)}$$

Taking log of both the sides of (1), we get

$$\log u = m \log x + n \log y + p \log z.$$

On differentiating, we get

$$\frac{1}{u} du = \frac{m}{x} dx + \frac{n}{y} dy + \frac{p}{z} dz$$

For the maxima and minima of u, we must have

$$du = 0$$

$$\Rightarrow \qquad \frac{m}{x} dx + \frac{n}{y} dy + \frac{p}{z} dz = 0 \qquad \text{...(3)}$$

Now, differentiating (2), we get

$$dx + dy + dz = 0. \qquad \text{...(4)}$$

Now, multiplying (3) by 1 and (4) by l, and equating the coefficient of dx, dy, dz to zero (after adding), we get

$$\frac{m}{x} + l = 0, \ \ \frac{n}{y} + l = 0 \ \text{ and } \ \frac{p}{z} + l = 0$$

which implies

$$x = -\frac{m}{l}, y = -\frac{n}{l}, z = -\frac{p}{l}$$

Putting the values of x, y and z in (2), we get

$$l = -\left(\frac{m+n+p}{a}\right)$$

therefore, we can say that, u is stationary when

$$x = \frac{am}{m+n+p}, y = \frac{an}{m+n+p}, z = \frac{ap}{m+n+p}$$

Now, we find the nature of this stationary value of u.

Let us regard x and y as independent variable and z is a function of x and y given by (2) [It is justify, because the variables x, y and z are connected by the relation (2), any two of them may be regarded as independent].

Now from (1), we get

$$\log u = m \log x + n \log y + p \log z$$

$$\therefore \qquad \frac{1}{u}\frac{\partial u}{\partial x} = \frac{m}{x} + \frac{p}{z}\frac{\partial z}{\partial x}$$

Now, differentiating (2) partially w.r.t x (treating y as constant), we get

$$1+\frac{\partial z}{\partial x}=0 \quad \Rightarrow \quad \frac{\partial z}{\partial x}=-1$$

Put this value in (5), we get

$$\frac{1}{u}\frac{\partial u}{\partial x}=\frac{m}{x}-\frac{p}{z}$$

$$\Rightarrow \quad \frac{1}{u}\frac{\partial^2 u}{\partial x^2}-\frac{1}{u^2}\left(\frac{\partial u}{\partial x}\right)^2=-\frac{m}{x^2}+\frac{p}{z^2}\frac{\partial z}{\partial x}=-\frac{m}{x^2}-\frac{p}{z^2}$$

At stationary point $\frac{\partial u}{\partial x}=0$

Therefore, $\quad \frac{1}{u}\frac{\partial^2 u}{\partial x^2}=\frac{-m}{x^2}-\frac{p}{z^2}$

$$\Rightarrow \quad \frac{\partial^2 u}{\partial x^2}=u\left[-\frac{m}{x^2}-\frac{p}{z^2}\right]=-x^m y^n z^p\left[-\frac{m}{x^2}-\frac{p}{z^2}\right]$$

which is negative for the obtained values of x, y and z.

Hence, at the stationary point, u is maximum and maximum value is

$$=\left(\frac{am}{m+n+p}\right)^m\left(\frac{an}{m+n+p}\right)^n\left(\frac{ap}{m+n+p}\right)^p$$

EXAMPLE 5. *Find the maximum and minimum value of $u=\dfrac{5xyz}{(x+2y+4z)}$ subject to the condition $xyz = 8$.*

SOLUTION. Here, we have

$$u=\frac{5xyz}{(x+2y+4z)} \qquad \qquad \dots(1)$$

The variables x,y,z are connected by the relation

$$xyz=8. \qquad \qquad \dots(2)$$

From (1) and (2), we get

$$u=\frac{40}{(x+2y+4z)} \quad \Rightarrow du=\frac{-40}{(x+2y+4z)^2}(dx+2dy+4dz)$$

For the maxima or minima of u, we must have $du=0$

$$\Rightarrow \qquad dx+2dy+4dz=0 \qquad \qquad \dots(3)$$

From (2), we get

$$\log x+\log y+\log z=\log 8.$$

On differentiating, we get

$$\frac{1}{x}dx+\frac{1}{y}dy+\frac{1}{z}dz=0 \qquad \qquad \dots(4)$$

Now, multiplying (3) by 1, (4) by l, adding and then equating to zero the coefficients of dx, dy and dz, we get

$$1+\frac{l}{x}=0,2+\frac{l}{y}=0,4+\frac{l}{z}=0$$

Now using (2), we get $\quad l=-4$

$\therefore \quad u$ is stationary at the point given by $x=4, y=2, z=1$.

Regard x and y as independent variables and z is a function of x and y given by (2).

From (1)

$$\frac{\partial u}{\partial x} = -\frac{40}{(x+2y+4z)^2}\left[1+4\frac{\partial z}{\partial x}\right]$$

From (2), we get

$$\log x + \log y + \log z = \log 8$$

$$\therefore \qquad \frac{1}{x}+\frac{1}{z}\frac{\partial z}{\partial x}=0$$

$$\Rightarrow \qquad \frac{\partial z}{\partial x}=-\frac{z}{x}$$

$$\therefore \qquad \frac{\partial u}{\partial x}=-\frac{40}{(x+2y+4z)^2}\left[1-4\frac{z}{x}\right]$$

$$\Rightarrow \quad \frac{\partial^2 u}{\partial x^2}=\frac{80}{(x+2y+4z)^3}\left[1+4\frac{\partial z}{\partial x}\right]\left[1-4\frac{z}{x}\right]-\frac{40}{(x+2y+4z)^2}\left[\frac{4z}{x^2}-\frac{4\partial z}{x\partial x}\right]$$

Now using $x=4, y=2, z=1$. We get $\frac{\partial^2 u}{\partial x^2}=-ve$

$\therefore$ u is maximum at the point given by $x=4, y=2, z=1$.

The maximum value is given by $u=\frac{5\times4\times2\times1}{(4+2\times2+4\times1)}=\frac{40}{12}=\frac{10}{3}$.

EXAMPLE 6.

SOLUTION.

In a plane triangle ABC, find the maximum value of u = cos A cos B cos C.

Here, we have $u = \cos A \cos B \cos C$ ...(1)

Since, we know that the sum of the angles of a triangle is always 180°.

$\therefore$ The variables A, B and C are connected by the relation

$$A + B + C = \pi \qquad ...(2)$$

From (1), we get

$$\log u = \log \cos A + \log \cos B + \log \cos C$$

$$\Rightarrow \qquad \frac{1}{u}du = -\tan A\, dA -\tan B\, dB - \tan C\, dC.$$

For the maxima and minima of u, we must have $du=0$

$$\Rightarrow \qquad \tan A\, dA + \tan B\, dB + \tan C\, dC = 0 \qquad ...(3)$$

Also from (2), $\qquad dA + dB + dC = 0$ ...(4)

Now, multiply (3) by 1, (4) by l, adding, and equating the coefficients of dA, dB and dC to zero, we get

$$\tan A + l = 0$$
$$\tan B + l = 0$$
$$\tan C + l = 0$$

$$\Rightarrow \qquad l = -\tan A = -\tan B = -\tan C \Rightarrow A = B = C.$$

Now from (2), $A = B = C = \frac{\pi}{3}$ *i.e.,* the triangle is equilateral.

Now to show that the stationary value of u given by $A = B = C = \frac{\pi}{3}$ is maximum.

Let C be a function of A and B, regarding A and B as independent variables. From (1),

$$\log u = \log \cos A+\log \cos B+\log \cos C$$

$$\Rightarrow \qquad \frac{1}{u}\frac{\partial C}{\partial A}=-\tan A-\tan C\frac{\partial C}{\partial A}$$

Now, differentiating (2), partially w.r.t. A, we get

$$1 + \frac{du}{dA} = 0 \quad \Rightarrow \quad \frac{\partial C}{\partial A} = -1$$

$$\therefore \quad \frac{1}{u}\frac{\partial u}{\partial A} = -\tan A + \tan C$$

$$\Rightarrow \quad \frac{1}{u}\frac{\partial^2 u}{\partial^2 A} - \frac{1}{u^2}\left(\frac{\partial u}{\partial A}\right)^2 = -\sec^2 A + \sec^2 C . \frac{\partial C}{\partial A} = -\left(\sec^2 A + \sec^2 C\right)$$

At stationary point $\frac{\partial u}{\partial A} = 0$

$$\because \quad \frac{\partial^2 u}{\partial^2 A} = -u\left(\sec^2 A + \sec^2 C\right) = -\text{ve} \ \text{ for } A = B = C = \frac{\pi}{3}.$$

Hence, u is maximum at $A = B = C = \frac{\pi}{3}$ and the maximum value is given by

$$u = \left(\cos\frac{\pi}{3}\right)^3 = \left(\frac{1}{2}\right)^3 = \frac{1}{8}.$$

EXERCISE 10.8

Using Lagrange's method of undetermined multipliers:

1. Find the maximum and minimum values of

$$\frac{x^2}{a^4} + \frac{y^2}{b^4} + \frac{z^2}{c^4}$$

where $lx + my + nz = 0$ and $\frac{x^2}{a^2} + \frac{y^2}{b^2} + \frac{z^2}{c^2} = 1$.

2. Find the maximum and minimum values of

$$f = a^2 x^2 + b^2 y^2 + c^2 z^2$$

where $x^2 + y^2 + z^2 = 1$ and $lx + my + nz = 0$.

3. Show that the maximum and minimum values of $u = x^2 + y^2 + z^2$ subject to the conditions

$$px + qy + rz = 0 \text{ and } \frac{x^2}{a^2} + \frac{y^2}{b^2} + \frac{z^2}{c^2} = 1$$

are given by $\frac{a^2 p^2}{u - a^2} + \frac{b^2 q^2}{u - b^2}$.

4. Find the minimum value of $u = x + y + z$ subject to the condition $\frac{a}{x} + \frac{b}{y} + \frac{c}{z} = 1$.

5. Find the minimum value of $u = x^2 + y^2 + z^2$, subject to the condition $ax + by + cz = p$.

6. Find the minimum value of $x + y + z$ where $xyz = c^3$.

7. Find the extreme values of $x^p y^q z^r$ subject to the condition $\frac{a}{x} + \frac{b}{y} + \frac{c}{z} = 1$.

8. Show that the maximum and minimum values of the radii vectors of the sections of the surface

$$(x^2 + y^2 + z^2)^2 = \frac{x^2}{a^2} + \frac{y^2}{b^2} + \frac{z^2}{c^2}$$

by the plane $\lambda x + \mu y + \nu z = 0$ are given by

$$\frac{a^2\lambda^2}{1 - a^2 r^2} + \frac{b^2\mu^2}{1 - b^2 r^2} + \frac{c^2\nu^2}{1 - c^2 r^2} = 0$$

9. Find the stationary points of the function $u = ax^p + by^q + cz^r$ subject to the condition $x^l + y^m + z^n = k$.

10. If two variables x and y are connected by the relation $ax^2 + by^2 = ab$, show that the maximum and minimum values of the function $u = x^2 + y^2 + xy$ will be the roots of the equation $4(u-a)(u-b) = ab$.

11. Prove that of all rectangular parallelopipeds of the same volume, the cube has the least surface.

12. Prove that if $x + y + z = 1$, $ayz + bzx + cxy$ has an extreme value equal to

$$\frac{abc}{2bc + 2ca + 2ab - a^2 - b^2 - c^2}$$

Also, prove if a, b, c are all positive and c lies between $a + b - 2\sqrt{ab}$ and $a + b + 2\sqrt{ab}$ this value is true maximum and that if a, b, c are all negative and c lies between $a + b \pm 2 \sqrt{ab}$. It is true minimum.

13. Find the maximum value of u, when
$$u=\sin x \sin y \sin z$$
and x,y,z are the angles of a triangle.

14. Find the triangle of maximum area inscribed in a circle.

15. Prove that the rectangular solid of maximum volume which can be inscribed in a sphere is a cube.

16. Find a plane triangle ABC such that
$$u=\sin^a A \sin^b B \sin^c C$$
has maximum value.

17. Find the rectangular parallelopiped of maximum volume that can be inscribed in the ellipsoid
$$\frac{x^2}{a^2}+\frac{y^2}{b^2}+\frac{z^2}{c^2}=1$$

18. Divide a number n into three parts x, y, z such that $ayz+bzx+cxy$ shall have maximum

or minimum and determine which it is.

19. Prove that a rectangular solid of maximum volume which can be inscribed in a sphere is a cube.

20. Find the maximum or minimum value of $x^p y^q z^r$ subject to the condition
$$ax + by + cz = p+q+r.$$

21. Show that the maximum and minimum value of $u=ax^2+by^2+cz^2+2fyz+2gzx+2hxy$ subject to the conditions $lx+my+nz=0$ and $x^2+y^2+z^2=1$ are given by the equation
$$\begin{vmatrix} a-u & h & g & l \\ h & b-u & f & m \\ g & f & c-u & n \\ l & m & n & o \end{vmatrix}=0$$

22. Show that of the perimeter of a triangle is constant, its area is maximum when it is equilateral.

ANSWERS

1. The maximum and minimum values of the given function is given by the equation
$$\frac{l^2 a^4}{a^2 u-1}+\frac{m^2 b^4}{b^2 u-1}+\frac{n^2 c^4}{c^2 u-1}=0$$

2. The maximum and minimum values of the given function is given by the equation
$$\frac{l^2}{u-a^2}+\frac{m^2}{u-b^2}+\frac{m^2}{u-c^2}=0$$

4. Stationary points are $x=\sqrt{a}\left(\sqrt{a}+\sqrt{b}+\sqrt{c}\right), y=\sqrt{b}\left(\sqrt{a}+\sqrt{b}+\sqrt{c}\right), z=\sqrt{c}\left(\sqrt{a}+\sqrt{b}+\sqrt{c}\right)$ minimum
value is $\left(\sqrt{a}+\sqrt{b}+\sqrt{c}\right)^2$.

5. Minimum value is $\frac{p^2}{\left(a^2+b^2+c^2\right)}$.   **6.** u is minimum at the point $x=y=z=c$. Value is $=3c^4$.

7. u is stationary when $\frac{px}{a}=\frac{qy}{b}=\frac{rc}{c}=p+q+r$, Minimum value is $\frac{a^p b^q c^r}{p^p q^q r^r}\left(p+q+r\right)^{p+q+r}$.

9. Stationary points are given by $\frac{x^{p-1}}{l/pa}=\frac{y^{q-m}}{m/qb}=\frac{z^{r-n}}{n/rc}$

13. u is maximum, when $x=y=z=\frac{\pi}{3}$. Maximum value is $\frac{3\sqrt{3}}{8}$.   **14.** Equilateral.

16. u is maximum when, A, B, C are given by $\frac{\tan A}{a}=\frac{\tan B}{b}=\frac{\tan C}{c}$.

17. Stationary points are $x=\frac{a}{\sqrt{3}}, y=\frac{b}{\sqrt{3}}, z=\frac{c}{\sqrt{3}}$, Maximum value $=\frac{8abc}{3\sqrt{3}}$.

REVIEW QUESTIONS AND ARCHIVE

1. Define the following :

   (i) Limit   (ii) Repeated limit

   (iii) Continuity   (iv) Differentiability

2. Show that $(|x|+|y|+|z|)$ is continuous but not differentiable at (0,0,0).

3. Show that for $k>0$, $\{|x+y|+(x+y)\}^k$ is everywhere differentiable on the finite xy-plane.

4. Show that $f(x,y)=\sin x \sin y \sin(x+y)$ has minima at $(0,0)$ and maxima at $\left(\dfrac{\pi}{3},\dfrac{\pi}{3}\right)$ in the positive quadrant when $x^2+y^2\le\dfrac{\pi^2}{2}$

5. Show that $f(x,y)=(y-x)^4+x^4$ has minimum at origin.

6. Show that the function $f(x,y,z)=(x+y+z)^3-12(x+y+z)-24xyz^3$ has maxima at $(-2,-2,-2)$ and minima at $(2,2,2)$.

7. If $lx+my+nz=d$, show that
$$\min\{(x-a)^2+(y-b)^2+(z-c)^2\}$$
$$=\frac{(al+bm+cn-d)^2}{l^2+m^2+n^2}.$$

8. Show that the function $f=x^2+y^2+2z^2$ when $xyz^2=1$ has four stationary points given by $(1,1,\pm1),(-1,-1,\pm1)$ giving a minima at each of these points.

9. Show that the largest length of the semi-axis of the ellopsoid $ax^2+by^2+cz^2+2dxy+2exz+2fyz=1$ is given by the largest real root of the equation.

$$\begin{vmatrix} a-1/r^2 & d & e \\ d & b-1/r^2 & f \\ e & f & c-1/r^2 \end{vmatrix}=0$$

10. Show that $xy+yz+zx$ has no extreme value when it is considered as a function of the independent variables x, y, z but it has a maximum value when $ax+by+cz=1$ where a,b,c are positive and
$$2(ab+bc+ca)>(a^2+b^2+c^2).$$

OBJECTIVE EVALUATION

▶ FILL IN THE BLANKS

1. A function of two variables x and y may be writen as _____ .

2. $\lim\limits_{(x,y)\to(a,b)} f(x,y)$, if exists, is _____ .

3. If a function $f(x, y)$ is totally differentiable, then the partial derivatives f_x and f_y _____ .

4. If $u=f\left(\dfrac{y}{x}\right)$, then $x\dfrac{\partial u}{\partial x}+y\dfrac{\partial u}{\partial y}=$ _____ .

5. If u is a homogeneous function of x and y of degree n, then $x\dfrac{\partial u}{\partial x}+y\dfrac{\partial u}{\partial y}=$ _____ .

6. The statement given in (5) is known as _____ .

7. If u is a function of variables x and y, and x and y both are the functions of the variable t, then u is said to be _____ .

8. If $f_{xy}(a, b)$ and $f_{yx}(a, b)$, both are continuous, then f_{xy} _____ .

▶ **TRUE/FALSE**

1. The limit of a function is unique. **(T/F)**

2. $\lim\limits_{(x,y)\to(1,1)} f(x,y)$ does not exist, where

   $f(x,y) = (x^2 + 2y)$. **(T/F)**

3. The two iterated limits obtained by reversing the order of limits, are always equal. **(T/F)**

4. The function $f(x, y) = xy$ is not continuous at $(2, 3)$. **(T/F)**

5. In case of a function of two variables, the continuity is a necessary condition for differentiability. **(T/F)**

▶ **MULTIPLE CHOICE QUESTIONS (CHOOSE THE MOST APPROPRIATE ONE)**

1. Simultaneous limits are also called :
   (a) repeated limit   (b) double limit
   (c) both (a) and (b)   (d) none of these

2. Iterated limits are also called :
   (a) repeated limit   (b) double limit
   (c) both (a) and (b)   (d) none of these

3. The limit $\lim\limits_{\substack{x\to a \\ y\to b}} f(x,y) = \lim\limits_{(x,y)\to(a,b)} f(x,y)$ is

   called :
   (a) iterated limit
   (b) simultaneous limit
   (c) both (a) and (b)
   (d) none of these

4. The limit, $\lim\limits_{x\to a}\left[\lim\limits_{y\to b} f(x,y)\right]$ or $\left[\lim\limits_{y\to b} f(x,y)\right]$

   is called:
   (a) iterated limit
   (b) simultaneous limit
   (c) both (a) and (b)
   (d) none of these

5. The simultaneous limit, $\lim\limits_{\substack{x\to a \\ y\to b}} \dfrac{xy^3}{x^2+y^6} =$

   (a) 0   (b) 2
   (c) 1   (d) does not exist

6. The value of $\lim\limits_{(x,y)\to(0,0)} \dfrac{x^2-y^2}{x^2+y^2} =$

   (a) 0   (b) 1
   (c) 2   (d) does not exist

7. Let $f : R^2 \to R$ be defined by $f(x, y)=x^2+y^2$
   then value of $\lim\limits_{(x,y)\to(0,0)} f(x,y) =$

   (a) 1   (b) 0
   (c) 2   (d) does not exist

8. The value of $\lim\limits_{(x,y)\to(0,0)} \dfrac{2x^3-y^3}{x^2+y^2} =$

   (a) 1   (b) 0
   (c) 2   (d) does not exist

9. The value of $\lim\limits_{(x,y)\to(0,0)} \dfrac{xy^2}{x^2+y^4} =$

   (a) 1   (b) 2
   (c) 0   (d) does not exist

10. If $f(x,y) = \dfrac{x^2y^2}{x^2y^2+(x-y)^2}$, where

    $x^2y^2+(x-y)^2 \neq 0$. Then $\lim\limits_{x\to 0}\left[\lim\limits_{y\to 0} f(x,y)\right] =$

    (a) 1   (b) 2
    (c) 0   (d) does not exist

ANSWERS

▶ **FILL IN THE BLANKS**

1. $f(x, y)$ 2. unique   3. both exist and equal   4. 0   5. *nu*
6. Euler's theorem   7. composite function   8. f_{yx}

▶ **TRUE/FALSE**

1. T   2. F   3. F   4. F   5. T

▶ **MULTIPLE CHOICE QUESTIONS**

1. (b)   2. (a)   3. (b)   4. (a)   5. (d)   6. (d)   7. (b)   8. (b)   9. (d)
10. (a)

⬛⬛◻   \ CHAPTER SUMMARY /   ⬛⬛◻

➲ If the simultaneous limit exists then two repeated limits if they exists are necessarily equal but converse is not true.

➲ If the repeated limits are not equal, the simultaneous limit can not exist.

➲ If f_x exists throughout a nbd of a point (a,b) and $f_y(a,b)$ exists then for any point $(a+h,b+k)$ of this neighbourhood
$$f(a+h,b+k)=f(a,b)+hf_x(a+\theta h,b+k)$$
$$+k[f_y(a,b)+\eta]$$
where $0<\theta<1$ and η is a function of k, tending to zero with k.

➲ A sufficient condition that a function f be continuous at (a,b) is that one of the partial derivatives exists and is bounded in a neighbourhood of (a,b) and the other exists at (a,b).

➲ A sufficient condition that a function be continuous in a closed region is that both the partial derivatives exist and bounded throughout the region.

➲ A function which is differentiable at a point possesses the first order partial derivatives at that point and necessarily continuous at that point.

➲ If (a, b) be a point of domain of a function f such that
  (i) f_x is continuous at (a, b)

  (ii) f_y exists at (a, b)
  then f is differentiable at (a, b)

➲ A function f is differentiable at (a, b) if f_x exists and f_y is continuous at (a, b) i.e., one of the partial derivatives is to be continuous and the other merely to exist at that point.

➲ (Young's theorem): If f_x and f_y are both differentiable at a point (a, b) of the domain of function f then $f_{xy}(a, b) = f_{yx}(a, b)$.

➲ (Schwarz's theorem): If f_y exists in a certain nbd of a point (a, b) of the domain of f and f_{yx} is continuous at (a, b) then $f_{xy}(a, b)$ exists and is equal to $f_{yx}(a, b)$.

➲ If f_{xy} and f_{yx} are both continuous at (a, b) then $f_{xy}(a, b) = f_{yx}(a, b)$.

➲ A necessary condition for $f(x, y)$ to have an extreme value at (a, b) is that $f_x(a, b) = 0$, $f_{yx}(a, b) = 0$ provided they exist.

➲ The necessary condition for a function $f(x, y)$ to be an extremum at (a, b) are that $f_x(a,b)=0$, $f_y(a,b)=0$.

➲ Lagrange's method of undetermined multipliers is used to find the extreme values of a function of three or more variables when the variables are not independent but are connected by some relation. It gives only the extreme points and not distinguish whether the point is a maxima or minima.

⬛⬛◻   \ FOR ADVANCED LEARNERS /   ⬛⬛◻

⇒ The usual metric, product and the postman metric on $\boldsymbol{R}^2$ are equivalent.

⇒ It is well known that for function of a single variable, derivability implies continuity, but for functions of two variables, the situation is rather different i.e. A function $f : D \to \boldsymbol{R}$ may possess partial derivative at a point but may not be continuous at that point.

⇒ A function may possesses a directional derivatives in every direction at a point but may fail to be continuous at that point.

⇒ Lagrange's method of undetermined multipliers yields only stationary values. However, in most cases one can find from other considerations as to whether the stationary values so obtained is a maximum or minimum.

❋❋❋❋

CHAPTER 11

Cantor's Theory of Real Numbers

11.1 INTRODUCTION

In this chapter we shall discuss about a very important concepts namely Cantor's theorem of Real numbers. We shall begin by defining a relation in the set of all Cauchy sequences of rational numbers and showing that this is an equivalence relation. We shall also define addition and multiplication of Cantor's numbers. Finally, we shall discuss the completeness and order completeness of Cantor's numbers.

11.2 CAUCHY SEQUENCE OF RATIONAL NUMBERS

Definition 1. *Let sequence $<s_n>$ of rational numbers is said to be Cauchy if for positive rational number ε, there exists a positive integer m such that*

$$|s_p - s_q| < \varepsilon \quad \forall \, p, q \geq m$$

Definition 2. (Bounded sequence). *A rational sequence $<s_n>$ is said to be bounded if there exists a rational number k such that $|s_n| \leq k \,\forall\, n$.*

Defintion 3. (Convergent sequence). *A rational sequence $<s_n>$ is said to be converge to a rational number l, if for given a positive rational number ε there exists a positive integer m such that $|s_n - l| < \varepsilon \,\forall\, n \geq m$. The number l is called the limit of the sequence and can be written as*

$$\lim s_n = l \text{ or } s_n \to l$$

Definition 4. *If $<a_n>$ and $<b_n>$ are two rational sequences, then the sequence $<c_n>$ where $c_n = a_n + b_n$ is called the **sum sequence** of $<a_n>$ and $<b_n>$ and is denoted by $<a_n + b_n>$.*

Definition 5. *If $<a_n>$ and $<b_n>$ are two rational sequences, then the sequence $<c_n>$ where $c_n = a_n \cdot b_n$ is called the **product sequence** of $<a_n>$ and $<b_n>$ and is denoted by $<a_n \cdot b_n>$.*

Definition 6. *If $<a_n>$ is a **Cauchy sequence of rational numbers** which does not converge to zero and $a_n \neq 0 \,\forall n$ then the reciprocal sequence $<\dfrac{1}{a_n}>$ is also a Cauchy sequence of rational numbers.*

FACTS : TO THE POINT

▶ A Cauchy sequence of rational numbers satisfies the necessary condition of convergence in Q, but it may not necessarily converge to a limit in Q.

▶ Every Cauchy sequence of rational numbers is bounded.

▶ If $<a_n>$ and $<b_n>$ are Cauchy sequences then
  (i) $<a_n \pm b_n>$ is also a Cauchy sequence.
  (ii) $<a_n \cdot b_n>$ is also a Cauchy sequence.
  (iii) $<\dfrac{1}{a_n}> \; (a_n \neq 0 \,\forall n)$ is also a Cauchy sequence.

▶ If $<a_n>$ is a Cauchy sequence of rational numbers which does not converge to zero, then only a finite number of elements of $<a_n>$ can be zero.

▶ If $<a_n>$ and $<b_n>$ are Cauchy sequences of rational numbers, $b_n \neq 0$ for any n and $<b_n>$ does not converge to zero, then the quotient sequence $<\dfrac{a_n}{b_n}>$ is also a Cauchy sequence of rational numbers.

▶ A sequence of rational numbers tends to a unique limi, if exist.

ILLUSTRATIONS

- The rational sequence $<a_n>$, where $a_n = 1$ for all n is a Cauchy sequence.

- The rational sequence $<b_n>$ where $b_n = (-1)^n \cdot \dfrac{1}{n}$ is a Cauchy sequence.

- The sequence $<a_n> = <(-1)^n>$ is not a Cauchy sequence.

- If $a_n = \dfrac{(-1)^{n+1}}{n}$ then $\lim a_n = 0$.

- If $<r^n>$ is a rational sequence such that $0 < r < 1$ then $\lim r^n = 0$.

11.3 A NECESSARY CONDITION FOR CONVERGENCE

If $<s_n>$ be a sequence of rational numbers which converges to some rational number then given any rational number $\varepsilon > 0$ there exists a positive integer m such that for every pair of positive integers p, q both greater than m we have

$$|s_p - s_q| < \varepsilon$$

PROOF. Let $\lim s_n = l$, $l \in \boldsymbol{Q}$

Then given $\varepsilon > 0 \ \exists \ m \in \boldsymbol{N}$ such that $|s_n - l| < \varepsilon/2 \ \forall \ n \geq m$

In particular, if $p, q \in \boldsymbol{N}$ and $p, q > m$, then

$$|s_p - l| < \varepsilon/2 \text{ and } |s_q - l| < \varepsilon/2$$

Then we have

$$|s_p - s_q| = |s_p - l + l - s_q| \leq |s_p - l| + |s_q - l|$$

$$< \frac{\varepsilon}{2} + \frac{\varepsilon}{2} = \varepsilon$$

11.4 CANTOR'S NUMBERS

In this section, we shall discuss about the concept of Cantor's numbers and related properties. Throughout this section we denote the set of all sequences of rational numbers by F_Q.

Definition 1. Let $<a_n>$ and $<b_n>$ be two Cauchy sequences of rational numbers such that $<a_n - b_n>$ converges to 0, then $<a_n>$ is said to be equivalent to $<b_n>$ i.e.

$$<a_n> \sim <b_n> \Leftrightarrow \lim(a_n - b_n) = 0$$

Definition 2. Let F_Q be the set of all sequences of rational numbers. In F_Q, if $\sim$ is the relation defined by the condition

$$<a_n> \sim <b_n> \text{ if and only if } \lim(a_n - b_n) = 0$$

Then each equivalence class $[a_n]$ where

$$[a_n] = \{<x_n> \in F_Q : <x_n> \sim <a_n>\}$$

is called a Cantor's number i.e. t is a Cantor's number then

$$t = \{<x_n> \in F_Q : <x_n> \sim <a_n>\}$$

☛ REMARK

➥ The set of Cantor's numbers is denoted by $\boldsymbol{R_c}$.

THEOREM I. **The relation $\sim$ in the set of all Cauchy sequences of rational numbers such that $<a_n - b_n>$ converges to 0 iff $\lim(a_n - b_n) = 0$ is an equivalence relation.**

PROOF. Let us suppose that, for any Cauchy sequence $<a_n>$ in $\boldsymbol{Q}$, $\lim(a_n - b_n) = 0$.

   (i) Since, clearly $<a_n> \sim <a_n> \quad \Rightarrow \quad \sim$ is reflexive relation.

   (ii) Let $<a_n>$ and $<b_n>$ be Cauchy sequences in $\boldsymbol{Q}$

      Then $\lim(a_n - b_n) = 0 \quad \Leftrightarrow \quad \lim(b_n - a_n) = 0$

So, $<a_n> \sim <b_n>$ $\Leftrightarrow$ $<b_n> \sim <a_n>$

$\Rightarrow$ The relation $\sim$ is symmetric.

(iii) Let $<a_n>$, $<b_n>$ and $<c_n>$ be Cauchy sequences in Q such that

$$<a_n> \sim <b_n> \text{ and } <b_n> \sim <c_n>$$

$\Rightarrow \lim(a_n - b_n) = 0$ and $\lim(b_n - c_n) = 0$

Consider $\lim(a_n - c_n)$

$$= \lim(a_n - b_n + b_n - c_n)$$
$$= \lim(a_n - b_n) + \lim(b_n - c_n)$$
$$= 0$$

$\Rightarrow$ $<a_n> \sim <c_n>$

$\Rightarrow$ $\sim$ is transitive relation.

Hence, from (i), (ii) and (iii) above we conclude that the relation $\sim$ is equivalence relation.

> **FACTS : TO THE POINT**
>
> ▶ If $[a_n] \neq [b_n]$, then $<a_n>$ is not $\sim <b_n>$.
> ▶ If $<a_n> \sim <b_n>$ and $\lim a_n = l$, then $\lim b_n = l$.
> ▶ If $<a_n> \sim <b_n>$ and $<a_n>$ does not converges in Q. Then $<b_n>$ also does not converge.

11.5 ALGEBRA OF CANTOR'S NUMBERS

THEOREM 1. (Equality of Cantor's numbers). *Two Cantor numbers $t_1 = [a_n]$ and $t_2 = [a'_n]$ are equal if and only if $<a_n> \sim <a'_n>$.*

PROOF. We have $t_1 = t_2$ $\Leftrightarrow$ $[a_n] = [a'_n]$

$\Leftrightarrow$ $<a'_n> \in [a_n]$

$\Leftrightarrow$ $<a'_n> \sim <a_n>$

$\Leftrightarrow$ $<a_n> \sim <a'_n>$

THEOREM 2. (Addition of Cantor Numbers). *Let $<x_n>$, $<y_n>$, $<a_n>$ and $<b_n>$ be the elements of F_Q, such that $<x_n> \sim <a_n>$ and $<y_n> \sim <b_n>$ then $<x_n + y_n>$ and $<a_n + b_n>$ are also the elements of F_Q and $<x_n + y_n> \sim <a_n + b_n>$.*

PROOF. Let $<x_n>$, $<y_n> \in F_Q$

$\Rightarrow$ $<x_n>$ and $<y_n>$ are Cauchy sequence in Q.

$\Rightarrow$ $<x_n + y_n>$ is Cauchy in Q.

In a similar way, we can say that $<a_n + b_n>$ is Cauchy in Q.

Since, $<x_n> \sim <a_n>$ and $<y_n> \sim <b_n>$ therefore,

$$\lim(x_n - a_n) = 0 \text{ and } \lim(y_n - b_n) = 0$$

Consider $\lim((x_n + y_n) - (a_n + b_n)) = \lim(x_n - a_n) + \lim(y_n - b_n)$

($\because$ limit of the sum of two functions is equal to the sum of their limits)

$$= 0 + 0$$
$$= 0$$

Hence, $<x_n + y_n> \sim <a_n + b_n>$.

THEOREM 2. *There exists a mapping $f : R_c \times R_c$ such that for every pair of Cantor numbers.*

$$t_1 = [a_n], \ t_2 = [b_n]$$
$$f(t_1, t_2) = t_3 \text{ where } t_3 = [a_n + b_n]$$

PROOF. Let $[t_1, t_2] \in R_c \times R_c$ be arbitrary and $t_1 = [a_n]$, $t_2 = [b_n]$ for some sequences $<a_n>$ and $<b_n>$ in F_Q.

We can uniquely defined the sum sequence $<a_n + b_n>$ in F_Q. Therefore, $[a_n + b_n] \in R_c$.

Now, let $t_3 = [a_n + b_n]$ and let us set $f(t_1, t_2) = t_3$.

Let $t_1 = [a'_n]$ and $t_2 = [b'_n]$ for some other sequence $<a'_n>$ and $<b'_n>$ in F_Q.

Now, $[a'_n] = [a_n]$ and $[b'_n] = [b_n]$

$\Rightarrow \quad <a'_n> \sim <a_n> \text{ and } <b'_n> \sim <b_n>$

$\Rightarrow \quad <a'_n + b'_n> \sim <a_n + b_n>$

$\Rightarrow \quad [a'_n + b'_n] = [a_n + b_n]$

Thus, f is well defined.

Hence, f is a mapping from $\mathbf{R}_c \times \mathbf{R}_c$ into $\mathbf{R}_c$.

THEOREM 3. **(Multiplication in $\mathbf{R}_c$).** *Let $<x_n>$, $<y_n>$, $<a_n>$ and $<b_n>$ are elements of F_Q such that $<x_n> \sim <a_n>$ and $<y_n> \sim <b_n>$, then $<x_n y_n>$ and $<a_n b_n>$ are the elements of F_Q and $<x_n y_n> \sim <a_n b_n>$.*

PROOF. Since, $<x_n>, <y_n> \in F_Q \quad \Rightarrow \quad <x_n y_n> \in F_Q$

Similarly, $<a_n>, <b_n> \in F_Q \Rightarrow \quad <a_n b_n> \in F_Q$

Further, since $<x_n>$ and $<b_n>$ are Cauchy sequence $\exists\ m_1, m_2 \in Q$ such that

$$|x_n| < m_1,\ |b_n| < m_2 \quad \forall n \in \mathbf{Z}^+$$

Also, since $<x_n> \sim <a_n>$ and $<y_n> \sim <b_n>$

$\Rightarrow \quad \lim(x_n - a_n) = 0$ and $\lim(y_n - b_n) = 0$

$\Rightarrow \quad$ For given $\varepsilon > 0\ \exists\ m_1, m_2 \in \mathbf{Z}^+$ such that

$$|x_n - a_n| < \frac{\varepsilon}{2m_1} \quad \forall n \geq m_1 \quad \text{and} \qquad |y_n - b_n| < \frac{\varepsilon}{2m_2} \quad \forall n \geq m_2$$

Choose $m = \max\{m_1, m_2\}$, we have

$$|x_n y_n - a_n b_n| = |x_n(y_n - b_n) + b_n(x_n - a_n)| \quad \forall n \geq m$$

$$\leq |x_n||y_n - b_n| + |b_n||x_n - a_n| \leq m_1 \cdot \frac{\varepsilon}{2m_1} + m_2 \cdot \frac{\varepsilon}{2m_2} = \varepsilon$$

Hence, $<x_n y_n> \sim <a_n b_n>$.

THEOREM 4. *There exists a mapping $f : \mathbf{R}_c \times \mathbf{R}_c \to \mathbf{R}_c$ such that for every pair of Cantor numbers.*

$$t_1 = [a_n],\ t_2 = [b_n]$$

$$f(t_1, t_2) = t_3 \text{ where } t_3 = [a_n b_n]$$

PROOF. Let $(t_1, t_2) \in \mathbf{R}_c \times \mathbf{R}_c$ be arbitrary.

Also, let $t_1 = [a_n]$ and $t_2 = [b_n]$ for some sequences $<a_n>$ and $<b_n>$ in F_Q.

$\to \quad <a_n b_n>$ is uniquely determined in F_Q.

$\Rightarrow \quad t = [a_n, b_n]$ is an element of $\mathbf{R}_c$.

Now, define $f(t_1 t_2) = t_3$

Let $t_1 = [a'_n], t_2 = [b'_n]$ for some other sequences $<a'_n>$ and $<b'_n>$ in F_Q.

$$[a'_n] = [a_n] \text{ and } [b'_n] = [b_n]$$

$\Rightarrow \quad <a'_n> \sim <a_n>$ and $<b'_n> \sim <b_n>$

$\Rightarrow \quad <a'_n b'_n> \sim <a_n b_n>$

$\Rightarrow \quad [a'_n b'_n] = [a_n b_n]$

$\Rightarrow \quad f$ is well defined.

Hence f is a mapping of $\mathbf{R}_c \times \mathbf{R}_c$ into $\mathbf{R}_c$.

FACTS : TO THE POINT

▶ $t_1 \oplus t_2 = t_2 \oplus t_1 \quad t_1, t_2 \in \mathbf{R}_c$

▶ $t_1 \oplus (t_2 \oplus t_3) = (t_1 \oplus t_2) \oplus t_3$

▶ There exists a Cantor's number $\tilde{0}$ such that $t_1 \oplus \tilde{0} = \tilde{0} \oplus t_1$ for $t_1 \in \mathbf{R}_c$

▶ $\tilde{0}$, the additive identity is always unique.

▶ $t_1 \odot t_2 = t_2 \odot t_1$

▶ $t_1 \odot (t_2 \odot t_3) = (t_1 \odot t_2) \odot t_3$

▶ There exists a Cantor number $\tilde{1}$ such that

$$t_1 \odot \tilde{1} = \tilde{1} \odot t_1 = t_1 \quad \forall t_1 \in \mathbf{R}_c$$

▶ The multiplicative identity $\tilde{1}$ is unique.

▶ $t_1 \odot t_3 = t_2 \odot t_3 \Rightarrow t_1 = t_2$ provided $t_3 \neq 0$.

11.6 ORDER IN $\mathbf{R}_c$

In this section, we shall discuss about the field $(\mathbf{R}_c, \oplus, \odot)$ of Cantor numbers. For this we have the following definitions.

Definition 1. (Positive sequence). *A rational sequence $<s_n>$ is said to be positive if there exist $c \in Q^+$ and $m \in Z^+$ such that $s_n > c \; \forall \, n \geq m$.*

Definition 2. (Cantor Positive Number). *A Cantor number α is said to be positive if for every sequence $<s_n> \in \alpha$, $<s_n>$ is a positive sequence in **Q**.*

Definition 3. *Let α, β be the Cantor numbers then α is said to be greater than β if $\alpha \ominus \beta \in R_c^+$. It can be written as $\alpha \oslash \beta$.*

Definition 4. (Modulus function). *The modulus of a Cantor number α, denoted by $|\alpha|$ can be defined as follows*

$$|\alpha| = \begin{cases} \alpha & if \quad x \oslash \tilde{0} \; or \; x = \tilde{0} \\ -\alpha & if \qquad \quad x \otimes \tilde{0} \end{cases}$$

Hence, $R_c^+ = \{ \alpha \in R_c : \alpha \text{ is positive} \}$

☛ Rᴇᴍᴀʀᴋ

⟹ For any positive sequence in **Q**, we can find a positive rational number c such that all except a finite number of elements of the sequence are greater than c.

TʜᴇᴏʀᴇᴍI. **Let $<a_n>$ and $<b_n>$ be positive sequences in Q then $<a_n + b_n>$ and $<a_n b_n>$ are also positive sequence in Q.**

PʀᴏᴏF. It is given that $<a_n>$ is a positive sequence in **Q**, so by definition $\exists \, c \in Q^+$ and $m_1 \in Z^+$ such that

$$a_n > c \quad \forall n \geq m_1 \qquad \qquad \qquad \text{...(1)}$$

Similarly, for $<b_n>$, $\exists \, d \in Q^+$ and $m_2 \in Z^+$ such that

$$b_n > d \quad \forall n \geq m_2 \qquad \qquad \qquad \text{...(2)}$$

Let $m = \max\{m_1, m_2\}$

Then we have $a_n > c$, $b_n > d \; \forall \, n \geq m$

$\Rightarrow \quad (a_n + b_n) > c + d$ and $a_n b_n > c \cdot d$

Hence, $<a_n + b_n>$ and $<a_n b_n>$ are positive sequences in **Q**.

Tʜᴇᴏʀᴇᴍ 2. **If $<a_n>$, $<b_n> \in F_Q$ such that $<a_n>$ is a positive rational sequence and $<b_n> \sim <a_n>$ then $<b_n>$ is also a positive sequence in Q.**

PʀᴏᴏF. It is given that $<a_n>$ is a positive rational sequence. Therefore, by definition $\exists \, c \in Q^+$ and $m \in Z^+$ such that

$$a_n > c \; \forall \, n \geq m_1$$

$$\Rightarrow \qquad a_n - \frac{c}{2} > \frac{c}{2} \qquad \qquad \qquad \text{...(1)}$$

Further, since $<b_n> \sim <a_n>$ and $\frac{1}{2} c \in Q$, therefore there exists $m \in Z^+$ such that

$$|b_n - a_n| < \frac{1}{2} c \quad \forall n \geq m_2$$

$$\Rightarrow \qquad a_n - \frac{1}{2} c < a'_n < a'_n + a_n + \frac{1}{2} c \qquad \qquad \qquad \text{...(2)}$$

Let $m = \max\{m_1, m_2\}$

Then from (1) and (2) we conclude that

$$\frac{1}{2} c < b_n < a_n + \frac{1}{2} c$$

$\Rightarrow \quad \exists \, \frac{1}{2} c \in Q^+$ and $m \in Z^+$ such that $b_n > \frac{1}{2} c \; \forall n \geq m$.

Hence, $<b_n>$ is also a positive sequence in **Q**.

Tʜᴇᴏʀᴇᴍ 3. **Let $\alpha, \beta \in R_c^+$ then $\alpha \oplus \beta$ and $\alpha \odot \beta$ also in R_c^+.**

PROOF.   It is given that $\alpha, \beta \in R_c^+$.

Now, $\alpha \in R_c^+$

$\Rightarrow$ $\exists$ a Cauchy sequence $<x_n>$ in α such that $<x_n>$ is a positive sequence in $\boldsymbol{Q}$.

and $\beta \in R_c^+$

$\Rightarrow$ $\exists$ a Cauchy sequence $<y_n>$ in β such that $<y_n>$ is a positive sequence in $\boldsymbol{Q}$.

From above, we conclude that $<x_n + y_n>$ and $<x_n y_n>$ are positive sequence in $\boldsymbol{Q}$.

Now, $\alpha \oplus \beta = [x_n] \oplus [y_n] = [x_n + y_n]$

and $\alpha \odot \beta = [x_n] \odot [y_n] = [x_n \cdot y_n]$

Hence, we conclude that $\alpha \oplus \beta \in R_c^+$ and $\alpha \odot \beta \in R_c^+$.

THEOREM 4.   **(Law of Trichotomy).** *If $\alpha \in R_c$ then one and only one of the following is true.*

    (i) $\alpha \in R_c^+$ ***(ii)*** $\alpha = \tilde{0}$ ***(iii)*** $-\alpha \in R_c^-$

PROOF.   Let $\alpha = [a_n]$ so that $<a_n> \in F_Q$ and $<a_n> \in \alpha$

If $\alpha \neq \tilde{0}$ then $<a_n>$ is not equivalent to $<0_n>$ where $0_n = 0 \; \forall \; n$.

Thus, $<a_n>$ does not converge to zero. So there exists $\lambda \in \boldsymbol{Q}^+$ and $m_1 \in \boldsymbol{Z}^+$ such that

$$|a_n| > \lambda \text{ for all } n \geq m_1$$

Since, $<a_n>$ is a Cauchy sequence so by definition for given $\lambda > 0 \; \exists \; m_2 \in \boldsymbol{Z}^+$ such that

$$|a_p - a_q| < \frac{\lambda}{2} \quad \forall p, q \geq m_2$$

$\Rightarrow$
$$a_q - \frac{\lambda}{2} < a_p < a_q + \frac{\lambda}{2} \quad \forall p, q \geq m_2$$

Let $m = \max\{m_1, m_2\}$. Then we have

$$a_m - \frac{\lambda}{2} < a_p < a_m + \frac{\lambda}{2} \quad \forall p \geq m$$

Then we have the following cases:

CASE I.   If $a_m > 0$ then $a_m > \lambda$, thus $a_p > \frac{\lambda}{2} \quad \forall p \geq m$

    $\Rightarrow$ $<a_n>$ is a positive sequence.

    Hence, $\alpha \in R_c^+$.

CASE II.   If $a_m > 0$ then $-a_m > \lambda$ or $a_m < -\lambda$ and therefore

$$a_p < -\frac{\lambda}{2} \quad \forall p \geq m$$

$\Rightarrow$ $-a_p > \frac{\lambda}{2} \quad \forall p \geq m$

$\Rightarrow$ $<-a_n>$ is a positive sequence.

$\Rightarrow$ $-\alpha \in R_c^+$.

Hence, we conclude that at least one of the three alternative holds. Now, it remains to **prove** that atmost one of the three alternatives is true.

If $\alpha = \tilde{0}$ then $<a_n> \sim <0_n>$

$\Rightarrow$ Given $c > 0 \; \exists \; m \in \boldsymbol{Z}^+$ such that $|a_n| < c \; \forall \; n \geq m$

$\Rightarrow$ There is no $c \in \theta^+$ such that for some $m \in \boldsymbol{Z}^+$, either $a_n \geq c \; \forall \; n \geq m'$

$\Rightarrow$ $-a_n \geq c \; \forall n \geq m'$

So, if $\alpha = \tilde{0}$ then neither $\alpha \in R_c^+$ nor $-\alpha \in R_c^+$

and if $\alpha \neq \tilde{0}$ then if possible, let $\alpha \in R_c^+$ and $-\alpha \in R_c^+$.

Then for some $c, c' \in \boldsymbol{Q}^+ (c < c')$ and $k, k' \in \boldsymbol{Z}^+$ we have $a_n \geq c \; \forall n \geq k$

and $\qquad -a_n \geq c' \quad \forall n \geq k'$

Let $n = \max\{k, k'\}$, then

$\qquad 0 < c' \leq -a_n \leq -c < 0$, which is a contradiction

Thus, $\alpha \in R_c^+, -\alpha \in R_c^+$ is not true.

Hence, the result.

11.7 CANTOR RATIONAL AND IRRATIONAL NUMBER

Let $<a_n>$ be a Cauchy sequence of rational numbers and $\lim a_n = r$, a rational number then $[a_n]$ is called Cantor rational number.

☛ **REMARK**

⟹ If a Cauchy sequence of rational number converges to a limit in **Q**, then the Cantor number determined by the sequence is called a Cantor rational number.

THEOREM 1. *If $r \in Q$, then there exists a Cauchy sequence of rational numbers converging to r.*

PROOF. Let $<s_n>$ be a rational sequence where $s_n = r \ \forall n$. Clearly, $<s_n>$ is Cauchy because

$$|s_p - s_q| = 0 \ \forall p, q$$

Here, $<s_n>$ is convergent also.

THEOREM 2. *If $<a_n>$ and $<b_n>$ are Cauchy sequences in Q both converging to the same limit r then $<a_n> \sim <b_n>$.*

PROOF. It is given that

$$\lim a_n = r = \lim b_n$$

$\Rightarrow \qquad \lim a_n - \lim b_n = 0$

$\Rightarrow \qquad \lim(a_n - b_n) = 0$

Hence, $<a_n> \sim <b_n>$.

THEOREM 3. *To every rational number r, there corresponds a unique Cantor rational number.*

PROOF. Let $r \in Q$ and $<a_n>$ be a Cauchy rational sequence such that $\lim a_n = r$.

$\Rightarrow \quad [a_n]$ is a Cantor rational number that corresponds to r.

If $[a_n']$ be another Cantor rational number that corresponds to r, then $<a_n'>$ is a Cauchy sequence such that $\lim a_n' = r$.

Further, $<a_n>$ and $<a_n'>$ both converge to r.

Therefore, $<a_n> \sim <a_n'>$.

Hence, $[a_n] = [a_n']$.

11.7.1 CANTOR RATIONAL NUMBER CORRESPONDS TO r

Let $r \in Q$, then the Cantor rational number that corresponds to r will be denoted by r^* and the set of all Cantor rational number will be denoted by Q*.

So, $Q^* = \{r^* : r^*$ is a Cantor Rational number$\}$

☛ **REMARK**

⟹ If $\alpha \in R_c$, but $\alpha \notin Q^*$ then α is called Cantor irrational number.

PROPERTIES

1. The set of all Cantor irrational number is $R_c \sim Q^*$.
2. There exists a function $\alpha : Q \to Q^*$ such that $\alpha(a) = a^* \ \forall \ a \in Q$.
3. The function $\alpha : Q \to Q^*$ such that for every $a \in Q$, $\alpha(a) = a^*$ is univalent and onto.

THEOREM 1.   *The mapping σ of Q onto Q^* such that*
$$\sigma(a) = a^* \in Q^* \ \forall \ a \in Q$$
is an isomorphism of Q onto Q^*, which preserves addition, multiplication and order.

PROOF.   Let $<a_n>$ be a Cauchy sequence in Q, let $\lim a_n = a$, $a \in Q$. Then clearly,
$$a^* = [a_n] = \sigma(a) \qquad \ldots(1)$$
Further, let $<b_n>$ be a Cauchy sequence in Q and let $\lim b_n = b$, $b \in Q$.
Then
$$b^* = [b_n] = \sigma(b) \qquad \ldots(2)$$
From (1) and (2), we have
$$\lim <a_n + b_n> = a + b$$
and $\lim <(a_n b_n)> = a \cdot b$
Therefore, $(a + b)^* = [a_n + b_n]$ and
$$(ab)^* = [a_n b_n] \qquad \ldots(3)$$
Now, $\sigma(a + b) = (a + b)^*$
$$= [a_n + b_n] = [a_n] \oplus [b_n]$$
$$= a^* \oplus b^* = \sigma(a) \oplus \sigma(b)$$
and $\sigma(ab) = (ab)^*$
$$= [a_n b_n] = [a_n] \odot [b_n]$$
$$= a^* \odot b^* = \sigma(a) \odot \sigma(b)$$
Also, $a > b$ in Q $\Leftrightarrow$ $a - b > 0$ in Q
$$\Leftrightarrow \ [a_n - b_n] \in R_c^+$$
$$\Leftrightarrow \ [[a_n] \oplus [-b_n]] \in R_c^+$$
$$\Leftrightarrow \ a^* \ominus b^* \in R_c^+$$
$$\Leftrightarrow \ a^* \oslash b^*$$
$$\Leftrightarrow \ \sigma(a) \oslash \sigma(b)$$
Hence, we conclude that $\sigma : Q \to Q^*$ is an isomorphism preserving addition, multiplication and order.

FACTS : TO THE POINT

▶ If $\varepsilon > 0$ be a Cantor number then there exists infinitely many rational numbers x such that $0 < x < \varepsilon$.

▶ If $\xi > 0$ be a Cantor number then there exists a rational number η such that $\xi < \eta$.

▶ If $\xi > 0$ be a Cantor number then there exist infinitely many rational numbers less than ξ and infinitely many rational number greater than ξ.

▶ Between any two positive Cantor numbers α, β, there exists a rational number r such that $\alpha < r < \beta$.

▶ Between any two Cantor numbers α, β with $\alpha < \beta$ there exists a rational number r such that $\alpha < r < \beta$ in R_c.

▶ Between any two Cantor's numbers there exists infinitely many rational numbers.

▶ Between any two Cantor's numbers there exists an irrational number.

▶ If α is any Cantor irrational number and x is a Cantor rational number then $\alpha + x$ and αx are both Cantor irrational.

SOLVED EXAMPLES

EXAMPLE 1.   *If $\varepsilon > 0$ be a Cantor number, then show that there exists a rational number x such that $0 < x < \varepsilon$ in R_c.*

SOLUTION.   Let $<s_n>$ be a Cauchy sequence in Q such that $[s_n] = \varepsilon$. Now since $\varepsilon > 0$, so $<s_n>$ is a positive sequence in Q. Therefore, for some $\delta \in Q^+$ and $m \in Z^+$, $s_n > \delta \ \forall \ n \geq m$.

Let $x \in Q$ such that $0 < x < \delta$ or $\delta' = (\delta - x) \in Q^+$

Now, for $n \geq m$, $a_n - x > \delta - x = \delta' > 0$ in Q. Therefore, $<a_n - x>$ is a positive sequence in Q. Thus $[a_n - x] > 0$ in R_c or $\varepsilon - x > 0$ is in R_c or $0 < x < \varepsilon$ where $x \in Q$.

EXAMPLE 2.   *Show that for each pair of positive Cantor number α, β there exists a positive integer n such that $n\alpha > \beta$.*

SOLUTION.   If $\alpha \geq \beta$ then $n = 2$ will suffice. If $\alpha > \beta$ then there exist rational numbers a and b such that $0 < a < \alpha < \beta < b$.

The system Q being Archimedian, there exists $n \in Z^+$ such that $na > b$. Hence, for this n, $n\alpha > na > b > \beta$ or $n\alpha > \beta$.

☞ REMARK

➠ **(Archimedian Property of R_c).** For each pair of positive Cantor number α, β there exists a positive integer n such that $n\alpha > \beta$.

11.8 CONVERGENCE OF CANTOR NUMBERS

Definition 1. (Convergence in R_c). *A sequence $<s_n>$ of Cantor number converges in R_c if there exists a Cantor number l such that for each Cantor number $\varepsilon > 0$ there exists $m \in Z^+$ such that*

$$|s_n - l| < \varepsilon \ \forall \ n \geq m$$

Here, the number l, if exists is called the limit of $<s_n>$.

Definition 2. (Cauchy sequence in R_c). *A sequence of Cantor number $<s_n>$ is said to be Cauchy sequence in R_c if to each Cantor number $\varepsilon > 0$ there exists $m \in Z^+$ such that*

$$|s_p - s_q| < \varepsilon \ \forall \ p, q \geq m$$

THEOREM 1. *If l is a Cantor number and $<s_n>$ is a Cauchy sequence in Q such that $<s_n> \in l$ then $\lim s_n = l$ in R_c.*

PROOF. Let $\varepsilon > 0$ be a Cantor number then clearly there exists $e \in Q$ such that $0 < e < \varepsilon$. Now, since $<s_n>$ is Cauchy in Q so by definition for given $e > 0 \ \exists \ m \in Z^+$ such that

$$|s_p - s_q| < e/2 \ \forall \ p, q \geq m$$

Now, consider for each $p \in Z^+$, the sequence $<y_q>$ where $y_q = e - |s_p - s_q|$. Since there exists a positive rational number e and a positive integer m such that $\forall \ p, q \geq m$

$$y_q = e - |s_p - s_q| > \frac{1}{2}e$$

$\Rightarrow$ For each $p \geq m$, the sequence $<y_q>$ of rational numbers is a positive sequence. Further, let $[y_q] = \eta$ then $<y_q>$ being a positive sequence, we have $\eta > 0$ in R_c.

So, $e - [|s_p - s_q|] > 0$ in R_c.

Finally, for each $p \geq m$, we have

$$|s_p - l| = |s_p - [s_q]| = |[s_p] - [s_q]|$$
$$= |[s_p - s_q]| = [|s_p - s_q|]$$
$$< \varepsilon < e$$

Hence, $\lim s_p = l$ and $l \in R_c$.

THEOREM 2. *Every Cauchy sequence of Cantor numbers converges to a limit in the set R_c of Cantor numbers.*

PROOF. Let $<s_n>$ be a Cauchy sequences in R_c then given $\varepsilon > 0 \ \exists \ m_1 \in Z^+$ such that

$$|s_p - s_q| < \frac{\varepsilon}{3} \ \forall p, q \geq m_1 \qquad \qquad \text{...(1)}$$

Now, we prove that for each $p \in Z^+$, there exists a rational number a_p such that

$$|s_p - a_p| < 1/p$$

Further, let $<r_{n(q)}>$ be a Cauchy sequence in Q such that

$$<r_{n(p)}> \in s_p$$

$\Rightarrow \quad \lim r_{n(p)} = s_p$

$\Rightarrow \quad$ For given $\left(\dfrac{1}{p}\right) > 0 \ \exists m_2 \in Z^+$ such that

$$|s_p - r_{n(p)}| < \frac{1}{p} \ \forall n \geq m_2$$

In particular, set $r_{m_2(p)} = a_p$ then we get

$$|s_p - a_p| < \frac{1}{p} \qquad \qquad \text{...(2)}$$

Now, consider the sequence $<a_p>$ and $<s_p>$. For each $\varepsilon > 0$ we can choose $p \in \mathbf{Z}^+$ and $m_3{}' \in \mathbf{Z}^+$ such that

$$\frac{1}{p} < \frac{\varepsilon}{3}$$

and $\qquad |s_p - a_p| < \dfrac{1}{p} < \dfrac{\varepsilon}{3} \quad \forall p \geq m_3$ ...(3)

Define $m = \max\{m_1, m_2, m_3\}$, then

$$|a_p - a_q| = |a_p - s_p + s_p - s_q + s_q - a_q|$$
$$\leq |a_p - s_p| + |s_p - s_q| + |s_q - a_q|$$
$$< \varepsilon/3 + \varepsilon/3 + \varepsilon/3 \qquad \text{(From (1), (2) and (3))}$$
$$= \varepsilon \ \forall \ p, q \geq m \qquad\qquad\qquad ...(4)$$

$\Rightarrow \quad <a_p>$ is a Cauchy sequence in $\mathbf{Q}$.

Hence, $[a_p] = l$ is a Cauchy number and $\lim a_n = l$

Further, given $\varepsilon > 0 \ \exists \ m_4 \in \mathbf{Z}^+$ such that

$$|a_p - l| < \frac{2}{3}\varepsilon \quad \forall p \geq m_4 \qquad\qquad ...(5)$$

Let us define $m_0 = \max\{m_3, m_4\}$ then for all $p \geq m_0$

$$|s_p - l| = |s_p - a_p + a_p - l|$$
$$\leq |s_p - a_p| + |a_p - l|$$
$$< \varepsilon/3 + 2\varepsilon/3 \qquad\qquad \text{(Using (3) and (5))}$$
$$= \varepsilon$$

Hence, $\lim s_n = l$ and $l \in \mathbf{R}_c$.

11.9 ORDER COMPLETENESS OF CANTOR NUMBER SYSTEM

In this section, we shall discuss the bounds and completeness of Cantor number system.

The important definitions are given below:

Definition 1. (Upperbound). *Let X be a subset of $\mathbf{R}_c$. A Cantor number u is called an upperbound of X if every member of X is less than or equal to u.*

Symbolically *u* is an upperbound of *X* if and only if $x \in X \Rightarrow x \leq u$.

☛ REMARK

⇒ If a set *X* of $\mathbf{R}_c$ has an upper bound then the set *X* is said to be bounded above.

Definition 2. (Supremum). *Let $X \subseteq \mathbf{R}_c$ be a bounded above set and the set U of upperbounds of X has a least element. Then this least element of U is called the supremum or least upper bound of X.*

THEOREM I. **A non-empty set of Cantor numbers has at most one supremum.**

PROOF. Let if possible s_1 and s_2 be two supremum of a non-empty set $S \subset \mathbf{R}$.

Now, since s_1 is the supremum of S and s_2 is the upperbound then $s_1 \leq s_2$. ...(1)

Similarly, if s_2 is the supremum of S and s_1 is the upperbound then $s_2 \leq s_1$. ...(2)

From (1) and (2), we conclude that

$$s_1 = s_2$$

Hence, supremum of a non-empty set of Cantor number is unique.

☛ REMARK

⇒ Every non-empty subset of Cantor numbers that is bounded above has a supremum.

REVIEW QUESTIONS AND ARCHIVE

1. Prove that the rational sequence $<a_n>$ where $a_1 = 1$ and $a_n = a_1 + \dfrac{a_2}{2} + \dfrac{a_3}{3} + \ldots + \dfrac{a_{n-1}}{n-1}$ is a Cauchy sequence.

2. For Cantor number t, prove that
   (i) $|t| = \max\{t, -t\}$
   (ii) $|-t| = |t|$

3. Prove that if $t > 0$ is a Cantor number, then there exists infinitely many rational numbers less than t and infinitely many rational number greater than t.

4. Write a short note on Cantor's theory of real numbers.

5. Show that any two complete ordered field are order isomorphic.

CHAPTER SUMMARY

⮩ Let sequence $<s_n>$ of rational numbers is said to be Cauchy if for positive rational number ε, there exists a positive integer m such that $|s_p - s_q| < \varepsilon \; \forall \, p, q \geq m$.

⮩ **(Bounded sequence).** A rational sequence $<s_n>$ is said to be bounded if there exists a rational number k such that $|s_n| \leq k \; \forall \, n$.

⮩ **(Convergent sequence).** A rational sequence $<s_n>$ is said to converge to a rational number l, if for given a positive rational number ε there exists a positive integer m such that $|s_n - l| < \varepsilon \; \forall \, n \geq m$. The number l is called the limit of the sequence and can be written as $\lim s_n = l$ or $s_n \to l$.

⮩ If $<a_n>$ and $<b_n>$ are two rational sequences, then the sequence $<c_n>$ where $c_n = a_n + b_n$ is called the sum sequence of $<a_n>$ and $<b_n>$ and is denoted by $<a_n + b_n>$.

⮩ If $<a_n>$ and $<b_n>$ are two rational sequences, then the sequence $<c_n>$ where $c_n = a_n \cdot b_n$ is called the product sequence of $<a_n>$ and $<b_n>$ and is denoted by $<a_n \cdot b_n>$.

⮩ If $<a_n>$ is a Cauchy sequence of rational numbers which does not converge to zero and $a_n \neq 0 \; \forall n$ then the reciprocal sequence $<\dfrac{1}{a_n}>$ is also a Cauchy sequence of rational numbers.

⮩ Let $<a_n>$ and $<b_n>$ be two Cauchy sequences of rational numbers such that $<a_n - b_n>$ converges to 0, then $<a_n>$ is said

to be equivalent to $<b_n>$ i.e. $<a_n> \sim <b_n>$ $\Leftrightarrow \lim(a_n - b_n) = 0$.

⮩ Let F_Q be the set of all sequences of rational numbers. In F_Q, if $\sim$ is the relation defined by the condition $<a_n> \sim <b_n>$ if and only if $\lim(a_n - b_n) = 0$. Then each equivalent class $[a_n]$ where $[a_n] = \{<x_n> \in F_Q : <x_n> \sim <a_n>\}$ is called a Cantor's number i.e. t is a **Cantor's** number then $t = \{<x_n> \in F_Q : <x_n> \sim <a_n>\}$.

⮩ **(Positive sequence).** A rational sequence $<s_n>$ is said to be positive if there exist $c \in Q^+$ and $m \in Z^+$ such that $s_n > c \; \forall \, n \geq m$.

⮩ **(Cantor Positive Number).** A Cantor number α is said to be positive if for every sequence $<s_n> \in \alpha$, $<s_n>$ is a positive sequence in Q.

⮩ Let α, β be the Cantor numbers then α is said to be greater than β if $\alpha \ominus \beta \in R_c^+$. It can be written as $\alpha > \beta$.

⮩ **(Modulus function).** The modulus of a Cantor number α, denoted by $|\alpha|$ can be defined as follows

$$|\alpha| = \begin{cases} \alpha & \text{if} \quad x \odot \tilde{0} \text{ or } x = \tilde{0} \\ -\alpha & \text{if} \quad x \oslash \tilde{0} \end{cases}$$

Hence, $R_c^+ = \{\alpha \in R_c : \alpha \text{ is positive}\}$

⮩ **(Convergence in R_c).** A sequence $<s_n>$ of Cantor number converges in R_c if there exists a Cantor number l such that for each Cantor number $\varepsilon > 0$ there exists $m \in Z^+$ such that $|s_n - l| < \varepsilon \; \forall \, n \geq m$. Here, the number l, if exists is called the limit of $<s_n>$.

⊃ **(Cauchy sequence in R_c).** A sequence of Cantor number $<s_n>$ is said to be Cauchy sequence in R_c if to each Cantor number $\varepsilon > 0$ there exists $m \in Z^+$ such that

$$|s_p - s_q| < \varepsilon \ \forall \ p, q \geq m$$

⊃ **(Upperbound).** Let X be a subset of R_c. A Cantor number u is called an upperbound of X if every member of X is less than or equal to u.

⊃ **(Supremum).** Let $X \subseteq R_c$ be a bounded above set and the set U of upperbounds of X has a least element. Then this least element of U is called the supremum or least upper bound of X.

\FOR ADVANCED LEARNERS/

➡ If α is a Cantor number and $<a_n> \in \alpha$ be a positive sequence in Q, then every sequence $<a'_n> \in \alpha$ is also a positive sequence in Q.

➡ There is a subset R_c^+ and R_c to be called positive class in R_c such that

(a) For each $\alpha \in R_c$, one and only one of the following holds

    (i) $\alpha \in R_c$ (ii) $\alpha = \tilde{0}$

    (iii) $-\alpha \in R_c^+$

(b) If $\alpha, \beta \in R_c^+$ then

         $\alpha \oplus \beta \in R_c^+$ and $\alpha \odot \beta \in R_c^+$

➡ The smallest subfield of a complete ordered field F is called the rational subfield of F and denoted by $Q(F)$.

➡ If $(F, +, \cdot, >)$ be a complete ordered field and $Q(F)$ its rational subfield then for each $a \in F$

$$a = \sup\{r \in Q(F) : r \leq a\}$$

➡ Any two complete ordered fields are isomorphic.

❋❋❋❋

S.No.	Metric Space	Separable	Complete	Totally bounded	Compact
1.	$X = \mathbf{R}$, $d(x, y) = \lvert x - y \rvert \ \forall \ x, y \in X$	✓	✓	✗	✗
2.	$X = [0, 1]$, $d(x, y) = \lvert x - y \rvert \ \forall \ x, y \in X$	✓	✓	✓	✓
3.	$X = {]}0, 1{[}$, $d(x, y) = \lvert x - y \rvert \ \forall \ x, y \in X$	✓	✗	✓	✗
4.	The complete plane, $X = \mathbf{C}$, $d(z_1, z_2) = \lvert z_1 - z_2 \rvert \ \forall \ z_1, z_2 \in \mathbf{C}$	✓	✓	✗	✗
5.	The Euclidean space, $X = \mathbf{R}^n$, $d(x,y) = \left(\sum_{i=1}^{n} \lvert x_i - y_i \rvert^2 \right)^{1/2} \ \forall x, y \in \mathbf{R}^n$	✓	✓	✗	✗
6.	$X = \mathbf{Z}$, $d(x, y) = \lvert x - y \rvert \ \forall \ x, y \in \mathbf{Z}$	✓	✓	✗	✗
7.	The real Hilbert space l_2: $X = \{<x_n> : x_n \in \mathbf{R}\} \ \forall \ n \in \mathbf{Z}^+$, Σx_n^2 is convergent $d(x,y) = \left(\sum_{n=1}^{\infty} \lvert x_n - y_n \rvert^2 \right)^{1/2}$ $\forall \mathbf{x} = <x_n>, \mathbf{y} = <y_n> \in Y$	✓	✓	✗	✗
8.	The Hilbert cube $X = \{< x_n > : \lvert x_n \rvert \le \frac{1}{n}\} \ \forall n \in \mathbf{Z}^+$ $\Sigma \lvert x_n \rvert^2$ is convergent $d(x,y) = \left(\sum_{n=1}^{\infty} \lvert x_n - y_n \rvert^2 \right)^{1/2}$ $\forall \mathbf{x} = < x_n > \in X, \mathbf{y} = < y_n > \in X$	✓	✓	✓	✓
9.	The space l_p $(p > 1)$ $X = \{< x_n > : x_n \in \mathbf{R}\} \ \forall n \in \mathbf{Z}^+$ $\sum_{n-1}^{\infty} \lvert x_n \rvert^p$ is convergent $d(x,y) = \left(\sum_{n=1}^{\infty} \lvert x_n - y_n \rvert^p \right)^{1/p}$ $\forall \mathbf{x} = < x_n > \in X, \mathbf{y} = < y_n > \in X$	✓	✓	✗	✗
10.	Frechet Space $X = \{<x_n> : x_n \in \mathbf{R}\} \ \forall \ n \in \mathbf{Z}^+$ $d(x,y) = \sum_{n=1}^{\infty} \frac{2^{-n} \lvert x_n - y_n \rvert}{(1 + \lvert x_n - y_n \rvert)}$ $\forall \mathbf{x} = < x_n > \in X, \mathbf{y} = < y_n > \in X$	✓	✓	✓	✓

11.	The space of all convergent real sequences with the supremum metric $d(x, y) = \sup|x_n - y_n|$.	✓	✓	✗	✗
12.	The space of all real sequences, converging to zero with supremum metric $d(x, y) = \sup|x_n - y_n|$.	✓	✓	✗	✗
13.	The space $B(0, 1)$ of all bounded real valued functions on $[0, 1]$ with the supremum metric i.e. $X = \{f : [0, 1] \to \boldsymbol{R} : f \text{ is bounded}\}$ $d(f, g) = \sup_{a \leq t \leq 1} |f(t) - g(t)| \quad \forall f, g \in X$	✗	✓	✗	✗
14.	The space $C(0, 1)$ of all continuous real valued function on $[0, 1]$ with the supremum metric i.e. $X = \{f : [0, 1] \to \boldsymbol{R} : f \text{ is continuous}\}$ $d(f, g) = \sup_{0 \leq t \leq 1} |f(t) - g(t)| \quad \forall f, g \in X$	✓	✓	✗	✗
15.	Uncountable discrete metric space	✗	✓	✗	✗

INDEX